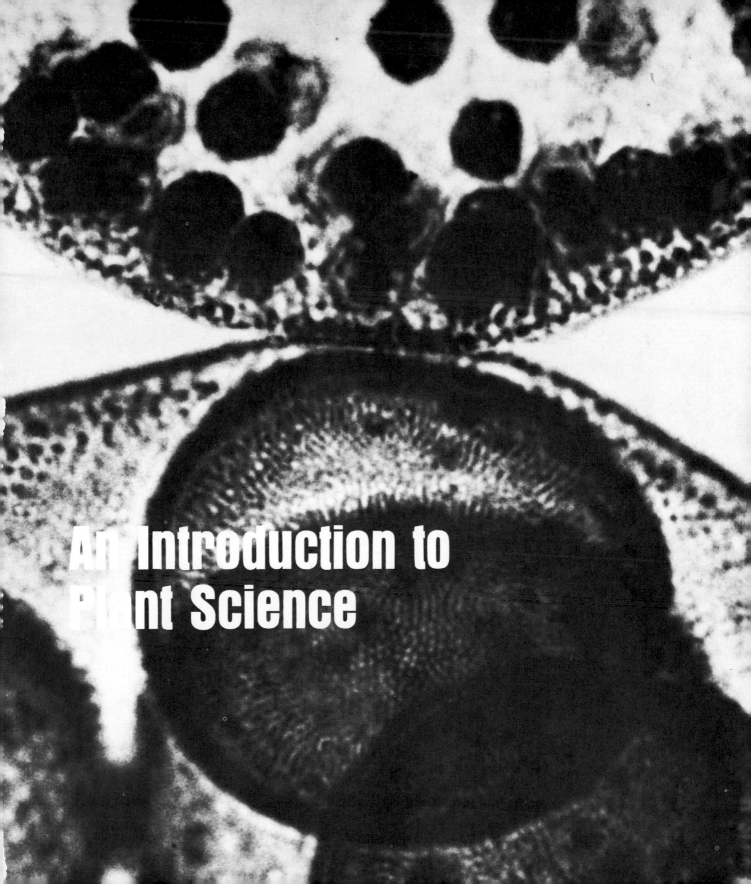

An Introduction to
Plant Science

THE MACMILLAN BIOLOGY SERIES
GENERAL EDITORS: *Norman H. Giles and John G. Torrey*

Gerard J. Tortora BERGEN COMMUNITY COLLEGE

Donald R. Cicero JERSEY CITY STATE COLLEGE

Howard I. Parish JERSEY CITY STATE COLLEGE

Plant Form and Function

AN INTRODUCTION TO PLANT SCIENCE

The Macmillan Company/Collier-Macmillan Limited, London

First Printing

Library of Congress catalog card number: 76-77510

THE MACMILLAN COMPANY
COLLIER-MACMILLAN CANADA, LTD., TORONTO, ONTARIO

PRINTED IN THE UNITED STATES OF AMERICA

Preface

Plant Form and Function is designed for use in introductory plant science courses at the college level. Botany, like most other scientific disciplines, is a rapidly advancing field of knowledge. In the last few decades the study of plants has moved from gross macroscopic investigations to highly sophisticated ultramicroscopic analyses. At one time, the plant scientist was concerned primarily with examining the structures and functions of plants as a whole in their environment. As better tools and techniques became available, plant anatomy and physiology were subjected to closer scrutiny, and with the further advance of instrumentation, the study of plants revealed successively lower levels of structural and functional organization. The botanist of today, utilizing the electron microscope along with various analytical tools of the chemist and physicist, interprets the form and activities of plants in terms of their constituent chemical substances. Essentially, the integration with botany of chemistry and physics, as well as other fields of science, has rapidly moved the frontiers of plant research from the macroscopic level to the ultramicroscopic level of organization.

In consideration of the recent explosion of botanical knowledge as a result of studies in electron microscopy and intensive biochemical research, this textbook attempts to maintain a balance between the molecular approach, which is concerned with form and function at the cellular level, and the traditional approach, which is basically descriptive and morphological. The result is not dichotomous because both aspects of botany now are interrelated and equally important for a total understanding of modern plant science.

The organization of *Plant Form and Function* is unlike that of most introductory botany textbooks.

The three principal areas of concentration are (1) the structural organization of plants, (2) the functional activities of plants, and (3) the diversity of plants.

After an introductory chapter on the general nature of plant science and the importance of plants to man, the various levels of plant organization, from molecules to organisms, are analyzed; in considering this hierarchy of structural organization, we use an angiosperm as a representative form because of its familiarity to most students. At the molecular level, the physical, chemical, and biological properties of the primary cellular building units are examined. The basic concepts of chemistry are outlined in an elementary and simplified manner in Chapter 2. We feel that such a full, chapter-length treatment not only reinforces previously learned chemical principles, but also provides a new learning experience for those who have never taken chemistry. In this way, all students will have the background necessary to the comprehension of the concepts of molecular biology treated in later chapters. At the microscopic level of structure, a generalized plant cell is analyzed with emphasis on newly acquired knowledge of cellular ultrastructure as revealed by the electron microscope. This is followed by a macroscopic presentation of plant tissues and organs, both vegetative and reproductive, and their relationships to the structurally integrated plant body. The advantage of studying form as a unifying concept with only limited reference to function is that the student is directed primarily to one aspect of botany without being confused by others.

A separate chapter entitled "The Integrated Plant Body: Form and Function" is a review of the various levels of structural organization and their relation to the physiological activities of plants, the second major area of concentration. Although structurally diverse, plants are functionally similar in that all undergo a myriad of biochemical reactions collectively termed metabolism. After a cursory survey of these metabolic activities and their relationship to form, the student is introduced to the principal activities of the functioning plant. Again with an angiosperm as the example, the plant is viewed in its physical environment with respect to metabolite availability, processes of metabolite procurement, and the distribution of metabolites to sites of metabolic reactions. This is followed by a fairly comprehensive treatment of the more important physiological activities of plants. Among the topics discussed are enzymes, energy relations, the synthesis of plant materials, and the breakdown of plant materials. Entire chapters have been devoted to enzymes as catalysts of plant metabolism; anabolic reactions such as photosynthesis and protein synthesis; and catabolic pathways, including digestion, respiration, and fermentation. We feel that the chemical tone of these chapters is justified because much of the preparatory material is presented in Chapter 2. The functioning plant is next treated with respect to growth and developmental phenomena, which are direct outcomes of its metabolic activities, and to heredity mechanisms, which are progenitors of metabolism.

The final area of concentration—the diversity of plants that inhabit our globe—is treated broadly. No attempt has been made to discuss all plant groups, so the chapters dealing with the plant kingdom are fairly brief. They are, however, comprehensive enough to encourage interested students to pursue a more detailed study of the plant kingdom. In keeping with the theme of the book, selected plant groups are analyzed in terms of form and function with special emphasis on adaptive characters, evolutionary relationships, patterns of reproduction, and economic importance. As a general orientation to plant diversity, a survey of the major plant divisions is presented in Chapter 13, in which the divisions are grouped into a modern, natural scheme based upon the most recent morphological and phylogenetic studies. Following this, the origin of living forms and the evolutionary history of plants in time are discussed in a single chapter, so that both chemical and biological evolution may be viewed as a continuous process. The major groups of plants, from Monera to higher vascular plants, are studied in an evolutionary-taxonomic framework in Chapters 15–20. The book concludes with a study of plant interactions with the environment and the principal plant formations.

Supplementary learning aids are employed throughout the book. These include a large number of electron micrographs with accompanying diagrammatic representations, a current list of suggested readings at the end of each chapter, chapter summaries, and selected questions and problems. The usefulness of the textual material is further augmented by an appendix, which contains a detailed classification of plants, and a glossary. We have also prepared a laboratory manual, correlated with the textbook, entitled *Plant Form and Function in the Laboratory*.

We wish to express our grateful appreciation to many botanists and biologists, both friends and colleagues, for their pointed comments, helpful suggestions, and new insights. Among those who read portions of an earlier draft of the text and offered their suggestions were Dr. Norman H. Giles of Yale University, Dr. John G. Torrey of Harvard University, Dr. Richard A. Goldsby of Yale University, Dr. David W. Bierhorst of Cornell University, Dr. David L. Dilcher of Indiana University, Dr. David S. Hauser of the State University of New York at Albany, Dr. Joseph Becker of Montclair State College, Dr. Rhoda Love of Lane Community College, and Dr. William J. Oostenink of Colgate University. We incorporated many of the changes they suggested, but any errors are our own.

The enormous burden of typing and retyping the manuscript has been carried by our two fine secretaries, Miss Diane Barbieri and Miss Maryjane Cappiello.

Finally, we wish to thank Mr. William D. Eastman, of The Macmillan Company, and his staff for the courtesy and efficiency with which the project has been handled since its inception. Mr. Eastman provided much guidance, inspiration, and encouragement in seeing this project to completion.

Paramus, N.J.   G. J. T.

Jersey City, N.J.   D. R. C.

Jersey City, N.J.   H. I. P.

Contents

1
An Introduction to Plant Science

INTRODUCTION

The land and waters of the earth contain a vast assemblage of plants upon which all other living forms are directly or indirectly dependent. Green plants are particularly important in this regard because they contain chlorophyll and possess the biological mechanism to provide the energy required by all living organisms for growth, maintenance, and reproduction. Through the process of photosynthesis, green plants are capable of capturing the free energy of solar radiation and converting it into usable chemical energy in the chemical bonds of food molecules. In this process green plants utilize water obtained from the soil or from the medium in which they are immersed and carbon dioxide from the atmosphere to produce food and oxygen. These food molecules in turn provide the energy for nongreen plants, animals, and man.

Plants have always provided mankind with many useful products beyond the primary necessities of food. In 1814 Thomas Jefferson noted the importance of plants to society with the words: "Botany, I rank with the most valuable sciences whether we consider its subjects as furnishing the principal substances of life to man and beast, delicious varieties for our tables, refreshments from our orchards, the adornment of our flower borders, shade and perfume of our groves, materials for buildings or medicaments for our bodies." In the last few decades, there has been an explosion of knowledge in the plant sciences. Not only have botanists learned more about the structure of plants, their activities, and their responses in different environments, they have also learned how to alter, and even control, the growth and behavioral patterns of plants. The impact of the application of this knowledge is readily apparent if one considers improvements in crop growth that have resulted from the development of new and better plants; the control of diseases caused by plants; the production of new and more effective herbicides; the development of antibiotics and other drugs from plants; and the manufacture of new fibers, gums, essential oils, and resins.

A BRIEF SURVEY OF THE PLANT WORLD

Plants exhibit considerable diversity in form, size, pigmentation, habitat, and activities. They range in size from microscopic bacteria, some of which are 1/50,000 inch in diameter, to giant sequoias and other trees which grow more than 350 feet high and may weigh more than 1000 tons. It has been estimated that the General Sherman tree, a giant California sequoia, may be over 3500 years old and that some bristlecone pines and Mexican cypresses may be several hundred years older. This means that the tallest, heaviest, and oldest living organisms are plants.

Plant form is closely related to environmental conditions. There is little resemblance between plants found in dry areas and those found in swampy regions. The same comparison can be made between plants found in tropical and polar climates. Other factors which determine the forms of vegetation found in a given area include soil factors, wind conditions, and the duration of sunlight. For example, the coastal plain of the Atlantic and Gulf states is characterized by extensive pine forests. The sandy soil of this coastal plain does not hold water very well. It can support a

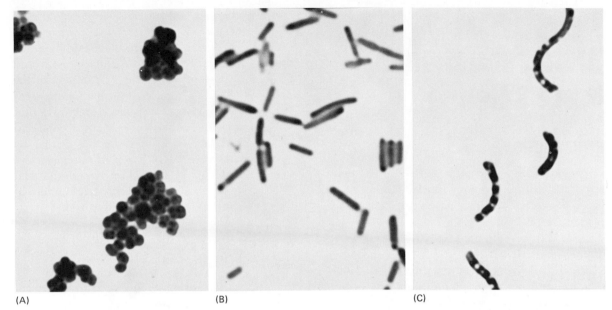

(A)   (B)   (C)

Figure 1-1. Bacteria. These primitive and extremely small plants appear as (**A**) spherical, (**B**) rod, and (**C**) spiral forms. (Courtesy of Carolina Biological Supply Company.)

pine forest, but broad-leaved trees grow better in a soil with more humus. In frost-free tropical rain forests, such as those in Mexico and the West Indies, the complexity of vegetation is unmatched. A trip to a local botanical garden will help to illustrate the enormous diversity which characterizes the plant world.

The origin of plants and, indeed, of the first living forms continues to be an unsettled question. Fossil data indicate that plants appeared over 2 billion years ago and that, through the continual process of evolution, many plants have originated and become extinct. The thousands of plants that surround us and are so vital to our existence are the products of this evolutionary process.

The most primitive (in terms of structural complexity) and smallest plants are the bacteria and algae. Bacteria (Figure 1-1) are well known because of the diseases they cause and the role they play in the decay of foods and dead organic material. The vast majority of bacterial species,

however, are either harmless or useful and aid man in innumerable ways. The algae, on the other hand, include the simplest photosynthetic plants, and some of them presumably resemble the first plants that developed on the earth. The term algae refers, not to a single group of plants, but to a large number of species that are differentiated primarily on the basis of their pigmentation, chemistry, and form. For example, the blue-green algae are the most primitive of all the algal groups and appear as single cells, colonies, or filaments in freshwater habitats (Figure 1-2A). Green algae are considerably more complex and diverse than the blue-green types and exhibit a wider variety of form (Figure 1-2B). Other groups of algae include brown algae (Figure 1-2C), often called kelps; euglenoids (Figure 1-2D), organisms that possess some animal-like characteristics; xantho-

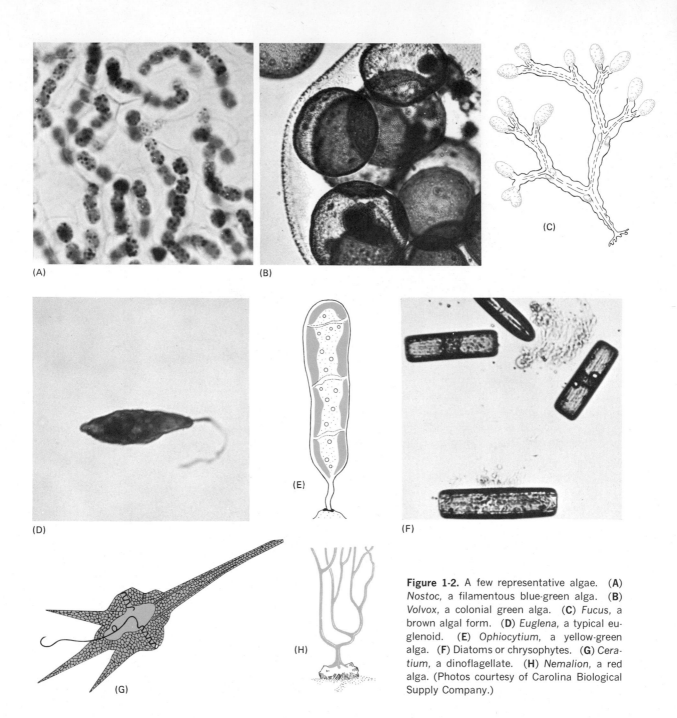

Figure 1-2. A few representative algae. (**A**) *Nostoc*, a filamentous blue-green alga. (**B**) *Volvox*, a colonial green alga. (**C**) *Fucus*, a brown algal form. (**D**) *Euglena*, a typical euglenoid. (**E**) *Ophiocytium*, a yellow-green alga. (**F**) Diatoms or chrysophytes. (**G**) *Ceratium*, a dinoflagellate. (**H**) *Nemalion*, a red alga. (Photos courtesy of Carolina Biological Supply Company.)

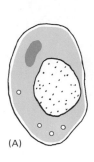

(A)

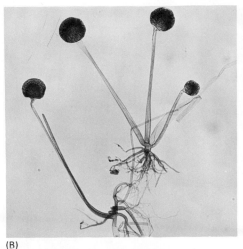

(B)

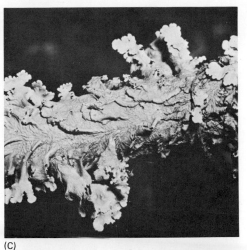

(C)

(D)

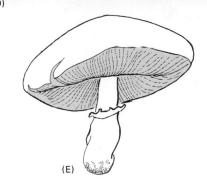

(E)

Figure 1-3. Representative fungus plants. (**A**) *Saccharomyces* (yeast). (**B**) *Rhizopus* (bread mold). (**C**) Lichens. (**D**) *Amanita* (poisonous mushroom). (**E**) *Agaricus* (edible mushroom). (Photos courtesy of Carolina Biological Supply Company.)

phytes (Figure 1-2E), mostly one-celled plants, some of which are yellow-green in color; chrysophytes (Figure 1-2F), which make up most of the freshwater and marine plankton; pyrrophytes (Figure 1-2G), or dinoflagellates; and red algae (Figure 1-2H), plants that in some ways resemble the brown algae.

Fungus plants, like the algae, are also primitive on the scale of plant evolution and seem to have had a development somewhat parallel with the algae. It is interesting to note that a lichen is a dual organism formed from an association of a fungus and an alga. Unlike the algae, the fungi are nonchlorophyllous and live on dead or decaying organic matter. Familiar fungus plants include yeasts, mushrooms, bracket fungi, puffballs, gill fungi, and bread molds (Figure 1-3). Although some fungus plants are associated with disease, others are important in the baking industry, in the production of alcohol, and in the manufacture of antibiotics.

(A)

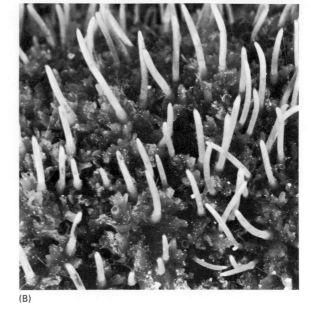

(B)

The relatively simple plants described thus far, although numerous and diverse, are not as familiar as other members of the plant world. To most of us, the term plant usually denotes some familiar, green, leafy land plant. Land plants evolved considerably later than algae, probably from a green algal ancestor. In the course of this transition from the aquatic to the terrestrial habitat, it appears that two distinct lines of evolution emerged. One of these was characterized by the presence of vascular or conducting tissue and supporting tissues; these plants reached considerable complexity in form and function. In the second line of evolution, no specialized vascular or supporting tissue developed, and these plants gave rise to no others.

Modern descendants of this latter line of evolution are the liverworts, hornworts, and mosses. Collectively, they are referred to as bryophytes (Figure 1-4), and they are the most primitive of

(C)

Figure 1-4. Representative bryophytes. (A) *Marchantia*, a liverwort. (B) *Anthoceros*, a hornwort. (C) Moss plants. (Photos courtesy of Carolina Biological Supply Company.)

the green land plants. Individually bryophyte plants are small and inconspicuous and often appear growing in clusters on the soil, on rocks, and on trees.

Tracheophytes, the vascular plants, are the largest group of plants on earth. The origin of

(A)

(B)

(C)

Figure 1-5. Tracheophytes. **(A)** *Psilotum,* a very primitive vascular plant. **(B)** *Lycopodium,* a club moss. **(C)** *Equisetum,* or horsetails. **(D)** A typical fern plant. (Photos courtesy of Carolina Biological Supply Company.)

these plants, as documented by the fossil record, dates back to over 400 million years ago. One of the oldest and simplest land plants is an organism called *Psilotum* (Figure 1-5A). For many millions of years plants such as these, as well as club mosses (Figure 1-5B), horsetails (Figure 1-5C), and ferns (Figure 1-5D), formed a conspicuous portion of the earth's vegetation. In time, giant club mosses and tree-sized ferns evolved, and it was from these dense forests that many coal deposits were formed (Figure 1-6).

Later in evolutionary time, a few vascular plants began to produce seeds. Among the earliest seed plants that have survived to the present are the conifers, or cone-bearing plants. Familiar conifers (Figure 1-7) include pine, spruce, hemlock, and fir trees. Following the conifers, the

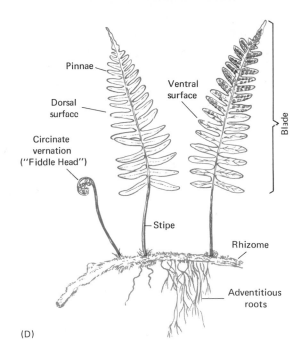

Pinnae

Dorsal surface

Ventral surface

Circinate vernation ("Fiddle Head")

Blade

Stipe

Rhizome

Adventitious roots

(D)

Figure 1-6. A Carboniferous swamp forest. (Courtesy of Field Museum of Natural History, Chicago, Illinois.)

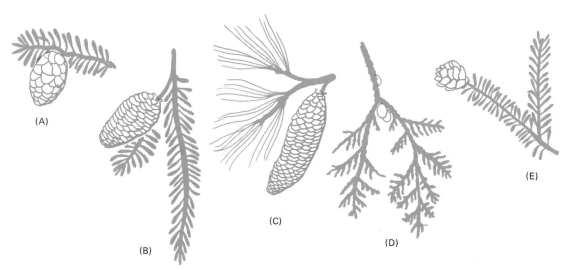

Figure 1-7. Leaf structure and cones of some representative conifers. (**A**) *Picea* (spruce). (**B**) *Abies* (fir). (**C**) *Pinus* (pine). (**D**) *Cedrus* (cedar). (**E**) *Tsuga* (hemlock).

flowering plants made their appearance. Like the conifers, they were seed bearers, but instead of producing exposed seeds in cones, they evolved flowers for reproduction (Figure 1-8). Once a flower completes its reproductive activities, it forms seeds enclosed in fruits. Flowering plants are not only the highest forms of plant life, but the most diversified and widespread. Much of the success of these plants is due to the evolution of the flower.

This brief survey of the plant kingdom has shown that some plants may lack roots, stems, and leaves; others are nongreen; some do not contain supporting and conducting tissues; some do not form seeds; others have exposed seeds; and some plants have flowers, from which seeds enclosed in fruits develop. Inasmuch as more than half of all living plants are flowering plants and these are the most conspicuous and familiar ones, approximately the first half of the book is devoted to an analysis of their form and function.

THE SYSTEMATIC STUDY OF PLANTS

Biology, the science of life, was first named and systematically studied by the Greeks more than 2000 years ago. They combined the words *bios* (life) and *logos* (the study of) to form the word biology. *Botany* and *Zoology* constitute two broad subdivisions of biology. Each of these areas of investigation has developed various specialized branches of study in response to new tools and techniques of research and to ever-expanding knowledge and accumulated data. Although each area of concentration is highly specialized, there is still a great interdependence among them.

Taxonomy (Systematic Botany)

Early botanical studies were represented mainly by somewhat random plant collections in which newly discovered plants were catalogued and named. In fact, until about a century ago, the

Figure 1-8. Representative flowering plants. **(A)** *Lilium rubescens* (ruby lily). **(B)** *Chrysanthemum leucanthemum* (daisy). **(C)** *Hibiscus palustris* (rose mallow).

(A)   (B)   (C)

principal botanical activity was the identification and classification of plants. This field of specialization is known today as *plant taxonomy,* or *systematic botany.*

Once a new plant is discovered, the professional taxonomist attempts to classify it with plants that have already been identified and classified. Whenever possible the taxonomist classifies plants on the basis of their reproductive structures (for example, flowers and cones), because these are less subject to environmental change than other plant parts. The ultimate goal of the taxonomist is to construct a classification scheme that indicates natural relationships (evolutionary) among the various plant groups. After a particular plant is classified, it is designated with an official name according to international rules of nomenclature. The enormous number of known named plants and the many new plants still being discovered each year necessitate some type of organizational pattern for designating specific names. The European white water lily has been given over 240 different common names, which certainly emphasizes the need for a uniform pattern of scientific nomenclature.

In recent years, the utilization of computers by taxonomists has had an enormous impact on plant classification. *Numerical taxonomy,* as this new field has been called, consists of quantifying basic taxonomic features which can be translated into computerized information. For example, a character called "hairiness of a leaf" might be coded in the following manner: hairless, 0; sparsely haired, 1; moderately haired, 2; densely haired, 3. In this way, all taxonomic characteristics can be converted to numerical values and arranged in a data matrix. Similar techniques are being employed in other areas of science: to classify soils, diseases, psychological types, and archeological artifacts. In addition, automatic data processing will greatly simplify and speed up the storage and retrieval of taxonomic information in the immediate future.

Employing the principles of evolution, computerized procedures have been useful in piecing together the characteristics of extinct plants and animals to gain a clearer picture of them. Also, it is possible, using the techniques of numerical taxonomy, to project present species of flora and fauna into the future and to speculate on the evolutionary outcome say 1000 years from now.

Morphology

It would seem logical that studies of individual plants should continue once the task of collecting and cataloguing was well underway. Actually, the next phase in the development of the science of botany was just that—the study of individual plant structures. This area of investigation is called *morphology* and deals specifically with the form and structure of plants, the relationships of the structures to each other, and the comparison of structural resemblances among plants (both living and fossil plants) to show natural relationships.

The morphology of plants may be studied at different levels of organization. A household geranium plant may be used to illustrate this point. A study of the *external morphology* of the plant is concerned with a macroscopic analysis of the arrangement, structure, and development of organs. In this regard the plant morphologist examines the form, vein patterns, and arrangement of leaves on the axis of the plant; the patterns of branching of the root and shoot (stem and leaves) system; and the structure of reproductive organs (flowers, fruits, and seeds).

At a more detailed level of structure the internal aspects of the organs of the geranium may be studied. This type of morphological study is referred to as *anatomy.* The anatomist dissects and prepares the organs for microscopic analysis and notes the distinctive feature of the tissues comprising the organs and their relationship to each other.

The science of *cytology* is the study of the basic units of structure (cells). At this minute level of morphological investigation, the cytologist employs highly specialized microscopic techniques, including electron microscopy, to elucidate the form and structural design of cells. An enormous amount of data has been amassed from cytological studies in the last few decades, and the data have greatly modified and clarified present concepts of cell ultrastructure.

Physiology

Next in the historical succession of plant studies emerged an interest in plant activities. This study, referred to as *plant physiology,* is concerned with the function of different plant organs, tissues, and cells, and with the development of this field came a variety of new laboratory techniques. Plants were no longer simply observed and described only in terms of structure and form; they were experimentally controlled with respect to various life processes and a variety of plant activities were investigated. Among those areas subjected to intensive investigation were the determination of mineral requirements, the processes involved in the procurement of materials from the environment, the mechanisms by which materials are transported within the plant body, the utilization of these materials by various cells, the identification of internal systems of control and regulation, the processes by which foods are synthesized and stored, the processes by which energy is liberated from food molecules and converted into usable energy, the mechanisms involved in reproduction, and the interrelations between the functioning plant and its physical environment. Research of this type, as well as many other types, eventually led scientists away from a natural history approach of the study of plants to an analysis of the dynamics of plant activities which characterize the present approach to modern botany.

Specialized Fields

With the establishment of taxonomy, morphology, and physiology other fields of specialization soon developed. One of these is *plant ecology,* the study of the interactions of plants with each other, with animals, and with the physical environment. The ecologist attempts to identify the various influences that affect plant growth, development, reproduction, and distribution in a particular geographic area. Ecological studies are concerned with soil factors, light, and water relations; the organisms of the soil that contribute to its fertility; and the animals which utilize the plants. Although ecological studies date back more than half a century, there has been a rapid and extensive development of the field during the last two decades, especially in the areas of population dynamics and the organization of populations into communities and ecosystems.

Plant pathology, which deals with plant diseases, was originally limited to studies relating to bacteria and parasites. Oddly enough, the significance of plant diseases in relation to the physiological conditions which they cause was not initially considered. In fact, cooperation between plant pathologists and other specialists was a much later development. This segregation between each of the new fields of investigation was a tremendous hindrance to the development of botany.

One comparatively new field of investigation which has developed within the present century is *plant breeding.* The scientific aspects of plant breeding derive from *genetics,* the study of plant heredity and variation. Plant geneticists study the ways in which plants transmit characteristics from one generation to another and the mechanisms which are involved in these transmissions. Not only have the principles of genetics been employed to develop superior varieties of plants, they have also been used to interpret the origin and evolution of living forms. In recent years, geneticists have ascertained the physical basis of inheritance, the manner in which genes act, the mechanisms by which genes replicate themselves, and the nature and causes of mutations.

Each of the abovementioned areas of study is broad enough to be considered in the context of either botany or zoology. In this book, as these areas are considered in more detail, emphasis will be placed on plants, but not to the exclusion of the rest of the living world and the environment.

Some branches of botany that are more exclusively identified with the plant world and are treated as such are *phycology,* the study of algae; *mycology,* the study of fungi; *agronomy,* the science of field crop production; *floriculture,* the study of ornamental plants; *horticulture,* the study of orchards, gardens, and greenhouse crops; *hydroponics,* the science dealing with growing plants in soilless media; and *economic botany,* the study of plants in relation to man.

Even these branches of botany, as specialized as they appear to be, are further subdivided by the research botanist. For example, some botanists may direct their research efforts within the narrow boundaries of selected plant diseases, such as those which infect tobacco or wheat plants. Such a specialty, within the more general field of plant pathology, represents a typical pattern of research as it is currently being conducted in the field of experimental botany.

PLANTS AND HUMAN AFFAIRS

Values of Plants

Plants and Food. All living organisms are chemical machines which require a steady and reliable supply of energy and nutrients. Radiant energy from the sun constitutes the principal source of available energy on earth. An all-important link between the energy provided by the sun and the world of living things is established by photo-synthetic activities of green plants. Photosynthesis is regarded as the most extensive chemical process on earth. Each year green plants convert some 200 billion tons of carbon from atmospheric carbon dioxide to carbohydrate. This amounts to approximately 100 times the combined mass of all products that man produces in a year.

Obviously, plant carbohydrates alone cannot provide all the needed nutrients to sustain life. In fact, additional chemical reactions take place within plants which transform the initial products of photosynthesis to form fats, proteins, vitamins, and other vital chemical compounds. Indeed, the link with the sun and subsequent chemical activity of green plants provides the total energy and nutritional needs for nearly all life on earth.

From the beginning man has always depended upon plants for food, shelter, clothing, fuel, and medicine. Today, in addition to providing a large variety of foods for millions of people, the field of agriculture furnishes many raw materials, such as wood, fibers, rubber, gums, resins, oils and waxes, and numerous other commercial products. Agriculture is one of the largest and most important industries in the world.

Foods of Early Civilizations. The history of the civilized world has been significantly influenced by certain plants which supply man with food. In fact, civilization began to take shape when man first began to systematically plant crops. This probably occurred between 13,000 and 9000 B.C. in the humid areas of southern Asia. During this period both root crop and cereal crop plants were already in evidence.

Whereas root crops are notably high in carbohydrate content, cereal crops are significantly high in protein value. Cereals also contain a reasonable balance of the essential human nutritional requirements. It should not be surprising that hardly any major civilization, including present ones, has ever existed without a great dependence upon at least one of the cereal crops. Cereals, such as corn, wheat, rice, and barley, have some basic advantages over most plant foods in that they are relatively easy to store and also provide a high yield per acre. The cereal crops provided early civilizations with the leisure time necessary to transmit and expand their knowledge. Perhaps it was the availability of this leisure time, not known to earlier nonagrarian nomads, which provided the impetus toward a more advanced civilization.

Probably the first cereals to be used were the "millets" (*Panicum* and *Pennisetum*), which first appeared near the mountainous terrain in tropical and subtropical regions. It is interesting to note that, in addition to the grain, other parts of these plants were utilized in the construction of "straw houses," which provided limited shelter for these primitive peoples. Millets were basically crops of the tropical regions and not suited for more temperate climates. With the spread of early Neolithic agriculture northward to Europe, other cereal crops grew better under the different climatic conditions to the north and they soon became the main sources of grain. Many of these cereal and root crop plants which were cultivated thousands of years ago are still in use today. A diet consisting only of grains and roots can become very monotonous, as one can readily imagine. It was for this reason that spices such as vanilla, ginger, cloves, cinammon, and pepper became so much in demand that at one time they were used as a substitute for money. The voyages of Columbus, Magellan, and Vasco da Gama during the fifteenth and sixteenth centuries were basically quests for spices rather than a desire to discover new lands. This is but one example which illustrates how the course of history has been significantly altered by certain plants.

Food and Modern Civilization. Although there are many unfortunate people in the world today who still subsist entirely on foods such as rice and wheat (and even these are often in short supply), those who are more fortunate can select from a wide variety of fruits and vegetables as well as nuts, cereals, and spices. Of course, the human diet often includes the edible flesh of swine, cattle, poultry, and fish. These are all produced at the expense of cereals, grains, grasses, legumes, algae, and other plants upon which these animals feed.

As the world population continues to increase at an unprecedented rate, an adequate supply of food will become more critical with each passing year. Many scientists predict that by the year 2000 the world population will increase to approximately 6 billion people. More conservative estimates place the annual population increase at about 1.8 per cent. This means that there is more than one extra mouth to feed every second. Continued improvement of public health measures has resulted in a decline of the death rate. Consequently, the rate of increase in population will rise. If human reproduction were to continue at its present rate, in about 1000 years the weight of the people on the earth would approximate the weight of the planet itself. Clearly, the population explosion cannot continue unabated for an indefinite period.

It is alarming to read almost daily suggestions for increasing food production "to keep up with the growing population," but only infrequently and usually in obscure places to come across authors who recognize the obvious fact that it is impossible to provide food for a population that continues to grow exponentially as ours is now doing. For example, suggestions have been proposed for using bacteria, fungi, or yeasts to convert petroleum directly into food for man. This process is touched upon again later in this chapter, and it is superficially attractive because it appears to be more efficient than first feeding the petroleum to refineries and then to tractors and other machines which eventually deliver food to us. It is, however, a melancholy fact that the metabolism of bacteria, fungi, and yeasts does not generate oxygen.

Plants and Oxygen.* The amount of oxygen in the atmosphere was negligible before the origin of living organisms that could carry on photosynthesis of the type characterizing green plants, and oxygen would disappear from the atmosphere now through natural geological processes if all green plants should be killed. The combination of green plants and oxidizing organisms, including animals, apparently became very efficient at taking oxygen from the atmosphere and returning it at equal rates so that the amount present remained virtually constant. Photosynthesis stops, of course, during the hours of darkness, and on land areas in high latitudes it practically stops during the winter. It continues in low latitudes (although often greatly reduced by seasonal drought) and in the oceans, and, fortunately, atmospheric circulation patterns are such that we have not yet had to be concerned that man would run out of oxygen to breathe at night or in winter.

However, in the United States alone, we are annually removing, largely by paving, 1 million acres from the cycle of photosynthetic productivity. We do not know to what extent we are inhibiting photosynthesis in either freshwater or marine environments by pollution. There appears to be no way for us to escape our dependence on green plants and no way for us to survive except by halting population growth.

Wood and Wood Derivatives. There are a number of direct uses of wood, which include lumber for construction and manufacture, as well as indirect uses such as paper, tannins, oils, rayon, resins, tars, charcoal, wood alcohol, acetic acid,

---

*This section was taken entirely from "Can the World Be Saved?" by LaMont C. Cole, published in *Bioscience,* July 1968.

and wood gas. The United States is the greatest per capita user of wood in the world (approximately 228 cubic feet per person). Forest conservationists estimate that about six times more timber is cut in this country than is replaced by normal growth. Unless more effective methods are employed to conserve our existing forest resources there will be inevitable and irreversible degradation of these resources.

Plants and Drugs. One of the most familiar drug plants is the opium poppy (*Papaver somniferum*), the source of opium. Among the derivatives of opium which are used as pain-killers is morphine. The European foxglove (*Digitalis purpurea*), another familiar plant, is the source of digitalin, a heart stimulant derived from the leaves of the plant. Cinchona trees, especially *Cinchona ledgeriana,* were found to produce quinine, which became the specific cure for malaria.

Numerous other drugs of plant origin have proved helpful in the treatment of various mental disorders. The much-discussed hallucinogenic agent LSD is derived from the fungus *Claviceps purpurea.* This drug is extremely dangerous when not used under strict medical supervision. Along with other hallucinogens such as peyote (from the cactus *Lophophora williamsii*), LSD has been investigated for its possible value in the treatment of certain mental diseases.

Table 1-1 lists other examples of well-known plant drugs, their sources, and principal uses.

Miscellaneous Products. A variety of other products which can be traced to plants include fibers (cotton, linen, jute, hemp, and ramie); beverages (tea, coffee, cocoa, and alcoholic beverages); gums (mucilages, soothing agents in cough medicines, stiffening agents in candy, ice creams, and meringues); essential oils (soaps, insect repellents, and perfumes); plant fats (coconut oil, olive oil, palm oil, and lubricants); and latex products (rubber for tires, tubes, and hoses).

TABLE 1-1. Plant Drug Sources and Principal Uses

Drug	Plant Source	Principal Use(s)
Atropine	*Atropa belladonna* (Belladonna)	Stimulant to sympathetic nervous system; dilates pupils
Cocaine	*Erythroxylon coca* (Coca plant)	Local anesthetic in dentistry
Curare	*Chondodenron tomentosum*	Muscle relaxant; treatment of rabies and lockjaw
Eucalyptus	*Eucalyptus globulus* (Eucalyptus tree)	Stimulant-antiseptic in nose and throat inflammations
Stramonium	*Datura stramonium* (Jimsonweed)	Asthma treatment
Penicillin	*Penicillium notatum* (a fungus)	Antibiotic
Reserpine	*Rauvolfia serpentina*	Treatment of mental disorders; sedative and depressant

Some Harmful Plants

Poison Ivy and Its Allies. Not all plants serve human needs. Some members of the plant community can cause illness and even death to man and domestic animals. Poison ivy (*Rhus radicans*) is one of the better known poisonous plants in America. When allergic individuals contact this plant (Figure 1-9) skin irritations characterized by water blisters and itching generally result. This condition will continue for approximately 1 week to 10 days unless medically treated.

Some people are apparently immune to the poison, a nonvolatile oil, but may lose their immunity with repeated exposure to the plant. Pacific poison oak (*Rhus toxicodendron*), more commonly found along the western coast of North America, produces similar effects when touched. Poison sumac (*Rhus vernix*) is also very poisonous to the touch. This smooth shrub is generally found in swampy areas and can be easily identified by its leaf structure and greenish-white berries.

Pollen and Hayfever. Wind-pollinated plants such as ragweed (*Ambrosia*) can cause hayfever to large numbers of people. Oaks, cottonwoods, sagebrush, and certain grasses also produce large quantities of pollen, which can be irritating to the mucous membranes of eyes, nose, and respiratory tract, often accompanied by fever and asthma.

Figure 1-9. *Rhus radicans* (poison ivy). The fruits of this plant are pale greenish white and the leaflets appear in clusters of threes.

Poisonous Plants. Some plants are poisonous when taken internally. For example, poison hemlock (*Conium masculatum*), a carrot-like noxious weed which is found in many parts of the United States, is extremely toxic to man and animals. A derivative of this plant was used by the Athenians to execute Socrates (469–399 B.C.). Poison hemlock sometimes poisons children who make whistles from its stem. Also, members of the nightshade family (Solanaceae), including henbane, belladonna, and jimsonweed, contain certain poisonous substances some of which can be used pharmacologically. Tobacco (*Nicotinana tabacum*), also a member of the nightshade family, has recently been linked with respiratory and cardiovascular disorders.

Many species of mushrooms produce toxic substances which, if eaten, can poison human beings and animals. Because it is very difficult to distinguish poisonous mushrooms from the edible variety, it is extremely unwise for the nonexpert to sample wild species of mushrooms. One species of mushrooms commonly known as the "death angel" can be fatal if eaten in sufficient quantity.

Some plants possess certain mechanical structures, such as spiny bristles, which can be harmful to domestic animals. The unicorn plant and the buffalo bur are examples of plants which when eaten by grazing animals will cause internal bleeding along the digestive tract.

Weeds. Botanically speaking, weeds are undesirable plants that compete with economically valuable plants for water, space, sunlight, and soil nutrients. Obviously, if their growth goes unchecked, weeds can have harmful effects on the growth of useful plants. A large variety of weed killers are currently being employed along with other techniques to inhibit or prevent the growth of weeds. One widely used weed killer, a synthetic hormone-like chemical compound called 2,4-D (2,4-dichlorophenoxyacetic acid), has been quite effective in eliminating certain weeds from lawns and farmlands.

Parasitic Plants. There are a variety of plants which parasitize healthy living organisms and as a result cause a tremendous amount of disease and often death. Many people have seen mistletoe at work sapping the energy of healthy plants. They possess modified roots which enter the tissues of the host and absorb plant nutrients. Other plants which have a long history of causing huge annual economic losses because of their parasitic activities belong to the non-chlorophyll-bearing group represented by bacteria, slime molds, and fungi.

PLANT SCIENCE TODAY

There is an enormous amount of research currently being done in the field of plant science, largely because of its close relationship to human affairs. However, research in the field of botany, like all scientific inquiry, is not limited to the immediate problems which confront the human race. Considerable research is being done solely to learn more about the plant world, without regard to socioeconomic factors. The future needs of mankind are often served by theoretical research not pointed toward a specific purpose. However, because of the gravity and immediacy of the problems, many active areas of plant study center around the supply of food.

Agricultural Research

Chemical Fertilizers. In a world in which the daily net increase of people exceeds 100,000, there exists an obvious need to seek new ways to produce food. One method involves the production and more effective distribution of chemical fertilizers. Each year there are about 30 million tons of basic plant nutrients—nitrogen, potassium, and phosphorus—added to the soil. At least four times this amount will be needed by the year 2000. This presents an enormous technological problem. In addition, there is a basic research problem which involves the complex process by which plants utilize soil nutrients. Many scientists around the world are engaged in the crucial problem of how to increase crop yield by the use of chemical fertilizers. Other related research includes better irrigation techniques, plant-breeding experiments, the use of growth-promoting substances, and techniques for combating destructive insects and parasitic fungus plants.

Saltwater Irrigation. Approximately one third of the land surface of the earth is occupied by arid and semiarid areas. Recent experimentation with saltwater irrigation indicates that many of these areas can be made agriculturally productive. For example, a surprisingly large number of plants can thrive in sandy soils when irrigated with salt water. It has long been believed that land plants, particularly angiosperms, have a very low tolerance for seawater. However, when planted in sandy soil and irrigated with seawater, certain cereals, vegetables, and flowers grow fairly well. This area of research offers an exciting challenge for future investigators.

Protein from Petroleum. Recently, biologists have discovered a process for converting petroleum to protein. It has been known for some time that certain molds commonly grow on petroleum and petroleum products. In 1952 Felix Just, a German biologist, reported that he succeeded in growing yeast plants on a waxy substance. This experiment led a team of French scientists to set up a pilot plant to determine whether petroleum, a chemical compound similar to wax, could support the growth of yeast. The objective of the research was to find a new source of protein. Yeast was grown on petroleum, dried, and purified in a form which contained more than 50 per cent high-quality, edible protein.

Algae as a Source of Food. Algae are a broad group of aquatic plants which play a major role in the economy of nature. Experiments have been conducted for some time to determine the feasibility of using algae as a food source. Although the food value of algae may be quite high, the present cost of production is impractical. The peoples of oriental countries, especially Japan and China, have supplemented their diets with marine algae for many years. The Japanese cultivate and harvest almost 100 species of various seaweeds. One of the most commonly consumed is the red alga *Porphyra*. Research is continually being conducted on colloidal extracts of marine algae which can be used as nutritional supplements.

Some species of blue-green algae such as *Tolypothrix tenuis* can add needed nitrogen to the soil, thus increasing its fertility. In some experiments done with rice crops, it has been demonstrated that a 20 per cent increase in yield can be accomplished by inoculating the soil with blue-green algae. In addition to increasing nitrogen content, blue-green algae will bind soil particles together, promoting maintenance of a high water content and a reduction in soil erosion.

Perhaps the time has come when population increases require that scientists find improved methods for cultivating blue-green algae along with other useful algae. *Nostoc* is a genus of algae which has proved to be quite palatable and possesses high protein content. This plant is eaten in parts of China and South America. Actually, most species of algae are unpalatable and serve best as a food for fish, poultry, and cattle destined for human consumption.

Other Plant Research

In addition to the application of plant research to food production and agriculture, an enormous amount of research is being conducted in the areas of molecular botany and plant physiology.

Botany at the Molecular Level. The development of highly sophisticated tools of research and the integration of the principles of chemistry and physics with those of botany have enabled research scientists to view and analyze the structure and activities of plants at exceedingly minute levels of organization. Such research comprises a comparatively new field of investigation termed *molecular botany.* Among the principal objectives of molecular botany are (1) to determine the nature of the structures and particles within plant cells, and (2) to ascertain the mechanisms by which different kinds of particles contribute to the efficiency of cellular activities and therefore to the well-being of the entire plant. Molecular botany is concerned with form and function at the cellular level.

One of the broad topics currently under investigation in the field of molecular botany is metabolism, the sum of all chemical reactions that occur in the cell. Plant scientists are attempting to determine how these reactions are directed and controlled and how they are related to each other and to the total functioning of the cell. Much of the knowledge concerning photosynthesis, respiration, enzymes, and DNA has been gained from such studies of metabolism.

Molecular botanists are also investigating the phenomena of growth and development. They are actively engaged in studies relating to mechanisms by which plant cells increase in size, to processes by which cells elongate, and to factors involved in the progressive change of one cell type into another.

Another area of inquiry is reproduction at the cellular level. Scientists have already determined that DNA is the hereditary material and they have proposed models to suggest how it is duplicated and then passed on to cells of a new generation. A great deal of emphasis is being placed on the manner in which genes exert control over cellular activities and the expression of genetic characteristics.

Inasmuch as an understanding of molecular botany is necessary for an understanding of modern plant science, entire chapters in this book have been devoted to such topics as metabolism, enzymes, photosynthesis, respiration, growth and development, and heredity.

Physiological Investigations. One of the most exciting and dynamic fields of current botanical research is plant physiology. The plant physiologist analyzes the component activities of the plant and attempts to relate them to the total functioning plant. In this way, the plant is viewed as a functionally integrated organism in constant interaction with its environment.

Plant physiologists are attempting to determine the kinds and amounts of substances needed for plant growth, the processes involved in procuring and transporting them, and where and how these materials are used by the plant. The manner in which growth and development are controlled by the plant is also an area of intensive research.

Physiological investigations are also underway to determine the effects of environmental factors on plant growth. Chief among these factors are light and temperature. Light, for example, is associated with responses such as flowering, and tropisms; temperature is related to seed germination.

The vast majority of plant growth and developmental responses appear to be the result of interactions between substances produced by the plant (hormones) and certain environmental factors. Many physiological investigations are attempting to ascertain the exact relationship between these interactions.

One of the most fascinating areas of current physiological research is that of determining how plants measure the passing of time based upon a period of approximately 24-hour intervals. Such cycles are called *circadian rhythms*. Investigations conducted by J. Woodland Hastings of Harvard University have shown that the single-celled marine alga *Gonyaulax polyedra* performs three separate functions that are related to three separate circadian rhythms: (1) photosynthesis, which is most efficient in the middle of the day; (2) cell division, which is specifically delimited to the hours just prior to dawn; and (3) bioluminescence (the emission of light by living organisms), which reaches a peak of brightness in the middle of the night.

Some plants, such as the common bean plant, raise their leaves during the day and lower them to the stem at night. These movements, called "sleep" movements, can be recorded by connecting the tip of a leaf to a pen by a thread, which records the movements on a rotating drum (Figure 1-10). These "sleep" movements occur at the same time intervals even when the plants are kept in darkness and continue until the plants die.

Circadian rhythms typically occur at intervals that are slightly longer or slightly shorter than 24 hours. Inasmuch as they do not occur on a regular 24-hour basis, it has been suggested that they are not dependent on any outside factor but are governed by factors within the organism itself. These factors, whatever their nature, represent the time-keeping machinery of the cell and are referred to as a *biological clock*. The biochemistry of the biological clock is unknown and the elucidation of its mechanism is one of the most challenging of all biological problems.

The list of topics currently being investigated

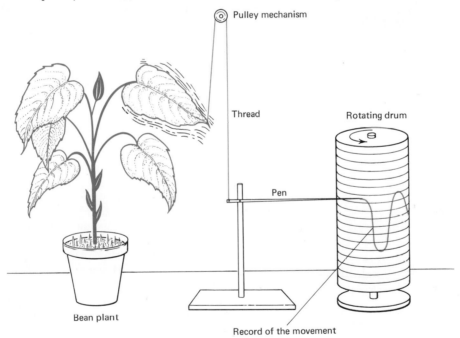

Figure 1-10. Procedure for recording the "sleep" movements of a bean plant. (Fig. 5.15 from *The Green Plant* by Arthur W. Galston. © 1968 by Prentice-Hall, Inc., Englewood Cliffs, N.J. Reprinted with permission.)

by plant scientists is quite extensive and with each passing year there are new problems and new challenges. Within the limits of this book, only a few of these areas of research will be discussed, some more superficially than others. Inasmuch as developments in the field of biochemistry during the last few decades have had an enormous impact on the study of botany, there is hardly a single area of plant science which has not been affected by biochemical research. In this regard, many areas of botany are biochemical in tone, and emphasis has been placed in this book on the chemistry of plant activities. Much of the necessary chemical background is presented in Chapter 2 so that the concepts of plant physiology treated in later chapters should be comprehended easily.

CHAPTER IN RETROSPECT

Plants are ubiquitous in their distribution and are the most conspicuous aspect of our outdoor environment. The biological importance of green plants lies chiefly in the fact that, through the process of photosynthesis, they can provide the energy requirements for practically all life on earth. From an historical point of view, civilization probably began when man developed methods to cultivate crop plants and ceased to wander about in constant search for food. The cereal crops (corn, wheat, rice, and barley) provided early man with the necessary leisure time to transmit and expand his knowledge.

There exists a huge variety of plant forms, which include algae, fungi, bryophytes, club mosses, horsetails, ferns, evergreens, and flowering plants. Most plants, including the majority of economically valuable ones, are members of the flowering (angiosperm) group.

The study of plants is subdivided into a variety of separate areas, such as taxonomy, morphology, anatomy, cytology, physiology, ecology, pathology, and genetics. Recent studies in each of these subjects have contributed greatly to the rapid growth of botany during the last few decades. Some areas of intensive research include saltwater irrigation, new sources of food, chemical fertilizers, molecular botany, and plant physiology.

Plants are essential to the survival of man, as they furnish food, oxygen, healing medicines, wood for construction and manufacture, and fibers for clothing. A few plants (poison ivy, ragweed, and poison hemlock) can cause illness and even death to both man and domestic animals.

Questions and Problems

1. Are the major plant groups (algae, fungi, bryophytes, gymnosperms, and angiosperms) represented within a 25-mile radius of where you live? List as many as you can and briefly describe the environmental setting in which each lives.
2. How have each of the specialized branches of botany contributed to the growth of this science?
3. Discuss the relationship between the availability of food and the advancement of civilization.
4. How have plants, other than crop plants, contributed to the general well-being of the human race? Compare the number of harmful and beneficial plants.
5. Briefly discuss several methods for increasing the supply of crop plants.
6. Compare recent botanical investigations with those of earlier botanists. What factors have contributed to the enormous amount of research currently being conducted in the field of plant science?
7. Describe how certain plants can be harmful or even fatal to man and domestic animals.
8. Assuming a world population increase of 100,000 daily, how many people would inhabit the earth 500 years from now? Discuss the problem of population increase in terms of (a) food supply, (b) air and water pollution, (c) plant research, (d) transportation, and (e) number of people per square mile of the earth's surface.

Suggested Supplementary Readings

Anderson, E., *Plants, Man, and Life*. Boston: Little, Brown & Company, 1952.

Baker, H. G., *Plants and Civilization*. Belmont, Calif.: Wadsworth Publishing Co., Inc., 1965.

Barry, J. M., *Molecular Biology: Genes and the Chemical Control of Living Cells*. Englewood Cliffs, N.J.: Prentice-Hall, Inc., 1964.

Campbell, N., *What Is Science?* New York: Dover Publications, Inc., 1952.

Cole, L. C., "Can the World Be Saved?", *Bioscience*, July 1968.

Davies, J. T., *The Scientific Approach*. New York: Academic Press, Inc., 1965.

DeBusk, A. G., *Molecular Genetics*. New York: The Macmillan Company, 1968.

Gabriel, M. L., and S. Fogel (editors), *Great Experiments in Biology*. Englewood Cliffs, N.J.: Prentice-Hall, Inc., 1955.

Hill, A. F., *Economic Botany*. New York: McGraw-Hill, Inc., 1952.

Hutchinson, J., and R. Melville, *The Story of Plants and Their Uses to Man*. London: P. R. Gawthorn, 1948.

Knobloch, I. W. (editor), *Selected Botanical Papers*. Englewood Cliffs, N.J.: Prentice-Hall, Inc., 1963.

Norman, A. G., "The Uniqueness of Plants," *American Scientist*, vol. 50, pp. 436–449, 1962.

Schery, R. W., *Plants for Man*. Englewood Cliffs, N.J.: Prentice-Hall, Inc., 1952.

Steere, W. C. (editor), *Fifty Years of Botany*. New York: McGraw-Hill, Inc., 1958.

Steiner, R. F., and H. Edelhoch, *Molecules and Life*. Princeton, N.J.: D. Van Nostrand Company, Inc., 1965.

Wald, G., "Innovation in Biology," *Scientific American*, September 1958.

Went, F., and the Editors of Life, *The Plants*. New York: Time-Life Books, 1963.

2
The Chemical Organization of Plants

INTRODUCTION

Decades ago, botanists analyzed plants in terms of gross macroscopic examinations. With the advent of better tools and techniques of investigation, plants were subjected to more intensive study. Perhaps the greatest influence on plant science in recent years has been the implementation and integration of the principles of chemistry and physics with those of botany. No science is an isolated branch of study; all are integrated in some manner. Today, the modern botanist is examining smaller and smaller components of plant organization and is interpreting the structures and activities of plants in terms of their constituent chemical components.

During the last century biologists have clearly established the concept that all living forms consist entirely of chemicals. Moreover, many data also support the hypothesis that life originated from chemicals present in the primitive atmosphere. Under the proper conditions of temperature, pressure, concentration, and other environmental factors, simple chemicals gradually combined into larger and larger units, each representing a more complex level of organization. After millions of years of chemical evolution, extremely large molecules arose which possessed those attributes ascribed to living forms. Once living things originated, in the form of cells, biological evolution became a dominant theme.

From the first cells thus produced, multicellular organisms, including plants, evolved. Many of these plants, conspicuous in our present-day flora, are composed of organs such as roots, stems, leaves, flowers, fruits, and seeds. These organs, in turn, are composed of tissues of many types, and an analysis of these tissues would reveal their distinct cellular nature.

Individual plants do not grow, develop, adapt, and reproduce as isolated organisms. They always function as part of a much larger environment consisting of other plants and animals as well as components of the nonliving world. In this regard, individual plants inhabit the earth as part of higher organizational levels in constant interac-

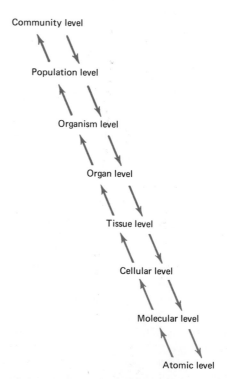

Figure 2-1. The structurally organized levels of multicellular plants. As the level progresses from atomic to community, there is an increase in the number of parts, the degree of complexity, and the energy requirements.

tion with the physical environment. The organizational hierarchy, of which multicellular plants are a part, may be represented by the following sequential levels from simplest to most complex: atomic, molecular, cellular, tissue, organ, organism, population (group of similar plants), and community (aggregate of different types of plants). These organizational levels are shown in Figure 2-1.

THE ATOMIC AND ELEMENTAL LEVELS OF PLANT ORGANIZATION

All forms of matter consist of a limited number of building blocks referred to as chemical **elements.** These fundamental substances are unique in that they cannot be decomposed into simpler substances by ordinary chemical reactions. At present there are 104 different chemical elements, of which 92 are naturally occurring and 12 are synthetic, or man-made elements. Iron, oxygen, aluminum, hydrogen, sodium, silver, and carbon are some familiar examples of elements. For convenience and consistency, elements are designated by letter abbreviations called *chemical symbols.* These symbols are usually derived from the first letter or two of the Latin or English name of the element (see Table 2-1).

Various chemical analyses of plants have revealed the presence of approximately 40 elements. Of these, however, only 15 are considered to be essential for plant growth and development. The key elements which play a salient role in the structural organization and functional activities of plants are carbon, hydrogen, oxygen, nitrogen, sulfur, and phosphorus. Carbon, hydrogen, and oxygen constitute approximately 99 per cent of the total weight of plants, and these elements are supplied from the atmosphere and water. Other essential plant elements include potassium, calcium, magnesium, and iron, and these are ob-

TABLE 2-1. Chemical Elements Commonly Found in Plants

Element	Chemical Symbol	Atomic Number
Elements essential for plant growth and development		
Hydrogen	H	1
Boron	B	5
Carbon	C	6
Nitrogen	N	7
Oxygen	O	8
Magnesium	Mg	12
Phosphorus	P	15
Sulfur	S	16
Potassium	K	19
Calcium	Ca	20
Manganese	Mn	25
Iron	Fe	26
Copper	Cu	29
Zinc	Zn	30
Molybdenum	Mo	42
Other elements		
Sodium	Na	11
Aluminum	Al	13
Silicon	Si	14
Chlorine	Cl	17

tained primarily from the soil in which plants live. Still others found in minute quantities, such as boron, copper, manganese, molybdenum, and zinc, are also derived from the soil environment. Table 2-1 lists some of the common chemical elements found in plants along with their respective chemical symbols and atomic numbers.

Atomic Structure

Every chemical element found in plants, and in all nature as well, consists of distinct particles called **atoms,** the smallest particles of elements which enter into chemical reactions. It has been estimated that the largest atoms are less than 1/50,000,000 inch in diameter and the smallest, those of the element hydrogen, are less than

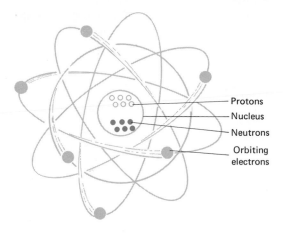

Figure 2-2. Atomic structure. An atom of an element consists of a relatively heavy, compact, centrally located nucleus and lighter particles called electrons that orbit about the nucleus at varying distances from its center.

1/250,000,000 inch in diameter. The degree of smallness of atoms is reflected by the fact that if 100 million (10^8) were placed end to end they would measure about 1 inch in length.

Essentially, an atom is composed of two basic portions, a centrally located *nucleus* and particles that orbit about the nucleus called *electrons* (*e*) (Figure 2-2). The nucleus of an atom is fixed, or nonreactive, in that it does not take part in chemical reactions. Within it are contained positively (+) charged particles called *protons* (*p*) and uncharged particles termed *neutrons* (*n*). The nucleus as a unit, therefore, bears a net positive charge. Both neutrons and protons have approximately the same mass (weight), being about 1890 times heavier than electrons. The electrons orbit about the nucleus in more or less definite regions called *shells* or *energy levels*. The charge on electrons is negative (−) and in all atoms the number of protons is equal to the number of electrons. In fact, an element is a substance whose atoms all

contain the same number of protons and the same number of electrons. The total positive charge on the nucleus equals the total negative charge of the electrons, so each atom is an electrically neutral unit. The number of electrons ranges from 1, in the case of hydrogen atoms, to more than 100 in some of the largest known atoms, such as mendelevium, which has 101 electrons. Atoms of elements are often listed in terms of *atomic number*, that is, the number of positive charges on the nucleus (see Table 2-1).

Electronic Configurations

For purposes of simplicity, somewhat at the expense of accuracy, it will be assumed that electrons occupy a definite orbit at a fixed distance from the nucleus and that these orbits form shells or energy levels. It will also be assumed, in the same context, that the innermost shell, or first energy level, designated as the *K* shell, can accommodate a maximum of 2 electrons; the next shell, the *L* shell, or second energy level, a maximum of 8 electrons; and the third shell, the *M* shell, or the third energy level, a maximum of 8 electrons if there are no more energy levels within a given atom. The *N, O,* and *P* shells, or fourth, fifth, and sixth energy levels, respectively, can each accommodate 18 electrons, although some variations exist with respect to this generalization. In terms of the positions of these shells relative to the nucleus, the *K* shell is closest and the *P* shell is farthest away. The distribution or arrangement of the electrons of atoms in various shells is referred to as an *electronic configuration*. Figure 2-3 shows the electronic configurations for the atoms of a few elements found in plants.

The various chemical properties of atoms are largely a function of the number of electrons in the outermost shell. Chemical reactions take place in such a manner as to produce the most stable electronic configurations in the outermost shells

Element	Electronic distribution Shell			Diagrammatic Representation of Electronic Distribution	Number of Outer Shell Electrons
	K	L	M		
Hydrogen	1				1
Carbon	2	4		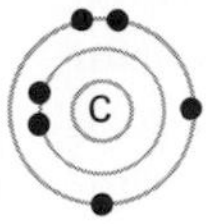	4
Nitrogen	2	5		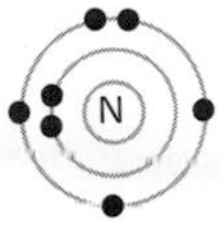	5
Oxygen	2	6		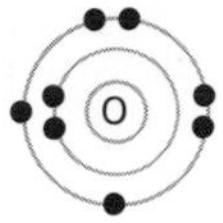	6
Magnesium	2	8	2	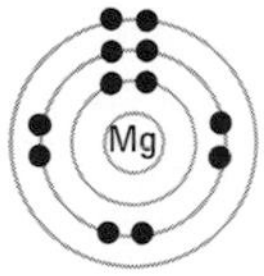	2
Phosphorus	2	8	5	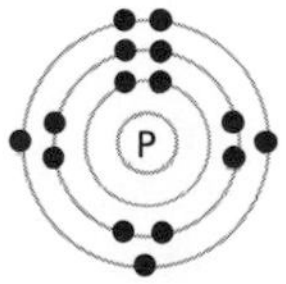	5
Sulfur	2	8	6	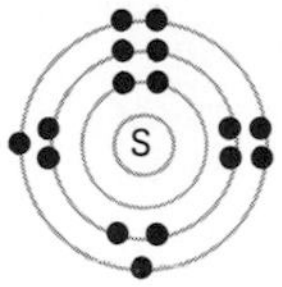	6

Figure 2-3. Electronic configurations for the atoms of a few elements found in plants.

of the reacting atoms. When this shell is completely filled in terms of the maximum number of electrons that it can hold, the atom does not react readily with other atoms and the atom is said to be stable, or *inert*. Helium (atomic number 2) and neon (atomic number 10) are examples of inert atoms that have filled outer shells. Helium has 2 electrons in the K shell, the outer shell for helium, and neon has 2 electrons in the K shell and 8 electrons in the L shell, the outer shell for neon.

Most atoms contain outer shells that are only partially filled. As a result they have unstable electronic configurations and they will tend to react with each other to attain stability. The degree of reactivity is dependent, at least in part, on the degree to which the outer shells are filled. Note the number of electrons in the outer shells for the atoms represented in Figure 2-3.

Valence and Chemical Bonding

When the outermost shell of an atom is only partially filled, it may be viewed as having either unfilled spaces or as having extra electrons in that shell. For example, an atom of oxygen ($K2$, $L6$) has two unfilled spaces in the L shell, whereas an atom of potassium ($K2$, $L8$, $M8$, $N1$) has 1 extra electron in the N shell. In order for these two atoms to gain stability, oxygen has to gain 2 electrons and potassium has to lose 1 electron. Atoms of elements combine so that the extra electrons in the outermost shell of some atoms can fill the spaces of the outermost shell of other atoms. The *valence*, or combining capacity, of an element is a numerical expression which represents the number of extra electrons or spaces in the outermost shell. For some elements the valence is a fixed value, whereas for others it may vary.

The atoms of most elements are capable of combining with each other to form more complex, stable aggregates called **molecules.** In the process

of combination the valence electrons in the outer shells of the combining atoms interact with each other in such a way that attractive forces are set up between the atomic nuclei involved. These attractive forces, which act to bind the nuclei together, are referred to as *chemical bonds*. In bond formation energy is required. As a result each chemical bond possesses a certain amount of potential chemical energy. Atoms of elements form only a specific number of chemical bonds; in this context, valence may also be viewed as the bonding capacity of elements. For example, hydrogen has a valence of 1, oxygen has a valence of 2, and carbon has a valence of 4. Essentially, this means that hydrogen can form one chemical bond, oxygen can form two, and carbon can form four chemical bonds with various atoms.

THE MOLECULAR LEVEL OF PLANT ORGANIZATION

As stated previously, atoms combine to form stable molecules. Basically, atoms form molecules and gain stability by completing the full complement of electrons in the outermost shells. For reasons far beyond the scope of this book, most, but not all, such molecules thus formed are stable. In general, atoms may gain stability by forming molecules in two ways: (1) by gaining or losing outer-shell electrons, and (2) by sharing outer-shell electrons. In both cases attractive forces are set up and chemical bonds are formed. When atoms gain or lose outer-shell electrons, the resulting chemical bond is referred to as an *ionic*, or *electrovalent*, *bond*. By contrast, when outer electrons are shared, the bond formed is called a *covalent bond*.

Ionic Bonding

To understand the concept of ionic bonding, consider first an atom of sodium. It has an atomic

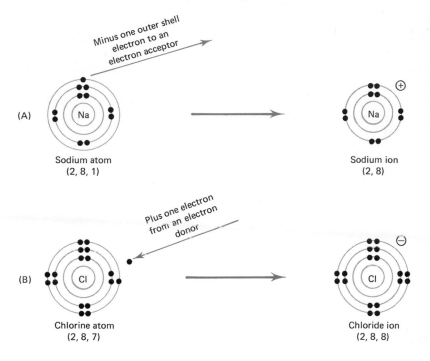

Minus one outer shell electron to an electron acceptor

(A)

Sodium atom
(2, 8, 1)

Sodium ion
(2, 8)

Plus one electron from an electron donor

(B)

Chlorine atom
(2, 8, 7)

Chloride ion
(2, 8, 8)

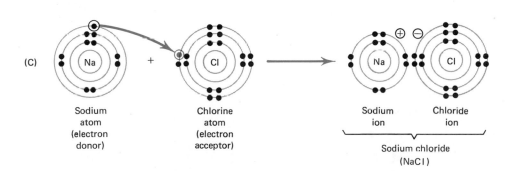

(C)

Sodium
atom
(electron
donor)

Chlorine
atom
(electron
acceptor)

Sodium
ion

Chloride
ion

Sodium chloride
(NaCl)

Figure 2-4. Ion formation and ionic bonding. **(A)** Sodium attains a stable configuration by acting as an electron donor, passing a single outer shell electron to an electron acceptor. **(B)** Chlorine acts as an electron acceptor in order to gain stability. **(C)** When a sodium ion and a chlorine ion combine, their nuclei are held together by an electrostatic attraction called an ionic bond. The substance formed from this attraction is sodium chloride, NaCl. Note that both the sodium ion and the chloride ion have stable electronic configurations.

number of 11 and a $2(K)$, $8(L)$, $1(M)$ electronic configuration. So that sodium can attain the stable electronic configuration $2(K)$, $8(L)$, the single electron in the M shell is passed on to an atom that would serve as an electron acceptor. Once it has lost the electron to gain stability, the sodium atom is left with one more proton than electrons and therefore has a net charge of $+1$. Any such charged atom or group of atoms is called an *ion* (Figure 2-4A).

Now consider an atom of chlorine. It has an atomic number of 17 and a 2(K), 8(L), 7(M) electronic distribution. In order for it to attain the stable electronic configuration 2(K), 8(L), 8(M), the chlorine atom accepts a single electron from an electron donor atom. In the process, as in the case of sodium, an ion is formed (Figure 2-4B).

The positively charged sodium ion and the negatively charged chloride ion tend to attract each other. This phenomenon may be explained on the basis that unlike charges attract each other. The resulting association, NaCl, or common table salt, is thus formed by the transfer of an electron from sodium to chlorine, and the ions are held together by the force of electrostatic attraction called an ionic bond (Figure 2-4C). Ionic compounds, such as sodium chloride, play an important role in the living economy and, as will be seen, many of the substances entering living plants are ions.

Covalent Bonding

Within living plants most atoms are bonded together to form molecules by sharing valence electrons. Once again, the most stable configura-

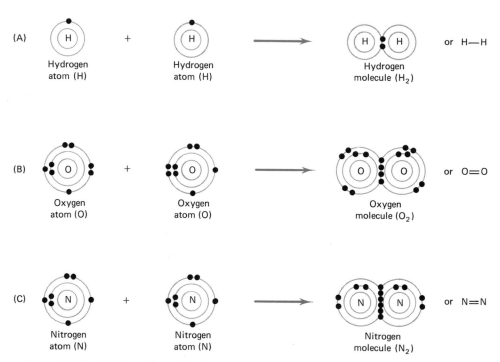

Figure 2-5. Covalent bond formation between atoms of the same element. (A) Single covalent bond between two hydrogen atoms. (B) Double covalent bond between two atoms of oxygen. (C) Triple covalent bond between two nitrogen atoms. Each covalent bond represents the sharing of two outer shell electrons. When a line (—) between atoms represents a single covalent bond, the above bondings would appear as shown to the right.

tion is one in which the outer shell contains the maximum number of electrons. This type of bonding, in which pairs of outer electrons of atoms are shared, is termed covalent bonding, and the bond shared between such atoms is called a covalent bond. Covalent bonds may be shared by atoms of the same element or by atoms of different elements. For example, two hydrogen atoms may be covalently bonded to form a molecule of hydrogen (H_2), in which each atom shares its outer electron with the other atom so that both have two electrons in the outer shell (Figure 2-5A).

In addition to the single covalent bond formed between one pair of electrons, atoms may also form double and triple covalent bonds. For example, when two atoms of oxygen bond together to form an oxygen molecule (O_2), they form a double bond, that is, a sharing of two pairs of electrons, because each oxygen atom needs two electrons to complete its outer shell (Figure 2-5B). Similarly, when two nitrogen atoms bond together to form a molecule of nitrogen (N_2), they form a triple covalent bond, because each atom needs three electrons to fill its outer shell (Figure 2-5C).

In the discussion of covalent bonds so far, single, double, and triple covalent bonds have been shown between similar atoms. Atoms of different elements also form covalent bonds (Figure 2-6). Consider the nature of covalent bonds formed between atoms in molecules of water (H_2O), carbon dioxide (CO_2), methane (CH_4), and ammonia (NH_3). In a molecule of water, each hydrogen atom forms a single covalent bond with the oxygen atom. By this type of sharing the oxygen atom has a full outer shell and each hydrogen atom also obtains a complete outer shell (Figure 2-6A). In the formation of carbon dioxide (Figure 2-6B), note that the carbon atom forms a double covalent bond with each oxygen atom. Methane (Figure 2-6C) represents a molecule in which the carbon atom forms a single covalent bond with each

hydrogen atom. Finally, a molecule of ammonia is formed (Figure 2-6D) by the nitrogen atom forming a single covalent bond with each hydrogen atom.

Note that in the formation of molecules of water, carbon dioxide, methane, and ammonia, as well as most other covalently bonded molecules, the component atoms have stable electronic configurations. In addition, it should be noted that paired electrons constitute a covalent bond. Finally, from these examples it becomes apparent that the number of covalent bonds formed by hydrogen is one, by carbon is four, by oxygen is two, and by nitrogen is three. Essentially, it is the numbers and types of atoms involved in molecule formation that determine whether a single, double, or triple covalent bond, or a combination of these, is formed.

Hydrogen Bonding

There is a third type of chemical bond, of special importance to all organisms, called a *hydrogen bond*. Hydrogen bonds usually occur between atoms of hydrogen and nitrogen and/or oxygen. Under certain circumstances when a hydrogen atom is chemically bonded to an atom of oxygen or a nitrogen atom, it may form, through a weak electrostatic attraction, a bridging bond to another nitrogen or oxygen atom in its vicinity.

An atom of nitrogen and an atom of oxygen are connected to a hydrogen atom with the hydrogen atom covalently bonded to either of the two atoms. Using (– – –) to represent hydrogen bonds, and (—) to represent covalent bonds, the following examples of hydrogen bonding may be considered:

$$O—H\text{-}\text{-}\text{-}O$$
$$O\text{-}\text{-}\text{-}H—N$$
$$N—H\text{-}\text{-}\text{-}N$$

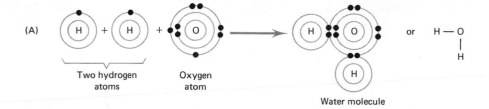

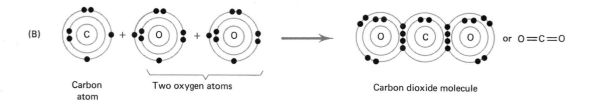

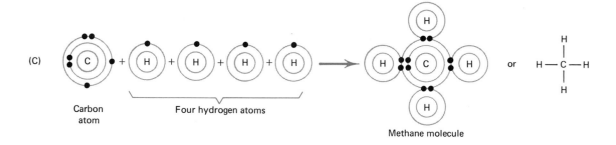

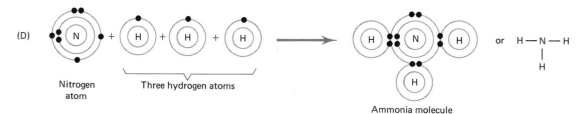

Figure 2-6. Covalent bond formation between atoms of different elements. (**A**) Water. (**B**) Carbon dioxide. (**C**) Methane. (**D**) Ammonia. Each of the molecules shown has a stable electronic configuration and each atom forms a limited number of covalent bonds with other atoms. Note bondings to the right as shown in Figure 2-5.

By way of comparison, hydrogen bonds are considerably weaker than either ionic or covalent bonds, being only about 5 per cent as strong. As a consequence, they are formed and broken relatively easily. This property is of special significance, because it accounts for the temporary bonding between certain atoms of large complex molecules such as proteins and nucleic acids.

INORGANIC COMPOUNDS FOUND IN PLANTS

It has been shown that atoms of the same element and atoms of different elements may combine chemically to form molecules. In essence, a vast array of molecules function as the structural building blocks of plants as well as the chemical functional units involved in the multiplicity of plant activities and chemical reactions. Any substance formed by the chemical combination of the atoms of two or more elements is called a **compound.** The compound thus formed will have different characteristics from those of the elements forming it. Whereas the fundamental unit of an element is an atom, the fundamental unit of a compound is a molecule. The compounds found in plants may be subdivided into two general types: (1) inorganic compounds, and (2) organic compounds. *Inorganic compounds* usually lack carbon, dissolve readily in water, resist decomposition (breakdown), undergo rapid chemical reactions, and are generally ionic in nature. *Organic compounds* are derived from living organisms and contain carbon as an essential constituent.

Inorganic Molecules and Ions

Water. An actively growing plant requires a wide variety of inorganic compounds for growth, repair, maintenance, and continuity. Water is one of the most abundant as well as the most important of these. Water is present in plants in amounts ranging from 5 to 90 or more per cent. In general, living plants contain 80 to 90 per cent water. Water is important to plants from a structural standpoint, from a chemical standpoint, and because of certain of its physical properties.

Structurally, water enters into the formation of various cellular structures such as cytoplasmic components, cell walls, and internal storage areas. Chemically, water also assumes a prominent role in that it serves as a medium of transportation for substances from the external environment and from one part of the plant to another. It is also the medium in which biochemical reactions in plants take place and is often a reactant or product of these reactions. Unless water is present in sufficient quantities, biochemical activities are inefficient. In a chemical sense, water is also a major source of hydrogen and oxygen which becomes incorporated into more complex molecules and an electron donor in oxidation–reduction reactions.

In addition to structural and chemical activities, water assumes another important role in plants, owing to its high specific heat. The *specific heat* is the amount of heat required to raise the temperature of 1 gram of a substance 1°C. A given quantity of water, as opposed to the same quantity of many other materials, requires a much greater amount of heat to increase its temperature. Without making a detailed exploration of the reasons this is so, it may be noted in passing that the additional structure introduced into aqueous systems by hydrogen bonding plays a major role in raising its specific heat. This high specific heat is one factor involved in preventing the destruction or dehydration of land plants that are exposed to direct solar radiation or aquatic plants that live on the surface of small bodies of water.

Minerals. From the soil higher plants obtain, besides water, inorganic substances that are essential for the formation of certain structural molecules as well as for the proper functioning of plant components. Much of the knowledge concerning what minerals are required by green plants has been obtained by growing plants in water cultures to which have been added appropriate inorganic substances in varying concentrations. According to this procedure, called *hydroponics,* if a particular element is omitted, poor growth and deficiency symptoms appear, indicating that the element in question can be considered essential. It was discovered that for plants to grow well the following elements are required in large amounts: nitrogen, potassium, calcium, phosphorus, sulfur, and magnesium. These six major elements, together with carbon, hydrogen, and oxygen, are the ones required in largest quantities by higher plants. Other elements, such as iron, boron, manganese, zinc, copper, and molybdenum, are also considered to be essential for the normal growth and development of plants, although they are required in extremely minute quantities. There is now evidence that some plants may also require varying quantities of yet other elements, such as cobalt, chlorine, sodium, silicon, iodine, aluminum, and vanadium. The exact role of these elements in plants is not completely understood.

Solutions and Ionization. The necessary mineral elements as well as all chemical substances that enter plants from the external environment are in solution in water. Two common examples of a solution are sugar in water and salt in water. A *solution* is a homogeneous mixture of two or more components, that is, a mixture in which the components cannot be distinguished because of their uniform distribution. The dissolved substances, sugar and salt, are referred to as *solutes,* and the water in which the solutes are dissolved is called the *solvent.*

Certain chemical compounds when placed in a water solution will *ionize;* that is, they will transfer one or more electrons from one atom to another.

In the process ions are formed. Some compounds ionize only partially, whereas others undergo more complete ionization. Consider the ionization of common table salt, sodium chloride (NaCl). This compound contains sodium (Na) and chlorine (Cl) and ionizes as follows:

$$NaCl \xrightarrow{\text{in water}} Na^+ + Cl^-$$

sodium   sodium   chloride
chloride

In this example an electron has been transferred from the sodium atom to the chlorine atom. This transfer accounts for the formation of the ions. Note that the water solution in which this ionization might occur is electrically neutral, because there are equal numbers of positively and negatively charged particles.

The mineral elements required by plants are always taken up in combined form and never as free elements in the soil solution in which plants live. Usually this combined form exists as ions of salts. A *salt* is a chemical compound that produces neither hydrogen ions (H^+) nor hydroxyl ions (OH^-) upon ionization. Common examples of salts which furnish ions for plant growth are potassium nitrate, KNO_3; calcium phosphate, $Ca_3(PO_4)_2$; ammonium sulfate, $(NH_4)_2SO_4$; and magnesium sulfate, $MgSO_4$. Upon ionization, these salts may provide plants with ions of nitrate, NO_3^-; ammonium, NH_4^+; sulfate, SO_4^{--}; phosphate, PO_4^{3-}; potassium, K^+; calcium, Ca^{++}, and magnesium, Mg^{++}. Nitrogen, the substance required in the largest amounts, is obtained from the soil as nitrate or ammonium ions. Sulfur is obtained as sulfate and phosphorus as phosphate. Other elements may be utilized as simple ions.

Acids and Bases. In addition to salts, two other classes of compounds also ionize in a water solution. These are acids and bases. There are a number of different definitions of acids and bases, but for the purposes of this textbook an *acid* may be defined as a substance that, upon ionization, yields hydrogen ions (H^+). A *base,* or alkali, may be defined as a substance that, upon ionization, produces hydroxyl ions (OH^-). The ionization of hydrochloric acid (HCl), an inorganic acid, and sodium hydroxide (NaOH), a base, are shown in Figure 2-7. A compound resulting from the chemical combination of an acid and a base is a salt (Figure 2-7C).

Plants contain all the necessary ions for the formation of most of the common inorganic acids and bases. Representative examples of the acids are hydrochloric acid (HCl), nitric acid (HNO_3), sulfuric acid (H_2SO_4), and phosphoric acid (H_3PO_4). Representative bases are sodium hydroxide (NaOH), potassium hydroxide (KOH), ammonium hydroxide (NH_4OH), and calcium hydroxide [$Ca(OH)_2$]. Note that in each of these examples the acids are capable of producing H^+ ions and the bases may furnish OH^- ions.

Hydrogen Ion Concentration (pH). The degree of acidity or alkalinity in plant systems is a critical factor in terms of both structural and functional integrity. The slightest deviation from the proper acidic or basic concentration can significantly alter the structure of large organic molecules or critically disturb the balance of certain plant chemical reactions. The degree of acidity or alkalinity of a given solution is expressed in terms of *pH,* a number which represents a measure of the relative acidity, that is, the concentration of H^+ ions, or alkalinity, the concentration of OH^- ions. The measure is graded on a scale from 0 to 14. A pH below 7 indicates an acid solution, one in which there are more H^+ than OH^- ions. A pH above 7 indicates an alkaline solution, one that contains more OH^- than H^+ ions. The midpoint, 7, represents neutrality, or

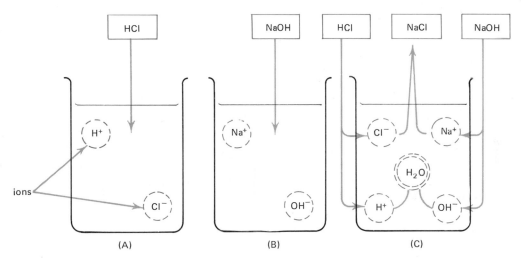

Figure 2-7. Acids, bases, and salts. **(A)** In a water medium, hydrochloric acid, HCl, produces H^+ ions. **(B)** Sodium hydroxide, NaOH, a base, produces OH^- ions. **(C)** NaCl, a salt, is formed by the interaction of an acid and a base.

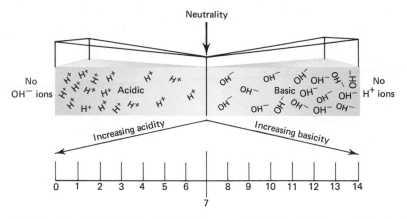

Figure 2-8. The pH scale. A pH below 7 indicates an acid solution, that is, a solution in which there are more H^+ ions than OH^- ions. The lower the numerical value, the higher the H^+ ion concentration. The opposite holds true for the concentration of OH^- ions. As the numerical expression increases, the OH^- ion concentration increases and the solution becomes increasingly basic. At pH 7, the concentrations of H^+ ions and OH^- ions are equal and the solution is neutral.

equal concentration of H^+ and OH^- ions. Water, for example, has a pH of 7 (Figure 2-8).

All chemical reactions taking place within living plants are limited in their ability to occur within fairly narrow ranges of pH on either side of neutrality. Any significant deviations from these ranges seriously disturb the structural framework and functional operations of plant cells. It becomes evident, in view of the fact that plants are constantly producing acids and bases as part of normal chemical reactions, that there must be some mechanistic control to maintain a constant pH.

In general, the maintenance of a constant pH is accomplished in two ways: (1) by the neutralization of acids and bases, and (2) by the existence of a buffer system. It will be recalled that acids and bases may combine to form salts, which upon ionization produce neither H^+ nor OH^- ions. Thus, if excessive H^+ or OH^- ions are introduced,

the H^+ ions may combine with the *anions* (negatively charged ions) of a base to form a salt and water. This process is termed *neutralization*. In this manner the H^+ ion concentration is kept relatively constant (see Figure 2-7).

A second way in which living matter is protected to some extent against pH change is by a *buffer system*. Such a system consists of an acid which ionizes slightly and the salt of that acid. Sodium bicarbonate ($NaHCO_3$), a salt of carbonic acid (H_2CO_3), is normally present in cells. This salt ionizes into sodium ions (Na^+) and bicarbonate ions (HCO_3^-). If excessive quantities of H^+ ions from the ionization of HCl were present in a cell, it should be expected that the solution would become more acid. Actually, there is only a negligible alteration of the pH, because the excess H^+ ions combine with the HCO_3^- ions of sodium bicarbonate to form carbonic acid. Inasmuch as carbonic acid ionizes very slightly, the

H$^+$ ions of the hydrochloric acid are removed from the solution by being "tied up" to the bicarbonate ions (Figure 2–9).

ORGANIC COMPOUNDS FOUND IN PLANTS

Whereas inorganic compounds, excluding water, constitute only 1 to 1.5 per cent of plant material, organic compounds constitute the bulk of the dry weight. This does not imply, however, that the two classes of compounds exist in plants as separate and distinct molecules based upon their relative proportions. On the contrary, inorganic molecules serve as raw materials for the synthesis of organic compounds, and they are interrelated through various chemical processes.

Organic compounds in plants are considerably larger and structurally more complex than inorganic molecules. Many organic molecules contain hundreds and even thousands of component atoms. The atoms found in the greatest abundance in these compounds are carbon, hydrogen, and oxygen. In addition, many contain nitrogen, phosphorus, and sulfur. To a lesser extent, other atoms, such as chlorine, magnesium, zinc, and calcium, are found as constituents of organic compounds. The principal organic compounds found in plants are carbohydrates, lipids, proteins, enzymes, nucleic acids, pigments, and hormones (growth regulators).

Carbohydrates

Carbohydrates represent a large group of organic compounds that play various significant roles in the functioning of plants. The importance of carbohydrates to plants is reflected by the fact that they comprise the bulk of the dry weight of the plant and are the basic molecules manufactured during the process of photosynthesis.

(A) Buffer system } Carbonic acid (H_2CO_3)—a slightly ionizing acid
Sodium bicarbonate ($NaHCO_3$)—a salt of a slightly ionizing acid

(B) Ionization of HCl and $NaHCO_3$ in solution }
$HCl \longrightarrow H^+ + Cl^-$
$NaHCO_3 \longrightarrow Na^+ + HCO_3^-$

(C) The buffering reaction }
$H^+ + HCO_3^- \longrightarrow H_2CO_3$
$Cl^- + Na^+ \longrightarrow NaCl$

Figure 2-9. Maintenance of a relatively constant pH by a buffer system. (**A**) Components of a buffer system. (**B**) In solution, hydrochloric acid, HCl, produces excess H$^+$ ions and sodium bicarbonate, $NaHCO_3$, ionizes into sodium ions, Na$^+$, and bicarbonate ions, HCO_3^-. (**C**) The buffering reaction. The H$^+$ ions that would normally make the solution more acidic by the lowering of the pH are combined with the bicarbonate ions to form carbonic acid and salt. Carbonic acid ties up the H$^+$ ions because it ionizes only very slightly. In this manner, excess H$^+$ ions are prevented from going into the solution. The principal mechanism of the buffering reaction is that bicarbonate ions tie up excess H$^+$ ions, sodium and chloride ions combine, and carbonic acid and a salt are produced.

Within living plants, carbohydrates are found as (1) structural components in the makeup of cell walls, (2) integral parts of the cytoplasm, (3) food storage products, (4) compounds which provide energy, and (5) raw materials for the synthesis of various other organic compounds.

Carbohydrates such as sugars and starches all contain carbon, hydrogen, and oxygen. Typically, the hydrogen and oxygen atoms are present in a 2:1 ratio, the same proportions as these atoms appear in water (H_2O). The molecular formula for carbohydrates is $(CH_2O)_n$, where n may represent any number, although there are certain exceptions to this rule. A *molecular formula* is a symbolic expression of the specific number and types of atoms or groups of atoms present in a given molecule.

Monosaccharides. The basic building blocks of carbohydrates are simple sugars, or **monosaccharides.** This class of carbohydrates consists of compounds containing three, four, five, six, or seven carbon atoms, the most important being the monosaccharides containing either five or six carbon atoms. Collectively considered, the names of monosaccharides end in -ose (meaning a sugar), and the prefix designates the number of carbon atoms present (see Table 2-2).

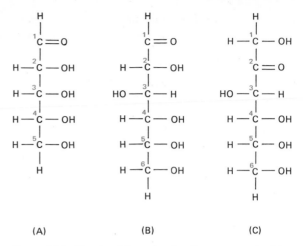

Figure 2-10. Structural formulas for the monosaccharides **(A)** ribose, **(B)** glucose, and **(C)** fructose. Ribose contains five carbon atoms while glucose and fructose each contain six carbon atoms. For simplicity and for reference, the carbon atoms are numbered.

It is often convenient to designate chemical compounds by a *structural formula,* that is, the arrangement of atoms within the molecule. In addition, for simplicity and from a reference standpoint, the carbon atoms of organic compounds are numbered. The structural formulas for the monosaccharides ribose (a pentose), glucose (a hexose), and fructose (a hexose) are shown in Figure 2-10. By comparing the molecular formulas for glucose and fructose, it should be noted that both monosaccharides contain 6 carbon atoms, 12 hydrogen atoms, and 6 oxygen atoms. The chemical properties for these compounds, however, are quite different. Any such molecules with similar chemical compositions but with different arrangements of the same atoms, and therefore different chemical properties, are referred to as *isomers,* and the phenomenon is called *isomerism.*

Some monosaccharides, such as glucose and fructose, assume a molecular configuration resem-

TABLE 2-2. Classification of Monosaccharides by Number of Carbon Atoms and Molecular Formulas

Sugar	Number of Carbon Atoms	Molecular Formula $(CH_2O)_n$	Value for n
Triose	3	$C_3H_6O_3$	3
Tetrose	4	$C_4H_8O_4$	4
Pentose	5	$C_5H_{10}O_5$	5
Hexose	6	$C_6H_{12}O_6$	6
Heptose	7	$C_7H_{14}O_7$	7

bling a ring-shaped arrangement of atoms. This form, called a *ring form,* is predominant when these compounds are in solution in plant cells. By comparison, the *straight-chain,* or linear, arrangement of atoms for molecules of glucose and fructose along with ribose are shown in Figure 2-10. Essentially, when atoms of molecules assume a ring form, specific carbon atoms are joined to each other via an "oxygen bridge." In the case of glucose, the first and fifth carbon atoms are bridged together, whereas in fructose the bridge is between the second and sixth carbon atoms (Figure 2-11).

Of all plant monosaccharides, glucose and fructose are most common. **Glucose** occurs in sweet fruits, honey, and fruit derivatives and is probably a universal sugar of plant cells in the free state (uncombined) or in combination with other chemical compounds. During photosynthesis, glucose is the major product of the reaction and it serves as the principal energy-supplying molecule within plant cells as well as being one of the molecules involved in building up protoplasm. In fact, many carbohydrates are converted into glucose before they are utilized as energy-producing molecules or molecules involved in synthetic reactions.

Fructose, or fruit sugar, is found in sweet fruits and honey together with glucose. To a somewhat lesser extent, fructose is utilized in plants in much the same manner as glucose.

One other group of monosaccharides that should be mentioned at this point is that of the five-carbon sugars, or *pentoses.* Three pentoses commonly found in plants as components of other molecules are *ribose, deoxyribose,* and *ribulose.* These sugars are important structural components of more complex molecules and assume a number of significant roles in various plant processes.

Disaccharides. Within plant cells two monosaccharides may combine with each other, and the resulting molecule is called a **disaccharide,** a sec-

ond class of carbohydrates. In the process of disaccharide formation, a molecule of water is lost and the reaction is termed *dehydration synthesis.* An example of this type of reaction, in which the disaccharide maltose is formed from two glucose molecules, may be represented as follows:

$$C_6H_{12}O_6 + C_6H_{12}O_6 \longrightarrow C_{12}H_{22}O_{11} + H_2O$$
$$\text{glucose} \qquad \text{glucose} \qquad \text{maltose} \qquad \text{water}$$

The chemical combination of two glucose molecules (monosaccharides) produces the disaccharide **maltose,** or malt sugar. Note that the molecular formula for disaccharides, as represented by maltose, is $C_{12}H_{22}O_{11}$. In a similar manner, the chemical combination (dehydration synthesis) of molecules of glucose and fructose will produce the disaccharide, **sucrose** (common table sugar). This sugar is found in abundance in plant cells and is widely distributed in nature.

The basic mechanism of disaccharide formation involves the union of smaller molecules, with the loss of water, to form a larger, more complex molecule. Many other organic compounds in plant cells, besides carbohydrates, are synthesized in a similar manner. It should be noted that dehydration synthesis is only one mechanism by which large molecules are made from smaller ones. Other processes involve more than the mere elimination of water.

It is also possible for large molecules to be broken down into smaller, simpler ones with the addition of water. This reverse chemical process is called **hydrolysis,** or **digestion.** A molecule of sucrose, for example, may be hydrolyzed into glucose and fructose, and a molecule of maltose may undergo hydrolysis into two glucose molecules. Figure 2-12 represents the dehydration synthesis and hydrolysis of the disaccharides maltose and sucrose.

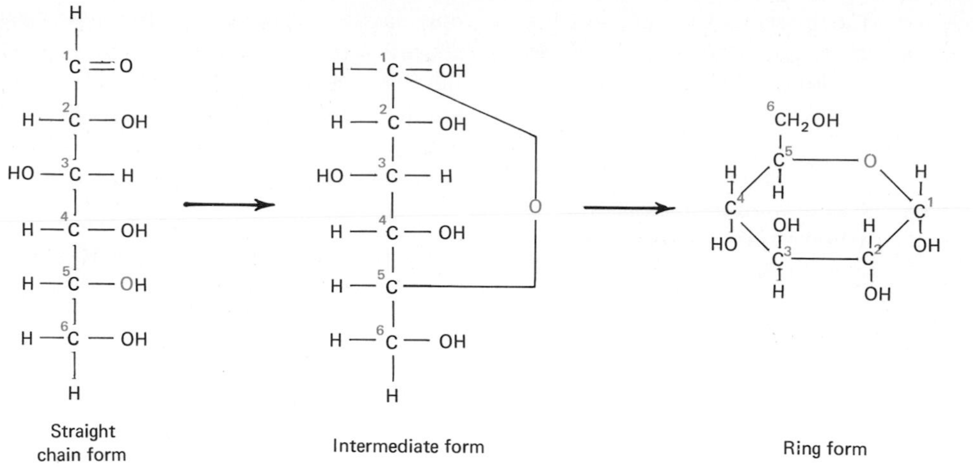

Figure 2-11. Ring forms for (**A**) glucose and (**B**) fructose. Note the relative positions of the numbered carbon atoms as each molecule assumes a ring configuration as well as the oxygen bridge between carbon atoms.

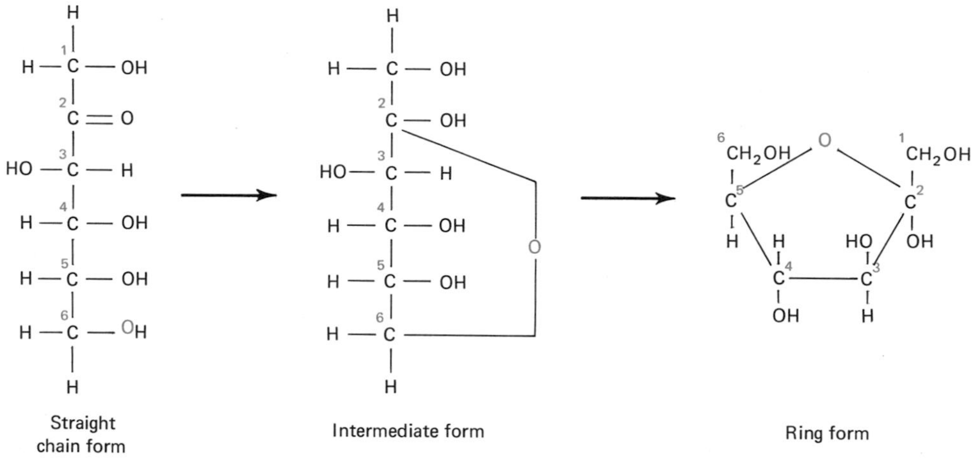

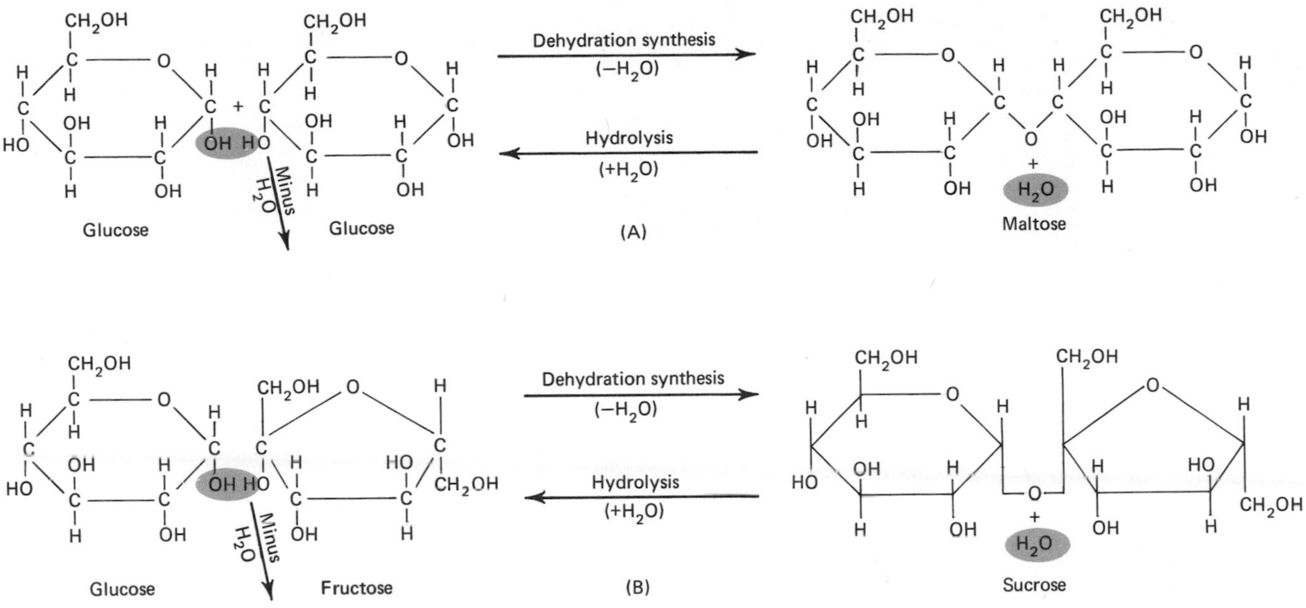

Figure 2-12. Dehydration synthesis and hydrolysis of (**A**) maltose and (**B**) sucrose. The synthetic reaction involves, among other mechanisms, the removal of water, whereas the hydrolysis reaction involves the addition of water. These reactions may be viewed as reverse chemical processes.

Polysaccharides. The dehydration synthesis of three or more monosaccharides results in the formation of a third class of carbohydrates called **polysaccharides.** Polysaccharides have the molecular formula $(C_6H_{10}O_5)_n$. Like disaccharides, they can be broken down into their constituent sugars by hydrolysis. Unlike either monosaccharides or disaccharides, they are not usually soluble in water and they lack the characteristic sweetness of sugars. Chief among the important plant polysaccharides are starches, cellulose, and inulin.

Starches are found in most green plants, particularly more complex land plants, as the principal food reserve or storage product. Typically, starches are stored in seeds, stems, leaves, and roots of plants in the form of *starch grains.* Starch grains consist of concentrically oriented layers of starch molecules. The concentric ring arrangement of starch molecules is due to the fact that during the day, when the plant is most active, starch is produced and deposited more densely than at night, when the plant is less active. Starch grains vary greatly in size and number in different plants, but the consistency of form and shape is adequately diagnostic for identifying certain plants (Figure 2-13).

Chemically, starches consist of a long series of glucose molecules bonded to each other. As a result of the manner in which this bonding occurs, starch molecules reveal two distinguishable components: amylose and amylopectin. The *amylose* portion consists of unbranched coiled chains of glucose molecules and may contain from 300 to 1000 such glucose units in a single, long molecule. *Amylopectin,* on the other hand, exhibits a fairly high degree of branching. Although the per-

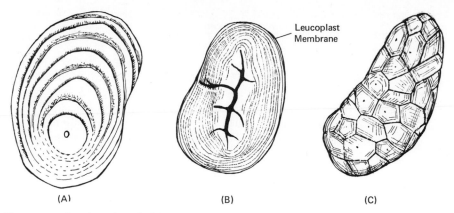

Figure 2-13. Starch grains of (**A**) potato, (**B**) bean, and (**C**) rice. Note the concentric layering effect due to the deposition of starch molecules. The outermost margin represents the membrane of the leucoplast.

centages of amylose and amylopectin differ widely in starches from different kinds of plants, most plants contain from 20 to 30 per cent amylose (Figure 2-14). The repetition of many simple, smaller molecules, such as glucose, into larger, chainlike molecules, such as starch, is called *polymerization,* and the molecule thus formed is termed a *polymer.* In all cases, polymerized molecules are fairly large.

Cellulose is an important structural component of the cell walls of higher plants. The widespread distribution of cellulose would rank it as one of the most abundant organic compounds on earth. Generally, it is found in combination with other compounds in plants, such as lignin; however, one source of pure cellulose is cotton fiber, a product of cotton plants. Other varieties of pure cellulose include flax and hemp, the raw materials of textiles. Chemically, cellulose is a polymer of 1000 or more glucose molecules. Like amylose, it is unbranched, but, unlike amylose, it is not coiled. Thousands of cellulose molecules are chemically bonded to each other in an intricate, organized network to form plant cell walls. The basic physical properties of cell walls—tensile strength, elasticity, and plasticity—are directly dependent upon the amount and arrangement of the cellulose molecules. Figure 3-13 represents the organizational development of cellulose into cell walls.

A third important polysaccharide is **inulin,** a fructose polymer. In certain plants, such as dandelion, goldenrod, and Jerusalem artichoke, inulin is accumulated as storage product instead of starch. In these plants the inulin is stored in underground portions, although, in some cases, it may be stored in leaves.

Lipids

Lipids constitute a second major group of organic compounds found in plants. Like carbohydrates, they are composed of atoms of carbon, hydrogen, and oxygen. They differ from carbohydrates, however, in that some may contain nitrogen and phosphorus, and the proportionate number of oxygen atoms is fewer. Most are insoluble in water. Lipids may be subdivided into three classes: (1) true fats, (2) waxes, and (3) phospholipids, or lipids containing phosphorus.

True Fats (Oils). **Fats,** the simplest and most abundant plant lipids, are food storage products of seeds and serve as a source of energy during the early stages of plant growth and development. Assuming equal quantities of a fat and a carbohydrate, fats would provide twice as much energy for various plant processes. Other plant organs, such as roots, stems, and leaves, also contain fats (about 5 per cent of the total dry weight). Fats are economically important to man as a source of nutrition from soybeans, peanuts, corn, and other fruits and seeds. Fats are found within various plant cells in the form of droplets or globules, owing to their insolubility in water.

Chemically, a molecule of fat is composed of two structural components: an alcohol called **glycerol** and a group of compounds collectively designated **fatty acids.** A single molecule of glycerol consists of three carbon atoms to which there are attached three *hydroxyl* or alcohol (OH) groups. The structural formula for glycerol is shown in Figure 2-15A.

Fatty acids are compounds with even numbers of carbon atoms, ranging from 12 to 26 in plants.

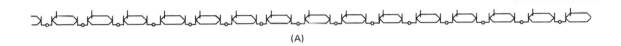

(A)

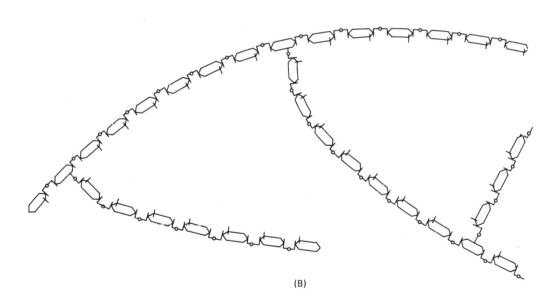

(B)

Figure 2-14. Arrangement of glucose molecules in (**A**) amylose and (**B**) amylopectin. Amylose is an unbranched, coiled, helical-shaped molecule, whereas amylopectin, by contrast, is highly branched.

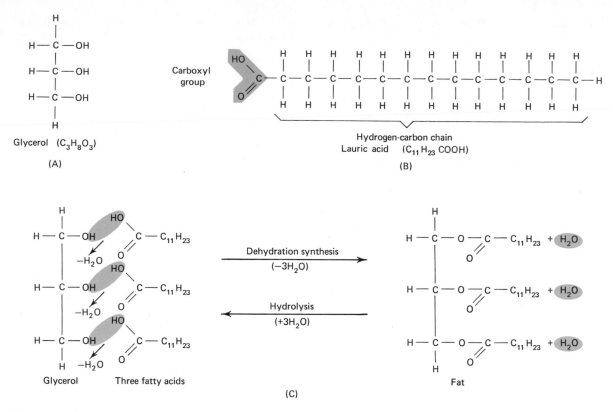

Figure 2-15. Structural formulas for **(A)** glycerol and **(B)** lauric acid, a fatty acid. **(C)** The chemical combination of a molecule of glycerol with three fatty acid (lauric) molecules forms a molecule of fat and three molecules of water (dehydration synthesis), and the addition of three water molecules to a fat forms glycerol and three fatty acids (hydrolysis).

All contain the *carboxyl* or acid group (—COOH), a characteristic of organic acids. *Lauric acid* ($C_{12}H_{24}O_2$), a common fatty acid of plants, contains 12 carbon atoms in a single chain. The structural formula for this fatty acid is shown in Figure 2-15B.

A single molecule of fat is formed when a molecule of glycerol combines with three fatty acid molecules. In the reaction, three molecules of water are removed. In the reverse reaction, hydrolysis, a fat molecule may be broken down into its component fatty acid and glycerol molecules. Figure 2-15C represents the synthesis and hydrolysis of a fat.

The three fatty acids involved in the formation of fats may be the same or they may be different. For example, three molecules of fatty acid A may combine with a glycerol molecule, or one molecule each of fatty acids A, B, and C may unite with a molecule of glycerol. As a result of the number

TABLE 2-3. Some Common Fatty Acids Found in Plants

Name	Structural Formula	Occurrence in Nature
Caprylic	$C_7H_{15}COOH$	Coconut oil
Lauric	$C_{11}H_{23}COOH$	Laurel kernel oil
Myristic	$C_{13}H_{27}COOH$	Nutmeg butter
Palmitic	$C_{15}H_{31}COOH$	Palm oil
Stearic	$C_{17}H_{35}COOH$	Cocoa butter
Geddic	$C_{33}H_{67}COOH$	Ghedda wax

of possible combinations, there are a great many different plant fats. Some of the common plant fatty acids are listed in Table 2-3.

Waxes. **Waxes,** a second class of lipids, are chemically composed of fatty acids combined with long-chained alcohols with 24 to 36 carbon atoms, not a three-carbon alcohol as in the case of fats. Plant waxes are found in combination with other substances as components of cell walls. Suberin and cutin are two such components commonly associated with plant waxes. **Suberin** is a water-proofing material found in the plant cell walls of cork. **Cutin** is found covering the epidermis of stems, leaves, and fruits. Both substances prevent excessive water loss and provide a certain degree of mechanical protection.

Phospholipids. These lipids contain glycerol, fatty acids, and in addition atoms of phosphorus and nitrogen. Of the three classes of lipids, **phospholipids** are least abundant in plants. They are involved in providing a structural framework for various cell membranes and cellular organelles. Various chemical analyses and electron-micrograph studies strongly indicate that the phospholipids are arranged in a double layer in forming these various membranes. In contrast to fats, they are not stored as reserve materials. Two examples of plant phospholipids are *cephalin* and *lecithin.*

Proteins

Proteins are far more complex molecules than either lipids or carbohydrates. Many biologists consider proteins to be the fundamental molecules of all living organisms. The distinguishing characteristics of living forms are reflected in the organization of their protein molecules. In plants, proteins may serve as a source of food and are involved in the processes of growth, repair, synthesis, and the production of new cells.

All proteins contain carbon, hydrogen, oxygen, and nitrogen. In addition, some contain sulfur and phosphorus. Just as the monosaccharide is the building block of carbohydrates, and fatty acids and glycerol are the building blocks of fats, **amino acids** are the fundamental units of proteins. Each of the 20 amino acids found in proteins of living organisms contains one or more basic *amino groups* ($-NH_2$) and one or more acidic carboxyl groups ($-COOH$). The structural formula for **glycine,** the simplest amino acid, is shown in Figure 2-16A.

When amino acids chemically unite so that the carboxyl group of one attaches to the amino group of another, water is lost. This, once again, is a dehydration synthesis reaction, although many other factors are involved. The chemical bond formed between adjacent amino acids is called a *peptide bond.*

In the example shown in Figure 2-16B, glycine and alanine undergo dehydration synthesis. The carboxyl group of glycine yields the OH and alanine supplies the H for the formation of water. The peptide bond is formed between the carbon atom of glycine and the nitrogen atom of alanine. The resulting compound, because it consists of two amino acids joined together via a peptide bond, is referred to as a *dipeptide.* The addition of another amino acid to a dipeptide would form a *tripeptide,* three amino acids connected by two peptide bonds. Further additions of amino acids would

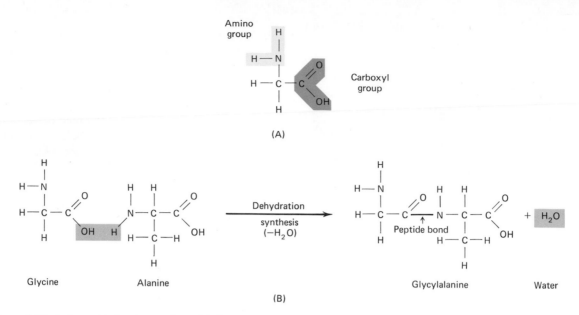

Figure 2-16. Amino acid structure and peptide bond formation. (A) Glycine, a representative amino acid, illustrates that all amino acids contain at least one amino (NH_2) group and at least one carboxyl (COOH) group. (B) When two or more amino acids are chemically united, a peptide bond is formed. In this example, the peptide bond is formed between glycine and alanine, and the resultant compound is glycylalanine, a dipeptide.

result in the formation of a long chainlike molecule referred to as a *polypeptide,* or a large protein molecule. Figure 2-17 lists some of the amino acids found in proteins with the standard abbreviated designations for each.

It should be obvious that an extremely large number of proteins is possible, owing to the number and sequences of amino acid linkages that can occur. The 20 amino acids found in plant proteins may be arranged in any order to form a sequence of hundreds of units in the formation of a single protein molecule. This situation is quite similar to having an alphabet of 20 letters with which to form words. Assuming that no prescribed sequence of letters exists to form these words, the number of possible words is exceedingly large.

The simplest plant proteins contain about 100 amino acids, whereas the average number of amino acids per protein molecule is 300 to 1000. For example, *zein,* a protein in corn, has the molecular formula $C_{736}H_{1161}N_{184}O_{208}S_3$, and *gliadin,* a protein in wheat, is represented as $C_{685}H_{1068}N_{196}O_{211}S_5$. The *molecular weight* (the relative weight of a molecule when the weight of

Figure 2-17. (Opposite) Structures for some amino acids formed in plant proteins. Almost all amino acids that form proteins have the same NH_2—CH—COOH unit at one end (unshaded). They differ, however, in the composition of the R groups (shaded). The particular R group gives each amino acid specific physical and chemical properties as well as definite bonding capacities with other amino acids.

Amino Acid	Structure	Standard Abbreviation	Composition of R group
Glycine		Gly	H
Alanine		Ala	CH_3
Valine		Val	$CH_3, CH_3.CH$
Tyrosine		Tyr	Ring form
Tryptophane		Try	Ring form
Aspartic Acid		Asp	Acidic (COOH)
Glutamic Acid		Glu	Acidic (COOH)
Lysine		Lys	Basic (NH_2)
Arginine		Arg	Basic (NH_2)
Cysteine		Cys	Sulfur containing
Methionine		Met	Sulfur containing

47

carbon is taken as 12) of some proteins is 500,000, compared to 342 for sucrose and 180 for glucose.

Levels of Structural Organization. It is generally recognized that plant proteins assume at least four levels of structural organization. These are referred to as primary, secondary, tertiary, and quarternary. The *primary structure* of a protein (Figure 2-18A) is the specific sequence of amino acids comprising the polypeptide chain or chains.

Analyses through chemical and physical studies indicate that few plant proteins exist in a straight-chain configuration as represented by the primary level of organization. In fact, the available data suggest that most proteins are coiled or twisted in a number of ways, the most common form being a spiral-shaped structure called a *helix*. The helix is probably produced by weak hydrogen bonds pulling adjacent portions of a polypeptide chain together. This three-dimensional coiling of a protein molecule represents a second level of protein organization referred to as a *secondary structure* (Figure 2-18B).

In addition to assuming straight-chain and helical configurations, plant proteins are also arranged into a third structural level of organization called the *tertiary level* (Figure 2-18C). This configuration is a result of the diversity of exposed chemical (R) groups comprising the amino acids. It can be noted by referring to Figure 2-19 that R groups are represented by ring structures (phenylalanine and tryptophane), acid groups (aspartic acid and glutamic acid), basic amino groups (lysine and arginine), and sulfhydryl (SH) groups (methionine and cysteine). This diversity permits a number of bonding situations to develop between adjacent R groups. Such interactions cause the protein molecule to bend or fold over itself, producing a characteristic tertiary structure.

All plant proteins exhibit a characteristic tertiary structure. Furthermore, individual polypeptide chains each having a tertiary structure may interact with one another to form a complex protein. The specific interaction between individual polypeptide chains is referred to as a quaternary structure. This structure, which represents the *quaternary level* of protein organization, is shown in Figure 2-18D.

Enzymes

Although enzymes are protein in nature, they will be considered as a separate class of organic compounds because of the vital functions they assume in living organisms. **Enzymes** are biochemical catalysts that increase the rates of chemical reactions or enable them to proceed more regularly. Many of these reactions are not only accelerated by enzymes but would not occur to any appreciable extent at the temperatures at which plants normally function. Although enzymes are intimately involved in the reactions they catalyze, they are essentially unchanged at the termination of the reaction. Inasmuch as enzymes are discussed extensively in Chapter 8, no further mention will be made of them at this point.

Nucleic Acids

Nucleic acids are extremely large organic molecules containing carbon, hydrogen, oxygen, nitrogen, and phosphorus. These compounds were so named because they were first discovered in the nuclei of cells. Like proteins, nucleic acids are long chainlike molecules composed of simpler units joined together by the removal of water at points of linkage between the units. These basic structural units of nucleic acids are called *nucleotides*. In the following discussion, the structure of two nucleic acids, **deoxyribonucleic acid (DNA)** and **ribonucleic acid (RNA),** will be examined.

A molecule of DNA is composed of repeating units of nucleotides. In this context, DNA is a polymer similar to cellulose and certain proteins.

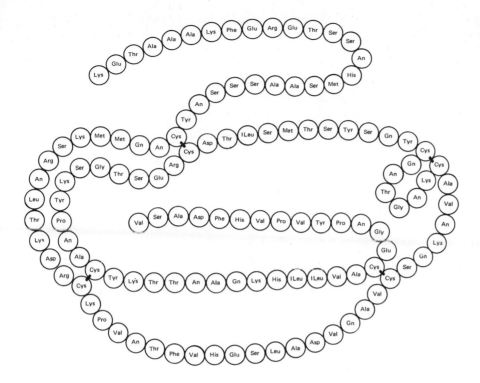

(A)

Figure 2-18. Levels of protein organization. (**A**) Primary structure—the sequence of amino acids in a polypeptide chain. In the example shown (ribonuclease, an enzyme), the whole sequence consists of 124 amino acids. (**B**) Secondary structure—the helical coiling of the polypeptide chain. (**C**) Tertiary structure—the bending or folding of the polypeptide chain over itself. (**D**) Quaternary structure—the relationship of the polypeptide chains of a complex protein to one another. The increasing levels of complexity are due to hydrogen bondings, as well as other chemical attractions between active R groups of certain amino acids within the polypeptide chains. (*A* and *B* from C. B. Anfinsen, *The Molecular Basis of Evolution,* New York: John Wiley & Sons, Inc., 1959. *C* and *D* from Conn and Stumpf, *Outlines of Biochemistry,* New York: John Wiley & Sons, Inc., 1963.)

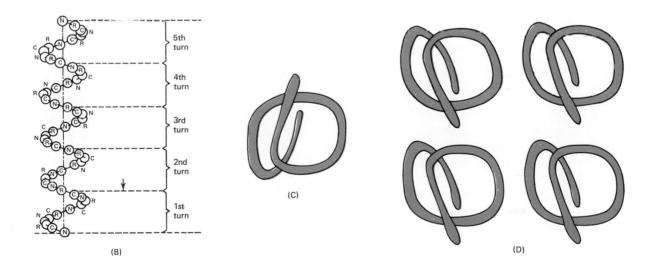

(B)

(C)

(D)

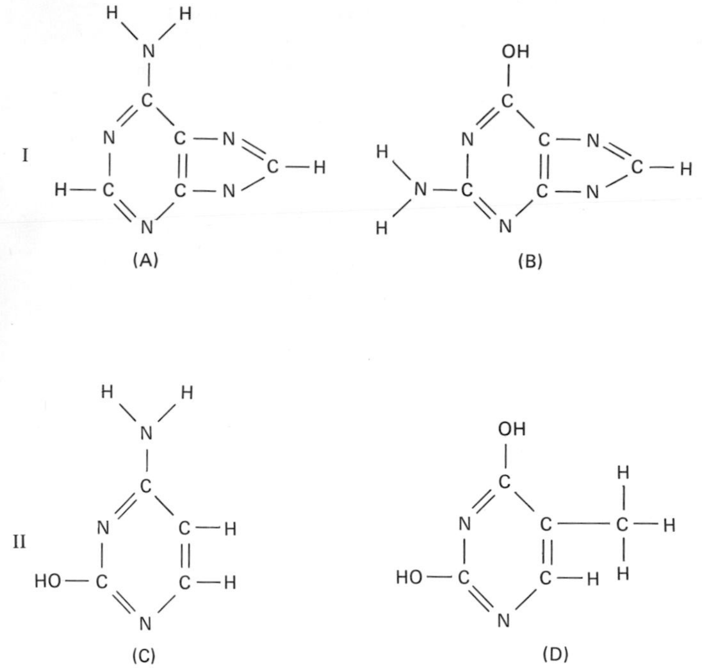

Figure 2-19. Structural formulas for the chemical compounds commonly found as components of nucleotides. *Above:* Nitrogenous bases. I. Purines, **(A)** adenine and **(B)** guanine. II. Pyrimidines, **(C)** cytosine and **(D)** thymine. *Below:* **(A)** Deoxyribose and **(B)** phosphoric acid (phosphate). The second carbon atom of deoxyribose lacks one oxygen atom as compared to ribose; thus it is designated deoxyribose.

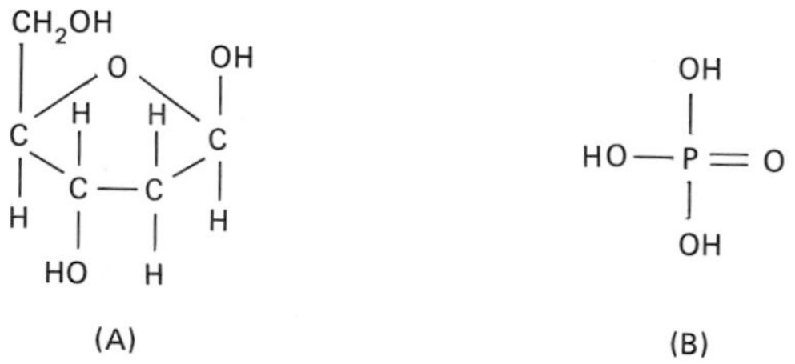

Each nucleotide consists of three distinguishable components: one nitrogenous base, one pentose sugar, and one phosphoric acid (phosphate) group.

Structure of Nucleic Acids. *Nitrogenous bases* are ring-shaped structures consisting of carbon, hydrogen, oxygen, and nitrogen atoms. They are further classified into two general types—*purines* and *pyrimidines.* The specific purines found in DNA are called *adenine* and *guanine;* the pyrimidines are called *thymine* and *cytosine.* The structures for these nitrogenous bases are shown in Figure 2-19.

The pentose (five-carbon sugar) found in DNA is called *deoxyribose.* This sugar is similar to ribose (Figure 2-10) except that it has one less oxygen atom attached to the second carbon atom. The structural formula of deoxyribose is represented in Figure 2-19 along with the structural formula of *phosphoric acid,* the third constituent of nucleotides.

If one of the purines or one of the pyrimidines is chemically bonded to deoxyribose and the sugar is then attached to a phosphate group, a nucleotide is formed. Each nucleotide thus formed is named for the constituent nitrogenous base. For example, a nucleotide containing adenine would be called *adenine nucleotide;* one containing thymine a *thymine nucleotide;* one containing cytosine a *cytosine nucleotide;* and one containing guanine a *guanine nucleotide* (Figure 2-20). These four nucleotides are the ones commonly found in DNA. Although the chemical components of DNA were known before 1900, it was not until 1953 that J. D. Watson and F. H. C. Crick presented a model for the arrangement of these components. Using data from various investigations, the following information concerning the structure of DNA was recorded:

1. Each DNA molecule consists of two strands, with cross members, twisted about each other in

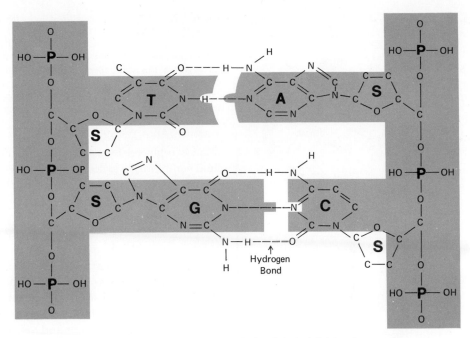

Key: G (guanine), C (cytosine), T (thymine), A (adenine), P (phosphate group)

Figure 2-20. Nucleotides commonly found in DNA. Shown are an adenine nucleotide, a thymine nucleotide, a guanine nucleotide, and a cytosine nucleotide. Each nucleotide consists of a nitrogenous base, a ribose sugar, and a phosphate group. When these nucleotides are assembled as components of a DNA molecule, the nitrogenous bases are held together by weak hydrogen bonds.

the form of a double helix (similar to a twisted ladder).

2. The uprights of the ladder are made of alternating phosphate and deoxyribose portions of a nucleotide.

3. The cross members, or rungs of the ladder, consist of paired purines and pyrimidines linked together by way of weak hydrogen bonds.

4. In the pairing of purines and pyrimidines, adenine always pairs off with thymine and cytosine always pairs off with guanine.

From these data and those of other investigators, Watson and Crick constructed a model of the probable structure for DNA (Figure 2-21).

The exact structure of RNA is much less clear than that of DNA. In some cases RNA consists only of a single strand of nucleotides, in others it is doubly stranded like DNA, and in still others it appears as a combination of both. Some data concerning RNA, however, have been ascertained. For example, it has been determined that RNA contains the pentose sugar ribose, not deoxyribose as found in DNA. Second, RNA does not contain the nitrogenous base thymine. Instead, the thy-

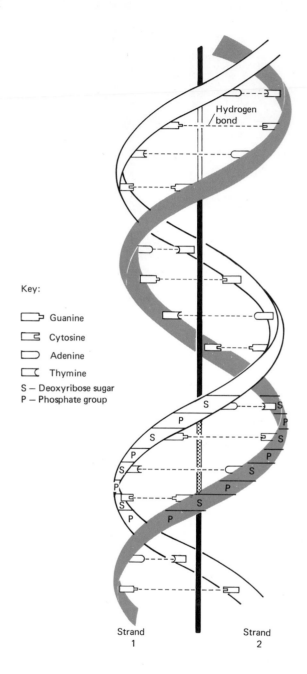

Figure 2-21. Schematic representation of a portion of a DNA molecule according to Watson and Crick. The entire molecule assumes a double-stranded helical configuration. The cross members represent paired purine and pyrimidine bases while the spiral backbone consists of alternating phosphate and sugar units. Note that adenine always pairs off with thymine, cytosine always pairs off with guanine, and the nitrogenous bases are attached to the sugar molecules.

Key:

Guanine

Cytosine

Adenine

Thymine

S — Deoxyribose sugar
P — Phosphate group

Hydrogen bond

Strand 1

Strand 2

mine is replaced by a pyrimidine called *uracil*. A final comparative aspect between DNA and RNA is that DNA is generally found in the nuclei of cells, whereas RNA may be found in both the nuclei or the cytoplasm, being more abundant in the cytoplasm.

Adenosine Triphosphate (ATP). In addition to the nucleotides that are found as components of DNA and RNA, one other should be mentioned at this point. This substance is **adenosine triphosphate (ATP).** ATP is a key organic compound that appears universally in plant and animal cells. The essential function of ATP is to store and carry energy for various cellular activities. Figure 2-22 shows the structure of this basically simple but extremely important molecule.

The major significance of ATP lies in the nature of the phosphate groups. The two terminal (end) chemical bonds that link the oxygen and phosphorus atoms are called *high-energy phosphate bonds*. To distinguish these from other chemical bonds, they are represented as wavy ($\sim$) lines. When these phosphate bonds are broken, they liberate two to three times as much energy as most chemical bonds. This energy is utilized by the cells to carry on basic life processes.

If one phosphate group is removed by splitting the terminal ($\sim$) bond, ATP becomes **adenosine diphosphate (ADP)** and energy is released. This reaction is represented as follows:

$$\text{ATP} \longrightarrow \text{ADP} + \text{P} + \text{energy}$$

adenosine   adenosine   phosphate
triphosphate   diphosphate

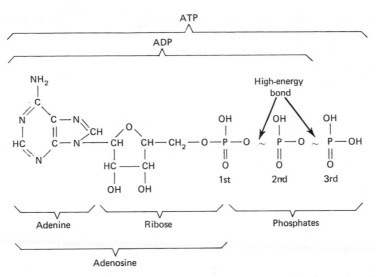

Figure 2-22. The structure of ATP. Adenosine triphosphate is composed of an adenosine unit (adenine + ribose) combined with three phosphate units. The first phosphate is attached to the ribose sugar by way of a low-energy bond. The second and third phosphates are attached by means of high-energy bonds. These high-energy bonds provide most of the chemical energy utilized by living cells. Compare the structure of ATP with that of the adenine nucleotide shown in Figure 2-20.

Plant cells are constantly using energy, which is supplied by the conversion of ATP into ADP, so the limited supply of ATP present in the plant at any given time would soon be completely consumed if provision were not made for its replenishment. This is accomplished when a phosphate group is added to ADP and a molecule of ATP is formed. Remember that when ATP is broken down, the reaction liberates energy. Conversely, when ATP is built up, the reaction requires the same amount of energy. This reaction might be represented as follows:

$$\underset{\substack{\text{adenosine}\\\text{diphosphate}}}{\text{ADP}} + \underset{\text{phosphate}}{\text{P}} + \text{energy} \longrightarrow \underset{\substack{\text{adenosine}\\\text{triphosphate}}}{\text{ATP}}$$

The energy needed to convert ADP and phosphate into ATP is derived from a long series of chemical reactions which will be considered later.

Pigments

Plants possess many striking external characteristics. These include variations in size, shape, color, growth patterns, and texture. Perhaps the most conspicuous and impressive of these characteristics is coloration. Because the substances that produce color in plants are intimately and critically associated with the vital physiological activities of plants, color is more than an aesthetic quality.

A cursory examination of plants in a limited geographic area would reveal a wide diversity of colors. The green leaves of oak trees, the red flowers of geranium plants, the yellow covering of pears, and the blue color of violets are but a few of the vast array of colors exhibited by different forms of vegetation. In fact, a detailed analysis of the entire plant kingdom would reveal all colors of the spectrum. Each of these colors is imparted by specific compounds called **pigments,** and the colors of various plants are the result of the presence of pigments in particular combinations and concentrations.

Chlorophylls. Apart from its chemical importance, **chlorophyll** is a pigment that imparts green color to the vast majority of plants. It is generally found in any part of the plant that is exposed to sunlight. Familiar examples are leaves, stems, and fruits. Within most plant cells, chlorophyll is contained in structures called **chloroplasts.** Although it is customary to refer to this green pigment as simply chlorophyll, actually there are about seven different varieties, of which **chlorophyll *a*** and **chlorophyll *b*** are found more commonly than any of the other types. The principal function of chlo-

rophyll is to trap the energy of sunlight and convert the solar energy into chemical energy during photosynthesis. The structural formulas for chlorophyll *a* and chlorophyll *b* are shown in Figure 2-23.

Carotenoids. Carotenoids constitute a number of pigments ranging in color from yellow to orange to red, the commonest color being yellow. As a group, carotenoids are found in chloroplasts with chlorophyll, and sometimes leaves become yellowish, or **etiolated,** if the carotenoids are more highly concentrated than the chlorophylls. Generally, carotenoids are subdivided into two classes: **carotenes** (orange) and **xanthophylls** (yellow), and as a rule both are found within the same plant cell. Although carotenoids are usually associated with chlorophylls in chloroplasts, they may be found in other pigment-containing structures, called

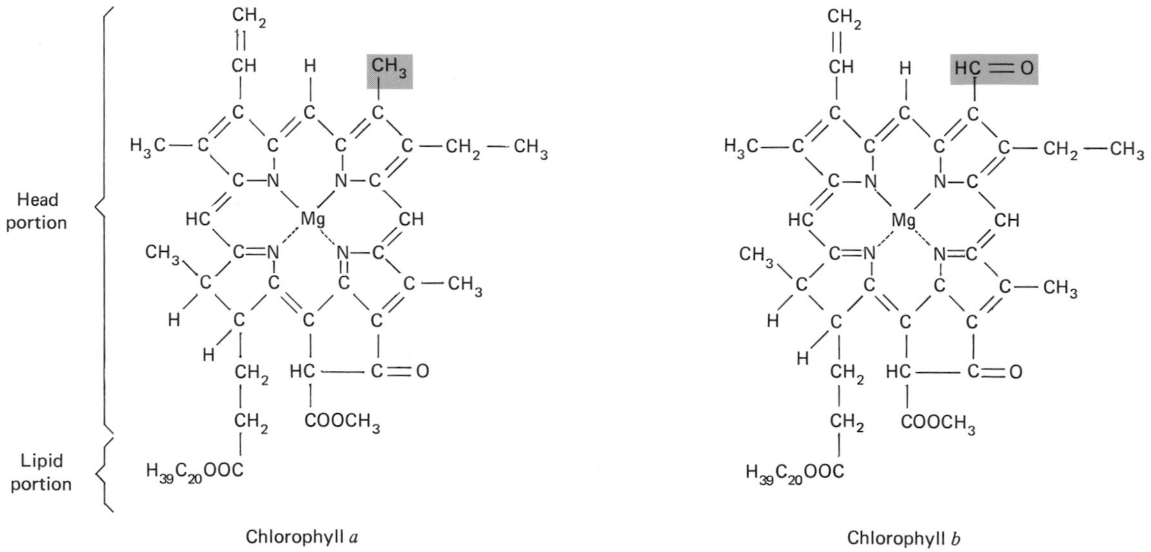

Figure 2-23. The chemical structures of chlorophyll *a* and chlorophyll *b*. The shaded areas represent the principal differences between the two molecules.

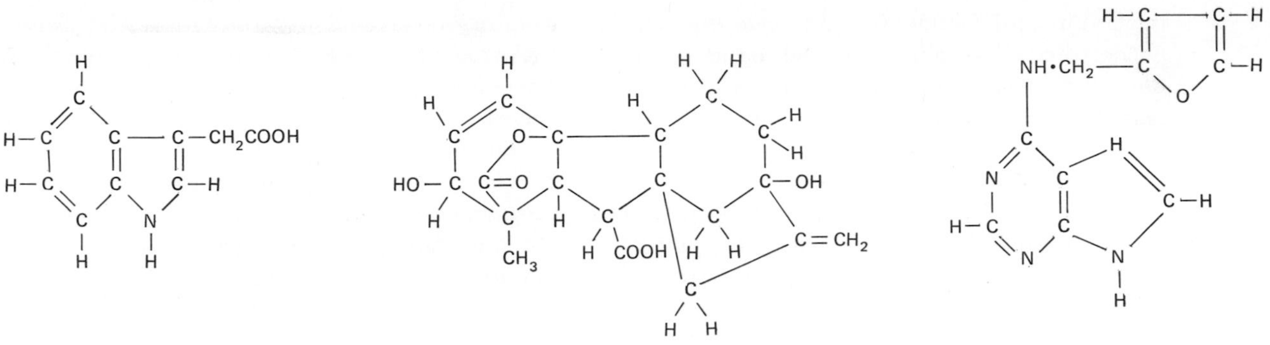

Figure 2-24. Structures of **(A)** indoleacetic acid, an auxin, **(B)** gibberellic acid, a gibberellin, and **(C)** kinetin, a cytokinin. Auxins, gibberellins, and cytokinins constitute the three principal classes of plant hormones, which assume various roles in growth and developmental phenomena.

chromoplasts. Chromoplasts are commonly found in flowers, seeds, fruits, and roots as well as in leaves of plants.

Anthocyanins. **Anthocyanins** are pigments that produce red, blue, and violet colors in plants. Whereas the chlorophylls and carotenoids occur in plastids, anthocyanins are found dissolved in certain portions of the plant cell such as vacuoles and may appear as crystalline bodies. Familiar plants and plant organs owing their distinctive colors to anthocyanins are radishes, garden beets, grapes, cherries, plums, poinsettias, geraniums, and tulips.

Hormones (Growth Regulators)

It is now recognized that the growth and development of plants are regulated by a variety of chemical substances called **hormones.** These compounds, active at very low concentrations, elicit profound effects on the physiological activities of the plant. The evidence suggests that they are synthesized in certain parts of the plant and dis-

tributed to other parts where they bring about certain growth and developmental responses.

Plant hormones are generally classified into three principal groups: (1) auxins, (2) gibberellins, and (3) cytokinins. Figure 2-24 shows the structural formulas of a representative hormone of each of these groups. **Auxins,** as a group, are involved in the growth (elongation) of cells, the dominance of vertical over lateral growth, the formation of roots in cuttings of economically useful plants, the fall of leaves and fruits, the production of seedless fruits, and the formation of wound tissue on injured or diseased plants. Many other functions have also been demonstrated by the action of auxins, and still others are suspected.

Gibberellins, although similar to some auxins in their biological activity, also exert other influences. For example, gibberellins promote the growth of dwarf plants, increase stem growth, promote seed germination, break the dormant period of seeds, and promote flowering under certain conditions in some plants. It appears that gibberellins and auxins act both independently and together, depending upon the species of the plant, the specific conditions under which the plant is growing, and the type of response being studied.

Cytokinins, as a group, have the capacity to stimulate the division and enlargement of cells, to initiate the growth of roots and stems, and to break seed dormancy. Much evidence supports the concept that the growth and development of the plant is under the influence of delicate balances and interactions among many different hormones. It is probably unlikely that a single hormone elicits a specific response without some degree of interaction with other hormones.

Other Organic Compounds

In addition to amino acids and fatty acids, there is another group of acids of considerable importance to plants. These are the so-called *"plant acids"* and, like other organic acids, they contain a characteristic carboxyl (—COOH) group. In general, plant acids contain six or less carbon atoms and most have from one to three carboxyl groups. Examples of plant acids are pyruvic, citric, succinic, and malic acids (Figure 2-25). Both *citric acid* and *malic acid* are present in considerable quantities in fruits and the leaves of succulent plants. They are also distributed, to a lesser ex-

tent, in other plant organs. *Pyruvic* and *succinic acids* are seldom present in appreciable quantities. These and other organic acids are intimately involved in plant activities which provide energy as well as other plant processes.

Other important organic compounds found in plants include vitamins, resins, gums, latex, and alkaloids. Some of these as well as other organic compounds will be treated later.

CHAPTER IN RETROSPECT

All living forms are composed entirely of chemicals. Of the 92 naturally occurring chemical elements, only 15 (hydrogen, boron, carbon, nitrogen, oxygen, magnesium, phosphorus, sulfur, potassium, calcium, manganese, iron, copper, zinc, and molybdenum) are considered to be essential to normal plant growth and development. The atoms of these elements, like all elements, are capable of combining to form larger stable aggregates (molecules). It is these larger aggregates of elements that act as the structural building blocks of plants (and animals) and also take part in multifarious chemical reactions which relate to specific plant functions. Substances formed by chemical combinations of atoms of two or more elements are called compounds. The compounds

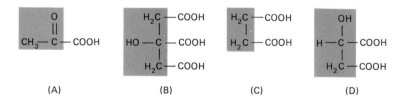

(A)   (B)   (C)   (D)

Figure 2-25. Structural formulas for **(A)** pyruvic, **(B)** citric, **(C)** succinic, and **(D)** malic acids. These "plant" acids, as well as others, are involved in a multiplicity of important biochemical reactions. The shaded areas represent differences in atomic structure among the molecules.

found in plants may be subdivided into two general types: (1) inorganic (water, sodium chloride, hydrochloric acid, ammonium hydroxide, and so on), and (2) organic (glucose, maltose, starch, cellulose, lauric acid, suberin, lecithin, glycine, DNA, ADP, and so on). Each of these compounds functions in specific ways, individually or collectively, to perform the highly complex task of maintaining a living system. For example, water, one of the most important inorganic compounds, serves as a medium for the transportation of substances needed within plants as well as a medium in which biochemical reactions take place. Cellulose, a very abundant organic compound, is an essential structural component of cell walls and is also found in combination with numerous other substances such as lignin. The DNA of the nucleus possesses the remarkable ability to produce an exact facsimile of itself. Considerable evidence exists to show that DNA is the substance of which genes are made and appears to work as the hereditary blueprint of all living things. Other organic compounds include plant pigments (chlorophylls, carotenoids, and anthocyanins), hormones (auxins, gibberellins, and cytokinins), enzymes, and vitamins.

To know the chemical basis for life, however, requires more than a knowledge of all the constituents and the myriad of chemical reactions in which they engage; there is also the need to know how chemical laws are translated into biological laws in the context of living organisms.

Questions and Problems

1. List some of the common chemical elements found in plants together with their respective chemical symbols. Which three elements constitute a major part of the total weight of plants?
2. Explain the nature of the ionic bond in a molecule of common table salt. Compare the bond in NaCl with that of O_2 and explain the difference in terms of electronic configurations.
3. List the major inorganic and organic compounds found in plants. What is the role of each of these compounds in the living plant?
4. When growing plants in water, what essential elements must be included? How would you devise an experiment to show the correct amount of essential elements to be included in a water culture (hydroponics)?
5. What is the effect of hydrogen ion concentration (pH) within the system of a plant? What mechanism exists for maintaining a proper pH within plants?
6. Compare the structural and molecular formulas for those monosaccharides, disaccharides, and polysaccharides which are normally found in plants. What advantage accrues from the use of a structural formula as contrasted with an empirical (molecular) formula? How are each of these three classes of sugars formed in plants?
7. How are amino acids involved in protein formation? Discuss the peptide bond in the formation of dipeptides, tripeptides, and polypeptides. What is meant by the primary, secondary, tertiary, and quarternary levels of protein organization?
8. Compare the chemical structure of RNA and DNA. Construct a three-dimensional model of DNA (optional).
9. How can you explain the differences in plant coloration in your area? (Consider the diversity of color in leaves, flowers, and fruits.)
10. How does the chemical organization of plants compare with humans? (Optional.) Are there any chemicals in plants that are not found in the environment? Describe the difference between the chemical organization of living and nonliving matter.

Suggested Supplementary Readings

Baker, J. J. W., and G. E. Allen, *Matter, Energy and Life.* Reading, Mass.: Addison-Wesley Publishing Company, Inc., 1965.

Baldwin, Ernest, *The Nature of Biochemistry.* New York: Cambridge University Press, 1967.

Bennett, T. P., and Earl Freiden, *Modern Topics in Biochemistry.* New York: The Macmillan Company, 1967.

Clevenger, Sarah, "Flower Pigments," *Scientific American,* June 1964.

Goldsby, R. A., *Cells and Energy.* New York: The Macmillan Company, 1967.

Kendrew, J. C., "The Three-Dimensional Structure of a Protein Molecule," *Scientific American,* December 1961.

Schmitt, F. O., "Giant Molecules in Cells and Tissues," *Scientific American,* September 1957.

Steiner, R. F., and Harold Edelhoch, *Molecules and Life.* Princeton, N.J.: D. Van Nostrand Company, Inc., 1965.

Stephenson, W. K., *Concepts in Biochemistry—A Programmed Text.* New York: John Wiley & Sons, Inc., 1967.

White, E. H., *Chemical Background for the Biological Sciences.* Englewood Cliffs, N.J.: Prentice-Hall, Inc., 1964.

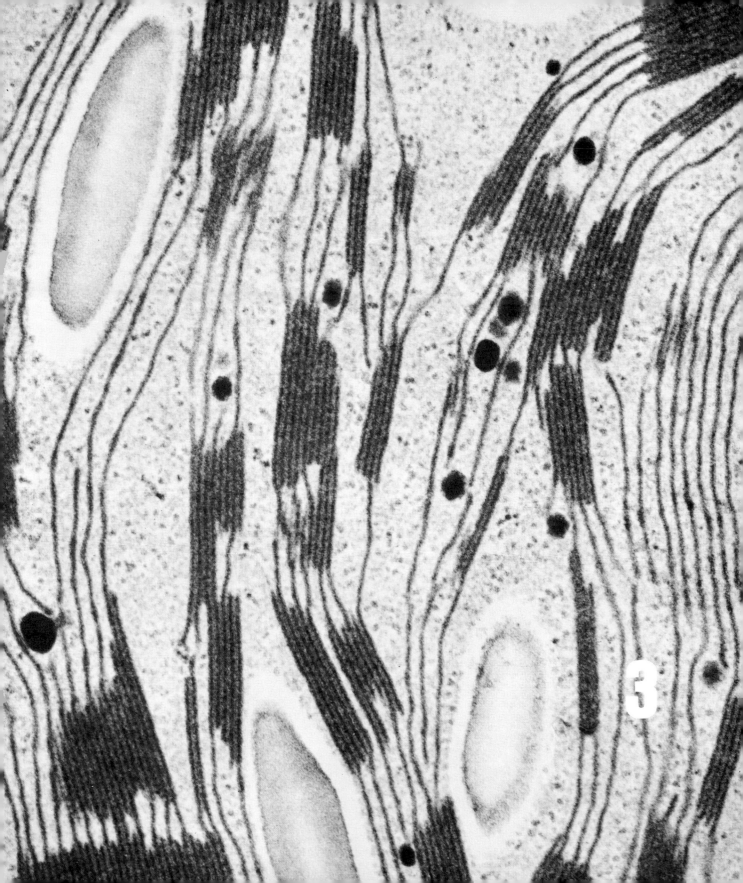

3
Microscopic Level of Plant Organization

INTRODUCTION

Prior to the seventeenth century, little was known about the structure of cells. Today it is universally accepted that cells are the units of life and that the forms and functions of these basic units are as diverse and as complex as life itself. The emergence of this concept, however, was not spontaneous. This view evolved from a long series of investigations which were undertaken utilizing newly discovered tools and techniques of observation and analysis.

DEVELOPMENT OF THE CELL CONCEPT

The Discovery of the Cell

Probably the most significant discovery leading to the development of the cell concept was the invention of the compound microscope by the Dutch lens maker Zacharias Janssen at the close of the sixteenth century. The instrument, however, did not lead at once to a systematic examination of the minute details of living things, because it was initially an object of curiosity. In the years that followed, the compound microscope and the fine art of lens grinding opened up the world of microscopic dimensions. Through the efforts of Anton van Leeuwenhoek (1632–1723) the science of microscopy advanced further (Figure 3-1).

One of the earliest contributions from microscopic examination was reported by Robert Hooke (1636–1700) in 1665. In that year he published a book entitled, *Micrographia: Or Some Physiological Descriptions of Minute Bodies Made by Magnifying Glasses.* In this publication, there appeared a sketch of a thin slice of cork which he had examined under the microscope (Figure 3-2). He described the cork as being composed of "little boxes" or "cells." The term "cell" as originally applied to dead plant material is still in use. Additional observations of living material revealed the presence of a liquid of some sort within the outer cell boundaries. The details and significance of the observations, however, were not understood. The important contribution made by Hooke was that he demonstrated that the substance of a plant is not homogeneous. It consists of units of varying sizes, shapes, and physical appearance.

In 1675 Marcello Malpighi, an Italian physician, published an account of his microscopic investigation of plants. This account consisted of a detailed description of the internal structure of plants, and, despite minor errors in observation, he provided investigators that followed with many clear and surprisingly accurate illustrations. The English microscopist Nehemiah Grew also made detailed studies of sections through roots and stems, the results of which he published in 1682.

In 1823 Robert Brown, a Scottish botanist, recognized for the first time the presence of a small spherical transparent object within each cell. He called this body the **nucleus** but did not concern himself with its exact nature or function. His second major contribution to cell theory appeared in a report published in 1828, *Microscopic Observations on the Pollen of Plants.* In this report he described the haphazard random movement he observed of pollen grains standing in water. This agitation of the pollen grains, called **Brownian movement,** is caused by the constant motion of water molecules too small to be seen even by microscopic observation. Such molecular motion, which is common

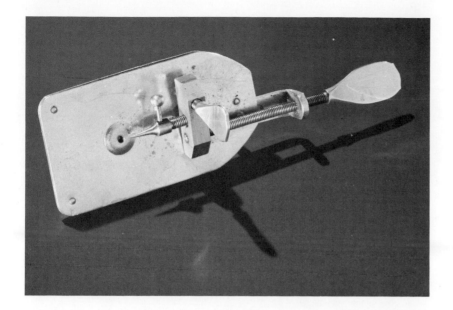

Figure 3-1. Leeuwenhoek's microscope. (Courtesy of Bausch & Lomb, Inc.)

to all matter, is extremely significant in bringing about chemical reactions.

As the magnifying power of the microscope was increased, investigators penetrated deeper into the structure of cells. The liquid observed by Hooke was subjected to closer scrutiny by Dujardin in 1835, at which time he described it as a jelly-like mass. In 1846 von Mohl called this viscous, jelly-like substance **protoplasm,** the substance now regarded as the physical and chemical basis of all life.

All these observations were consolidated by the German biologists Matthias Schleiden and Theodor Schwann in the early part of the nineteenth century. In 1838 Schleiden, a botanist, announced that the cell was the basic structural unit of all plants. One year later, Schwann, a zoologist, announced that all plants and animals consist of cells and that each cell contains a fluid substance (protoplasm) and a solid spherical inclusion, the nucleus. The cell, therefore, is the common denominator, the basic unit, of all living forms. Just as an atom is the smallest particle of an element capable of retaining its identity, the cell is the smallest unit of life. Subsequent investigations into the structure and functions of cells were based upon the cell concept as experimentally established by Schleiden and Schwann.

A Look Inside the Cell

Until a relatively few years ago, the concept of the cell was limited to what could be seen through a *compound microscope* employing visible light (Figure 3-3). Magnifications of 1000 to 2000 times, the practical limit for the light microscope, revealed the outer boundary, the protoplasm, and the nucleus of the cell. Even with the use of special staining techniques, smaller structures either appeared as minute, indiscrete dots or strands, or remained invisible. For many years the ability to observe submicroscopic structures was restricted by the limitations of the light microscope. These

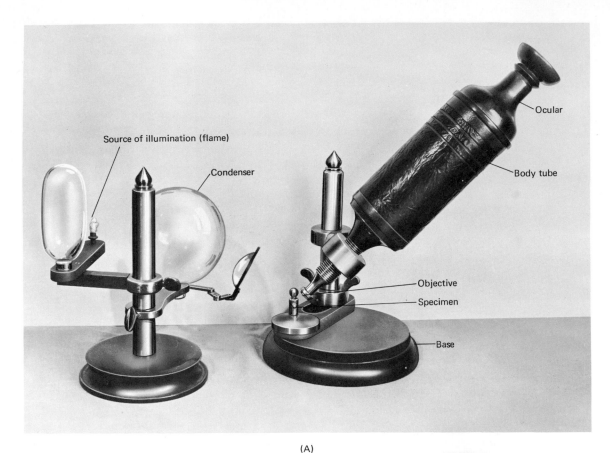

Source of illumination (flame)

Condenser

Ocular

Body tube

Objective

Specimen

Base

(A)

Figure 3-2. Hooke's microscope and his contribution to cell biology. **(A)** This microscope, considerably more sophisticated than the ones used by Leeuwenhoek, is basically similar in design to modern compound microscopes. (Courtesy of Bausch & Lomb, Inc.) **(B)** Cork cells as they appeared to Hooke. These cells demonstrated that the substance of plants is not homogeneous.

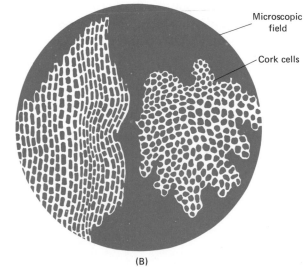

Microscopic field

Cork cells

(B)

limitations resulted from the relatively long length of the waves of the visible light employed in the instrument. As a result, cell structures smaller than a certain size could not be seen. Cell parts whose dimensions are less than 0.3 μ (micron) could not be resolved. A *micron*, symbolized μ, the micro-

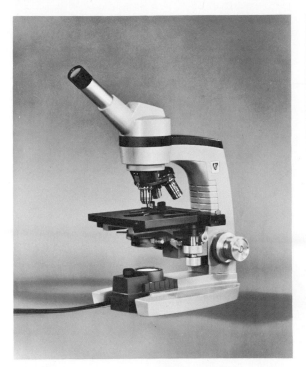

Figure 3-3. Photo of a modern compound microscope. Inasmuch as microscopes of this type use light, they cannot resolve cellular structures smaller than 0.3 μ. This limitation is the result of the relatively long length of visible light waves employed in the instrument. (Courtesy of American Optical Corporation.)

scopic unit of measurement, is equal to 0.001 millimeter (1/254,000 inch). Many cellular components are smaller than 0.3 μ, so an instrument was needed which utilized something shorter than a light wave as a source of illumination.

The most important breakthrough in the study of submicroscopic structure came in the 1930s with the advent of the first practical *electron microscope* (Figure 3-4). This instrument employs high electrical voltages to drive a beam of electrons through a vacuum to magnify the image of the object being studied. The electron microscope uses a beam of electrons instead of light. Because the

wavelength of electrons is about 1/100,000 that of the normal wavelength of white light, the electron microscope can distinguish much smaller objects (about 1/100 the size) than those revealed by the best light microscope. The upper limit of magnification possible with an electron microscope is of the order of 200,000 times.

Using a light microscope, the observer looks directly at the object. In electron microscopy, the observer views the object indirectly. The beam of

Figure 3-4. A modern electron microscope. Instead of light, this instrument uses a beam of electrons. Since the wavelength of electrons is about 1/100,000 that of the normal wavelength of white light, an electron microscope can magnify objects up to 200,000 times. This instrument permits scientists to see and photograph particles as small as 1/50,000,000 of an inch in diameter. (Courtesy of Radio Corporation of America.)

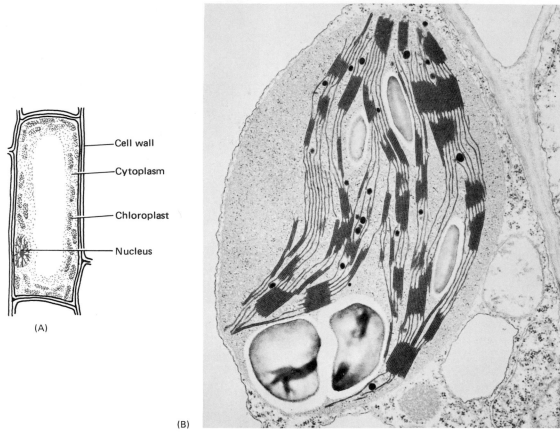

Cell wall

Cytoplasm

Chloroplast

Nucleus

(A)

(B)

Figure 3-5. Comparison between a diagrammatic representation of a cell as seen through a light microscope and an electron micrograph. **(A)** A single cell of *Elodea,* a water plant, containing chloroplasts. X6,000. The light microscope reveals the chloroplasts only as disc-shaped structures. **(B)** Electron micrograph of a chloroplast from the leaf of *Phleum pratense* (timothy grass). X20,600. In this micrograph the ultrastructure of a chloroplast is readily demonstrable. Close examination reveals numerous detailed anatomical features. (Courtesy of Dr. Myron C. Ledbetter, Brookhaven National Laboratory.)

electrons is directed through an object on to a screen or photographic plate. The plate is then developed, enlarged, and examined. Such a developed plate is called an *electron micrograph* (Figure 3-5). Inasmuch as the electron microscope reduces the observational area of the object being studied,

a unit of measurement called the *angstrom* is used. One angstrom, symbolized Å, is equal to 1/10,000 μ (1/2,540,000,000 inch).

Electron micrographs have elucidated many of the mysteries which have long been associated with the cell. They have revealed cellular struc-

Clamps to hold block

Paraffin ribbons with
cell sections in series

Paraffin block with
embedded cell

Blade

Rotary handle

Blade clamp

Figure 3-6. A rotary microtome. This instrument is used extensively in the preparation of cell sections for microscopic examination. First the cells are embedded in a paraffin block to hold them rigid when cut by the blade. The block is then pushed against the blade and thin sections, of predetermined thickness in microns, are cut. The cells are then stained and studied microscopically. (Courtesy of Ontario Region Department of Forestry and Rural Development, Sault Ste. Marie, Ontario, Canada.)

tures not even known to exist and have added a new dimension to the knowledge of the structural organization of the cell. Undoubtedly, the electron microscope, coupled with other tools and techniques of investigation, will reveal smaller and smaller components of cellular organization.

Preparing Cells for Study

The constant companion to all experimental light microscopy is an instrument called the *rotary microtome* (Figure 3-6). To prepare microscopic specimens for investigation it is first necessary to slice thin sections, usually not more than one or two cells in thickness, to enable light to pass through the object. A typical cell from the elongating zone of a stem or root has an approximate depth of 10 μ. Sometimes the use of an ordinary single-edge razor and a steady hand can be satisfactory in accomplishing sufficiently thin sections; however, for more precise work, a rotary microtome is indispensable.

A microtome and the butcher's meat-slicing machine operate on the same general principle. However, the degree of precision with a rotary microtome is remarkably high. A skilled technician operating a quality microtome can cut a well-prepared specimen into sections 1 μ thick.

When permanent slides are required, it is necessary to employ a rather complex technique, which usually involves four major steps: fixation, embedding, sectioning, and staining. *Fixation* is a special method of rapidly killing the cells, which is usually accomplished by employing chemical fixatives. Often a combination of formaldehyde, acetic acid, and alcohol is used as the chemical fixative. The method of *embedding* is accomplished by pouring melted paraffin wax around the specimen and allowing it to penetrate and harden. Synthetic plastics have recently been employed satisfactorily as a substitute for paraffin wax.

Sectioning embedded specimens requires the very sharp blade of a rotary microtome. Once sections have been made, further treatment involving stains and dyes is necessary to accentuate various structural features of cells.

Obtaining Information from Cells by Chemical Analyses

With the aid of the light and electron microscopes, the structural aspects of cells can be examined in detail. If, however, the functions of cells are to be analyzed, attention must be turned to the molecular level, because this is the level at which chemical reactions occur. Inasmuch as molecules cannot be observed directly, owing to their relatively small size, chemical information about cells can only be deduced. Among the many techniques utilized by biologists to obtain chemical information from cells are chromatography, centrifugation, isotopic traces, and electrophoresis.

Chromatography. The laboratory technique for separating mixtures of similar materials is called *chromatography*. This term literally means "writing with color." Chromatography is a technique for separating mixtures of similar materials based on their differential adsorption or partitioning between two phases. It is frequently employed to help separate and identify various amino acids and sugars found in living cells and tissues. An extensively used chromatographic technique is *paper chromatography* (Figure 3-7).

This technique employs the use of strips of filter paper upon which a small spot of unknown amino acids (or sugars) is placed. An organic solvent such as *n*-butanol is allowed to flow up the paper by capillary action. As the solvent flows past the sample spot, it carries the individual amino acid components along with it at various speeds, depending upon their solubility in the solvent and their affinity for the paper. With this process completed, the paper is removed and dried and made ready for observation by either ultraviolet light or by spraying the paper with a chemical reagent, which reacts with the components to form an identifiable colored compound. The mixture may be identified by comparison with pure standards (known compounds) run under identical conditions.

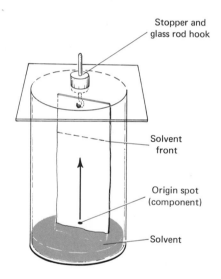

Stopper and glass rod hook

Solvent front

Origin spot (component)

Solvent

Figure 3-7. Paper chromatography apparatus. The solvent moves up the filter paper by capillary action taking the components along at varying rates. The components become visible after the paper is removed from the apparatus, dried, and sprayed with a special developer. (Adapted from Edward S. Lenhoff, *Tools of Biology.* New York: The Macmillan Company, © 1966.)

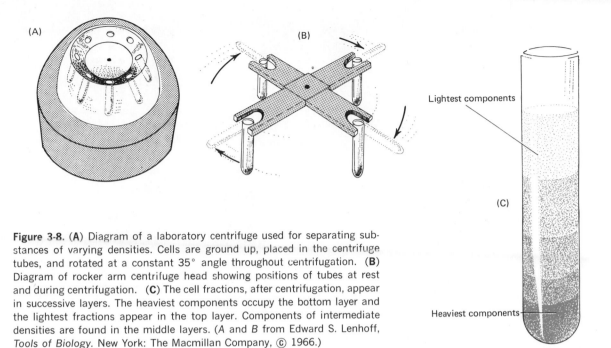

Figure 3-8. **(A)** Diagram of a laboratory centrifuge used for separating substances of varying densities. Cells are ground up, placed in the centrifuge tubes, and rotated at a constant 35° angle throughout centrifugation. **(B)** Diagram of rocker arm centrifuge head showing positions of tubes at rest and during centrifugation. **(C)** The cell fractions, after centrifugation, appear in successive layers. The heaviest components occupy the bottom layer and the lightest fractions appear in the top layer. Components of intermediate densities are found in the middle layers. (*A* and *B* from Edward S. Lenhoff, *Tools of Biology*. New York: The Macmillan Company, © 1966.)

Chromatographic techniques are also employed to separate plant pigments such as chlorophyll *a* from chlorophyll *b* and carotenes from other carotenoids. Numerous interesting chromatography experiments can be done with simple and inexpensive materials. The original technique of column chromatography developed by Tswett in 1910 has been greatly modified and refined to include two-dimensional chromatograms, thin-layer chromatography, and gas chromatography. Texts on chromatography should be consulted to appreciate the full range of this technique. Other means of separating and identifying organic compounds found in living systems have been developed, but their consideration is beyond the scope of this text.

Centrifuges. One of the common techniques for separating cellular components is *centrifugation.* The instrument used for separating substances of different densities is called a *centrifuge* (Figure 3-8). The cells are ground up, usually in a blender, so that the individual cells are broken. This mixture of water and cellular components is then placed in a centrifuge tube and the tube is inserted into the centrifuge apparatus. After rotating for about 20 minutes at very high velocities the mixture will separate into distinct layers. The bottom layer will contain the heaviest particles and the top layer will consist of the lightest particles. Intermediate layers will contain cellular components of varying densities. Extremely large and costly ultracentrifuges can attain rotary speeds up to 65,000 revolutions per minute and gravitational forces as high as 300,000 times gravity.

Isotopic Tracers. A method of biological investigation which was perfected soon after World War

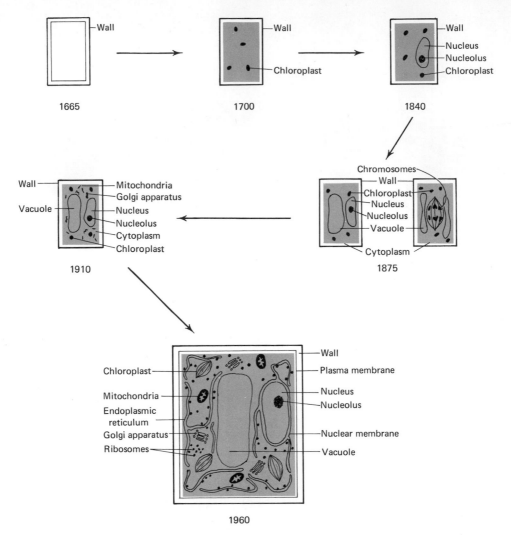

Figure 3-9. Development of knowledge about the structure of the cell from 1665 to 1960. (From *The Plant Cell* by William A. Jensen. ⓒ 1964 by Wadsworth Publishing Company, Inc., Belmont, California. Reproduced by permission.)

Figure 3-10. (Opposite) Diagrammatic representation of a composite plant cell. (**A**) Cell wall, (**B**) pit, (**C**) plasma membrane, (**D**) endoplasmic reticulum, (**E**) chromoplast, (**F**) chloroplast, (**G**) Golgi body, (**H**) starch grain, (**I**) ribosomes, (**J**) mitochondrion, (**K**) tonoplast, (**L**) vacuole, (**M**) crystal, (**N**) nuclear membrane, (**O**) chromatin, (**P**) nucleolus, (**Q**) karyolymph, and (**R**) cytoplasm. (Jack J. Kunz, Time-Life Books, from *The Plants*, ⓒ 1963 Time Inc.)

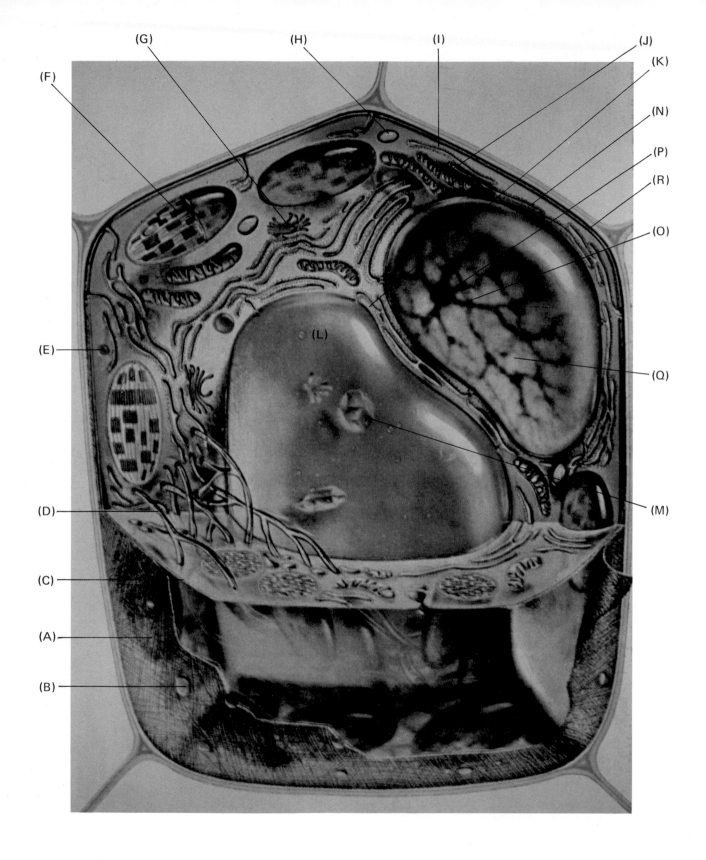

II involves the use of *isotopic tracers.* Atoms of the same element which differ in mass are called *isotopes.* This technique is somewhat technical and can only be mentioned here. It is worth noting that Melvin Calvin received the Nobel prize in 1961 for his outstanding work in tracing the pathway of carbon during the process of photosynthesis. Radioactive carbon (^{14}C) was substituted for ordinary carbon (^{12}C) and traced by means of its radioactivity. The experiments provided much of the detailed information about the intermediate reactions which occur during the process of photosynthesis. The usefulness of radioactive elements in such experiments is that the paths of individual molecules of the radioactive substance can be traced through a series of chemical reactions in a living organism.

Electrophoresis. Another method for separating materials makes use of an electrical field across a liquid containing charged particles. This technique, called *electrophoresis,* separates materials according to their electrical charge. Positively charged particles are attracted to the negative electrode and negatively charged particles to the positive electrode. The rate of movement of substances with the same kind of electrical charge varies according to the magnitude of the charge.

With the development of sophisticated instrumentation for observing the structure of cells and analyzing the chemical constituents of cells, biologists have greatly expanded their knowledge of cellular form and function. In the years to come, the mysteries of the cell will unravel one by one. In essence, to understand the cell is to understand the meaning of life itself. The development of the knowledge about cell structure from 1665 to the present is outlined in Figure 3-9.

THE GENERALIZED PLANT CELL

In a very broad sense, a **cell** may be defined as the basic structural and functional unit of living plants. Each plant cell represents a highly complex unit composed of molecules organized to form a series of component cell structures. Each of these components has a particular structural framework, chemical composition, and specialized function. In treating plant cells in this chapter, more than the usual number of electron micrographs have been included. For most of the cell parts illustrations have been designed that consist of a diagram of an entire plant cell, to show the location of the specific part under consideration and its anatomical relationship to other parts; an adjacent electron micrograph to show detailed ultrastructure; and an accompanying diagrammatic representation of the proposed chemical composition of the part based upon recent research. In this way, the cell components are seen as highly organized molecular units. Although the functions of these structures are discussed briefly, they will be analyzed in greater detail later.

Plant cells exhibit a great diversity in shape, size, and component parts, and to study plant cells implies an analysis of many different cell types. Therefore, attention will be directed first to a study of a generalized plant cell, which represents a composite of many plant cells (Figure 3-10). For convenience, a plant cell may be considered to consist of three general parts: (1) the cell wall, or outer covering; (2) the protoplast, or organized functional unit of the cell; and (3) ergastic substances, which are stored nutritive materials or products of cell activities. Of these three cellular parts, only the protoplast is living.

Cell Wall

The outer boundary of the plant cell, produced by the living protoplast, consists of a nonliving structure called the **cell wall.** The presence of cell walls is considered to be the outstanding structural characteristic that distinguishes plant cells from

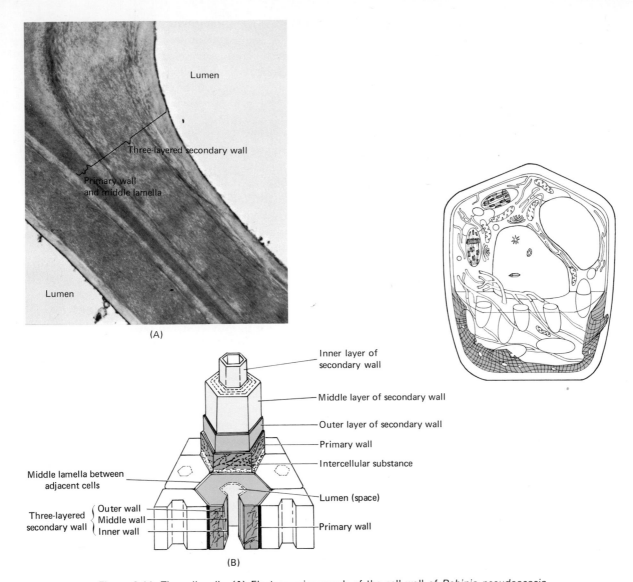

Figure 3-11. The cell wall. **(A)** Electron micrograph of the cell wall of *Robinia pseudoacacia* (black locust) showing primary and secondary cell walls. X16,300. (Courtesy of Dr. Myron C. Ledbetter, Brookhaven National Laboratory.) **(B)** Diagrammatic representation showing relationship of the middle lamella, primary wall, and secondary wall. (Modified from A. Wardrop and D. Bland, *Proceedings of the 4th International Congress of Biochemistry*, Vol. 2. New York: Pergamon Press, Inc. Reproduced with permission.)

animal cells. With relatively few exceptions, plant cells possess walls. These walls assume protective and supportive functions and in addition determine the shape and the texture of plant cells. On the basis of its development and structure, the cell wall may consist of three distinguishable parts: an intercellular substance called the middle lamella, a primary cell wall, and a secondary cell wall (Figure 3-11).

Parts of the Cell Wall. The **middle lamella** is a relatively thin layer of intercellular material formed between adjacent plant cells during division which persists after division (Figure 3-11B). The lamella is composed of viscous *pectic substances* (pectin), calcium, cellulose, and polymers of various types. In terms of consistency, the middle lamella, owing to the presence of pectin, is a viscous, jelly-like substance. Commercial preparations of pectin are obtained from the cell walls of various fruits and are added to jellies to ensure the "jelling" of these materials. In plant cells, the middle lamella serves as an intercellular cementing material.

The **primary cell wall,** found in all plant cells, is formed during the early stages of growth and development (Figure 3-11B). It is relatively thin, being about 1 to 3 μ thick. It is also quite plastic and capable of considerable extension as the cell increases in size. The cells of many fruits, roots, fleshy stems, and leaves possess only a primary cell wall and middle lamella.

The **secondary cell wall,** as the name implies, follows the primary wall in order of appearance and is found only in certain mature cells (Figure 3-11B). It forms after the primary cell wall ceases to grow. In many cell types, after secondary wall formation is completed, the living components of the cell die and disappear. The secondary wall may be about 5 to 10 μ thick, lending considerable strength and mechanical support to the cell. Plant products or derivatives containing these thickened cell walls, such as lumber and textiles, are much stronger and more durable than those containing only primary walls. An electron micrograph showing the primary and secondary walls from the cells of the black locust is shown in Figure 3-11A.

Pits, Primary Pit Fields, and Plasmodesmata. For the most part, the cell wall is a continuous covering. In some plants, however, especially multicellular ones, secondary walls are characterized by the presence of cavities or depressions called **pits** (Figure 3-12). Cells consisting only of primary walls contain somewhat similar depressions called **primary pit fields.** Pits and primary pit fields originate when the wall is formed unevenly, leaving depressions or extremely thin portions of wall material. These pitted areas greatly facilitate the movement of materials from one cell to another.

Pits are usually found in nonliving cells that are concerned with conduction and support, such as fibers and tracheids (Chapter 4). In some cases, thickened overhangs form around the pits. These pits are referred to as *bordered pits* (Figure 3-12C). In other cases, the overhanging borders are absent and the pits are known as *simple pits* (Figure 3-12B). With special staining techniques and high magnification many primary pit fields reveal exceedingly small openings (pores) in the membrane of living cells through which fine, delicate strands of cytoplasm connect adjacent cells. These cytoplasmic projections are termed **plasmodesmata.** Figure 3-12A shows an electron micrograph of plasmodesmata formed between two adjacent cells of a corn plant.

Chemical Composition of Cell Walls. The most abundant organic compound in most plant cell walls is cellulose. Other compounds include *hemicelluloses* and *pectic compounds,* substances structurally designated as polysaccharides. At maturity,

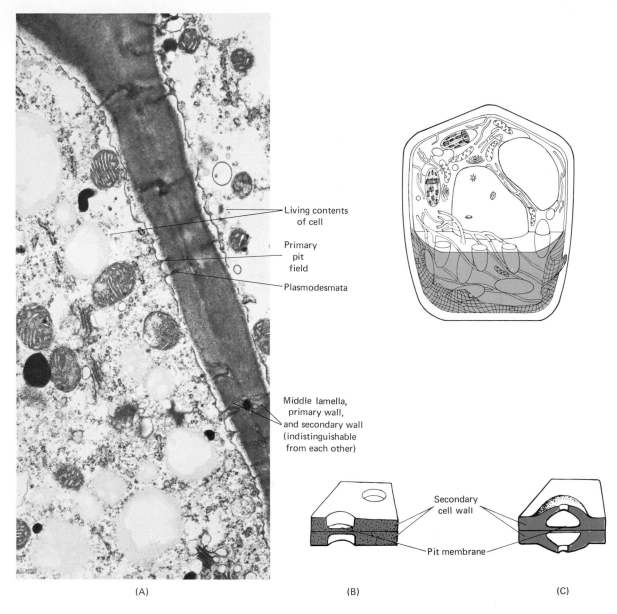

Living contents
of cell

Primary
pit
field

Plasmodesmata

Middle lamella,
primary wall,
and secondary wall
(indistinguishable
from each other)

Secondary
cell wall

Pit membrane

(A)   (B)   (C)

Figure 3-12. Primary pit fields, plasmodesmata, and pits. (**A**) Electron micrograph of primary pit fields and plasmodesmata of *Zea mays* (corn). X18,300. (Courtesy of Dr. Myron C. Ledbetter, Brookhaven National Laboratory.) (**B**) Diagram of a simple pit. (**C**) Diagram of a bordered pit.

if a secondary wall develops, another wall component, called lignin, appears. *Lignin,* a constituent of the cell walls of materials such as wood, affords a high degree of rigidity and strength. In addition, lipids are found as structural components of cell walls. Representative lipids include suberin, cutin, and waxes. These substances are usually abundant in cell walls found at the periphery of various plant organs.

The primary wall is composed chiefly of cellulose, hemicellulose, and pectic compounds, whereas the secondary cell wall contains cellulose, hemicellulose, and lignin as the chief constituents. The middle lamella, by comparison, contains pectic compounds as its main component.

The organizational pattern of cellulose into the formation of cell walls is shown in Figure 3-13. As previously stated, cellulose is a polymer of glucose molecules and, together with other cellulose molecules, forms a *micelle.* Micelles are in turn organized into larger bundles called *microfibrils.* The microfibrils are then arranged into layers forming the cell wall.

Protoplast

The **protoplast** of plant cells is a collective designation for the living, organized functional unit bounded by the cell wall. On the basis of organization and differentiation, the protoplast may be conveniently divided into three portions: the

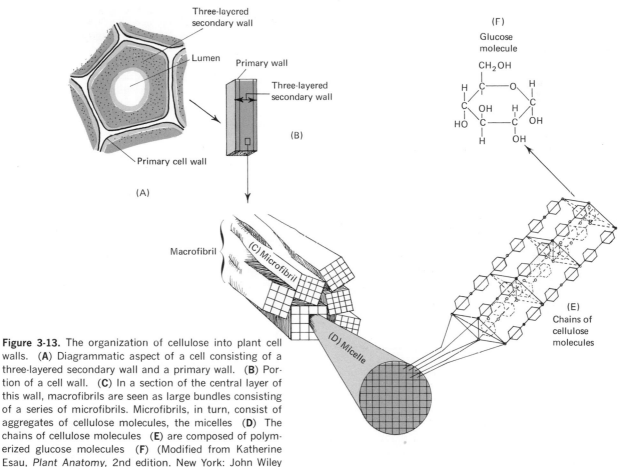

Figure 3-13. The organization of cellulose into plant cell walls. (**A**) Diagrammatic aspect of a cell consisting of a three-layered secondary wall and a primary wall. (**B**) Portion of a cell wall. (**C**) In a section of the central layer of this wall, macrofibrils are seen as large bundles consisting of a series of microfibrils. Microfibrils, in turn, consist of aggregates of cellulose molecules, the micelles (**D**) The chains of cellulose molecules (**E**) are composed of polymerized glucose molecules (**F**) (Modified from Katherine Esau, *Plant Anatomy,* 2nd edition. New York: John Wiley & Sons, Inc., © 1965.)

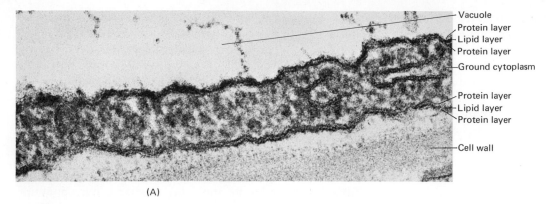

- Vacuole
- Protein layer
- Lipid layer
- Protein layer
- Ground cytoplasm
- Protein layer
- Lipid layer
- Protein layer
- Cell wall

(A)

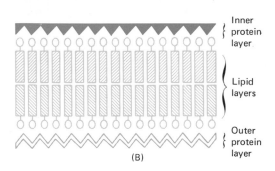

Inner protein layer

Lipid layers

Outer protein layer

(B)

Figure 3-14. The plasma membrane. **(A)** Electron micrograph of the plasma membrane of *Statice sinuata* (statice). X150,000. (Courtesy of Dr. Myron C. Ledbetter, Brookhaven National Laboratory.) **(B)** A proposed model of the orientation of molecules in the plasma membrane. The unit membrane consists of a bimolecular lipid layer covered on each side by a protein layer; as the diagram indicates, the outer and inner protein layers are probably not identical. (Modified after J. D. Robertson, "The Ultrastructure of Cell Membranes and Their Derivatives," *Biochemical Society Symposia,* No. 16, 1959, pp. 3–43.)

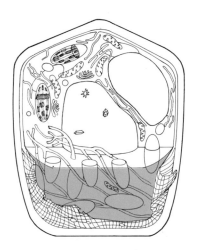

plasma (cell) membrane, cytoplasm, and organelles.

Plasma Membrane. The **plasma membrane** is a thin, flexible structure located beneath the cell wall, separating the wall from the cytoplasm (Figure 3-14A). It is differentiated from the cell wall on the basis of structure, chemical composition, and function. Whereas the cell wall is thick and rigid, the plasma membrane is relatively thin and flexible. Chemically, the cell wall is formed primarily from carbohydrates, whereas the membrane is composed essentially of lipids and proteins. Functionally, the cell wall is supportive in nature, whereas the membrane regulates the passage of various materials into and out of the cell. One of the distinctive properties of the cell membrane is termed *differential* or *selective permeability.* This is a phenomenon by which certain

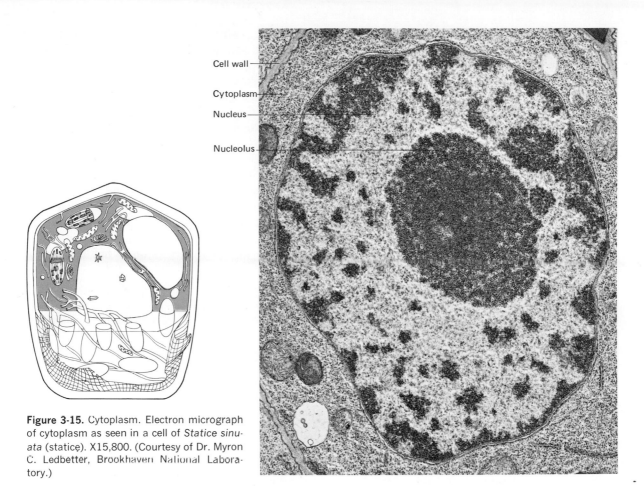

Cell wall

Cytoplasm

Nucleus

Nucleolus

Figure 3-15. Cytoplasm. Electron micrograph of cytoplasm as seen in a cell of *Statice sinuata* (statice). X15,800. (Courtesy of Dr. Myron C. Ledbetter, Brookhaven National Laboratory.)

substances freely enter and exit cells to the selective exclusion of other materials. An analysis of the importance of selective permeability will be given later.

Based upon interpretations of numerous electron micrographs and chemical analysis, a model has been proposed for the structure of the membrane (Figure 3-14B). Such a model, referred to as the *unit membrane,* shows that there is a bimolecular lipid layer covered on each side by a layer of protein molecules. Recent investigations have shown that the unit membrane structure is also characteristic of many other cellular membrane systems. Inasmuch as there is some morphological evidence indicating that the unit membrane concept does not sufficiently account for the structure of all membranes, the universality of this model should not be accepted.

Cytoplasm. **Cytoplasm** is a rather general term for the living matter or physical substance of the cell inside the plasma membrane and external to the nucleus (Figure 3-15). It is the ground substance, containing organelles, and, together with vacuoles, usually comprises the bulk of cell contents. Cytoplasm is commonly referred to as the physical basis of life because within it resides the structural integrity of the cell. It is best described as a fluid, semitransparent, viscous, elastic substance. Although apparently homogeneous, it varies from cells of one species of plant to another and even from one cell to another in plants of the same species. Chemically, 75 to 90 per cent of cytoplasm is water. The remainder is solid material in the form of assorted chemical compounds. Of the solid materials, more than one half is protein, with lipids, carbohydrates, acids, and other organic and inorganic compounds comprising the remaining portions. It should not be thought that cytoplasm is a random mixture of water and assorted chemical substances. On the contrary, it represents a complex and highly organized chemical and physical system of detailed ultrastructure capable of myriad energy-transfer reactions.

In a physical context, cytoplasm is neither a solid nor a liquid but a system composed of both solid and liquid constituents. The inorganic components of cytoplasm, as well as most carbohydrates, are soluble in water and are present as a true solution. The vast majority of organic compounds, specifically proteins and lipids, are present as colloids. True solutions and colloids may be differentiated on the basis of particle size.

If all the particles are small, such as those of inorganic salts or small organic molecules, the system is termed a *true solution*. A true solution is a homogeneous mixture of two or more components; that is, the particle sizes of the components are very small and cannot be distinguished from each other. In fact, the molecules are so small that they cannot be detected, even with the most powerful microscope. No matter how long a true solution stands, the dispersed particles will not settle. Two examples of true solutions are salt (NaCl) dissolved in water and sucrose ($C_6H_{12}O_6$) dissolved in water and in the cytoplasm of cells.

If the particles are somewhat larger than those in a true solution, ranging from 0.001 to 0.1 μ in diameter, the system is termed a *colloid*. Like a true solution, the particles of a colloid remain suspended. In the case of colloids, this is due to the fact that colloidal particles bear positive and negative electric charges; these unlike electric charges repel each other, keeping the particles suspended as well as apart from each other.

Cytoplasmic colloids undergo reversible *sol-gel* transformations. In the *sol state,* the colloidal particles are dispersed in a random fashion. In the *gel state,* the particles interact and form a network (Figure 3-16). Interconversions between sol and gel may be brought about by changes in temperature, concentration of various cytoplasmic components, and salt concentration, among others. Consider gelatin as an example; at high temperatures it is a sol (fluid). When the temperature is lowered, it is transformed into a gel (semisolid).

Cytoplasm is, then, both a colloid and a true solution. It differs from ordinary colloids, however, in that the colloidal components are not haphazardly arranged but systematically organized. It also differs from an ordinary colloid in that it contains specialized bodies called organelles.

One distinctive cytoplasmic characteristic that should be mentioned at this point is **cytoplasmic streaming (cyclosis),** which may be described as the flowing or streaming of cytoplasm within a cell in a continuous movement. Although there is no movement from one cell to another, many of the distinct bodies found in the cytoplasm, such as chloroplasts, are carried about much like logs in a stream. As shown in Figure 3-17, the movement of one portion of cytoplasm may be opposite to that of another; lower layers of the cytoplasm may flow in one direction while the upper layers flow

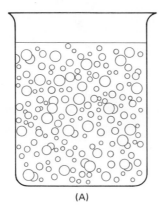
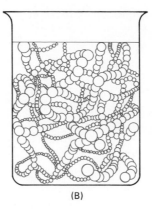

(A)   (B)

Figure 3-16. Sol and gel states of a colloid. **(A)** In the sol state the colloidal particles are randomly dispersed. **(B)** The particles in a gel interact and form a more orderly network. (Modified from M. S. Gardiner and S. C. Flemister, *The Principles of General Biology*, 2nd edition. New York: The Macmillan Company, © 1967.)

in the opposite direction. Such movement indicates great chemical activity and may be important in the transport of food and other materials.

Organelles. Organelles, or "little organs," represent the third subdivision of the protoplast. These are highly organized and specialized portions of the cytoplasm that assume various roles in cellular maintenance, growth, repair, and continuity. The chemical machinery by which cells perform various activities is localized within these organelles. In essence, the integration of biochemical processes into the structural framework of different organelles is the structural basis of cellular function.

Plastids. **Plastids** represent a variety of differentiated organelles embedded in the cytoplasm. These, along with cell walls, are truly distinctive plant structures. Microscopically, plastids reveal a number of sizes and shapes. Various forms exhibited by plastids are plates, discs, and spirals, among others. Plastids are believed to originate from minute, defined precursor structures called *proplastids.* These proplastids, typically found in young cells, are transmitted from one generation to another during cell division and are also capable of duplicating themselves in the mature plant cell. Plastids are commonly categorized into three distinct types on the basis of color and function.

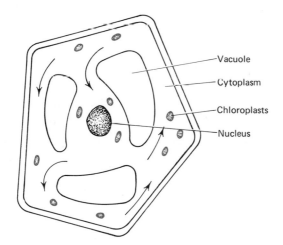

Vacuole

Cytoplasm

Chloroplasts

Nucleus

Figure 3-17. Cytoplasmic streaming. In the diagram shown note the movement of the various portions of the cytoplasm. Certain organelles such as chloroplasts are carried along in the streaming but the nucleus remains fixed.

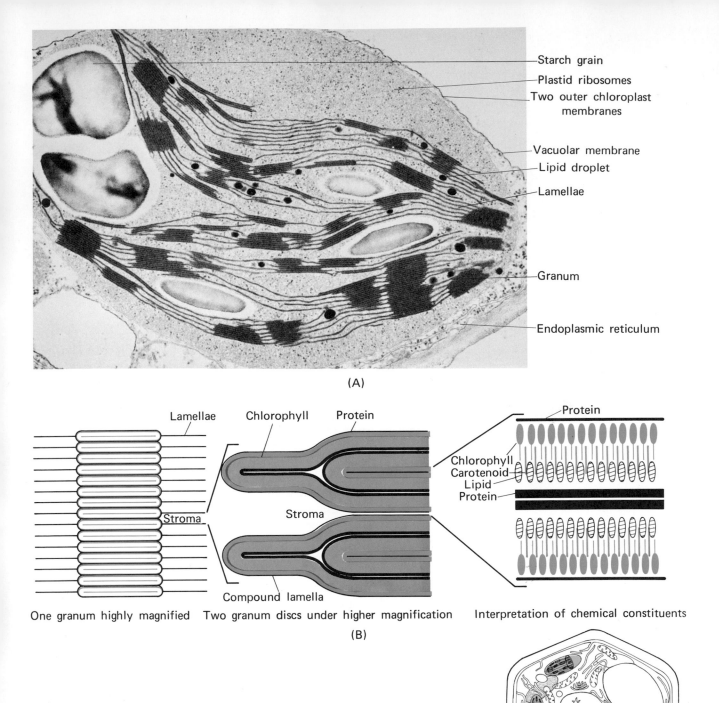

Starch grain
Plastid ribosomes
Two outer chloroplast membranes
Vacuolar membrane
Lipid droplet
Lamellae
Granum
Endoplasmic reticulum

(A)

Lamellae   Chlorophyll   Protein   Protein

Stroma   Stroma   Chlorophyll
                                                    Carotenoid
                                                    Lipid
                                                    Protein

One granum highly magnified   Compound lamella
                               Two granum discs under higher magnification   Interpretation of chemical constituents

(B)

Figure 3-18. The ultrastructure of a chloroplast. (**A**) Electron micrograph of a chloroplast from the leaf of *Phleum pratense* (timothy grass). X27,400. (Courtesy of Dr. Myron C. Ledbetter, Brookhaven National Laboratory.) (**B**) Diagrammatic representation of a single granum in successive magnifications showing possible chemical make-up and orientation. (Modified from A. J. Hodge, in J. L. Onceley et al. (eds.), *Biophysical Science—A Study Program.* New York: John Wiley & Sons, Inc., © 1959.)

These include chloroplasts, chromoplasts, and leucoplasts.

Chloroplasts are green plastids that owe their color to the presence of the pigment chlorophyll. In higher plants they are usually disc-shaped and range in size from 2 to 4 μ in diameter and 0.5 to 1 μ in thickness. About 20 per cent of the volume of leaf cells is occupied by chloroplasts. This is not surprising in view of the fact that leaves are the primary organs of photosynthesis. At high magnification, chloroplasts reveal an elaborate, highly organized internal structure (Figure 3-18A). Each consists of a double-layered bounding membrane that separates it from the cytoplasm. In addition, three internal structures are distinguishable. The first of these, a series of membranes that lie parallel to each other and extend the length of the chloroplast, are called *lamellae*. The second definitive structures are a series of multilayered, coin-shaped discs referred to as *grana*. The grana appear at various intervals throughout the lamellae and represent areas in which there is a high concentration of chlorophyll molecules. Each granum is believed to consist of layers of chlorophyll molecules associated with lipids sandwiched between layers of protein molecules (Figure 3-18B). The third component of a chloroplast is a clear, nonmembraneous region lacking chlorophyll and referred to as the *stroma*.

Although chlorophyll and carotenoids are the major chemical constituents of chloroplasts, chemical studies have also revealed the presence of lipids (lamellae), proteins (grana and enzymes), RNA, and DNA of a sort unique to chloroplasts and distinct from nuclear DNA and RNA.

Chromoplasts are plastids that owe their distinctive color to the presence of various carotenoids and are typically yellow, orange, or red. Although they are generally disc-shaped, some appear as angular, spherical, or spindle- or rod-shaped bodies (Figure 3-19; see also Figure 3-10).

Chromoplasts may be plastids in which chlorophyll disappears and becomes less dominant, or they may originate directly from proplastids, never having contained chlorophyll. For example, when bananas ripen, chloroplasts become chromoplasts as the chlorophyll breaks down and carotenoids become dominant. Another example is the ripening of fruits such as sweet peppers or tomatoes, in which chlorophyll is broken down and the carotenoid pigments predominate. A final example is the coloration in autumn leaves, in which carotenoids, originally present in leaves, become the dominant pigments as chlorophyll is broken down in the fall.

Leucoplasts, for the most part, are colorless plastids and are difficult to locate and identify unless cells containing them are specially stained. One such common stain is iodine, which has an affinity for the starch contained in the leucoplasts. These plastids are generally found in colorless leaf cells (variegated leaves), stems, roots, and other storage organs. Usually tissues not exposed to sunlight and underground portions of plants contain numerous leucoplasts. Functionally, leucoplasts serve as starch-forming and storage centers in plants (refer to Figure 2-13).

Mitochondria. A second group of organelles found universally in the cytoplasm of plant and animal cells consists of small spherical or rod- or filamentous-shaped bodies called **mitochondria** (Figure 3-20). They range in size from 0.5 to 1 μ in diameter and from 1 to 2 μ in length. The size, shape, and distribution of mitochondria are fairly constant in cells of the same type. However, cells of different types exhibit considerable variations.

Although mitochondria are rendered visible by using the ordinary light microscope and special staining techniques, very little detailed structure is revealed. Through the use of thin sectioning and the electron microscope mitochondria exhibit a fairly elaborate internal organization. Electron

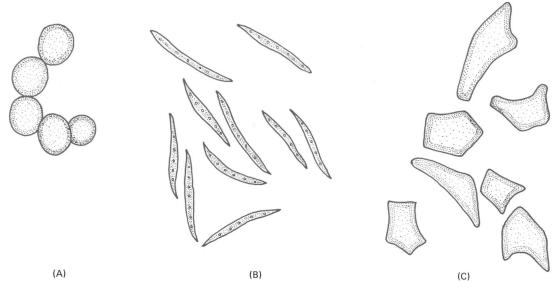

Figure 3-19. Diversity in chromoplasts. All of the chromoplasts shown are from the petals of flowers. (**A**) *Cucurbita* (squash). (**B**) *Tropaeolum* (nasturtium). (**C**) *Impatiens* (jewelweed).

micrographs have shown that each mitochondrion consists of a double membrane similar in structure to the plasma membrane. Chemically, the mitochondrial membrane, like the cell membrane, is lipid and protein in nature. The outer membrane forms a smooth surface separating the internal portions of the mitochondrion from the cytoplasm. The inner membrane is thrown into a series of folds that extend into the *matrix,* or center, of the mitochondrion. These folds, or *cristae,* assume various forms in the mitochondria of different cells.

As with other cellular organelles, mitochondria illustrate a high degree of correlation between structure and function. The cristae are arranged in a specific pattern in the form of uneven and incomplete folds. This pattern affords an enormous surface area for chemical reactions, a great deal of reaction space, and the possibility of molecules orienting themselves in a linear arrangement. Experimental evidence indicates that enzyme molecules, concerned with energy-releasing reactions, are arranged on the cristae. Organic acids, derived from the breakdown of more complex molecules, are further broken down into carbon dioxide and water as part of a stepwise series of reactions termed **respiration.** Each reaction is catalyzed by a specific enzyme, and many steps in the sequence release energy. Some of this energy is then stored as ATP and passes to other parts of the cell, where it is utilized. Because of the central role mitochondria play in cellular respiration, they are referred to as "powerhouses of the cell."

Endoplasmic Reticulum. Electron-microscope studies of cytoplasm have discredited the idea that cytoplasm is merely a viscous, homogeneous sub-

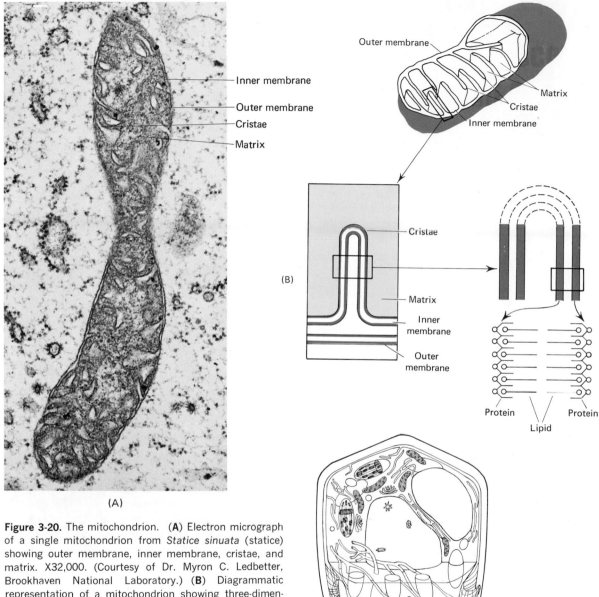

Labels in figure:

(A) — Inner membrane, Outer membrane, Cristae, Matrix

(B) — Outer membrane, Matrix, Cristae, Inner membrane; Cristae, Matrix, Inner membrane, Outer membrane; Protein, Lipid, Protein

Figure 3-20. The mitochondrion. (**A**) Electron micrograph of a single mitochondrion from *Statice sinuata* (statice) showing outer membrane, inner membrane, cristae, and matrix. X32,000. (Courtesy of Dr. Myron C. Ledbetter, Brookhaven National Laboratory.) (**B**) Diagrammatic representation of a mitochondrion showing three-dimensional structure, details of a single cristae, and interpretation of molecular orientation. (Modified from E. D. P. De Robertis, W. W. Nowinski, and F. A. Saez, *Cell Biology*, 4th edition. Philadelphia: W. B. Saunders Company, 1965.)

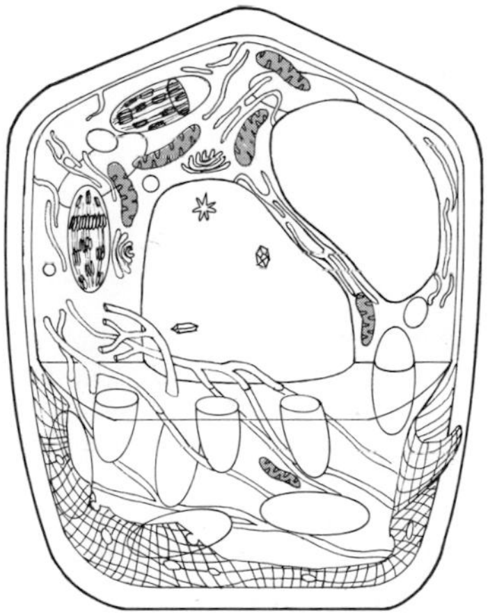

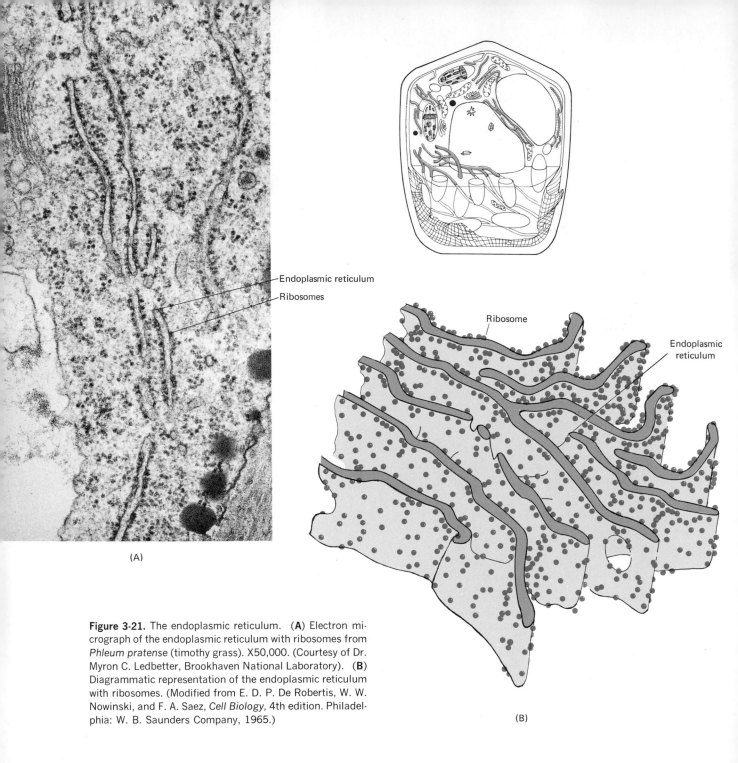

Endoplasmic reticulum

Ribosomes

(A)

Ribosome

Endoplasmic reticulum

Figure 3-21. The endoplasmic reticulum. (A) Electron micrograph of the endoplasmic reticulum with ribosomes from *Phleum pratense* (timothy grass). X50,000. (Courtesy of Dr. Myron C. Ledbetter, Brookhaven National Laboratory). (B) Diagrammatic representation of the endoplasmic reticulum with ribosomes. (Modified from E. D. P. De Robertis, W. W. Nowinski, and F. A. Saez, *Cell Biology,* 4th edition. Philadelphia: W. B. Saunders Company, 1965.)

(B)

stance. Rather, these studies have revealed that cytoplasm contains a complex system of double membranes called the **endoplasmic reticulum (ER)** (Figure 3-21). These membranes appear as an extensive branching network (reticulum) forming a continuous elaborate system. At times the ER appears in the form of long, narrow tubes, but more commonly it consists of folded sheets of membranes. The structural and functional relationship of the ER to other plant cell membranes, such as the plasma, nuclear, and vacuolar membranes, is still not entirely clear.

Consistent with other cellular membranes, the ER is composed of lipid-protein molecules. It is not fixed or static, but breaks up, re-forms, and regroups in response to cellular activities. The role of the endoplasmic reticulum is still a question of much debate. Among the possible functions ascribed to it are (1) a transport and storage system for proteins, (2) a connecting membrane between cells through plasmodesmata, and (3) at or near it ribosomes perform their function in protein synthesis.

Ribosomes. **Ribosomes** are small, spherical bodies composed of RNA and protein. The presence of ribosomes in cells of bacteria, yeasts, higher plants, and animals implies their universal distribution. These organelles, approximately 250 Å in diameter, may exist free in the cytoplasm or they may be attached to the ER (Figure 3-21). When associated with the ER, ribosomes are found only on the outer side of the tubular membrane structure and give the ER a rough or granular appearance. Functionally, ribosomes are the sites of protein synthesis. The mechanism of these protein-synthesizing reactions will be treated later.

Golgi Bodies. **Golgi bodies,** or **dictyosomes,** were first described by Camillo Golgi, an Italian physician, in 1898. Although once considered to be organelles of animal cells only, electron microscopy has demonstrated their presence in plant cells

as well (Figure 3-22). Structurally, dictyosomes reveal two components. First, a series of much flattened discs consisting of two membranes and, second, numerous spherical vesicles, arranged around the edges of the discs. Both the discs and vesicles vary in size and number in response to changes in cellular activities. Although the function of Golgi bodies is still obscure, some evidence indicates that they assume a role in synthesis of material for cell wall formation during cell division and after the cell has been formed. In this context it is believed that the vesicles migrate to the cell wall and discharge their contents (synthesized materials), which combine with other wall elements.

Nucleus. The **nucleus** of plant cells, which averages about 15 μ in diameter, assumes a spherical or oval shape and its contents are visibly more viscous than the cytoplasm (Figure 3-23). Because of this dense appearance, it is more conspicuous than other cellular organelles. A nucleus of characteristic appearance is universally present in living plant cells except in mature food-transporting cells, bacteria, and blue-green algae. It is centrally located and occupies a major proportion of the cell contents in young plant cells, whereas it is peripherally located and comprises a smaller proportion of the total contents of more mature cells. Most plant cells contain only one nucleus; however, in certain species of algae and fungi, cells contain two or more nuclei, a condition referred to as multinucleated, or **coenocytic.**

The nucleus is delimited from the cytoplasm in which it is embedded by a double membrane, the **nuclear membrane.** This structure resembles other cellular membranes both in appearance and chemistry. Electron micrographs reveal that the nuclear membrane communicates with the ER and is traversed by minute pores. These pores probably facilitate an exchange of materials between the nucleus and the cytoplasm.

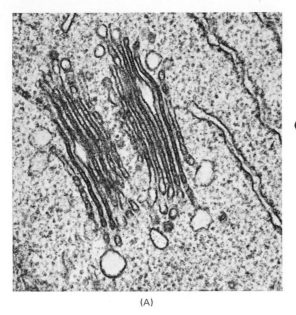

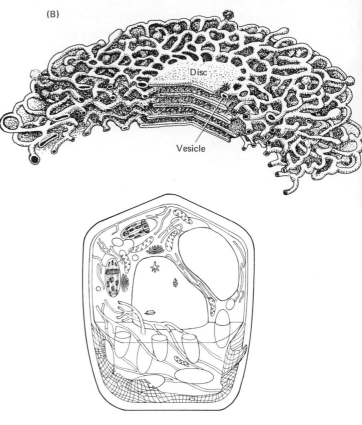

Figure 3-22. The Golgi bodies. (A) Electron micrograph of two Golgi bodies from *Potamogeton pectinatus* (pond weed) showing flattened discs and spherical vesicles. X52,000. (Courtesy of Dr. Myron C. Ledbetter, Brookhaven National Laboratory.) (B) Diagrammatic representation of a Golgi body. (After H. H. Mollenhauer and D. J. Morré, "Golgi Apparatus and Plant Secretory Processes," *Annual Review of Plant Physiology,* vol. 17, p. 27, 1966.)

The body of the nucleus within the nuclear membrane is composed of a gel-like nuclear cytoplasm called **nuclear sap** or **karyolymph,** one or more spherical bodies called nucleoli (nucleolus, singular), and a threadlike network referred to as chromatin.

Nucleoli are composed chiefly of protein and RNA and are believed to assume a role in protein synthesis. **Chromatin,** which is the genetic material, consists of DNA and protein. In a nondividing cell chromatin appears as a distended threadlike mass. When the cell is about to divide, the chromatin mass shortens and thickens to form rodshaped bodies called **chromosomes.**

Ergastic Substances

Ergastic substances, or **cell inclusions,** constitute a group of assorted materials that are produced as a result of cellular activities. These substances, much simpler in structure than organelles, may appear and disappear at different times in the life of the cell. Some are stored as reserve materials; others are products of various biochemical processes. Many assume characteristic forms in the cell that are readily visible.

Vacuoles. The most conspicuous inclusions of mature plant cells are vacuoles. They are almost as distinctive a feature of plant cells as are the cell

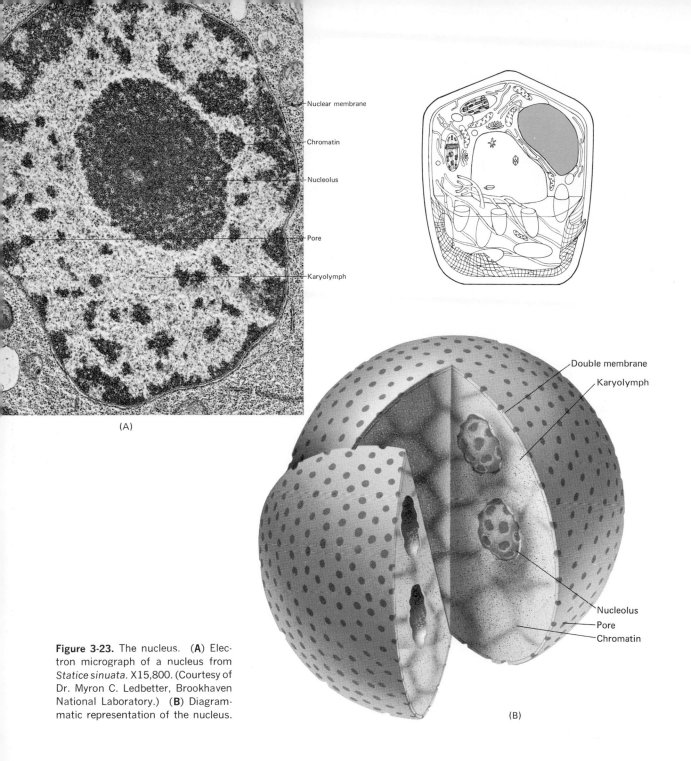

Nuclear membrane

Chromatin

Nucleolus

Pore

Karyolymph

(A)

Double membrane

Karyolymph

Nucleolus

Pore

Chromatin

(B)

Figure 3-23. The nucleus. **(A)** Electron micrograph of a nucleus from *Statice sinuata*. X15,800. (Courtesy of Dr. Myron C. Ledbetter, Brookhaven National Laboratory.) **(B)** Diagrammatic representation of the nucleus.

walls. A **vacuole** is a region containing a highly dilute solution of substances bounded by a membrane. The solution is termed *vacuolar sap* and the vacuolar membrane, a one-layered structure, is called the *tonoplast* (see Figure 3-10).

The major component of vacuoles is water. Some of the dissolved substances include atmospheric gases, salts, sugars, organic acids, soluble proteins, and certain pigments. The most prevalent pigments are **anthocyanins.** The red color of many flowers, such as roses, is a result of pigments being concentrated in the vacuoles in the cells of the flower petals.

Starch. *Starch grains* represent a second type of ergastic substance. Glucose, produced during photosynthesis, is converted into starch and stored in plastids in the form of starch grains (see Figure 3-10). Starch is generally stored in leucoplasts and chloroplasts.

Crystals. These inorganic deposits in plants consist chiefly of calcium salts, the most common *crystals* being those of *calcium oxalate.* The chemical union of oxalic acid (an organic acid) with calcium produces the calcium oxalate crystals. These crystals assume a number of forms in plants. Some are needle-like and are termed *raphides;* some form clusters and are designated *druses;* others appear as single crystals (Figure 3-24). In addition to calcium oxalate crystals, crystals of calcium carbonate, calcium sulfate, silica, and protein also occur.

Other Inclusions. Proteins, the major constituents of cytoplasm, may also occur as temporary ergastic substances. Two examples are *gluten* and *aleurone grains,* which are stored proteins found in the cells of seeds, such as wheat. Lipids also appear as inclusions being found as *oil globules* in the cytoplasm. Protein, lipid, and starch inclusions are formed when excess foods are produced by the plant cell and they may be hydrolyzed and utilized during periods of minimum food synthesis.

Less frequently found inclusions are tannins, resins, gums, and mucilages. It is believed that all are waste products of the physiological activities

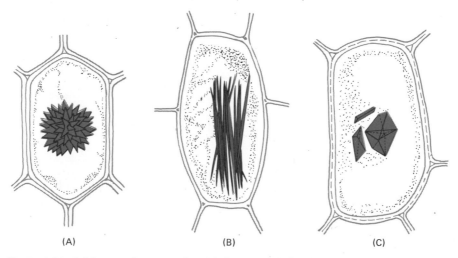

Figure 3-24. Calcium oxalate crystals. **(A)** Druses. **(B)** Raphides. **(C)** Rhombohedral and prismatic crystals.

TABLE 3-1. Principal Components of a Generalized Plant Cell

I. Cell wall	II. Protoplast	III. Ergastic substances
A. Middle lamella	A. Plasma membrane	A. Vacuoles
B. Primary wall	B. Cytoplasm	B. Starch grains
C. Secondary wall	C. Organelles	C. Crystals
	1. Plastids	D. Other: protein,
	2. Mitochondria	lipids, tannins,
	3. Endoplasmic	resins, mucilages,
	reticulum	gums
	4. Ribosomes	
	5. Golgi bodies	
	6. Spherosomes	
	7. Nucleus	
	a. Nuclear	
	membrane	
	b. Nucleoli	
	c. Karyolymph	
	d. Chromosomes	

of plant cells and probably protect the protoplast against desiccation, decay, and injury by animals. The components of a generalized plant cell are summarized in Table 3-1.

The Sizes of Plant Cells

In the preceding discussion, the dimensions, in microns and/or angstroms, for various cellular constituents were indicated. To gain some understanding of the diversity of plant cells, in terms of size, some representative examples will be considered.

With few exceptions, plant cells are too small to be seen with the unaided eye. Examination of living or killed plant cells with various types of microscopes reveals considerable variations in size. *Mycoplasma*, a bacterium-like organism, measures $0.1\ \mu$ in diameter as seen with an electron microscope. Other cells, particularly bacterial cells, range from 0.2 to $5\ \mu$ and are visible with the light microscope. For cells of most higher plants, the general range in diameter is 0.5 to $100\ \mu$, with the vast majority measuring from 5 to $20\ \mu$. Plant

cells several millimeters in length, such as fibers from flax and cotton hairs, are also found in nature.

DIVISION, ENLARGEMENT, AND DIFFERENTIATION OF CELLS

The organization of similar cells performing a specialized function results in the formation of a **tissue.** Tissues are in turn organized into higher structural levels called **organs.** Familiar plant organs are roots, stems, leaves, flowers, fruits, and seeds.

Higher plants are composed of a number of tissues and organs each contributing to the form and function of the total plant. These tissues originate from a group of unspecialized cells in the seed called *meristematic cells.* The formation of a mature, structurally and functionally integrated plant from a seed involves three phenomena: cell division, cell enlargement, and cell differentiation.

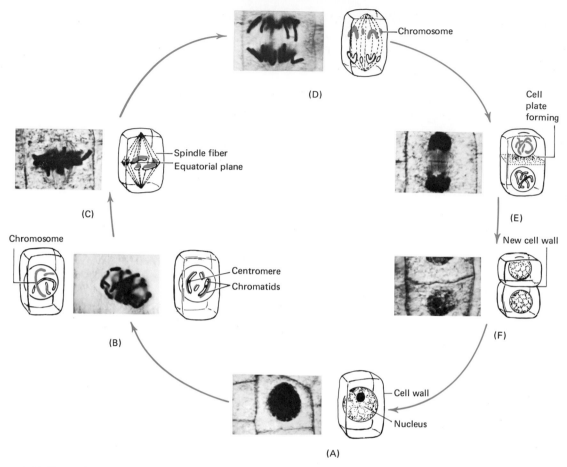

Figure 3-25. Photomicrographs and diagrammatic representations of root tip cells in the process of cell division. (**A**) Interphase. (**B**) Prophase (early and late). (**C**) Metaphase. (**D**) Anaphase. (**E**) Telophase. (**F**) New cells in interphase. (Photomicrographs courtesy of Carolina Biological Supply Company.)

Cell Division

Cell division is an important fundamental aspect of growth and reproduction in plants and consists of two more or less simultaneous processes. The first of these, **mitosis,** involves the replication of nuclear material into two separate and equal nuclei. The second process, **cytokinesis,** is concerned with the division of the cytoplasm into separate components with the formation of a new wall between the components.

Stages of Mitosis. When a cell is not dividing it is said to be in a stage called **interphase** (Figure 3-25A). Most of the time cells are in interphase. In this condition, the cell is physiologically active;

the nucleus is a clearly defined structure with the enveloping membrane; and nucleoli, chromatin, and karyolymph are clearly visible. Although mitosis is a continuous process, it is convenient to describe it in terms of stages or phases. These arbitrarily designated stages, in order of appearance, are prophase, metaphase, anaphase, and telophase.

The term **prophase** (Figure 3-25B) refers to the earliest recognizable stage of mitosis. It is characterized by a shortening and thickening of the chromatin material of the nucleus into discrete bodies called **chromosomes.** The chromosomes are doubled, with the two members sharing a point of attachment. Each of the halves is called a **chromatid** and the point of attachment is termed a **centromere.** The chromatids indicate that each chromosome has replicated itself. This duplication probably occurs during interphase. As prophase continues and terminates there is a sudden contraction of the nucleus, the nucleoli dissolve, and the nuclear membrane disappears. Then the chromosomes begin to move about the cytoplasm and eventually converge at the center of the cell, called the *equatorial plane.*

During **metaphase** (Figure 3-25C) fine radiating fibers become prominent at opposite ends, or poles, of the cell. These structures, believed to be composed of chainlike polypeptides, are called *spindle fibers.* The fibers attach themselves to the centromeres of the chromatids, which have now become oriented along the equatorial plane.

During **anaphase** (Figure 3-25D) the chromatids separate, forming two identical groups, each migrating to different poles in the cell. Once the separation is completed, each chromatid is considered to be a distinct chromosome. The contraction of the spindle fibers probably aids the migration of the two groups of chromosomes. At the termination of anaphase, one group of identical chromosomes is located at each pole of the cell.

Telophase (Figure 3-25E) is the last recognizable stage of mitosis. Essentially, the events of telophase are described as the reverse of the happenings of prophase. The chromosomes form a threadlike network (chromatin), the nuclear membrane and nucleoli again become apparent, spindle fibers disappear, and a new cell wall forms from a *cell plate.* The development of the new wall is one of the events associated with cytokinesis.

The salient features of mitosis may be summarized as follows: (1) it brings about an increase in the number of cells, (2) each new (daughter) cell has the same number and kind of chromosomes as the original (parent) cell, and (3) the daughter cells have the same hereditary material and genetic potential as the parent cell.

The most rapidly growing plant cells spend 75 to 80 per cent of their time in interphase. This means that 3 to 4 hours a day are spent in active nuclear division. The time required to complete a divisional cycle varies with the type of plant cell, its location, and the effect of certain factors such as temperature. In general, an increase in temperature increases the rate of mitosis. For example, at 59°F, the root cells of pea plants undergo a complete divisional cycle in about 25 hours. If the temperature is increased to 77°F, the cycle is completed in about 16 hours.

The duration of the time involved in the various stages of mitosis varies from one plant to another. Prophase is usually the longest, lasting from 1 to several hours. Metaphase is much shorter and requires about 5 to 15 minutes. The shortest stage is anaphase, which lasts only 2 to 10 minutes. Finally, the telophase may vary from 10 to 30 minutes. These figures indicate only the relative length of time for the various stages of mitosis and should not be interpreted as exact limits.

Cytokinesis. The term **cytokinesis** refers to the division of the cytoplasm and the formation of a new cell wall (Figure 3-25). It is closely associated with mitosis and usually immediately follows the division of the nucleus during late anaphase or telophase. As mitosis nears completion, granular material appears across the equatorial region. These particles become larger, more dense, and coalesce to form the cell plate, which extends across the cell from wall to wall. The cytoplasm is thus divided into two portions, each containing one of the new nuclei. The cell plate becomes the middle lamella of the new wall. The two proto-plasts of the resulting daughter cells deposit additional material on the middle lamella until the cell wall layers are formed. As indicated earlier, the vesicles of the Golgi bodies may be involved in the deposition of new wall materials (Figure 3-25F).

If a parent cell has a fixed amount of chromosomal material within the nucleus, and if this material is divided equally between daughter nuclei during mitosis, it becomes evident that the quantity of heredity material (DNA) would be reduced by half with each successive division. This constant reduction would seriously affect the physiological behavior of the nucleus, and in consequence the behavior of the cell, tissue, organ, and plant to which it belonged. It becomes apparent that there must be some mechanism by which DNA itself is quantitatively and qualitatively duplicated so that it will remain constant in daughter cells derived from parent cells.

It is generally believed that the double helix of DNA uncoils and forms a ladder-like structure during interphase (Figure 3-26). Further, it is assumed that the weak hydrogen bonds between purines and pyrimidines are broken, leaving two unpaired strands of nucleotides. As part of the raw materials stored in the cell, there are a variety of nucleotides, which, during interphase, fit into place in specific sequences on both unpaired strands until the double chain is once again completed. In the presence of an appropriate enzyme the hydrogen bonds once again join the nitro-genous bases, the enzyme catalyzes the formation of the sugar phosphate backbone, and two identical double helices stand where one stood before.

Cell Enlargement

In the process of cell division, one cell divides to form two. A closely related event is the process of *cell enlargement.* This is the dominant and most obvious phase of plant growth. At least three factors are involved in cell enlargement: (1) an intake of water, (2) an increase in the area of the cell wall, and (3) the deposition of new cell wall material to accommodate the enlargement of the cell. In addition, there is an increase in cytoplasmic content as a result of the production of more cellular proteins. The role of the plant hormone auxin is considered to be essential in the process of cell enlargement.

Cell Differentiation

The differentiation of cells, which occurs after or in conjunction with enlargement, involves several different types of structural and functional specializations. Essentially, *differentiation* is the transformation of apparently identical cells, produced by division, into a variety of highly specialized cells. Consider the formation of a mature plant from a single fertilized egg. The egg divides successively to form 2, 4, 8, 16, and so on, cells. The resulting cells are presumed to be genetically identical, yet, despite this hereditary consistency, the cells produced by later divisions give rise to a variety of tissues and organs that are structurally and functionally specialized.

As a multicellular plant matures, most of its cells differentiate and specialize in various ways. Some cells specialize to a minor extent and are capable of performing several comparatively simple functions. Others become highly specialized and are capable of performing one or two relatively complex functions. In either case, specialization in function is always based upon a specialization in structure. Cells may differentiate and become specialized with respect to size, shape, nature, and extent of the secondary wall, or with respect to cytoplasmic contents. A more detailed discussion of enlargement and differentiation is presented in Chapter 11.

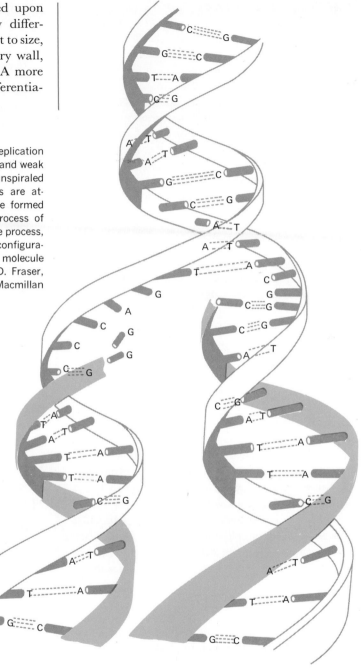

Figure 3-26. Diagrammatic representation of the replication of DNA. The double helical configuration uncoils and weak hydrogen bonds between nucleotides of the unspiraled strands break. New complementary nucleotides are attached at the proper sites. Hydrogen bonds are formed between complementary nucleotides and the process of duplication is completed. At the termination of the process, each molecule once again assumes the helical configuration. Note that each DNA strand of the original molecule has produced an exact replica of itself. (From D. Fraser, *Viruses and Molecular Biology.* New York; The Macmillan Company, © 1967.)

CHAPTER IN RETROSPECT

More than a century ago, Schleiden and Schwann discovered that the cell was the basic unit or common denominator of all living forms. Since that time, a considerable amount of additional information has been learned about cells, largely as a result of more precise instrumentation and the development of more sophisticated biochemical techniques. Considering that the average size of plant cells ranges between 5 and 20 μ, it is indeed remarkable that scientists have discovered so much about their exact nature. For instance, it is now known that the cell wall consists of a middle lamella as well as a primary and secondary cell wall. It is also known that the cell protoplast is made up of a plasma membrane, cytoplasm, and at least seven organelles. In addition, various ergastic substances such as vacuoles, starch grains, and crystals may appear and disappear at different times in the life of the cell. Each of these components possesses a specific structural framework, chemical composition, and specialized function.

The procedure by which plants mature from a seed involves three cytological phenomena: cell division, cell enlargement, and cell differentiation. The first of these, cell division, consists of the replication of nuclear and cytoplasmic material into two separate components with the formation of a new wall between the components. These smaller daughter cells will increase in size by the process of cell enlargement. The second and third processes, those of cell differentiation, occur after or in conjunction with cell enlargement and involve structural and functional specializations.

Questions and Problems

1. What is the relative significance of the discoveries of Schwann and Schleiden in 1838–1839?

2. Compare the magnifying power of an electron microscope with that of a compound microscope. How has each instrument contributed to a more exact knowledge of cell morphology?

3. In preparing microscopic slides, why is it necessary to cut a specimen extremely thin (about 10 μ)? Why are specimens embedded in wax as part of the procedure for preparing slides?

4. Discuss the mechanism of chromatography with respect to the separation and identification of various amino acids and sugars. How has this technique furthered our understanding of cellular makeup?

5. Briefly describe the function of the following cell structures: cell wall, secondary cell wall, pits and primary pit fields, plasma membrane, cytoplasm, chloroplasts, leucoplasts, mitochondria, ribosomes, and nucleus.

6. By way of a series of labeled diagrams, describe the various processes of cell division. What is accomplished as a result of cell division?

7. Briefly describe any chemical changes which may occur in DNA molecules during the process of mitosis.

8. Discuss the essential differences between cell enlargement and cell differentiation. Why is each process necessary in the growth and development of living organisms?

Suggested Supplementary Readings

Brachet, Jean, "The Living Cell," *Scientific American,* September 1961.

Brown, R., "The Plant Cell and Its Inclusions," in vol. 1A of *Plant Physiology,* edited by F. C. Steward. New York: Academic Press, Inc., 1960.

de Duve, Christian, "The Lysosome," *Scientific American,* May 1963.

Esau, Katherine, *Plant Anatomy.* New York: John Wiley & Sons, Inc., 1965.

Guthe, K. F., *The Physiology of Cells.* New York: The Macmillan Company, 1968.

Hokin, L. E., and M. R. Hokin, "The Chemistry of Cell Membranes," *Scientific American,* October 1965.

Jensen, W. A., *The Plant Cell.* Belmont, Calif.: Wadsworth Publishing Co., Inc., 1964.

Lenhoff, E. S., *Tools of Biology.* New York: The Macmillan Company, 1966.

Loewy, A. G., and Philip Siekevitz, *Cell Structure and Function.* New York: Holt, Rinehart and Winston, Inc., 1963.

Mazia, D., "How Cells Divide," *Scientific American,* September 1961.

Rich, Alexander, "Polyribosomes," *Scientific American,* June 1963.

Robertson, J. D., "Membrane of the Living Cell," *Scientific American,* April 1962.

Siekevitz, P., "Powerhouse of the Cell," *Scientific American,* July 1957.

Solomon, A. K., "Pores in the Cell Membrane," *Scientific American,* December 1960.

Swanson, C. P., *The Cell.* Englewood Cliffs, N.J.: Prentice-Hall, Inc., 2nd ed., 1964.

Taylor, J. H., "The Duplication of Chromosomes," *Scientific American,* June 1958.

Walker, J. J., "The Chloroplast and Photosynthesis—A Structural Basis for Function," *American Scientist,* vol. 47, pp. 202–215, 1959.

Wardrop, A. B., "Cell Wall Organization in Higher Plants," *Botanical Review,* vol. 28, pp. 241–285, 1962.

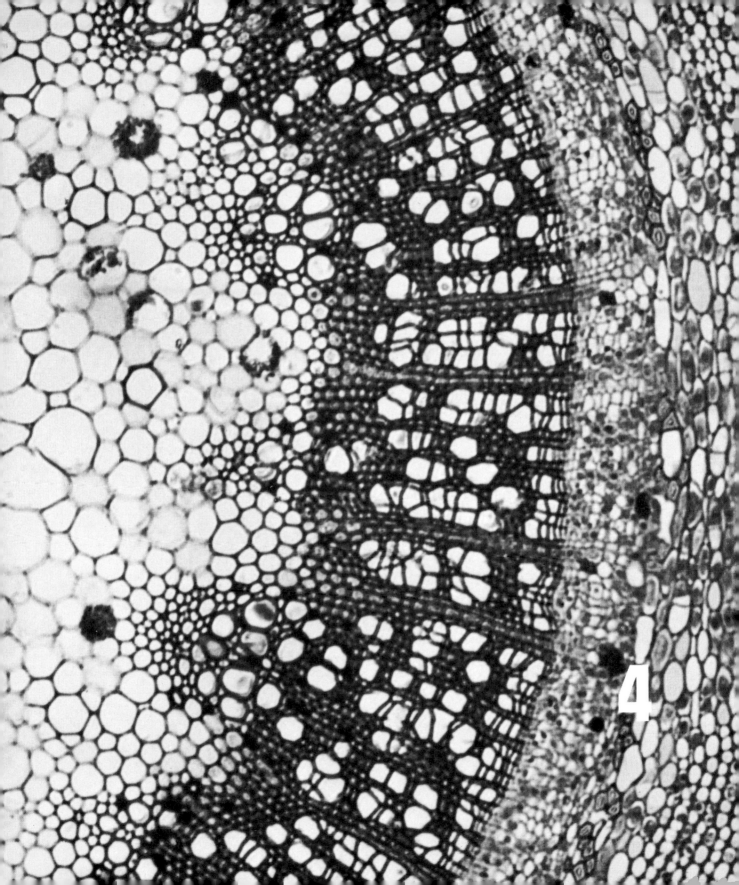

4
Macroscopic Level of Plant Organization: Vegetative Organs

INTRODUCTION

In Chapter 3 the cell was discussed as the basic structural and functional unit of plants. In this context, the cell was treated as a highly organized and integrated unit with each of its components having a particular structural organization, chemical composition, and specific function. By the processes of division, enlargement, and differentiation, cells become organized into tissues, and tissues in turn may form organs, thus producing a multicellular plant.

The complex nature of a multicellular plant implies a high degree of coordination and correlation among its component organs, tissues, and cells. Even though no structures or functions of an entire plant are isolated from all others, it is frequently helpful to separate individual portions or processes for the purpose of logical discussion and understanding. However, the interrelationships of processes and of structures and function of the entire plant must always be kept in mind.

In the following discussion of the tissues and organs of a plant, attention will be directed to those plants which are the most familiar. These plants are the **angiosperms,** or flowering plants. Angiosperms are subdivided into two groups, called *monocots* and *dicots*. Although both groups produce flowers, they can be distinguished on the basis of certain structural characteristics.

In this chapter, after an analysis of the nature and types of plant tissues, plant organs will be studied. The organs of an angiosperm are classified as vegetative or reproductive. *Vegetative organs* are those structures of the plant which are concerned with growth, maintenance, and development, such as roots, stems, and leaves. Those parts of the plant which are concerned with reproduction and the production of seeds are classified as *reproductive organs*. These would include flowers, which give rise to fruits, which in turn contain seeds. Figure 4-1 shows the various organs of an angiosperm.

TYPES OF PLANT TISSUES

Plant tissues, as noted earlier, are groups of similar cells performing a specialized function. All plant tissues are generally subdivided into two types: meristematic and permanent.

Meristematic Tissue

The continued growth and development of plants depend upon the mitotic activity of **meristematic tissues.** In such tissues, the cells are actively dividing and new cells are constantly being produced. Although meristematic cells exhibit considerable variation from one part of the plant to another, certain common features do exist. They tend to be small, and thin-walled, rich in cytoplasm, and have small, numerous vacuoles. New cells produced by a meristem are initially alike, but as they grow and mature, their characteristics slowly change as they become differentiated into cells of other tissues.

Meristematic tissues located near the tips of roots and in the buds of stems are called *apical meristems*. These meristems bring about an increase in length of cells after they are formed. **Vascular cambium** is a term applied to meristematic tissue found within the stems and roots of certain plants.

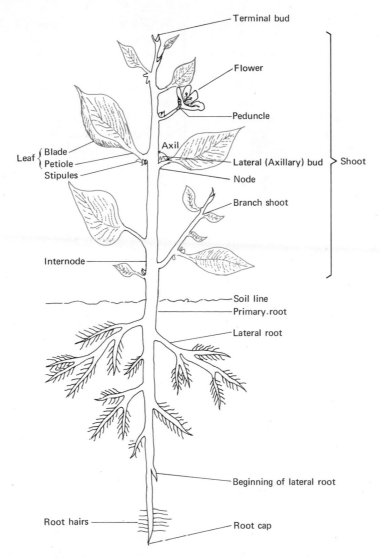

Figure 4-1. Diagrammatic representation of the principal external features of a flowering plant.

This tissue is considered to be a *lateral meristem,* because it causes lateral growth or an increase in diameter of a plant part. Another type of lateral meristem, called the *cork cambium,* usually develops if a plant is damaged in any way, although it is also quite common in uninjured woody angiosperms as well. It is located in the outer bark of trees and shrubs and produces cork.

Permanent Tissues

Permanent tissues, as contrasted with meristems, are stable in that they are no longer actively dividing. Although derived from meristematic regions, these tissues develop considerable structural and functional modification. Each type of permanent tissue is composed of differentiated cells that contribute to an efficient division of labor in the total functioning plant. For example, cells of conducting tissue are usually quite elongated and thick-walled, whereas storage cells are generally isodiametric and thin-walled. The term *isodiametric* is used to describe cell shape where length and width are essentially equal. Permanent tissues are divided into simple and complex types. *Simple permanent tissues* are composed of cells which are structurally and functionally similar, and *complex permanent tissues* are composed of several types of tissues which differ in structure and function. In the latter type of tissue, the cells are involved in a group of interrelated activities in which one function usually dominates.

Simple Permanent Tissue. The variety of cells that comprises simple permanent tissue is shown in Figure 4-2.

Epidermis. The **epidermis** is usually one cell layer in thickness and forms the surface layers of leaves, flowers, and young stems and roots. In older roots and stems of certain plants, the epidermis is often replaced by other tissues, such as cork. Epidermal cells are generally relatively flat, with large vacuoles and only a small amount of cytoplasm. Their outer walls, at least on the aerial portions of the plant, are covered by **cutin,** a waxy, waterproofing material. The continuous layer of cutin, termed the **cuticle,** aids in protection against water loss, mechanical injury, and invasion by disease organisms.

Epidermal tissues of aerial portions of plants sometimes give rise to single-celled or many-celled structures such as spines, hairs, or glands. These epidermal appendages are called **trichomes.** Some epidermal cells, especially those covering leaves and green stems, are specialized as guard cells. **Guard cells** are modified epidermal cells, often kidney-shaped, that regulate the size of tiny openings called **stomata.** Each stoma is actually a pore between two adjacent guard cells (Figure 4-2A). Stomata permit the exchange of gases between the atmosphere and internal tissues and also facilitate the evaporation of water.

Parenchyma. **Parenchyma** tissue occurs in roots, stems, and leaves. These cells are characterized by thin primary cell walls, lack of a secondary wall, and well-developed vacuoles surrounded by a peripheral distribution of cytoplasm. In addition, the cells are more or less spherical, although the shape is distorted by the pressure of surrounding cells so that they appear angular or box-shaped (Figure 4-2B). Parenchymatous tissue carries on a variety of functions. Most of the chloroplasts of leaves are in the cells of parenchyma tissue (for example, palisade and spongy mesophyll), and it is in these cells that photosynthesis occurs. Other parenchyma cells, found in roots and to a lesser extent in stems, are colorless and function chiefly in food and water storage.

Collenchyma. **Collenchyma** is composed of cells with unevenly thickened primary walls. The thickened areas are usually most prominent at the angles (corners) of the cells when viewed in cross section (Figure 4-2C). The principal function of collenchyma is mechanical support and strengthening. It is usually the first supportive tissue produced in young stems and leaves as well as a permanent tissue in mature plant organs. The combination of tensile strength and flexibility of these cells is ideally adapted to the support of growing organs.

Sclerenchyma. **Sclerenchyma** is a tissue characterized by thick cell walls. The cells have a definite

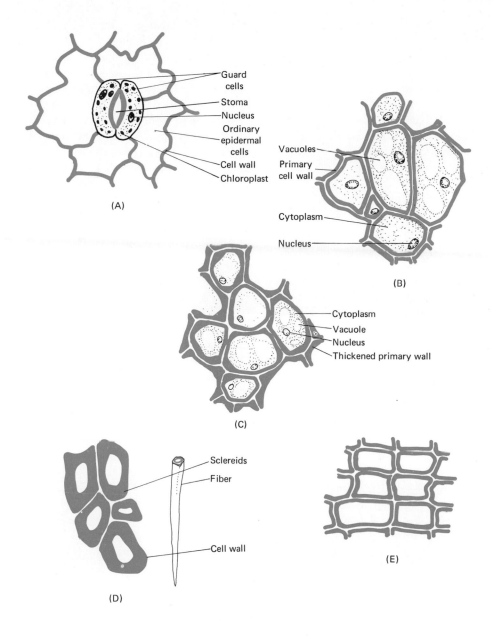

Figure 4-2. Cells from simple permanent tissue. **(A)** Epidermis. **(B)** Parenchyma. **(C)** Collenchyma. **(D)** Sclerenchyma (sclereids on left, fiber on right). **(E)** Cork.

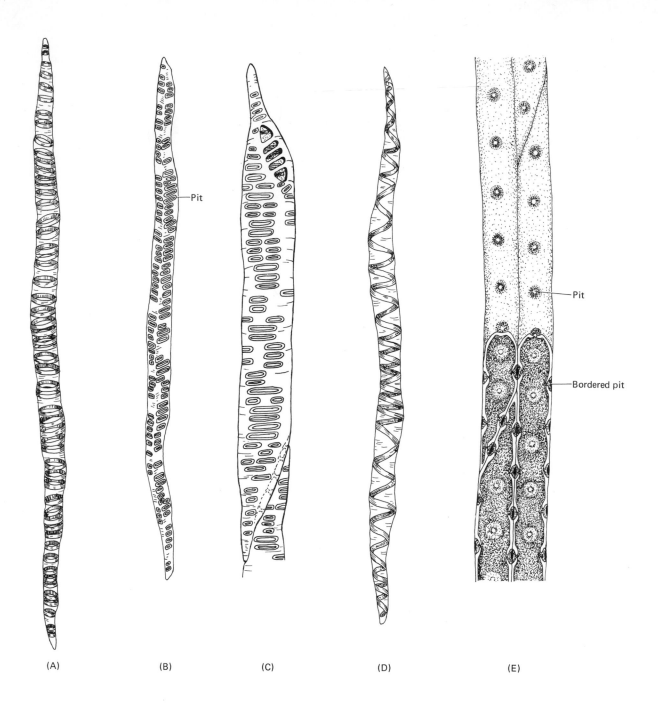

(A)   (B)   (C)   (D)   (E)

secondary wall that is often impregnated with lignin. With the appearance of lignin, the protoplast dies, so that at maturity there are no living contents. Sclerenchyma tissue reveals two types of cells which vary greatly in size and shape. These are fibers and sclereids (Figure 4-2D). **Fibers** are elongated sclerenchymatous cells with tapered ends. They are tough and strong, but flexible, and because of these properties and because they are aggregated into strands are important in the manufacture of rope, twine, and textiles from such plants as flax, jute, and hemp. **Sclereids,** or **stone cells,** are similar in strength to fibers but are characteristically different in shape. Typically, they are not elongated but rather are irregular in form (Figure 4-2D). Sclereids are common in the shells of nuts, the hard parts of seeds, and are scattered in the fleshy portions of certain hard fruits. The gritty texture of pears, for example, is due to small clusters of stone cells.

Cork. **Cork** is composed of cells which have thick cell walls impregnated with suberin, a waxy material (Figure 4-2E). The cells are dead at maturity and form a waterproofing layer of tissue. Cork is found as the outer layer of stems and roots of older woody plants as an epidermal replacement. The essential functions of cork are protection and water conservation. In the case of the cork tree, *Quercus suber,* a profuse amount of cork is produced which is removed and used commercially in the preparation of cork stoppers, life preservers, and insulating materials. As the cork cambium produces new cork it may again be removed from the tree in 3 to 4 years.

Complex Permanent Tissue. Complex permanent tissue, or vascular tissue, consists of more than one type of tissue and is a distinctive feature of higher plants. These tissues function as tubes or ducts within which water and dissolved substances (minerals and foods) are conducted from one part of the plant to another. The two principal types of vascular tissue are xylem and phloem.

Xylem. **Xylem,** or wood, is a vascular tissue that functions in the transport of water and dissolved substances, usually minerals, upward in the plant body. In addition, it serves as a supporting tissue. As a conducting tissue, it forms a continuous pathway through the roots, stems, leaves, flowers, and fruits. As with sclerenchyma, a number of types of cells are unique to xylem. These include tracheids, xylem fibers, and vessel elements.

Tracheids are long, slender cells, tapered at the ends, with well-developed lignified secondary walls (Figure 4-3). At maturity, the protoplasm within tracheids dies, leaving a space called a *lumen.* On the basis of the pattern of the secondary wall, tracheids are divided into several categories. Tracheids in which the secondary cell wall consists of a series of separate rings are called *annular tracheids* (Figure 4-3A). If the secondary wall forms a continuous spiral, a *spiral tracheid* results (Figure 4-3D). Tracheids in which the wall resembles the rungs of a ladder are designated as *scalariform tracheids* (Figure 4-3B), and those in which the secondary wall forms a continuous network are called *reticulate tracheids* (Figure 4-3C). Finally, tracheids consisting of a continuous secondary wall interrupted by pits are referred to as *pitted tracheids* (Figure 4-3E). As mentioned previously, pits facilitate the passage of materials from one cell to another.

Figure 4-3. (Opposite) Types of tracheids. Based upon the pattern of secondary wall thickening, tracheids are classified into five principal types: **(A)** annular, **(B)** scalariform, **(C)** reticulate, **(D)** spiral, and **(E)** pitted. (From V. A. Greulach and J. E. Adams, *Plants: An Introduction to Modern Botany.* New York: John Wiley & Sons, Inc., © 1967.)

Xylem fibers and vessel elements are cell types that are dead at maturity and represent two evolutionary modifications of tracheid cells. In the course of evolution, xylem fibers have assumed the function of support at the expense of conduction. **Xylem fibers** are longer than tracheids, having more extensive and heavier wall thickenings and lignification. In addition, they are more slender and tapering, with greatly reduced pits (Figure 4-4). A wood fiber is a good example of an evolutionary derivative of a tracheid in which the supportive function has become predominant.

Vessel elements, by comparison, are derived through evolutionary modifications involving slight shortening and widening of a tracheid with the loss or perforation of the end walls. The latter modification produces cells which are open and tubular in form (Figure 4-5). Note that vessel elements have similar wall patterns as exhibited by tracheids. Vessel elements, in terms of their modifications, have assumed the function of conduction at the expense of support. In the plant, vessel elements are arranged end to end with little or no overlapping. This structural orientation produces a continuous, open, pipelike structure called a **vessel.**

Phloem. Like xylem, **phloem** is a vascular tissue and is usually associated with xylem in various plant organs. Together they constitute the vascular system. Phloem functions particularly in the transport of dissolved organic substances in a process called **translocation.** Being a complex permanent tissue, phloem contains a variety of cell types. Always present in phloem are sieve cells. In addition, the phloem of higher plants also contains parenchyma cells and fibers.

The fundamental cells in phloem are **sieve cells.** These are long, slender, thin-walled cells, which at maturity retain their cytoplasm but lose their nucleus. Just as vessel segments in xylem form a continuous tubular network, the sieve cells in phloem

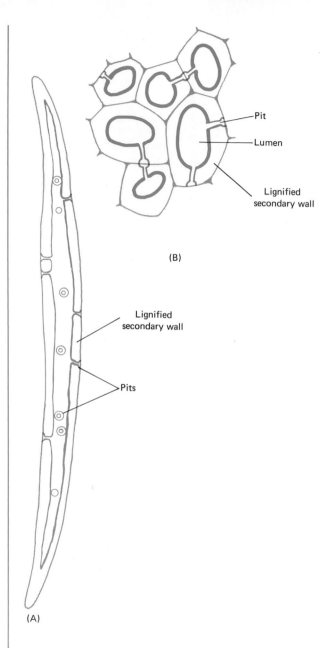

Figure 4-4. Xylem fibers. (**A**) Longitudinal aspect. (**B**) Xylem fibers in cross (transverse) section.

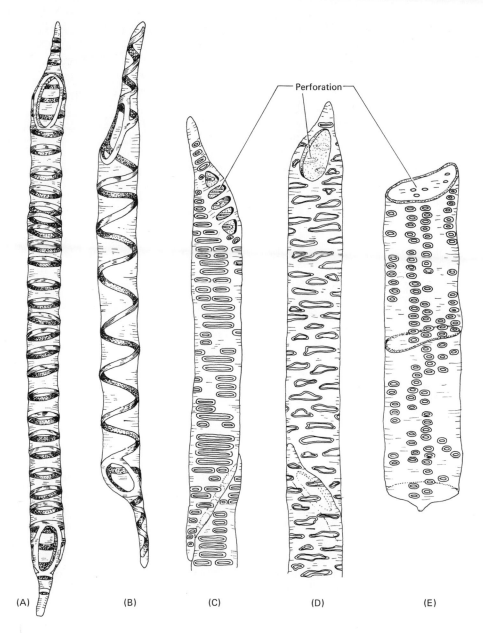

Perforation

(A) (B) (C) (D) (E)

Figure 4-5. Vessel segments found in xylem tissue are (**A**) annular, (**B**) spiral, (**C**) scalariform, (**D**) reticulate, and (**E**) pitted. (From V. A. Greulach and J. E. Adams, *Plants: An Introduction* to *Modern Botany,* New York: John Wiley & Sons, Inc., © 1967.)

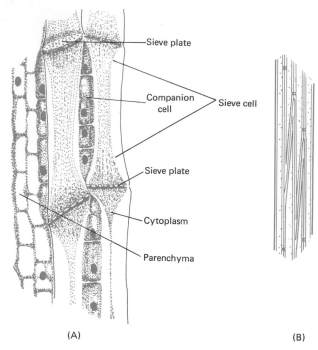

Sieve plate

Companion
cell

Sieve cell

Sieve plate

Cytoplasm

Parenchyma

(A)   (B)

Figure 4-6. Cells in phloem tissue. (**A**) Sieve cells containing cytoplasm. Note the adjacent companion cells with nuclei and the accompanying parenchyma cells. (**B**) Phloem fibers (modified sclerenchyma cells).

produce a similar structure called a **sieve tube.** The cytoplasm is continuous from one sieve cell to another within the sieve tube by means of **sieve plates** on the end walls (Figure 4-6A). These sieve plates are perforated end walls through which cytoplasm passes from one cell to another.

Ordinarily associated with sieve cells are modified parenchyma cells which retain cytoplasm and a functional nucleus. These cells are called **companion cells** (Figure 4-6A), and since there are cytoplasmic connections between sieve cells and adjacent companion cells, it is thought by some that the latter may aid the sieve cells in regulating conduction.

In addition to sieve cells and companion cells, the phloem of higher plants also contains parenchyma cells and fibers. The *parenchyma cells* function chiefly in food storage, and the **phloem fibers** (Figure 4-6B), modified sclerenchyma cells, afford strength.

The tissues of an angiosperm are summarized in Table 4-1.

VEGETATIVE ORGANS OF AN ANGIOSPERM

The Root

The **root** of an angiosperm is typically a subterranean organ which functions chiefly by anchoring the plant to the substrate and by absorbing and conducting water and nutrients from the soil. These absorbed materials are either utilized in the roots themselves or transported to other plant organs where they are used. In addition, some roots also serve as food storage organs.

Roots represent the first portion of a young growing plant to establish contact with the environment, in this case, the soil. Considered collectively, all the roots of a plant form the *root system.* The nature of the root system and its vertical and lateral extent are related to the species of plant, the amount of moisture in the soil, the soil temperature, and the physical properties of the soil. The root system generally equals or exceeds the portion of the plant growing above ground. In the sugar beet, for example, at the end of the first growing season, the root system may penetrate to a depth of 6 feet with a similar lateral spread but considerably less growth above ground.

Types of Roots. In terms of origin, roots may be classified into three main types: primary, secondary, and adventitious. The first root to emerge from a growing seed lengthens rapidly and pushes its way down into the soil. This root is called the

**TABLE 4-1. Classification of the Tissues of
an Angiosperm**

I.   Meristematic tissue—tissue of rapidly dividing cells

II.   Permanent tissue—matured tissue
        A. Simple permanent—consisting chiefly of one kind of cell
            1. Epidermis—protection
            2. Parenchyma—photosynthesis and storage
            3. Collenchyma—support
            4. Sclerenchyma—strengthening
            5. Cork—waterproofing and protection
        B. Complex permanent—consisting of several kinds of cells
            1. Xylem—conduction and support
            2. Phloem—conduction, storage, and support

primary root (Figure 4-7A). The structure originates from an embryonic root inside the seed called the **radicle.** After a short period of growth, *secondary, or lateral, roots* originate from the primary root, first near the top of the primary root and later farther down (Figure 4-7A). Some roots frequently arise from plant parts other than the primary or secondary roots. These roots are called *adventitious roots* (Figure 4-7B). Such roots may arise from stems and leaves. The rooting of stem cuttings on

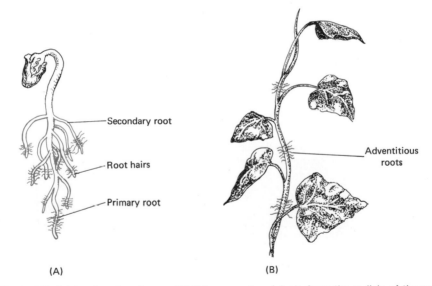

(A)   (B)

Figure 4-7. Origin of root systems. **(A)** Primary roots originate from the radicle of the seed, whereas secondary roots develop from existing primary roots. **(B)** Adventitious roots arise from vegetative organs such as stems and leaves.

(A)   (B)

Figure 4-8. Types of root systems. (**A**) Diffuse (fibrous) root system of a grass. (**B**) Taproot system of a carrot (*Daucus carota*).

such plants as geranium and *Coleus* depends upon the growth of adventitious roots, as does the rooting of African violet leaf cuttings.

As the root system continues to grow, there is a considerable degree of branching, penetration into the soil, and horizontal spreading of root branches. On the basis of the growth pattern that

results, root systems are classified into two types: diffuse (fibrous) root systems and taproot systems.

Diffuse root systems are characteristic of all cereals as well as other grasses. In such a root system, no one root is the largest; all are nearly equal in size (Figure 4-8A). Diffuse root systems serve as efficient organs for anchorage and absorption because of the enormous soil area that they occupy. The small, abundantly branched roots are intertwined throughout the soil particles and tend to anchor both the plant and the soil in place, thus preventing erosion.

In the *taproot system* such as that in beets, dandelions, and carrots, the primary root grows most rapidly, enlarges considerably, and remains the largest part of the underground system (Figure 4-8B). If the main root, or *taproot,* contains stored materials when the plant is mature, it is referred to as a *fleshy root.*

Structure of Dicot Roots. All root systems terminate in delicate white tips which are usually less than $\frac{1}{16}$ inch in diameter. Close examination of one of these tips reveals four more or less distinct regions: the root cap, meristematic region, enlargement region, and differentiation region. Figure 4–9 shows these regions in various aspects as they appear both externally and internally in a root tip.

Root Cap. The **root cap** is found on practically all roots except those of many water plants but is never found on any other plant organ. Microscopically, it appears as a thimble-shaped mass of parenchyma cells that protects the growing portion of the root tip (Figure 4-9A and B). As the root grows and the tip advances through the soil, the outer cells of the root cap are sloughed or rubbed off and are continually being replaced by divisions of the cells in the region directly behind it. Cells sloughed off the root cap lubricate the passage of the young root through the soil.

Meristematic Region. This region, also called the

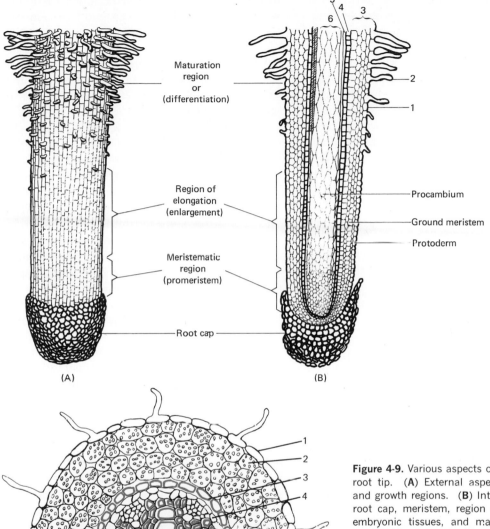

Maturation region or (differentiation)

Region of elongation (enlargement)

Meristematic region (promeristem)

Root cap

(A)

5 4 3
6

2

1

Procambium

Ground meristem

Protoderm

(B)

(C)

Figure 4-9. Various aspects of a young developing root tip. (**A**) External aspect showing root cap and growth regions. (**B**) Internal aspect showing root cap, meristem, region of enlargement with embryonic tissues, and maturation region with primary tissues: (1) epidermis, (2) root hairs, (3) cortex, (4) endodermis, (5) pericycle, and (6) vascular tissue (primary phloem, vascular cambium, and primary xylem). (**C**) Cross-sectional aspect through the region of maturation showing (1) epidermis, (2) cortex, (3) endodermis, (4) pericycle, (5) secondary root formation from pericycle, (6) primary xylem, (7) primary phloem, (8) vascular cambium, and (9) root hair.

promeristem (Figure 4-9A and B), is an apical meristem and is characterized by active cell division. The cells are small, cube-shaped, parenchyma-type cells with thin walls, dense cytoplasm, and large nuclei. The division of these cells, and consequently the addition of new ones, brings about lengthening of the root tip. Cells produced by the meristem are destined for one of three possible fates. Those toward the tip of the root become part of the root cap. The cells in the middle remain as cells of the meristem. Those cells at the upper portion enlarge and differentiate into permanent tissues. Such a distribution of cells allows the meristematic zone to remain fairly constant in size even though there is a continual production of new cells.

Enlargement Region. Those cells produced by and located directly above the meristem constitute the **enlargement region,** or region of cell elongation (Figure 4-9A and B). Although the regions are not sharply distinguishable from each other, close microscopic examination reveals two bases for delineation. First, the cells of the enlargement region tend to be elongated longitudinally as opposed to the cubical form of the meristematic cells. Second, three distinct immature permanent tissues can be differentiated. The tissues represent the first definitive changes in form from the cells of the meristem. These tissues are the **protoderm, ground meristem,** and **procambium** (Figure 4-9A and B). All are immature, meristematic tissues which will give rise to primary plant tissues. **Primary plant tissues** are all tissues that develop from the apical meristem and bring about an increase in length of a plant part.

Differentiation Region. In the **region of differentiation** (Figure 4-9A and B), cells of the protoderm, ground meristem, and procambium divide, enlarge, and differentiate sufficiently to become recognizable as primary tissues. These primary tissues include the epidermis, which is produced by the protoderm; the cortex and endodermis, which are derived from the ground meristem; and the vascular cylinder consisting of pericycle, primary phloem, vascular cambium, and primary xylem, all derivatives of the procambium. A cross section of a root tip through the region of differentiation reveals these differentiated primary tissues (Figure 4-9C).

Primary Tissues. The **epidermis** consists of a single layer of parenchyma cells and forms the outermost tissue of the root. In addition to producing root hairs, it affords protection for the underlying cells. The **cortex,** inside the epidermis, is composed of large, thin-walled, angular parenchyma cells. This tissue helps to transfer water and minerals from the root hairs to the xylem and also serves as a food storage tissue. The **endodermis** is a rather conspicuous tissue in roots consisting of a single layer of thick-walled cells and forms the inner boundary of the cortex.

Within the endodermis is the **vascular cylinder,** or **stele,** the principal strengthening and conducting portion of the root. The outermost tissue of the vascular cylinder is the **pericycle,** which consists of one or more layers of parenchyma cells. This tissue is extremely important because it produces the secondary roots and the cork cambium. The xylem occupies the center of the stele and is referred to as the *primary xylem.* As viewed in cross section, it is usually arranged in the form of a star, although many other arrangements also exist. In each of the angles formed by the star-shaped xylem, there are groups of smaller, thinner-walled cells comprising the *primary phloem.* In dicots, the phloem is separated from the xylem by layers of thin-walled undifferentiated parenchyma cells. One layer of these cells remains meristematic and is known as the vascular cambium. The cambium cells continue to divide and produce the secondary tissues. Most monocots lack a cambium. In treelike monocots such as *Aloe* and *Yucca,* there is a special

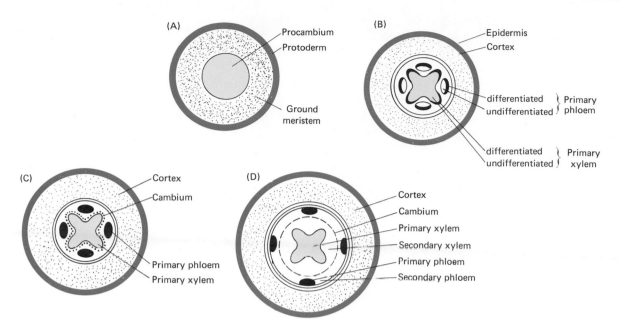

Figure 4-10. Secondary growth in a young root as seen in various cross-sectional aspects. (**A**) In a very early stage of development the three embryonic tissues are shown. (**B**) At a somewhat later stage of development the embryonic tissues have differentiated into primary tissues. (**C**) With the formation of primary tissues the vascular cambium becomes mitotic. (**D**) At a later stage of growth, the cambium produces secondary xylem and secondary phloem between the tissues of primary xylem and primary phloem. Note the relative distribution and spatial arrangement of the primary and secondary tissues in early and later stages of growth.

type of secondary growth. A cambium arises in the cortex and produces groups of cells on the inner side which develop into typical closed vascular bundles. On the other side parenchyma cells are produced.

Secondary Tissues. The roots of most trees, shrubs, perennial plants, and other dicots have secondary tissues as well as primary tissues. **Secondary tissues** are those that develop from a vascular cambium and cause an increase in diameter, or lateral growth, of a plant part. For this reason, a cambium is also refered to as a lateral meristem. The two principal types of secondary tissues in roots are those that arise from the vascular

cambium and those originating from the pericycle.

Of all the primary tissues, the vascular cambium and pericycle alone remain meristematic and continue to divide. The others are more or less specialized and no longer divide to produce new cells. When the cambium divides, new cells are added to both the primary xylem and the primary phloem. These new secondary tissues, *secondary xylem* and *secondary phloem,* grow at uneven rates, the former growing much more rapidly. As a result, the secondary xylem occupies the spaces between the radiating arms of the primary xylem, originally occupied by the primary phloem (Fig-

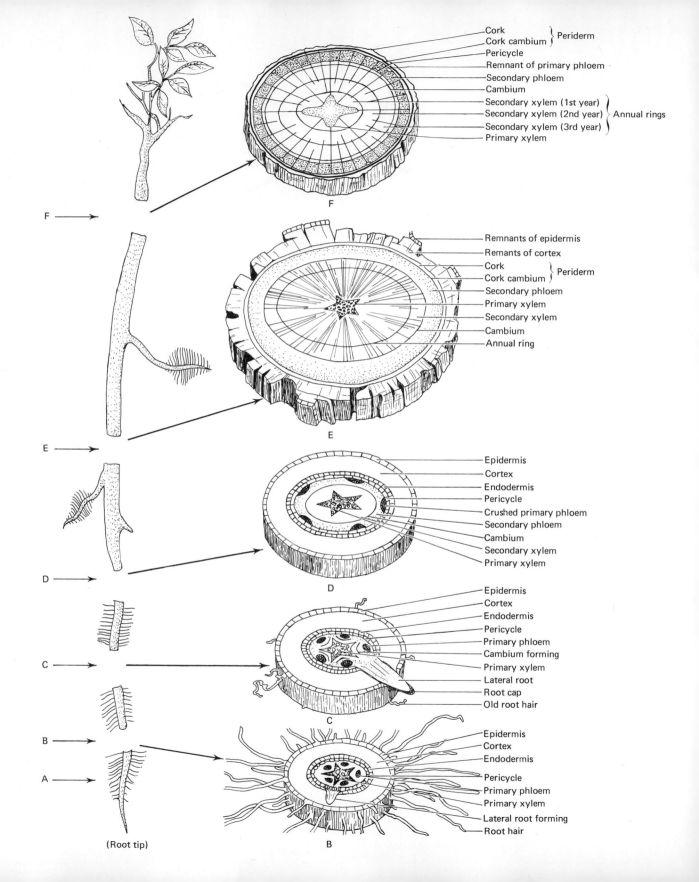

Cork ⎫
Cork cambium ⎬ Periderm
Pericycle
Remnant of primary phloem
Secondary phloem
Cambium
Secondary xylem (1st year) ⎫
Secondary xylem (2nd year) ⎬ Annual rings
Secondary xylem (3rd year) ⎭
Primary xylem

F

F

Remnants of epidermis
Remants of cortex
Cork ⎫
Cork cambium ⎬ Periderm
Secondary phloem
Primary xylem
Secondary xylem
Cambium
Annual ring

E

E

Epidermis
Cortex
Endodermis
Pericycle
Crushed primary phloem
Secondary phloem
Cambium
Secondary xylem
Primary xylem

D

D

Epidermis
Cortex
Endodermis
Pericycle
Primary phloem
Cambium forming
Primary xylem
Lateral root
Root cap
Old root hair

C

C

Epidermis
Cortex
Endodermis
Pericycle
Primary phloem
Primary xylem
Lateral root forming
Root hair

B

A

(Root tip)

B

ure 4-10). This pushes the phloem cells outward and the xylem region becomes circular. In a similar manner, the phloem also becomes a continuous cylinder of tissue external to the xylem and vascular cambium.

When the cells of the pericycle begin to divide, they produce a second lateral meristem referred to as the cork cambium, or *phellogen*. The cork cambium usually remains a single layer of cells, and new layers of cells are added both outwardly toward the epidermis and inwardly toward the phloem as it divides. Those produced outwardly are called cork, or *phellem*. Those developed inwardly are referred to as the *phelloderm*. Collectively considered, this entire new secondary tissue consisting of cork, cork cambium, and phelloderm is designated as the *periderm* (Figure 4-11E).

The cells of the phelloderm are parenchyma-like and remain alive and active. Those of the cork layer become suberized, thus waterproofed, and die at maturity. When the periderm is fully developed, the epidermis with its root hairs, and the cortex die, so that in older roots, the periderm eventually becomes the outermost layer (Figure 4-11F). At maturity, the periderm breaks, cracks, and furrows like the bark of the stems of trees. The entire sequence of primary and secondary tissue growth is summarized in Figure 4-11.

The Growth of Roots. As cells divide, enlarge, and differentiate, the root cap and apical meristem are forced farther and farther into the soil. A given cell starts as part of the apical meristem. As it ages, if it does not become a root cap cell, it then exists as part of the enlargement region and finally as a cell of the differentiation region. In effect, the cell has not changed position, it has merely changed in structure and function. Figure 4-12 represents the growth of a root tip in which a single cell is shown in various stages of development. In the example cited, the cell develops into a tracheid; for the sake of simplicity, other cells have not been shown. Bear in mind, however, that all cells which eventually become part of the tissues of a root undergo similar changes.

Root Hairs. In the region of differentiation there occurs a multitude of fine white hairs radiating outward from all sides of the root. These appendages are called **root hairs** (Figure 4-9). Each root hair is a projection of an epidermal cell and develops perpendicular to the cell from which it forms. It is not unusual to find 200 to 300 root hairs per square millimeter of epidermis in root tips. They are short-lived, maintaining existence for only a few days, or at most a few weeks. Root hairs are the principal structures performing the vital function of absorption.

Specialized Roots. The functions of absorption, conduction, anchorage, and storage are common to most root systems. There are some plants, however, in which at least some of the roots have become highly specialized in terms of one particular function. This function may be one commonly performed by roots or it may be an entirely different one. In either case, such specialized roots are greatly modified in structure (Figure 4-13).

Storage Roots. Most roots store at least some food in the form of carbohydrates, especially starch and sucrose. Sugar beets, for example, commonly store 15 to 20 per cent sucrose. One of the commonest types of modified root is the fleshy *storage root* found among taprooted plants. Carrots, beets, parsnips, radishes, and turnips are familiar examples (Figure 4-13A).

Aerial Roots. Although roots are generally under-

Figure 4-11. (Opposite) Regions of a root in various stages of growth. The figure on the left indicates the regions at which sections were made. (Adapted from Walter H. Muller, *Botany: A Functional Approach*, 2nd edition. New York: The Macmillan Company, © 1969.)

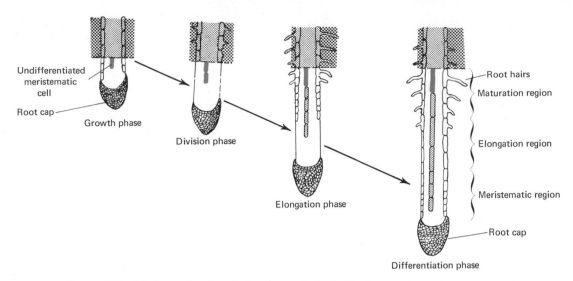

Undifferentiated
meristematic
cell

Root cap

Growth phase

Division phase

Elongation phase

Root hairs

Maturation region

Elongation region

Meristematic region

Root cap

Differentiation phase

Figure 4-12. Cellular maturation during root growth. In the figures shown the growth and development of an undifferentiated meristematic cell is traced as it elongates and specializes into a tracheid cell. Note that as the root tip grows downward the developing cell does not change position.

ground structures, in some species of plants they are found either partially or completely above the ground, in which case they are called *aerial roots*. Aerial roots are found in some plants such as corn, screw pine, and banyan trees, in which the roots grow from the stem into the soil. These aerial roots are called *prop roots* and they function by giving added support to the stem (Figure 4-13B). Other aerial roots are modified as *climbing roots* and serve to anchor stems to walls, fences, and trees along which they grow. Roots of this type are common to vines such as poison ivy, English ivy, trumpet creeper, and tropical orchids (Figure 4-13C).

Parasitic Roots. Specialized roots called **haustoria** develop along the stems of parasitic plants such as mistletoe and dodder whenever the stems come in contact with the plant on which the parasite is growing (Figure 4-13D). The common dodder sends out roots into the stem of the host plant which penetrate the vascular tissues and absorb water and nutrients. Whereas the leaves of mistletoe are green and photosynthetic, the dodder depends almost entirely upon the host plant for nutrition.

Economic Importance of Roots. First, it should be further emphasized that roots are necessary for the growth and development of higher plants. The principal root crops used as food by man are carrots, beets, turnips, radishes, parsnips, and sweet potatoes. Other roots, such as those from horseradish, licorice, tumeric, sassafras, and sarsaparilla, furnish spices and aromatic flavorings. Some roots such as those from madder and alkanna are important sources of dyes. Finally, the roots of other plants are used in pharmaceutical preparations. These include cortisone from Mexican yams and reserpine, a tranquilizer, from Rauwolfia. Among other medicinal plants are aconite, gentian, goldenseal, and valerian.

Stems

The **stem** is that part of the plant that rises above the ground and together with the leaves comprises the *shoot*. Although most stems are erect aerial organs, some remain underground, and others creep along the surface of the ground. The stem typically serves as a mechanical support for leaves, flowers, and fruits; a pathway for the conduction of materials; a site for food manufacture (in green stems); and as a potential reproductive structure.

Types of Stems. An examination of stems from various plants will reveal many differences in size, shape, color, and texture. In addition, the nature and arrangement of the internal tissues will exhibit considerable variations. The stems of angiosperms may be either herbaceous or woody. **Herbaceous** stems, such as those found in peas, beans, and corn, are generally soft and green and have little or no woody tissue. There is little growth in diameter and the plants are short-lived. In contrast, **woody** stems, such as those of oak, maple, and hickory trees, are hard, thick, long-lived, capable of increase in diameter, and consist mostly of wood (xylem). The outer surface of older, woody stems is rough and is covered with cork, the common bark of trees and shrubs. The anatomical distinction between a tree and a shrub is that in a tree the trunk rises some distance above

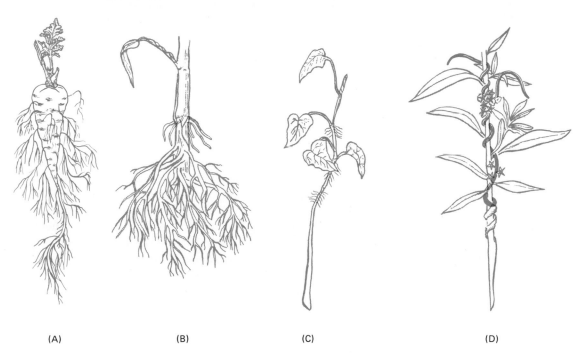

(A) (B) (C) (D)

Figure 4-13. Specialized roots. (**A**) Storage root (carrot). (**B**) Prop roots (corn). (**C**) Climbing roots (English ivy). (**D**) Parasitic roots (dodder).

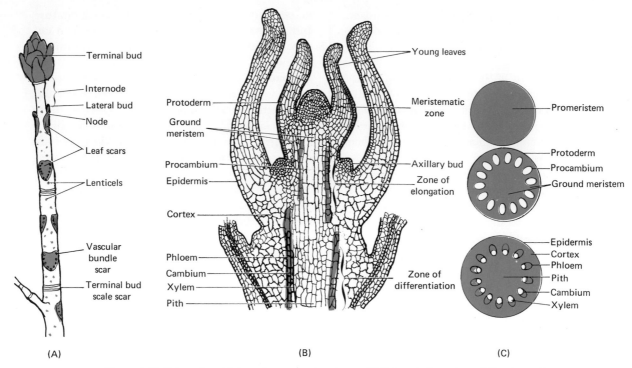

Figure 4-14. External and internal aspects of a stem. (A) External characters of the stem of a horse chestnut tree in the winter condition. (B) Internal aspect of the growth regions and tissues of a stem tip as seen in longitudinal section. (C) Transverse view of the same regions and tissues shown in (B).

the ground before it branches, whereas in a shrub, several stems of equal size arise at, or close to, the ground.

External Features of Stems. The twigs of woody plants in the winter condition afford excellent subjects for a study of the external characteristics of stems (Figure 4-14A). New stems and their leaves develop from **buds.** The top of a twig generally bears a large *terminal bud,* and at regular intervals along the sides of the stem are *axillary,* or *lateral buds.* At the base of each axillary bud is a scar that was made when a leaf fell from the twig. This is called the **leaf scar** and is usually circular, oval, or shield-shaped. The conducting tissue of the stem is continuous with that of the leaf, so this tissue also ruptures when the leaf falls, producing *vascular bundle scars* inside each leaf scar.

Leaves and buds are present at specific locations along the stem, each point of attachment being called a **node;** the distances between nodes are called **internodes.** The immature cells within buds are protected by a series of overlapping structures, the **bud scales,** which drop off as the new stem develops, and in doing so they leave scars called *bud-scale scars.* The portion of the stem between sets of terminal bud-scale scars is formed during one growing season. After a portion of a stem is 1 year old, there is no increase or decrease in the length of that portion. Tiny openings or pores called **lenticels** also appear on woody stems. They facilitate gaseous exchange between the interior of the stem and the atmosphere.

The arrangement of buds (or leaves) on a stem may be categorized into three types: opposite, alternate, and whorled. Ash and maple trees have buds opposite each other at each node, and the arrangement is called *opposite*. In the walnut twig, only one bud is present at each node, producing an *alternate* arrangement. The *whorled* arrangement, as exemplified by catalpa trees, consists of three or more buds at each node.

Growth and Structure of a Woody Dicot Stem.
Assume that an oak tree is 40 feet high, 1 foot in diameter, and the first branch is exactly 6 feet above the ground. After 10 years of growth, it will be found that the trunk is 16 inches in diameter and 50 feet high. However, the first branch will still be 6 feet above the ground. The obvious conclusion to be drawn is that stems grow both vertically and laterally at their tips, and regions below the tips grow only in diameter.

This growth process can be understood by first examining a longitudinal section through a dormant stem tip (Figure 4-14B). Growth in length of the stem is due to cell divisions in the apical meristem, which comprises the meristematic zone. In effect, the meristem is constantly being carried upward by the growth of new cells it produces. The region just behind the meristem, characterized by rapid cell enlargement, is called the enlargement region. In this region, three conspicuous embryonic primary tissues are produced. These are the protoderm, ground meristem, and procambium. As development of these tissues occurs in the differentiation zone, each produces specific primary tissues. The protoderm gives rise to the epidermis; the ground meristem produces the cortex and pith; and the procambium differentiates into the vascular cambium, primary

xylem, and primary phloem. The similarities that exist between the growth and development of primary tissues in roots and stems should now be obvious.

Primary Tissues. The young woody stem consists of primary tissues alone, and secondary tissues are not produced until after the primary tissues are established. All the primary tissues are produced by the apical meristem, after which cell enlargement and differentiation result in these tissues being part of the permanent tissues. The primary tissues are shown in Figure 4-14C in cross-sectional views through various growth regions. Each of these tissues will now be discussed from the outermost to the central area of the stem.

Epidermis. This tissue forms the surface protective layer and is one cell layer thick. The outer cell walls are frequently thick and covered with cutin. In older stems the epidermis is replaced by cork.

Cortex. This region, just under the epidermis, is composed chiefly of parenchyma cells with some collenchyma and sclerenchyma. The cortex, usually many cells deep, is a food storage tissue which degenerates in older stems.

Primary Phloem. This complex permanent tissue provides for the efficient conduction of materials and also assumes the functions of storage and support. The principal conducting cells are sieve cells, the storage cells are companion cells and phloem parenchyma cells, and the strengthening and supporting cells are phloem fibers. These fibers are long, narrow, vertically elongated, thick, and dead at maturity. Phloem fibers of flax and hemp plants are made into linen and rope, respectively.

Vascular Cambium. This tissue is immediately internal to the phloem and consists of a region of meristematic cells.

Primary Xylem. Like the primary phloem, the

Figure 4-15. Division and differentiation of a cell of the vascular cambium into secondary xylem and secondary phloem cells. (C) Vascular cambium cell; (X¹, X², X³) secondary xylem cells; (P¹, P²) secondary phloem cells. Note the relative proportions of xylem and phloem formed in the sequence.

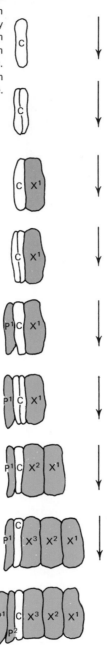

primary xylem is a complex permanent tissue and assumes a number of functions. Conduction is facilitated by tracheids and vessel elements. Xylem parenchyma cells and xylem fibers, similar to those found in primary phloem, assume the functions of storage and support.

Pith. This region is the central portion, or core, of the stem and contains parenchyma cells that function in storage.

Secondary Tissues. After the primary tissues are formed in a young dicot stem, little or no further lengthwise growth of the differentiated portion takes place. The terminal bud continues its growth through the growing season, each year forming a new stem segment with its primary tissues. In general, the primary tissues are established in the first year during the first few weeks and bring about lengthening of the stem. Also, in the first year, the stem begins to grow in diameter as a result of the continued division of the vascular cambium cells, which produce the secondary tissues.

As the cambium divides, cells produced toward the outer part of the stem develop into the secondary phloem, and those that are produced toward the inner portion of the stem form the secondary xylem. Usually more xylem than phloem is produced, and as the stem grows in diameter, more and more xylem is formed. The mechanism by which the cambium forms this uneven distribution of secondary phloem and secondary xylem cells is shown in Figure 4-15. The secondary tissues which reinforce or replace the primary tissues are the secondary phloem, secondary xylem, cork cambium, and cork.

Secondary phloem. The cells of this tissue are similar to those of the primary phloem except for two differences: The first is one of origin; the second is the presence of added cells that serve a specific function. As the stem increases in diameter, horizontal as well as vertical conduction

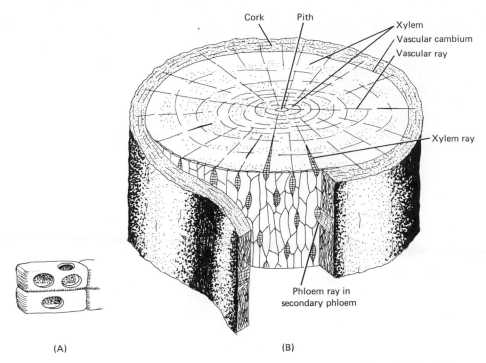

Cork   Pith   Xylem
Vascular cambium
Vascular ray

Xylem ray

Phloem ray in
secondary phloem

(A)   (B)

Figure 4-16. Lateral conducting elements. **(A)** Enlarged aspect of ray cells. **(B)** The relation-ship of phloem rays, xylem rays, and vascular rays as seen in a mature woody stem section.

is necessary. This specific function is accomplished by *phloem ray cells,* not found in primary phloem. These parenchyma-type cells are elongated hori-zontally (radially) and function in transporting materials across the stem (Figure 4-16A). The large simple pits in the walls of phloem ray cells greatly facilitate the horizontal conduction of materials. In the stem, these cells become oriented end to end, forming a continuous horizontal con-ducting structure called a phloem *ray* (Figure 4-16B).

Secondary Xylem. Here again, the secondary xylem differs from the primary xylem with respect to origin and the presence of new cells. *Xylem ray cells,* similar to those found in the phloem, are

attached end to end to the phloem rays and form *vascular rays,* which provide an excellent horizontal conducting system (Figure 4-16B).

The cambium continues to produce secondary xylem and secondary phloem year after year, so that the stem progressively increases in diameter. Xylem cells produced in the spring are relatively large, but as the season progresses, the cells are considerably smaller. As a result there is usually an abrupt contrast between the xylem formed during different growth seasons. These annular growth layers, which appear as a series of concen-tric rings when the stem is viewed in cross section, are called **annual,** or **growth, rings** (Figure 4-16B).

Cork Cambium and Cork. In addition to secondary

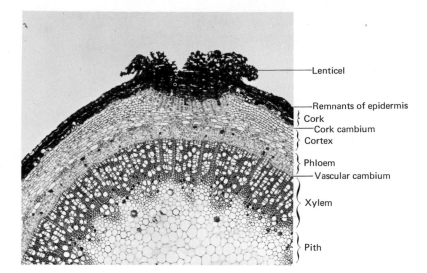

Figure 4-17. Cross-section of a portion of the stem of elder (*Sambucus*) with a lenticel. Note the secondary tissues. (Courtesy Carolina Biological Supply Company).

Lenticel

Remnants of epidermis
Cork
Cork cambium
Cortex

Phloem
Vascular cambium

Xylem

Pith

tissues produced by the vascular cambium, another group of secondary tissues is produced by the cork cambium (Figure 4-17). The cork cambium is a group of meristematic cells that originates from parenchyma cells of the outer cortex. As the cells of the cork cambium divide, the outer ones develop into cork cells and the inner ones give rise to phelloderm. The cork thus formed serves as a protective, waterproofing tissue.

Although cork forms in response to tissue damage, this is not an unusual occurrence. In fact, all woody dicots have external layers of cork by the second year of growth. The reason for cork production is that secondary tissue development brings about an increase in diameter while the epidermis and cortex no longer increase in size. This continued stress causes the epidermis and cortex to tear and rip, and, as a result of this damage, cork forms before the damage is apparent. Cork is impervious to most materials, so all cells external to it die as a result of a lack of nutrition. The **bark** of a tree is a collective designation for the secondary phloem, cork cambium, and cork.

Structure of Various Stem Types. The preceding discussion of stems was related to woody dicot stems. At this point, two other types will be mentioned to show basic internal differences. These are the herbaceous dicot and herbaceous monocot stems. Structurally, herbaceous stems differ from woody stems in having much less xylem and phloem tissue. The xylem and phloem that are present are found as long strands called **fibrovascular bundles,** which run longitudinally through the stem.

The *herbaceous monocot stem,* as represented by iris, orchid, lily, and corn plants, as well as sedges and grasses, contains scattered fibrovascular bundles (Figure 4-18A). In addition, there is no vascular cambium, no secondary tissues, and little increase in diameter. This is why herbaceous monocots are generally long and slender. The vast majority of the stem consists of parenchyma tissue with some collenchyma and sclerenchyma found near the epidermis for added support.

In the *herbaceous dicot stem,* the vascular tissue is arranged in discrete bundles which are concen-

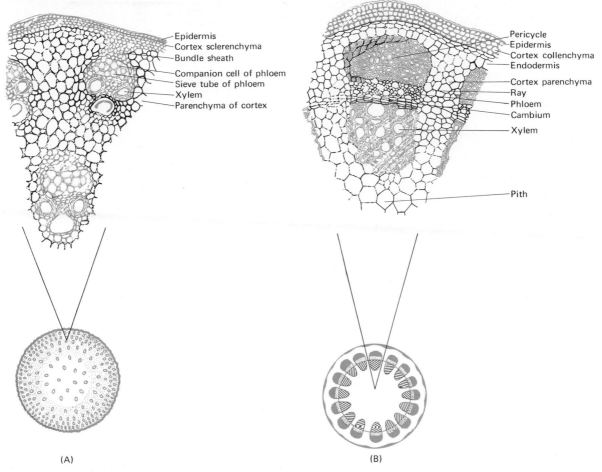

Figure 4-18. Comparison of herbaceous monocot and herbaceous dicot stems. **(A)** Herbaceous monocot stem showing arrangement of tissues and detailed structure of a fibrovascular bundle. **(B)** Herbaceous dicot stem showing orientation of tissues and detailed structure of a fibrovascular bundle.

trically oriented (Figure 4-18B). The cambium, located between the xylem and phloem, may be restricted to individual vascular bundles or may be continuous from one bundle to the next. Despite the presence of a cambium, secondary tissues are poorly developed and the stem remains nonwoody. The herbaceous dicot stem, such as found in tomato, buttercup, and bean plants, does not live through the winter, so that any growth in diameter is restricted to one season only.

Specialized Stems. The stems of many species of angiosperms, like roots, may be specialized in form and function and sometimes bear little resemblance to ordinary stems. Although most stems

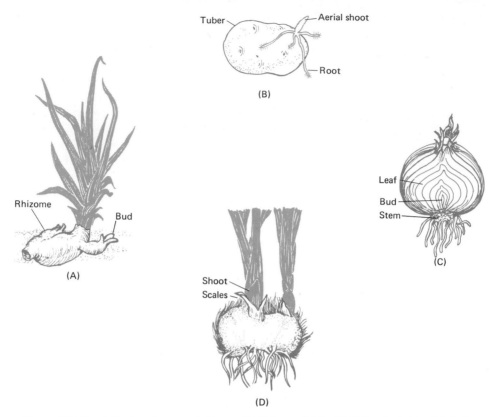

Figure 4-19. Specialized underground stems. (**A**) Rhizome (iris). (**B**) Tuber (potato). (**C**) Bulb (onion). (**D**) Corm (crocus).

are cylindrical, erect, and above ground, some exhibit prominent and unusual modifications (Figure 4-19).

Underground Stems. The four principal types of underground stems are rhizomes (root stocks), tubers, corms, and bulbs. The **rhizome,** or **rootstock,** is an underground, horizontal stem with nonchlorophyllous leaves which are generally reduced to a considerable extent. Some are relatively long, as in Kentucky bluegrass; others are thickened and fleshy, as in iris (Figure 4-19A). Some slender rhizomes have greatly thickened portions

at their ends referred to as **tubers.** An excellent example of a tuber is the Irish potato (Figure 4-19B). Tubers have little vascular tissue and a high proportion of parenchyma, in which food, usually starch, is stored.

Some plants, usually monocots, such as crocus and gladiolus, contain a short, stout underground stem called a **corm** (Figure 4-19D). These modified stems are usually broader than they are high and represent the enlarged bases of stems as contrasted with tubers, which are swollen tips of stems. Finally, the common onion (Figure 4-19C) is an example of a **bulb,** another specialized underground stem. It differs from the corm in that the principal storage organs are bulb scales, which surround the short, erect stem. Generally, a bulb stores sugar rather than starch.

Aerial Stems. Any stem growing above the ground is considered to be an **aerial stem.** One type of modified aerial stem is a **runner,** or **stolon.** It consists of a slender stem that grows horizontally along the surface of the ground and is exemplified by strawberry plants. Another modified aerial stem is a **tendril,** a slender curling structure that helps to support the plant. Some tendrils, as in the case of Boston ivy, have adhesive discs, but others, as in the case of the Virginia creeper, do not. One other fairly common type of modified aerial stem is the **thorn,** a typical structure found in hawthorns, rose bushes, and honey locusts.

Economic Importance of Stems. Stems, especially those modified as food storage organs, are an important source of food. Tubers, corms, and bulbs of many plants have long been eaten by man. The most important of these modified stems are potatoes. Another important food derived from a typical plant stem is sugar, extracted from the sap of sugar cane. Other modified stems important as food sources are the bulbs of onions, the corms of the taro, and the tubers of Jerusalem artichokes.

Other commercially important products derived from stems include (1) wood, a principal construction material; (2) wood products, such as tannins, wood pulp (source of newspaper); wood alcohol, tar, and oil; cellophane; and rayon; (3) fibers from the phloem and pericycle (linens, hemp); (4) cork from the bark of trees; (5) latex (rubber); and (6) various gums, resins, and balsams.

Leaves

Leaves are the characteristic photosynthetic organs of most higher plants. Another important leaf activity is **transpiration,** the loss of water vapor. Leaves, like roots and stems, vary in many respects. For example, leaves differ according to arrangement on the stem, vein patterns, form, structure, and size. Despite these variations, certain basic features are distinguishable.

External Features of Leaves. The leaf stalk, or **petiole,** is a continuation of the stem and distributes vascular tissue from the stem to the leaf (Figure 4-20). The flattened, expanded portion of the leaf is the **lamina,** or **blade,** which usually is green because of the presence of chlorophyll-containing cells. At the base of the petiole, small leaflike structures called **stipules** may be found along with **axillary buds.** These axillary buds may produce secondary branches, flowers, or both. The shape and form of the petiole may vary in different species of plants. In some, leaf petioles are absent, and the leaves are said to be **sessile.**

Vein Patterns. The blade of the leaf is strengthened by the presence of **veins,** which are composed chiefly of vascular tissue continuous with that of the petiole and the stem. These veins may be arranged in several ways and form two principal patterns. In *parallel venation,* a character-

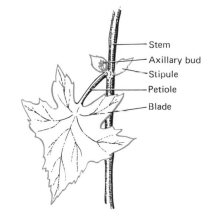

Figure 4-20. Principal parts of a typical leaf.

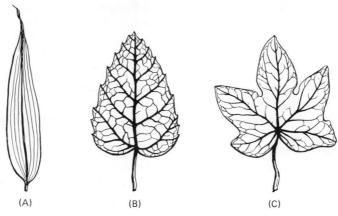

Figure 4-21. Types of vein patterns. (**A**) Parallel venation in bamboo leaf (*Bambusa*). (**B**) Pinnately net-veined leaf from grape (*Vitis*). (**C**) Dwarf ivy leaf (*Hedera*) showing palmately net-veined pattern. (From Walter H. Muller, *Botany: A Functional Approach,* 2nd edition. New York: The Macmillan Company, © 1969.)

istic vein pattern of monocots (for example, lily, onion, corn, and grasses), the principal veins run parallel to each other from the base to the tip of the blade (Figure 4-21A). In most leaves of this type the principal veins are interconnected by smaller, inconspicuous veins. *Net venation,* by contrast, is characteristic of dicots such as oak, maple, and sycamore trees and bean plants. In this case, the veins continuously branch, with one or more being prominent and the remaining ones forming a complete network throughout the blade. The main, or prominent, vein is referred to as the *midrib.* When there is a single midrib, from which other veins diverge, like the pinnae of a feather,

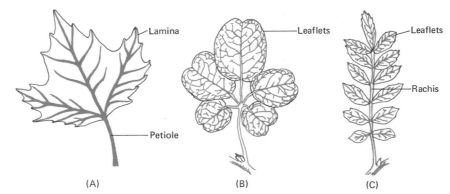

Figure 4-22. Types of leaves. (**A**) Simple leaf. (**B**) Palmately compound. (**C**) Pinnately compound.

the vein pattern is called *pinnately net-veined* (Figure 4-21B). When several principal veins diverge from the base of the blade, each in turn branching, the leaf is said to be *palmately net-veined* (Figure 4-21C).

Types of Leaves. The blades of some leaves are deeply indented at their margins or edges, while others are completely separated into individual parts called *leaflets.* If the blade remains in one piece, even though deeply lobed, the leaf is referred to as a *simple leaf* (Figure 4-22A). A complete segmentation of the blade into leaflets produces a *compound leaf.* Accordingly, compound leaves are of two types: *palmately compound* (Figure 4-22B), in which the leaflets are attached at a common point at the tip of the petiole, and *pinnately compound*

(Figure 4-22C), in which there is a principal central axis called the *rachis,* to which all the leaflets are attached.

Internal Structure of Leaves. The internal arrangement of the tissues of a leaf is commonly studied by microscopic examination of a cross section cut at right angles to the broad surface of the blade. Such a section reveals three distinguishable areas: (1) epidermis, (2) mesophyll, and (3) vascular tissue (Figure 4-23).

Figure 4-23. Diagram of a section view of a leaf showing internal structural composition. (From Walter H. Muller, *Botany: A Functional Approach,* 2nd edition. New York: The Macmillan Company, © 1969.)

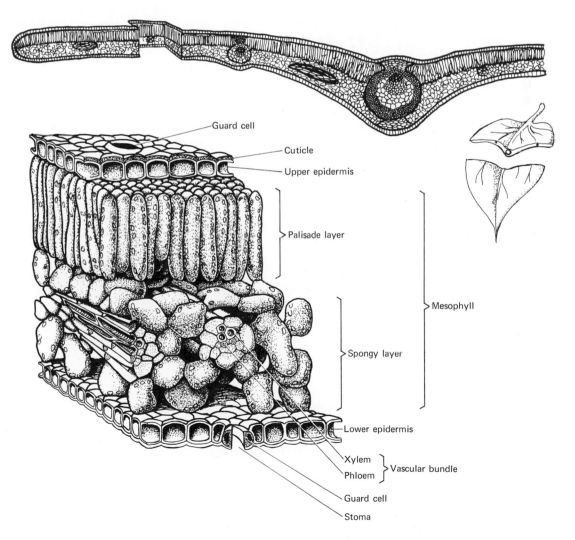

Guard cell

Cuticle

Upper epidermis

Palisade layer

Mesophyll

Spongy layer

Lower epidermis

Xylem

Phloem

Vascular bundle

Guard cell

Stoma

Epidermis. This tissue, in cross section, consists of a single layer of more or less rectangular-shaped cells lacking chloroplasts. In surface view, the epidermal cells are fairly irregular in form. The epidermis covers both surfaces of the leaf and is therefore subdivided into the upper and lower epidermis. The *upper epidermis* is covered by a cuticle which prevents drying out of the underlying tissues and protects them against injuries. The *lower epidermis,* and sometimes the upper, is perforated by **stomata** bordered by **guard cells.** In contrast to the cells of the epidermis proper, the guard cells contain chloroplasts.

In many plants, the epidermis of the leaf and stem bear various types of hairs called **trichomes** (Figure 4-24). These hairs are produced as a result of the outward growth either of a single epidermal cell or a group of such cells. They may be simple, unicellular epidermal outgrowths (Figure 4-24A)

or they may be complex, multicellular appendages. Sometimes they are living structures in the form of glandular cells that secrete oils and other substances (Figure 4-24B). The stinging hairs of nettle are of this type (Figure 4-24D). If touched, the end usually breaks off, leaving a sharp point which may penetrate the skin and inject an irritating poisonous fluid. Other epidermal hairs are dead and are filled with air. Trichomes of the common mullein (Figure 4-24C) are a representative example. These branched hairs commonly give the leaf a silky, wooly, or velvety appearance.

Mesophyll. Between the upper and lower epidermal layers, the **mesophyll** is found. On the basis of location and structure, the mesophyll is divided into two portions. The upper, rectangular-shaped compact rows of cells are called the *palisade mesophyll* (Figure 4-23). These cells are oriented at right

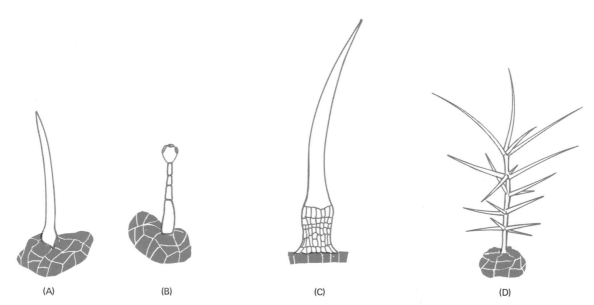

(A) (B) (C) (D)

Figure 4-24. Plant hairs. **(A)** Simple hair of a geranium plant. **(B)** Glandular hair of a geranium plant. **(C)** Branched hair of mullein. **(D)** Stinging hairs of nettle.

angles to the surface of the leaf. Below these, and extending to the lower epidermis, are irregularly shaped cells with abundant intercellular spaces. These layers of cells are designated the *spongy mesophyll* (Figure 4-23). Both layers of mesophyll cells are composed of parenchyma with numerous chloroplasts. The term *chlorenchyma* is applied to chlorophyll-bearing cells, whether they are found in the leaf or in other plant organs.

Vascular Tissue. Standing out rather conspicuously in the spongy mesophyll are the **veins** of the leaf (Figure 4-23). These veins, or **vascular bundles,** are specialized structures that function in conduction and support, being similar to the vascular tissue found in roots and stems. Each vein consists of three distinct tissues: an outer *bundle sheath* consisting of parenchyma cells, *xylem,* and *phloem.*

Leaf Abscission. The vast majority of broad-leaved trees and shrubs produce new leaves each spring and lose all of them in the fall. Plants which lose their leaves, such as oak, sycamore, and maple, are said to be **deciduous.** The separation, or **abscission,** of leaves is initiated by shortening of the day length or reduced water availability resulting in a decreased production of growth-promoting substances (auxins). Abscission is related to the development of a special layer of cells at the base of the petiole called the **abscission layer** (Figure 4-25). These parenchyma cells extend all through the base of the petiole, except through the vascular tissues. As the season progresses, the cell walls of the abscission layer become softened. This causes the cells to separate from each other, until finally the leaf is held only by the vascular tissue. Eventually, the vascular tissue is ruptured by wind action, the impact of rain, or by the formation of ice crystals, and the leaf falls. Along with abscission, a layer of cork cells which protects the exposed stem tissues develops just beneath the abscission layer.

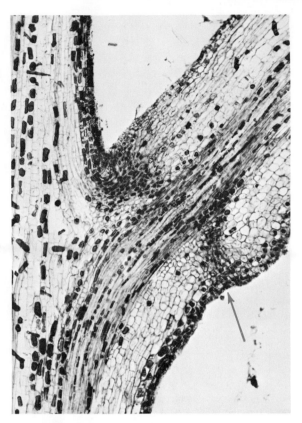

Figure 4-25. Leaf abscission. Longitudinal section of the base of the petiole of a *Coleus* leaf. The dark band of cells indicated by the arrow is the abscission layer. (Courtesy Carolina Biological Supply Company.)

In some species of plants, such as the white oak, it is not uncommon to find some of the dead leaves remaining on the tree throughout the winter months. This may be caused by the failure of the vascular tissue to break or by the failure of the abscission layer to develop. Plants that retain their leaves throughout the year are known as evergreens. Even evergreens eventually shed all their leaves but not all at one time. Common evergreens are rhododendrons, pines, spruces, and hemlocks.

(A)

(B)

(C)

Figure 4-26. Modified leaves. (**A**) Pitcher plant (*Sarracenia purpurea*). (**B**) Venus's-flytrap (*Dionaea muscipula*) in open and closed positions. (**C**) Sundew (*Drosera intermedia*). (Courtesy Carolina Biological Supply Company.)

Preceding abscission, there are many striking color changes in leaves. During the late spring and summer, leaves are green, owing to the presence of chlorophyll. With the coming of fall the temperature drops below the point necessary for chlorophyll formation and other pigments are revealed. Brilliant golden hues of *xanthophylls* and *carotenes* appear in the leaves of some maple, hickory and tulip trees. The cool weather and increased moisture also accelerates the production of the red pigment *anthocyanin*, which is revealed in the leaves of dogwoods, sweetgums, and varieties of maple trees. Brown leaf coloration results from the death of leaves accompanied by the production of *tannic acid.*

Specialized Leaves. Some leaves or parts of leaves are so highly specialized that they lose their identity as leaves. Angiosperms, in particular, provide many structural and functional specializations. Perhaps the most curious adaptation of leaves is found in the *insectivorous plants,* so-called because of their ability to attract, trap, and digest insects (Figure 4-26). Three representative examples are the pitcher plant, Venus's-flytrap, and the sundew. Insects crawl into the vaselike leaves of the pitcher plant (Figure 4-26A) and fall to the bottom, where they are trapped and digested. The outer ends of the leaves of the Venus's-flytrap (Figure 4-26B) resemble a spring trap. When insects alight on these leaves and touch the sensitive hairs, the two halves of the leaf close. The unique trapping device of the sundew (Figure 4-26C) consists of leaves that are covered with long, gland-tipped tentacles which curl about the insect and smear it with a mucilaginous, hydrolyzing secretion.

In some perennial monocots, leaves are modified as water or food *storage organs* (Figure 4-27A). The thick, fleshy bases of leaves that make up much of the onion bulb, for example, accumulate large quantities of stored food for use during the next

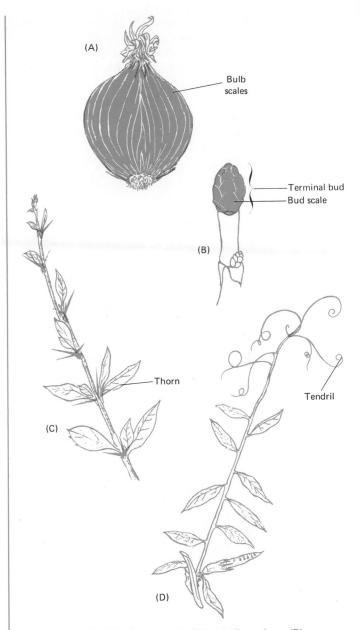

Figure 4-27. Modified leaves. (A) Onion bulb scales. (B) Bud scales protecting the terminal bud of a horse chestnut stem. (C) Barberry thorns. (D) Tendrils of vetch.

growing season. *Bud scales,* the enclosing scales of dormant buds, are greatly reduced, tough-textured, modified leaves that protect the meristematic tissue of the stem tips (Figure 4-27B). In the common barberry some of the leaves are specialized into *thorns* or *spines* (Figure 4-27C). In various species of cacti, all the leaves are spines. A final type of leaf modification is represented by *tendrils.* These are slender, elongated, threadlike structures which either twist around various supporting structures or contain small discs by which they adhere to supports. The tendrils of vetch represent this leaf modification (Figure 4-27D).

Economic Aspects. The leaves of many plants are utilized as a source of food. Some leaves are consumed directly. These include the leaves of cabbage, kale, Brussels sprouts, watercress, and spinach and the petioles of celery and rhubarb. Other leaves, or products of leaves, such as sage, bay, parsley, peppermint, thyme, and spearmint, are widely utilized for flavoring.

Some fibers, drugs, and waxes also come from leaves. The leaves of certain plants are a source of bowstring hemp, manilla hemp, sisal hemp, and New Zealand flax. Drugs obtained from leaves include *cocaines,* from the cocoa shrub, which relieves pain; digitalis, a heart stimulant; and belladonna, which is used to dilate the pupils of the eye. Most commercial waxes such as *carnauba wax* are obtained from the cuticles of leaves. Carnauba wax is an ingredient in automobile and floor waxes, cosmetics, carbon paper, and phonograph records. The uses of tea leaves and tobacco leaves are well known.

CHAPTER IN RETROSPECT

Roots, stems, and leaves of angiosperms are vegetative organs consisting of highly specialized tissues which perform specific functions within the plant body. For purposes of clarification, tissues are subdivided into two general types: (1) meristamatic tissue, in which cells actively divide, and (2) permanent tissue, in which there is relatively little cell division. The continued growth and development of plants depend upon the activity of meristamatic tissue, whereas permanent tissues (epidermis, parenchyma, xylem, phloem, and so on) provide a system for carrying out numerous plant activities. Some of these activities include food and water storage (parenchyma), mechanical support and strengthening (collenchyma), water conservation (cork), transport of water and mineral salts (xylem), and conduction of synthesized foods (phloem). Unlike stems, roots do not bear leaves or buds and represent the initial structures to establish contact with the environment, which, except for aerial and aquatic roots, is the soil. Angiosperm roots function chiefly to anchor the plant and to absorb water containing nutrient salts. Some roots become highly specialized in terms of a particular function; that is, storage roots (carrots, beets, parsnips) reserve food in the form of carbohydrates, aerial roots (poison ivy, tropical orchids, trumpet creeper) anchor stems to walls and trees and also permit exchange of gases, and parasitic roots (mistletoe, dodder) penetrate the vascular system of host plants and absorb water and nutrients.

Stems bear leaves and are usually erect aerial organs which serve to support other plant structures as well as to provide a pathway for the conduction of essential materials. Certain specialized stems—rhizomes, tubers, bulbs, runners, and tendrils—function in more specific ways and exhibit unusual structural modifications. Leaves are the characteristic photosynthetic organs of a majority of higher plants. The leaf can be viewed as several layers of actively photosynthesizing cells (the mesophyll) enveloped by a protective layer (the epidermis). Other important leaf activities include transpiration, respiration, and, in some cases, food and water storage.

It is the leaves, stems, and roots that provide the chief economic importance of plants. Foods, drugs, textiles, lumber, wood alcohol, latex, various gums and resins, dyes, and a host of other commercial products rank plants high as contributors to the economic welfare of man.

Review Questions and Problems

1. Define each of the following.

meristemic	taproot
permanent tissue	stele
parenchyma	cork cambium
collenchyma	prop root
sclerenchyma	lateral bud
xylem	node
phloem	pith
lumen	vascular ray
sieve cell	bark
adventitious root	stolon
stomata	chlorenchyma
abscission layer	

2. Distinguish between the vegetative and reproductive organs of angiosperms. List examples of each.
3. Discuss the pathway by which water and dissolved substances are transported from one part of the plant to another.
4. Compare roots and stems in terms of their specialized functions within plant systems.
5. List representative plants which possess the following types of specialized roots: (a) storage roots, (b) aerial roots, (c) prop roots, and (d) parasitic roots.
6. Compare the relative economic importance of roots, stems, and leaves.
7. What are the essential differences between woody dicot stems and herbaceous dicot stems?
8. Discuss the external and internal features of leaves, with specific mention of how each structure performs a specific task. How are leaves uniquely designed to accommodate photosynthetic activities?

Suggested Supplementary Readings

Ashby, E., "Leaf Shape," *Scientific American,* April 1949.

Cormack, R. G. H., "Development of Root Hairs in Angiosperms," *Botanical Review,* vol. 28, pp. 446–464, 1962.

Esau, Katherine, *Plant Anatomy.* New York: John Wiley & Sons, Inc., 1965.

Foster, A. S., and E. M. Gifford, Jr., *Comparative Morphology of Vascular Plants.* San Francisco: W. H. Freeman and Company, 1959.

Jacobs, W. P., "What Makes Leaves Fall?" *Scientific American,* November 1955.

Sachs, R. M., "Stem Elongation," *Annual Review of Plant Physiology,* vol. 16, pp. 73–96, 1965.

Sinnott, E. W., *Plant Morphogenesis.* New York: McGraw-Hill Book Company, Inc., 1960.

Waddington, C. H., "How Do Cells Differentiate?" *Scientific American,* September 1953.

Wigglesworth, V. B., "Metamorphosis and Differentiation," *Scientific American,* February 1959.

Williams, S., "Wood Structure," *Scientific American,* January 1953.

5

Macroscopic Level of Plant Organization: Reproductive Organs

THE NATURE OF PLANT REPRODUCTION

Reproduction may be viewed as a very specialized mechanism, common to all organisms, by which new individuals are produced. Reproduction is not only a means of species continuity, that is, the extension of individual plants from one generation to the next, it is also a process concerned with species maintenance. For example, as cells age, they tend to wear out. Through various synthetic processes parts of cells are replaced or, in some cases, entire living cells are replaced in response to wear and tear and death due to disease, accidents, or injuries. Species continuity is dependent on the addition of new cells as well as replacement. As noted earlier, growth of new tissues and organs in an individual plant is dependent upon the continual addition of cells.

Types of Plant Reproduction

Plants exhibit a wide variety of reproductive patterns which can be categorized into two principal types: asexual reproduction and sexual reproduction. The details of a number of these reproductive patterns will be given later as they apply to specific groups of plants. At this point, only their general nature will be discussed.

Asexual Reproduction. *Asexual reproduction* is a type of reproduction in which organisms multiply without the formation and fusion of specialized sex structures (gametes). Essentially, all asexual reproduction starts at the cellular level with cell division. Among living organisms cell division is a universal occurrence and is one of the most fundamental of all life processes. In single-celled plants it represents the reproduction of the entire organism, with the newly formed organism being genetically identical to the parent cell that produced it. In multicellular plants, cell division is the mechanism by which new cells are added or by which existing cells are replaced.

Among the common types of asexual reproduction in plants at the organismic level are fission, budding, fragmentation, spore formation, and vegetative reproduction. **Fission** is a method of reproduction in many simple single-celled plants in which the entire organism divides to form two (Figure 5-1A). Many unicellular plants such as bacteria and numerous algae reproduce by fission. **Budding** is the production by the parent organism of new individuals which initially appear as small outgrowths but later develop into mature organisms of full size. Yeast cells are common representative plants that reproduce by budding (Figure 5-1B).

Many forms of algae consist of threadlike colonies of cells (filaments) joined end to end. Quite often, these colonies of algal cells are broken apart by water currents, rain, wind, or contact with animals living in ponds. This mechanical separation of the multicellular plant body into segments, each capable of growth into a new colony, is called **fragmentation** (Figure 5-1C).

Spore formation, or **sporulation,** is a common asexual reproductive method among a number of plants. It consists of the production of a specialized reproductive cell or group of cells (spores) that separate from the parent plant and by repeated cell divisions grow into a complete new plant (Figure 5-1D). Spores are generally surrounded by

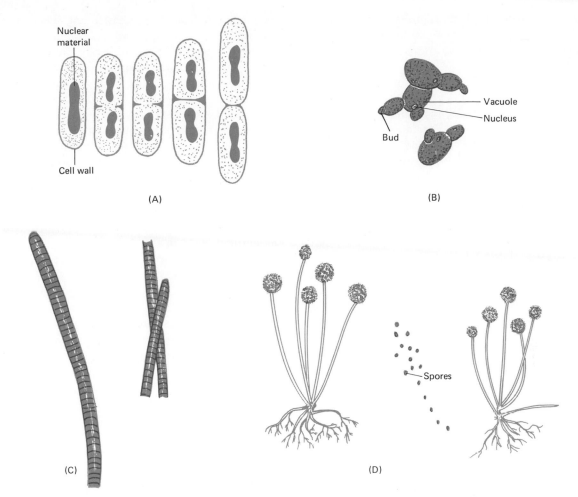

Nuclear material

Cell wall

(A)

Vacuole

Nucleus

Bud

(B)

(C)

Spores

(D)

Figure 5-1. Types of asexual reproduction. **(A)** Fission in bacteria. **(B)** Budding in yeast cells. **(C)** Fragmentation in a filament of blue-green algae. **(D)** Sporulation in a bread mold.

tough, protective coverings which enable them to withstand adverse environmental conditions. Sporulation is common among lower plants such as bacteria, algae, and fungi as well as more complex plants such as mosses and ferns. In fact, spore formation has even persisted as an important portion of the life cycle of flowering plants.

For the most part, asexual reproduction is more common to lower plants than to higher forms. However, higher plants do possess various means of asexual reproduction through **vegetative propagation.** Roots, stems, and leaves, collectively called vegetative organs, are capable of producing new individuals from the parts by natural vegetative propagation. Typical examples of this natural process are the production of bulbs, rhizomes, stolons or runners, shoots, and tubers (see Chapter 4).

Artificial vegetative propagation is a form of vegetative propagation developed by man which is of considerable economic importance. Four principal types of artificial vegetative propagation are cuttings, grafting, budding, and layering. **Cuttings** are small stems, usually with attached leaves, which have been removed from parent plants and placed in water or moist sand until roots develop. Once adventitious roots are produced, the cuttings are transplanted to soil. Geranium, rose, begonia, and ivy are commonly propagated in this manner.

Grafting is a horticultural practice in which a cutting from one plant is attached to the root or stem of another actively growing plant. This is accomplished by shaping and exposing the vascular cambium of the cutting called the *scion* and placing it so that it makes contact with the vascular cambium of the root or stem called the *stock*. Then the scion and stock are firmly bound together with a string or tape and are coated with grafting wax. In this way the graft is protected against damage and desiccation (Figure 5-2A).

A method of artificial vegetative propagation closely related to stem grafting is *bud grafting*. In this process, a stem apex or a bud, together with a small portion of the surrounding bark, is cut from the stem of a plant whose desirable fruit-bearing characteristics are to be propagated. The bud, which in this case is the scion, is inserted into an incision made in the bark of a stem (stock) and is secured with a special elastic tape (Figure 5-2B). Many varieties of fruit, including peaches, cherries, and grapes, are propagated by bud grafting.

In the process called *layering*, the end of a branch of a plant is bent so that it makes contact with the soil (Figure 5-2C). After the branch has taken root, it is cut from the parent plant. Then it is removed from the soil and transplanted. The process of layering is horticulturally employed for such plants as black raspberries, gooseberries, rhododendrons, roses, and other ornamental or food plants.

Sexual Reproduction. *Sexual reproduction* is defined as the fusion of the nuclei of specialized sex cells called gametes leading to the production of a fertilized egg called a zygote. Unlike asexual reproduction, sexual reproduction involves elaborate preparatory and specialized activities and structures. The principal preparatory activity involves a complex series of events associated with the chromosomes in the nucleus; the specialized activities are concerned with the production of sex cells followed by the union of the nuclei of these sex cells.

Meiosis. Within the nuclei of diploid plant cells are found pairs of threadlike bodies called **chromosomes.** The number of chromosomes within the nuclei of cells in the meristematic root or shoot is usually constant for each species of plant or animal. For example, the nuclei of meristematic cells of corn plants have 20 chromosomes and those of pine trees have 24; somatic cells of a housefly have 12, and somatic cells of a man have 46 chromosomes. If the nuclei of cells of these organisms could be examined during past or future generations, the chromosome numbers presumably would remain constant for each species.

Inasmuch as the nuclei of two sex cells are fused together during sexual reproduction and the chromosome number remains constant for all generations, a paradox is apparent concerning a numerically constant chromosome number. It would appear that the chromosome number would double for each succeeding generation. However, it does not double because of a special type of nuclear division called **meiosis.** This nuclear division occurs at some stage in the life cycle

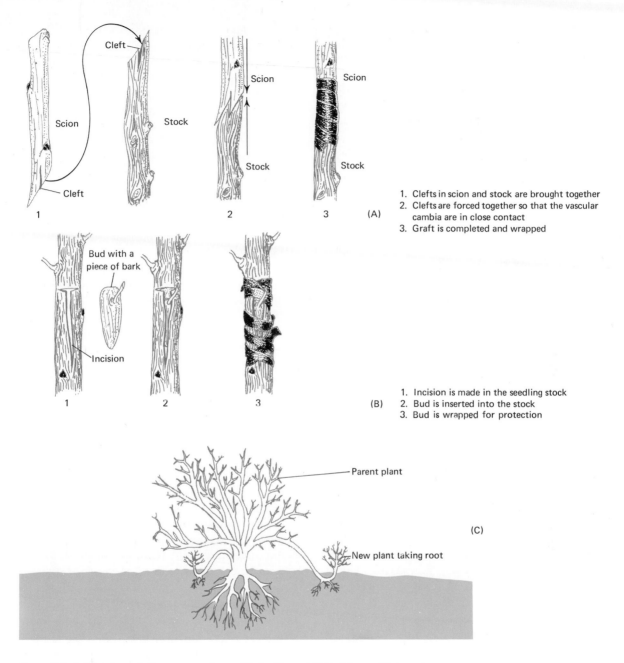

1. Clefts in scion and stock are brought together
2. Clefts are forced together so that the vascular cambia are in close contact
3. Graft is completed and wrapped

1. Incision is made in the seedling stock
2. Bud is inserted into the stock
3. Bud is wrapped for protection

Figure 5-2. Artificial vegetative propagation. **(A)** Grafting. **(B)** Budding. **(C)** Layering.

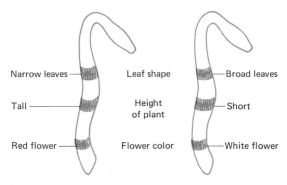

Narrow leaves — | Leaf shape | — Broad leaves

Tall — | Height of plant | — Short

Red flower — | Flower color | — White flower

Figure 5-3. Homologous chromosomes. Members of a pair of chromosomes are similar in size, shape, and the genetic traits that they control. They may differ, however, in terms of the degree to which a particular trait is controlled, as shown in the diagram.

prior to the formation of sex cells, so each of the gametes contains only one half as much hereditary material (chromosomes) as the parent cells from which they were formed. Thus, when two sex cells unite, each contributes one half the number of chromosomes, so that the resulting zygote has the normal number of chromosomes for that species.

It is by no means coincidental that most organisms have even numbers of chromosomes. The members of a pair of chromosomes are referred to as **homologous chromosomes** or homologs (Figure 5-3). Structurally, homologous chromosomes are very similar in size and shape and functionally they are alike concerning the hereditary characteristics that they control (for example, height, leaf shape, and flower color). However, homologs may be dissimilar in terms of the degree to which a particular hereditary trait is controlled (short–tall; narrow leaves–broad leaves; red flowers–white flowers). For example, the height of a pea plant is governed by genes (Chapter 12) at specific locations on homologous chromosomes, but one chromosome may have the gene for shortness while the other has the gene for tallness. In

a similar manner, considering the shape of leaves, one chromosome might contain the gene for broad leaves while the other contains the gene for narrow leaves, although both chromosomes are concerned with leaf shape.

To understand the numerical distribution and constancy of chromosomes during sexual reproduction, consider the normal chromosome number for cells of the shoot apex of corn, which is 20. This full complement of chromosomes is referred to as the **diploid** number and is commonly designated $2n$. In other words, the diploid, or $2n$, number of chromosomes for corn is 20. In cells resulting from meiosis, the chromosome number is reduced from $2n$ to n, the **haploid** number, which is one half (10) the full complement of chromosomes. Therefore, when meiosis occurs in cells of corn plants, each cell thus produced contains 10 chromosomes. Logically, when two such cells fuse to form a zygote during fertilization, the diploid number is restored. All organisms that reproduce sexually have both haploid and diploid phases, but in higher plants diploid cells are more numerous. Only sex cells of higher plants are haploid as a result of meiosis, and these are few in comparison to diploid vegetative cells.

In flowering plants there are two kinds of diploid **spore-mother cells,** that is, cells capable of undergoing meiosis. One kind is formed in the anthers, or male reproductive structures; the other is produced in the ovules, or female reproductive structures. Within the anther each diploid spore-mother cell forms four haploid spores (pollen grains or microspores). These pollen grains germinate and produce the pollen tube, within which the haploid gametes (sperms) are formed. In a similar manner, a spore-mother cell in an ovule forms four haploid spores, the megaspores. Three of these disintegrate; the fourth develops into the haploid embryo sac, which contains the haploid egg. At fertilization, the sperm and egg nuclei fuse and the diploid chromosome number is restored. The diploid chromosome number characterizes the embryo plant and the adult plant until, at maturity, it, in turn, forms spores from spore-mother cells preparatory to the production of gametes.

In other plant groups meiosis may occur at any time in the life cycle, depending on the type of plant. However, regardless of when or where meiosis occurs in the life cycle, the essential features of the process are the same. Meiosis starts with a cell containing the diploid number of chromosomes characteristic of the plant. The cell undergoes two mitosis-like divisions, but the chromosomes are duplicated only once. The net effect of this is the formation of four cells, each containing the haploid number of chromosomes. For purposes of simplicity, meiosis will be described below in terms of a starting diploid cell containing only four chromosomes.

Meiosis, although similar to mitosis (see Chapter 3), differs considerably in some important respects. Meiosis is typically characterized by two stages (I, II), the second generally following immediately after the first (Figure 5-4). The prophase of the first stage (prophase I) is similar to the prophase of mitosis, except that the homologous chromosomes become associated in pairs. This pairing, or *synapsis,* occurs throughout the length of the chromosomes (Figure 5-4A). In addition, during a late stage of prophase I, each chromosome doubles, so that each pair of homologous chromosomes consists of four **chromatids** (Figure 5-4B). Throughout the prophase, especially during the latter part, the chromosomes gradually thicken and shorten.

During metaphase I the chromosomes migrate toward the equatorial region of the cell (Figure 5-4C), and, at anaphase I, the paired chromosomes separate and move toward opposite poles of the cell (Figure 5-4D). Telophase I is the same as telophase in mitosis (Figure 5-4G). In some plants

a cell wall is formed between the two nuclei during telophase I, but in many plants wall formation is delayed until after four nuclei have been formed at the termination of stage II.

The second stage of meiosis generally follows quickly. During prophase II, the nuclear membranes disappear. The chromosomes already consist of two chromatids each and are ready to enter metaphase II. During metaphase II, the chromosomes usually orient themselves at right angles to that of the first division in the equatorial region (Figure 5-4F). At anaphase II, the chromatids separate and move toward opposite poles of each cell (Figure 5-4G). The last sequence of events consists of the formation of cell walls between migrating chromosomes followed by the passage of the four nuclei into telophase and interphase (Figure 5-4H). In summary, then, meiosis consists of two nuclear divisions (stages) and one duplication of chromosomes. Each of four cells thus formed contains one half the number of chromosomes of the original cell and all four are not genetically alike.

There are two rather significant results of meiosis. One is obviously the reduction of chromosome number from the $2n$ to n condition. The second result is that the cells produced in meiosis may be genetically unlike with respect to each other and with respect to the original cell that produced them. To illustrate the genetic variability of spores, assume that the starting cell consists of two pairs of chromosomes and that each pair of chromosomes governs the same genetic characteristic but to varying degrees. In other words, assume that one pair of homologous chromosomes is responsible for the expression of height. One member of the pair contains gene A for tallness and the other member of the pair contains gene a for shortness. Similarly, assume that the other pair of homologous chromosomes governs flower color. In this case one member of the pair contains gene B for

Prophase I. Homologous
chromosomes pair (synapsis)
(A)

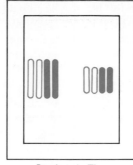

Prophase I. The
chromosomes split
into chromatids
(B)

Metaphase I. The chromosomes
migrate to the equator
(C)

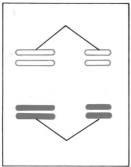

Anaphase I. Paired chromosomes
move toward opposite ends of cell
(D)

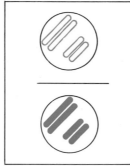

Telophase I
(E)

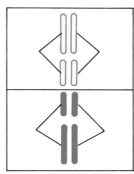

Metaphase II. Chromatids
orient at right angles
(F)

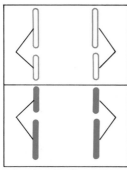

Anaphase II. Chromatids
separate and move toward
opposite cell poles
(G)

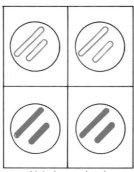

Meiosis completed
(H)

Figure 5-4. Meiosis. This process, an activity preparatory to sexual reproduction, consists of two nuclear divisions or phases (I and II) and one duplication of chromosomes. See text for amplification.

red flowers and the other member of the pair contains gene *b* for white flowers (Figure 5-5A).

During metaphase I of meiosis, these chromosomes and their respective genes may orient themselves in one of two ways. Either *AB* and *ab* or *Ab* and *aB* may result (Figure 5-5B). This chromosomal orientation is determined purely by chance, and the resulting cells produced after metaphase II of meiosis are also determined by this chance arrangement (Figure 5-5C). Note that the cells may contain genes *AB*, *ab*, *Ab*, or *aB*, that is, four different sets of genetic characteristics. Genetically, the first cell contains the genes for tall plants with red flowers, the second possesses a genetic expression for short plants with white flowers, the third for tall plants with white flowers, and the fourth for short plants with red flowers. Logically, the genetic combination achieved by nuclear fusion during fertilization will determine the genetic traits for height and flower color of the plant.

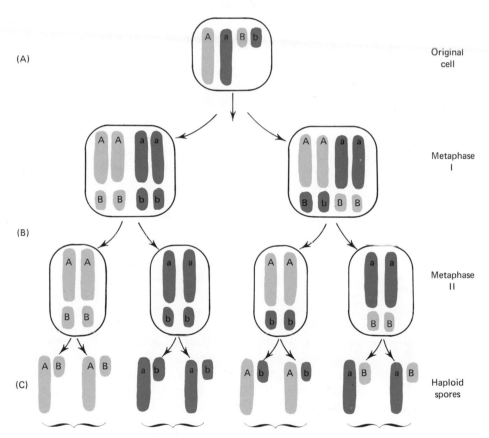

Figure 5-5. Formation of genetically dissimilar haploid cells during meiosis. Note the difference between the original cell and the four haploid spores that are produced.

The Cycle of Sexual Reproduction

It was noted earlier that the structures and activities of flowering plants may be divided into two rather distinct types: *vegetative* and *reproductive.* The vegetative organs (roots, stems, and leaves) are concerned primarily with the maintenance of the individual plant. The reproductive organs (flowers, fruits, and seeds) are concerned primarily with the propagation of the species. In most angiosperms, the orderly processes of vegetative development terminate in the formation of flowers. These specialized reproductive structures exist for only a short period of time, and certain parts of them develop into fruits containing the seeds. The seeds in turn produce a new generation of plants.

The basic steps involved in the life cycle of a flowering plant are (1) the production of specialized reproductive cells—the eggs, and pollen; (2) pollination, or bringing the pollen into close contact with the eggs; (3) fertilization, the union of the pollen and egg nuclei; (4) the development of fruits and seeds; (5) the dissemination of fruits and seeds; and (6) seed germination, the development of a seed into a mature plant. The mature plant, then, after a period of vegetative activity produces flowers, and the cycle is completed.

Flowers

The Structure of a Typical Flower

In the angiosperms, identification and classification of different plants is based largely on flower structure, because floral structure is remarkably stable within a particular species of plant. Consequently, the great morphological variation in flowers found among the different groups of flowering plants makes this structure a useful organ on which to base classification. A **flower** may be defined as a specialized, branched stem bearing lateral appendages. It consists typically of structures directly involved in reproduction and others which are only indirectly involved in reproductive activities. Flowers develop from buds, as do vegetative shoots, but unlike vegetative shoots there is no elongation of their internodes. The result of this growth pattern is that floral appendages are compressed together and are not distributed at intervals along the floral axis. Furthermore, once the floral organs are formed, there is no further apical growth. Some of the factors affecting floral initiation will be considered in Chapter 11.

A cursory survey of a limited number of flowers would indicate that there is an enormous diversity in floral structure. Yet the similarities are greater than the differences, because all flowers are based upon the same structural design (Figure 5-6). The flower stalk is a stem which supports the flower and is known as the **pedicel.** The terminal portion of the pedicel consists of an enlarged structure, the **receptacle,** to which the floral parts are attached. The most familiar types of flowers have four kinds of floral organs attached to the receptacle. These organs, always arranged in the same order, are attached at successively higher levels. The first of these organs are the **sepals.** These appendages, attached at the base of the flower, are the outermost and lowest parts and are usually green in color, closely resembling leaves. Collectively, the sepals are referred to as the **calyx.** The next appendages proceeding toward the center are the **petals,** collectively known as the **corolla.** The petals are usually variously colored or white, leaflike structures which frequently extend beyond the sepals. The **perianth** is a collective designation for the calyx and corolla. The third group of appendages consists of the **stamens,** collectively termed the **androecium,** which are attached just above the base of the petals. Each stamen, or male reproductive structure, consists of a slender elongated stalk, the **filament,** and a somewhat expanded lobed structure at its distal end, the

anther. The anthers contain pollen sacs in which large numbers of minute pollen grains are formed.

Attached to the uppermost site in the center of the flower is the female reproductive structure, the **gynoecium.** The basic structural unit of the gynoecium is the **carpel,** which is generally flask-shaped and is composed of a number of prominent regions. The swollen basal portion, the **ovary,** contains **ovules** which develop into seeds. The ovary is connected to a stalk-like **style,** which terminates in an expanded area, the **stigma. Pistil** is another term referring to the megasporangial part of the flower and although the abandonment of the term

has been advocated, it continues to be useful. In this regard, a gynoecium may contain only one carpel (*simple pistil*) as in the flowers of beans and buttercups (Figure 5-7A), or may be composed of two or more carpels fused together (*compound pistil*) characteristic of the flowers of carnations, tulips, lillies, and geraniums (Figure 5-7B,C). The tissue of the ovary to which the ovules are attached is called the **placenta** and the arrangement of ovules, that is, the position of the placentae, is termed *placentation.*

Sepals and petals are referred to as *accessory parts* of a flower because they do not play a direct role

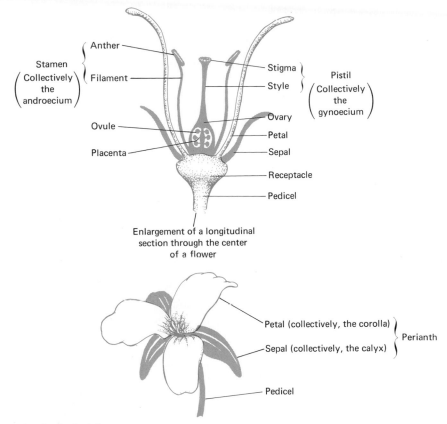

Stamen (Collectively the androecium)
Anther
Filament

Pistil (Collectively the gynoecium)
Stigma
Style

Ovule
Placenta

Ovary
Petal
Sepal
Receptacle
Pedicel

Enlargement of a longitudinal section through the center of a flower

Petal (collectively, the corolla)
Sepal (collectively, the calyx)
Perianth

Pedicel

Entire flower

Figure 5-6. The parts of a typical flower.

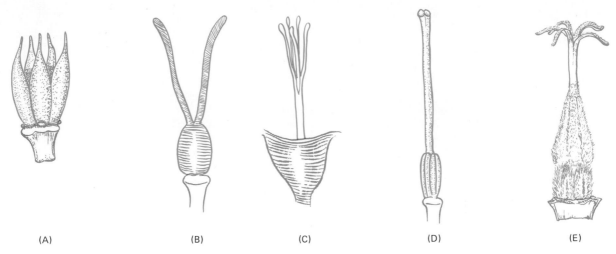

Figure 5-7. Types of pistils. (**A**) Simple pistil (stonecrop). (**B**) Compound pistil with two free styles (carnation). (**C**) Compound pistil with five styles (apple). (**D**) Compound pistil with one style (lily). (**E**) Compound pistil with five united carpels having free style tips (geranium).

in reproduction. Stamens and pistils, by contrast, are the *essential parts* of a flower because they are directly related to the reproductive process.

Variations in Floral Structure

A *complete flower* is one which possesses the four kinds of floral organs just described—sepals, petals, stamens, and carpels (see Figure 5–6). A familiar example is the lily. An *incomplete flower* is one in which one or more of the floral whorls are lacking. Examples of incomplete flowers include those of oats, anemones, clematis, and willows. In oat flowers, there are no sepals and petals; only stamens and carpels are present. Flowers of anemone and clematis, by contrast, have sepals, stamens, and carpels but no petals.

In many species (for example, tulips, roses, lilies, and orchids) stamens and a carpel are present in the same flower and the flower is referred to as a *perfect flower*. In others, however, stamens and

carpels are produced in separate flowers and the flowers are called *imperfect flowers*. Unisexual flowers which bear only stamens and not pistils are termed *staminate;* those which are carpel-bearing and do not contain stamens are called *pistillate*. When staminate and pistillate flowers are produced on the same plant, as in walnut, oaks, Indian corn, pumpkin, and numerous other species, the plant is said to be **monoecious.** A condition described as **dioecious** occurs when staminate and pistillate flowers are borne on separate plants. Examples of dioecious plants are hemp, willows, and asparagus. In the case of edible fruits which are borne on dioecious plants, both pistillate ("female") and staminate ("male") individuals must be close enough together for the pollen to be transferred. This fact must be considered in commercial plantings of dioecious plants; otherwise, fruits will not develop.

There are numerous other floral variations that

occur in addition to those relating to the presence or absence of accessory or essential structures. These include variations in the size and form of petals, the relative position of floral organs, the degree of fusion among various floral structures, the symmetry of the flower, and differences in the number of parts. Such modifications, as well as others, not only contribute to success in pollination but also reveal certain evolutionary relationships among flowering plants, because variation is a principal feature used in classification.

Most flowers show a definite numerical arrangement of parts. In the monocots (Monocotyledonae) the flowers generally have their parts in three or multiples of three (see Figure 20-12). In the dicots (Dicotyledonae) flower parts are usually in fours and fives (see Figure 20-12). Monocots and dicots are also differentiated on the basis of seed structure, leaf anatomy, and stem structure.

In some flowers the petals of the corolla are of similar shape and size and radiate from the center of the flower. Such flowers in which the petals are equidistant from each other and are built on a circular plan are called *regular flowers* and are said to have *radial symmetry*. Roses, buttercups, and petunias are examples of radially symmetric flowers. In others, such as snapdragons, mints, orchids, and sweet pea, the petals vary in size, do not radiate from the center, and are not equidistant from each other. These flowers, in which the left and right half are mirror images, are termed *irregular flowers* and the symmetrical pattern is called *bilateral symmetry*.

The fusion of floral organs is another important modification found in flowers. In the tulip tree flower all parts are separate and distinct; each sepal, petal, stamen, and carpel is attached at its base to the receptacle. Such flowers are referred to as *apopetalous*. In many flowers, however, there is a fusion among members of the same or different whorls. Such a union of floral parts, either partial or complete, is called *coalescence*. Coalescence may involve the fusion of sepals (mints), the union of petals (African violet), the fusion of stamens (cotton), and the union of carpels (lily).

In some kinds of flowers the floral organs are arranged spirally on the receptacle. In the magnolia flower, for example, the stamens are of indefinite number and are attached in a *spiral* fashion to the axis. In an iris flower, by contrast, the parts are arranged in distinct whorls or cycles upon the receptacle. In such a flower the stamens are definite in number (three), and all are attached at the same level. The *cyclic* arrangement of floral parts is more common than the spiral configuration in angiosperms.

In some plants the calyx, corolla, and stamens originate below the ovary, and the ovary is termed *superior*. Such flowers are called *hypogynous* and are represented by those of buttercups, grasses, and morning glories. In flowers of apple, rose, and cherry plants, the sepals, petals, and stamens are inserted on an urnlike rim and the receptacle forms a cuplike structure around the superior ovary. Such flowers are referred to as *perigynous*. In still other flowers, such as dandelion, honeysuckle, and apple, the ovary is enclosed by the receptacle and the other floral parts are attached above the ovary. These flowers are termed *epigynous* flowers and the ovary is referred to as *inferior*.

The evolutionary significance of the floral modifications discussed in this section is treated in more detail in Chapter 20.

Inflorescences

In some angiosperms, individual flowers are quite large and are borne singly on the pedicel. Such flowers are termed **solitary** flowers. In most angiosperms, the flowers are small, but together they form a rather conspicuous cluster called an

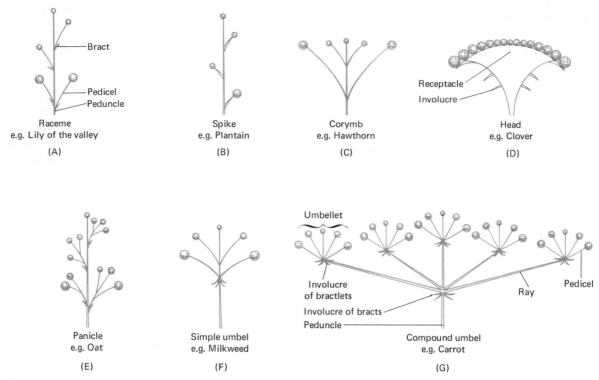

Figure 5-8. Common types of inflorescences. See text for amplification.

inflorescence. The more commonly observed types of inflorescences are shown in Figure 5-8.

A very simple type of inflorescence is the **raceme** (Figure 5-8A), which is characteristic of the flowers of such plants as snapdragons, currants, radishes, and hyacinths. In this arrangement, individual flowers are borne on short branches called *pedicels,* which are approximately equal in length. The pedicels arise from a main axis, the *peduncle,* and the flowers are spaced at regular intervals. Tiny leaflike structures, called *bracts,* are often found at the base of each pedicel.

Another type of simple inflorescence is a spike (Figure 5-8B). **A spike** is similar to a raceme, except that the flowers are attached directly to the pe-

duncle and are not borne on pedicels. Common plantain, cattail, and gladiolus illustrate the spike.

Another variation of the raceme is the **corymb** (Figure 5-8C), in which the lower flowers have longer pedicels than the upper ones. The inflorescence is more or less flat-topped or round-topped in appearance. The corymb is represented by such flowers as cherry, candytuft, and hawthorne.

A disc-shaped inflorescence with much reduced or totally absent pedicels is termed a **head** (Figure 5-8D). A head resembles a very short, dense spike. The individual flowers are quite small and are packed together over the surface of the disc. The crowded flowers lack pedicels; they are borne on a disclike expansion at the tip of the stem. Often the flower clusters are subtended by a whorl of bracts (*involucre*). Sunflower, dandelion, aster, and chrysanthemum flowers have heads.

A branched inflorescence which bears pedicelled flowers with the upper branches progressively shorter is termed a **panicle** (Figure 5-8E). A panicle is a branched raceme in which each branch of the main axis bears more than one flower. Panicles are characteristic of the flowers of grapes, bluegrass, lilacs, and oats.

An inflorescence known as a **simple umbel** (Figure 5-8F) is characterized by a much reduced main axis in which all the pedicels appear to arise from a central point. Milkweed flowers possess a simple umbel.

A final type of inflorescence is the **compound umbel** (Figure 5-8G), which is actually a series of simple umbels arranged in such a manner that the branches (*rays*) of the cluster grow from the apex of the main axis. Most members of the carrot family produce inflorescences of the compound-umbel type.

Reproductive Activities of Flowers

Development of Pollen. A cross section of a young anther reveals that it is composed of four pollen sacs, or **microsporangia** (Figure 5-9A).

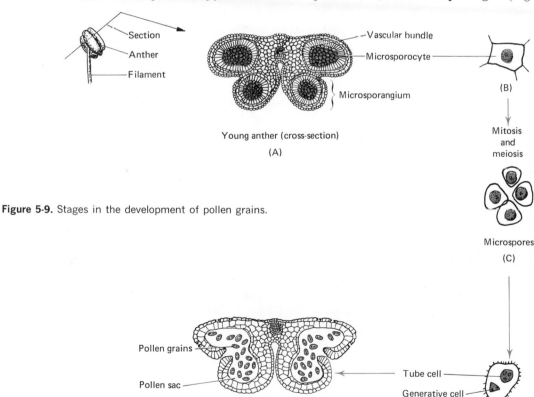

Figure 5-9. Stages in the development of pollen grains.

Within each microsporangium there is a cluster of cells, the *microsporocytes.* These cells are characterized by their large size, abundant cytoplasm, and prominent nuclei (Figure 5-9B). Concomitant with the growth of the anther, the microsporocytes (spore-mother cells) undergo a series of divisions (meiosis) in which the chromosome number of each microsporophyte is reduced by half. In addition to a reduction in chromosome number, each microsporocyte undergoes divisions and gives rise to four cells, the **microspores** (Figure 5-9C). Essentially, then, each microsporophyte, which contains the full chromosome number (diploid) of the species of plant involved, produces four microspores containing only one half the chromosome number (haploid). Each microspore becomes a pollen grain.

The microspores undergo further mitotic division into two daughter nuclei (Figure 5-9D). There are no walls to separate these nuclei, so the nuclei are retained within the wall of the microspore. One of these cells, the **tube cell,** is larger; the other cell, the **generative cell,** is smaller; later it divides once again to form two sperm. Once the tube cell and the generative cell are formed, the entire structure is referred to as a **pollen grain,** or young *male gametophyte.* At about this stage of development, the wall between the microsporangia disintegrates, resulting in the formation of two pollen sacs (Figure 5-9E). With the maturity of the flower, the anther splits along its long axis and the pollen grains are liberated.

Development of Ovules. The activities leading to the development of ovules are not unlike those which lead to the production of sperm (Figure 5-10). The ovule, one of many which may be produced in an ovary, first appears as a projection on the ovary wall. This protuberance consists of a layer of tissue, one to several cells in thickness, called the **nucellus,** which encloses the megasporocyte, or spore-mother cell (Figure 5-10A). The nucellus is the wall of the **megasporangium.** As growth and development proceed, the nucellus is raised on a stalklike structure, and one or two protective layers, the **integuments,** grow from the base of the nucellus and completely surround it except for a small opening, the **micropyle** (Figure 5-10B to D).

The single, large, diploid megasporocyte undergoes meiosis and forms four haploid cells, the **megaspores.** Of these, only one persists; the other three disintegrate (Figure 5-10B). The remaining megaspore undergoes successive mitotic divisions, which result in eight haploid nuclei within the enlarged megaspore. During the development of these eight nuclei, the megaspore gives rise to an oval-shaped structure, the **embryo sac** (Figure 5-10E). Three of the nuclei migrate toward the micropylar end of the embryo sac, three move in the opposite direction, and two remain near the central region. Membranes then develop around all nuclei except those that are centrally located.

Of the three cells at the micropylar end, one is the egg; the other two are referred to as *synergids.* The three cells at the opposite end of the embryo sac are termed the *antipodals* and the two nuclei in the center are called *polar nuclei* (Figure 5-10F). The fully developed embryo sac, composed of the polar nuclei, antipodals, synergids, and egg, together with the enveloping nucellus, integuments, and stalk, constitutes the mature ovule.

Pollination. **Pollination** refers to the deposition of pollen grains on the stigma of the female reproductive organ. The surface of the stigma is covered with a sticky secretion so that pollen grains will adhere more readily to its surface and therefore increase the chance that pollination will occur. Among the agents that facilitate pollination are wind currents, gravity, water, insects, birds, and other animals. Many flowering plants are *cross-pollinated;* that is, pollen is transferred from the anther of one plant to the stigma of the flower

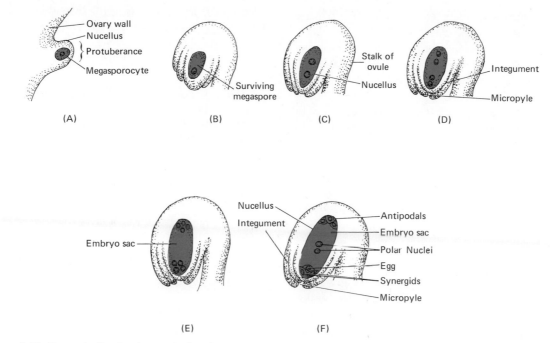

Figure 5-10. Stages in the development of ovules.

of another. Other angiosperms are *self-pollinated,* a process in which pollen is transferred from the anther to the stigma of the same flower or another flower on the same plant. Cross-pollination occurs mainly by wind and insects. The most common insect pollinators are moths, bees, and butterflies. Flowers which are pollinated by insects generally possess large and conspicuous petals, secrete aromatic substances or nectar, and frequently have relatively small stigmas which lack the characteristic hairs and bristles of most wind-pollinated flowers. In addition, wind-pollinated flowers produce copious amounts of pollen to compensate for the large number of pollen grains which fail to find receptive stigmatic surfaces.

Once a pollen grain is deposited on the stigma, it germinates (Figure 5-11). In this process, the pollen grain produces a cytoplasmic tube, the *pollen tube,* which grows (digests) through the tissues of the stigma, style, and ovary toward the ovule. The pollen tube produces extracellular enzymes which hydolyze the tissues of the style. In this manner, energy for growth is provided. At this stage of development, the generative cell divides mitotically and produces two male gametes, the **sperm.** The sperm move down the pollen tube, the tip of the tube enters the embryo sac, bursts, and its contents are discharged.

Fertilization. Within the embryo sac, one of the sperm nuclei fuses with the egg nucleus, forming a diploid zygote (Figure 5-12). At about the same time, the other sperm nucleus fuses with the two polar nuclei (triple fusion) and the result is a primary *endosperm nucleus.* This nucleus also under-

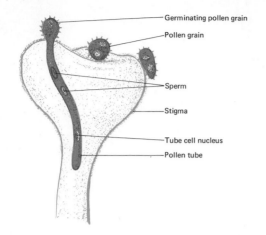

Figure 5-11. Germination of a pollen grain on the surface of the stigma.

goes a series of mitotic divisions and gives rise to a multicellular food storage tissue, the **endosperm**. Subsequently the zygote develops by many successive mitotic divisions into a multicellular embryo, or rudimentary plant. The antipodal cells and synergid cells disintegrate after fertilization. In addition, the tube cell also disappears. The term *double fertilization* is applied to the union of egg and sperm nuclei together with the fusion of the second sperm nucleus with the polar nuclei. This phenomenon is unique to the angiosperms.

SEEDS

Seeds are the characteristic reproductive structures of angiosperms. The success of the flowering plants has been due largely to the evolution of a mechanism for protecting the new generation within the old. This mechanism is accomplished by seed production and subsequent germination. A **seed** consists of an immature plant surrounded by a quantity of a stored food available for its early nourishment and a protective coat derived from the hardened surface layers of the integument. More simply, seeds are ripened ovules.

Development of the Seed

After fertilization has occurred, the zygote begins to develop into an embryo, or young plant, within the embryo sac (Figure 5-13). After several divisions, the zygote produces a row of cells, the

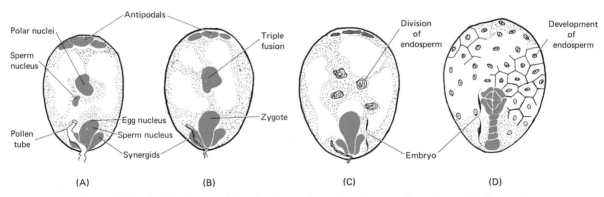

(A)   (B)   (C)   (D)

Figure 5-12. Fertilization, embryo development, and endosperm formation. (A) Egg and sperm fusion, sperm nucleus approaching polar nuclei, and disintegration of the pollen tube. (B) Sperm nucleus and polar nuclei have fused to form a triploid endosperm nucleus. (C) Division of the endosperm nuclei and the division and differentiation of the zygote into an embryo. (D) A later stage of endosperm and embryo development.

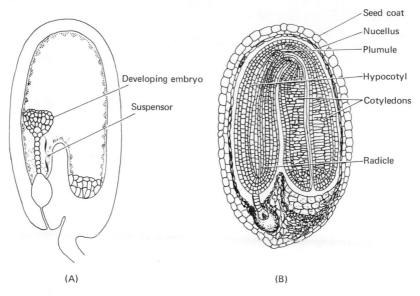

Figure 5-13. Development of the embryo. (A) Early stage of embryo development. (B) Mature embryo.

proembryo, which push into the embryo sac. The uppermost cell of the proembryo divides longitudinally and transversely and eventually produces most of the embryo. The remaining cells, called the *suspensor,* push the embryo into the endosperm so that it can derive nutrition for further growth and development (Figure 5-13A). In time the suspensor disintegrates. The mature embryo consists of a short axis with one or two cotyledons (Figure 5-13B). **Cotyledons** are food storage organs which absorb food from the endosperm and which function as leaves after germination. Monocots produce only one cotyledon, whereas dicots give rise to two. In addition to cotyledons, the mature embryo also contains a **plumule,** which gives rise to the shoot; the **radicle,** which becomes the root system; and the **hypocotyl,** a region between the radicle and the cotyledons which elongates during germination.

The seed is surrounded by the **testa,** or seed coat, which develops from the integuments of the ovule (Figure 5-14). Although integumentary growth practically closes the micropyle, its location may still be visible as a minute pore. Associated with this pore is a scar, the **hilum,** which indicates the point of attachment of the seed to the ovary.

Dispersal of Seeds

The dissemination of seeds serves as the mechanism for distribution of plants over a relatively wide geographic region. As with pollen, the principal agencies of seed dispersal are wind, animals, and water. A number of seeds are sufficiently light to be carried great distances by wind currents. Wind-dispersed seeds usually possess devices such as wings, tufts, or hairs which make air travel easier (Figure 5-15). Some seeds and fruits float on water and frequently drift many miles before they again become land-based. In some cases, the fruit may be eaten by animals and the enclosed seeds may survive the digestive process. Frequently, seeds possess spines, hooks, or a sticky secretion which enables them to adhere to the fur of mammals or the feathers of birds. Of course, man himself is a great disseminator of seeds, especially those which he finds economically worthwhile.

Germination of Seeds

The length of time a seed may remain viable varies from a few weeks (silver maple) to more than 1000 years (Indian lotus), depending largely upon environmental conditions and the particular species. Seed **germination** is the resumption of the growth of the embryo after a period of much-reduced metabolic activity. The external conditions required for germination are usually an adequate supply of water, available oxygen, and a favorable temperature. As water is absorbed by the seed there is an increase in seed volume and a general softening and weakening of the seed coats; as a result the developing embryo emerges. The most characteristic physiological effect of seed germination is the increased rate of respiration. Prior to germination, the seed respires at an almost imperceptible rate.

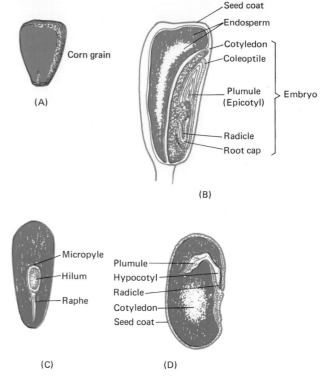

Figure 5-14. Structure of typical monocot and dicot seeds. (**A**) Corn grain, external view. (**B**) Corn grain in longitudinal section showing parts of the embryo and seed. (**C**) Garden bean, external view. (**D**) Garden bean in longitudinal section showing internal structure of embryo and seed.

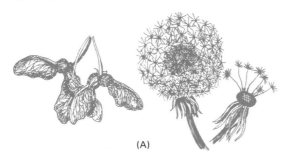

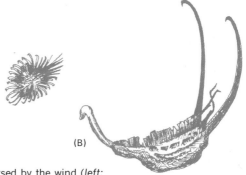

Figure 5-15. Adaptations for seed dispersal. (**A**) Plants that are dispersed by the wind (*left:* maple; *right:* dandelion). (**B**) Plants that are dispersed by animals (*left:* cocklebur; *right:* unicorn).

There is a dramatic increase in the rate of cell division in the meristematic region of the plumule and radicle just after germination takes place. From the time that the radicle emerges from the seed to the time that it can carry on an existence independent of the stored food reserves in the seed, a plant is known as a seedling.

Corn (a monocot) and lima bean (a dicot) represent the two typical patterns of seedling development. In the germination of corn (Figure 5-16A), the cotyledon absorbs digested food from

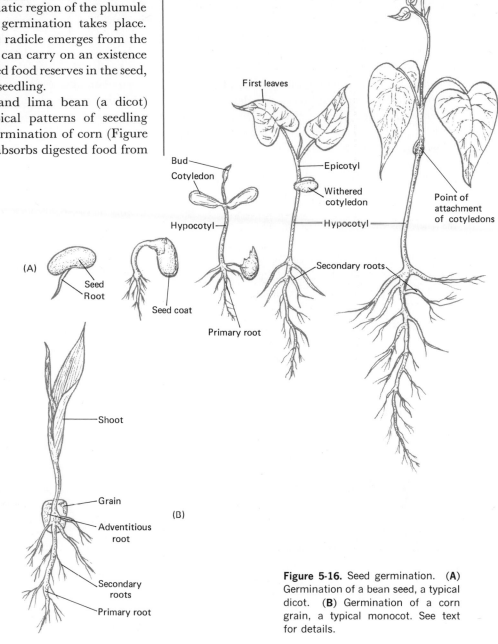

Figure 5-16. Seed germination. **(A)** Germination of a bean seed, a typical dicot. **(B)** Germination of a corn grain, a typical monocot. See text for details.

the endosperm and transfers it to the growing regions. The outside appearance of the radicle from the corn seed is quickly followed by the emergence of the plumule sheath or coleoptile. The hypocotyl does not elongate but remains with the cotyledon in the soil. Actually the radicle becomes the primary root and persists for a short period until the adventitious roots develop. The plumule elongates, passing through the coleoptile, then develops into a vegetative shoot which becomes the independent plant. By contrast, in lima bean germination (Figure 5-16B) the radicle emerges first and the hypocotyl ultimately grows above the soil. The early growth of the cells of the bean hypocotyl is unequal on the two sides of the axis resulting in the formation of a hypocotyl hook. This hook soon straightens and forces the cotyledons out of the seed coats and above the soil. At this time, the plumule undergoes rapid cell division and develops into the leafy shoot. When the food in the cotyledons has been used up, they die and fall off. The lowermost portion of the stem is the hypocotyl.

FRUITS

Development of the Fruit

The botanist describes a **fruit** as a structure composed of one or more ripened ovaries with or without seeds, together with any other accessory flower parts which may be associated with ovaries. Such a definition includes grains of wheat and corn, pea pods, tomatoes, cucumbers, chestnuts, and many other structures, which, in familiar usage, are not regarded as fruits. It is commonly thought that a structure must be edible to be considered a fruit, but the ovaries of a large number of flowers develop into fruits that cannot be eaten. The question of whether or not a plant food should be called a fruit or a vegetable may be decided on the basis of the structural origin. If the structure develops from a floral ovary, it is a fruit; if it represents some other plant structure, it is usually a vegetable.

In most species pollination is necessary, not only for fertilization and the development of the embryo plant and seeds, but also for the development of fruits. If the stigma is not pollinated and fertilization does not take place, the flower usually withers and drops from the plant. Failure to set fruit may also result from unfavorable environmental conditions such as late frosts in the spring. Pollination normally prevents the abscission of the flower and initiates the development of the tissues of the fruit itself. Evidence indicates that hormones are present in pollen grains and initiate the development of the ovary (see Chapter 11).

Without pollination, the egg cannot be fertilized. The development of fruits without pollination and fertilization is termed **parthenocarpy,** and such fruits are usually seedless. In some plants fertilization may occur, but the ovules fail to develop into mature seeds even though fruit develops normally. Natural parthenocarpy is found in certain kinds of citrus fruits, pineapples, bananas, cucumbers, and seedless grapes. Artificial parthenocarpy has been induced in some plants by treating them with dilute solutions of growth substances. Among the plants that have been induced to form seedless fruits by artifical parthenocarpy are watermelon, summer squash, cucumber, tomato, and holly.

Fruit Structure

The main structure of a fruit wall is the **pericarp,** which was formerly the wall of the immature ovary. The pericarp is differentiated into three morphologically distinct layers: the exocarp, mesocarp, and endocarp. The **exocarp** constitutes the outermost layer of cells and is sometimes

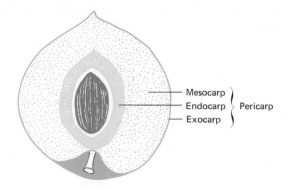

Figure 5-17. Diagrammatic representation of the pericarp showing its three subdivisions as they appear in a peach.

characterized by epidermal hairs and stomata. The **mesocarp** is the middle layer of tissue, which varies in thickness, and is often composed of parenchyma and vascular tissue. Of the three tissues, the internal **endocarp** is the most variable in structure, texture, and thickness. For example, in such fruits as the cherry and peach, the endocarp is the stony pit which encloses the seed (Figure 5-17).

Kinds of Fruits

Inasmuch as fruits exhibit enormous diversity it is difficult to group all of them into even the most detailed schemes of classification. Nevertheless, fruit classification is usually based upon the following criteria: (1) the structure of the flower from which the fruit develops, (2) the number of ovaries which comprise the fruit, (3) the number of carpels in each ovary, (4) the type of pericarp (fleshy or dry), (5) the extent to which the pericarp splits (dehisces) at maturity, (6) the nature of splitting if it occurs, and (7) the role of the sepals or receptacle in fruit formation. On the basis of these characters, fruits are usually classified into three main groups: simple, aggregate, and multiple. The following outline includes the more important or more familiar fruits in each of these groups (Figure 5-18).

I. **SIMPLE FRUITS.** Simple fruits are those derived from the ovary of one pistil, either simple or compound. Maturation of the pericarp is quite variable.

A. **DRY FRUITS.** In dry fruits as the pericarp ripens it becomes leathery, woody, or papery and is frequently hard when mature.

1. **INDEHISCENT FRUITS.** These fruits may consist of one or more carpels and do not split open when seeds are released.

a. **ACHENE.** This is a one-seeded fruit in which the seed is attached at only one place to the pericarp. Examples: buttercup, buckwheat, and sunflower.

b. **GRAIN** (caryopsis). These are also one-seeded fruits, but the seed is completely fused to the pericarp. Examples: corn, wheat, oats, barley, and other representatives of the grass family.

c. **SAMARA** (key fruit). These fruits are basically winged achenes in which the wings develop from the pericarp. Examples: maple, elm, and ash.

d. **NUT.** This is similar to an achene except that the pericarp becomes hard throughout. Examples: chestnut, acorn, walnut, and hickory.

2. **DEHISCENT FRUITS.** These fruits split open along one or several lines at maturity and the seeds may be dispersed. Typically more than one seed is present.

a. **LEGUME.** In this fruit the single carpel dehisces along two lines at maturity. Examples: Bean, pea, and most members of the pea family.

I. Simple fruits

  A. Dry fruits

    1. Indehiscent fruits

      a. Achene

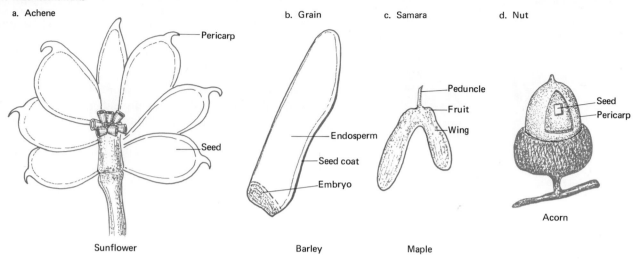

b. Grain   c. Samara   d. Nut

Pericarp

Peduncle
Fruit
Wing

Seed
Pericarp

Seed

Endosperm

Seed coat

Embryo

Sunflower   Barley   Maple   Acorn

    2. Dehiscent fruits

      a. Legume   b. Follicle

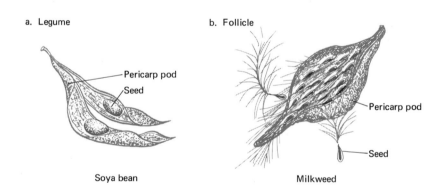

Pericarp pod
Seed

Pericarp pod

Seed

Soya bean   Milkweed

      c. Capsule   d. Silique

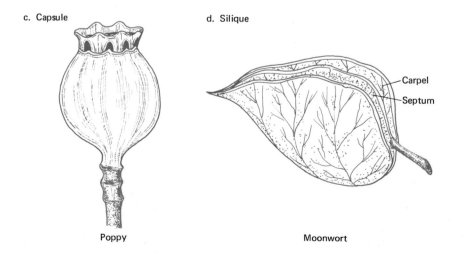

Carpel
Septum

154

Poppy   Moonwort

B. Fleshy fruits
  1. Derived from pericarp only
    a. Berry

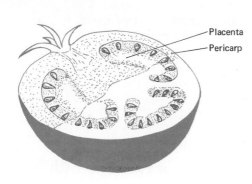

Placenta
Pericarp

Tomato

    b. Drupe

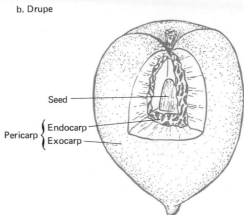

Seed

Pericarp { Endocarp
           Exocarp

Peach

  2. Derived from pericarp plus floral parts
    a. False berry

Scar left following abscission
of floral structures

Banana

    b. Pome

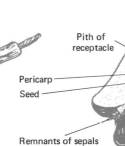

Pith of
receptacle

Pericarp
Seed

Remnants of sepals

Cortex of
receptacle

Apple

II. Aggregate fruits

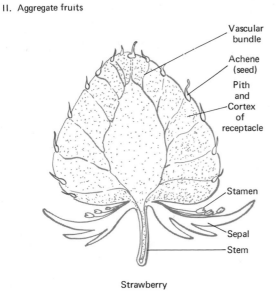

Vascular
bundle

Achene
(seed)

Pith
and
Cortex
of
receptacle

Stamen

Sepal

Stem

Strawberry

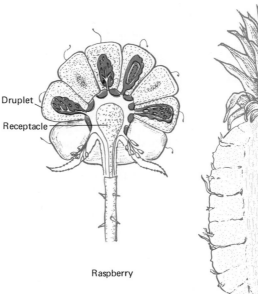

Druplet

Receptacle

Raspberry

III. Multiple fruit

Individual
fruit

Bract

Pineapple

b. FOLLICLE. A single carpel dehisces longitudinally along one line at maturity, Examples: milkweed, peony, and larkspur.

c. CAPSULE. Two or more carpels are present and dehiscence is variable. Examples: yucca, cotton, lily, orchid, and snapdragon.

d. SILIQUE. Two carpels separate at maturity, leaving a thin partition between. Examples: members of the mustard family.

B. FLESHY FRUITS. At least a part of the pericarp is soft and fleshy.

1. Fruits derived from the pericarp only.

a. BERRY. The entire pericarp is fleshy and often edible. Examples: tomato, grape, date, avocado, and citrus fruits.

b. DRUPE. The pericarp is divided into the exocarp (skin), mesocarp (fleshy pulp), and endocarp (stone or pit). Examples: peach, plum, cherry, olive, and apricot.

2. Fruits derived from the pericarp plus certain parts of the flower.

a. FALSE BERRY. Similar to the berry except that parts of the flower comprise the fruit. Examples: cucumber, squash, banana, cataloupe, and cranberry.

b. POME. The fleshy parts are all derived from the flower and enclose the portions produced by the pericarp. Examples: apple and pear.

II. AGGREGATE FRUITS. These fruits are formed by a single flower containing several separate pistils on an enlarged receptacle. In some aggregate fruits, each ovary develops into a small drupe. Common examples are raspberries and blackberries. In other aggregate fruits, each ovary develops into an achene and the receptacle develops into the fleshy part of the fruit. The strawberry is a typical example.

III. MULTIPLE FRUITS. These are formed from the flowers of an inflorescence and remain together as a single mass. Examples: pineapple, fig, and mulberry.

ECONOMIC IMPORTANCE OF FLOWERS, SEEDS AND FRUITS

Flowers

In addition to the essential role of providing the mechanism for sexual reproduction, flowers are universally admired for their beauty and their production has become a leading industry around the world. Thousands of species and varieties have been cultivated as ornamentals. A large number of flowers possess desirable fragrances which are used in making perfumes (rose, lavender, jasmine, lily of the valley, carnation, and many others).

Only a few flowers are used directly as food by man. For example, the globe artichoke (an unopened floral head) possesses certain edible portions, including the receptacle. Broccoli (a cluster of flower buds) and cauliflower (a head of abortive flowers) are popular "vegetables" in most parts of the world. Not so widely popular is the use of roses and other flowers in making jellies (rose petal jelly) and also to garnish salads and other special dishes. We do not often think of bee honey as a flower product, although it does come from the nectar of flowers. Honey is a wholesome and satisfying food. Cloves (unopened flower buds of the clove tree) provide a valuable spice and source of clove oil. The dried stigmas of the saffron crocus are the source of a yellow dye used to color and flavor foods and various medicinal preparations. Some insecticides have been developed from certain flowers. For example, pyrethrum powder is a useful insecticide obtained from a few species of chrysanthemum. The flowers of hops (*Humulus lupulis*) are especially valuable in the brewing industry. Numerous yellow glandular hairs of hops found on the bracteoles secrete *lupulin*, and to this the plants owe their economic value.

Seeds

Man and wild animals depend heavily on food stored in seeds. Perhaps the grass family (Gramineae) is the most important economically, because it includes such staple foods as wheat, corn, rice, barley, rye, oats, sorghum, millet, and buckwheat. The bean family (Leguminosae) ranks a close second because it includes such popular foods as peas, beans, lentils, soybeans, and peanuts. Seeds also make up a large portion of the food eaten by domestic animals from which we depend for milk, eggs, and meat. There are a few spices and condiments which are of seed origin, that is, coriander, anise, dill, caraway, nutmeg, and mustard. Seeds from coffee plants (*Coffea*), cocoa (*Theobroma cacao*), and cola (*Cola nitida*) are sources of popular beverages.

Fats and oils from a large variety of seeds are rapidly becoming among the most valuable commercial products in the world. Oils obtained from soybean, coconut, flax, tung, cotton, olive, peanut, castor bean, sesame, and oil palm are widely used in the food industry as well as in the manufacture of paints, varnishes, linoleum, printers' ink, soap, artificial leather, and insulating materials.

Fibers which surround cotton seeds still serve widely in the textile industry despite a number of synthetic fibers which have recently come on the market. Kapok seeds (*Ceiba pentandra*) yield fibers which are used in life preservers and as stuffing for mattresses and furniture.

The medical sciences employ a variety of plant seeds. For example, chaulmoogra oil, from the seed of this name, is the source of a medicine which is employed in the treatment of leprosy. Strychnine, curare, and castor oil are also much used as medicines.

Of course, it should not be overlooked that seeds are the chief method for the propagation of both food and ornamental plants.

Fruits

The food value of fruits to man and wildlife is well known and requires little commentary. However, man cannot eat all fruits and vegetables because many are indigestible and some are poisonous. A number of edible fruits are rich in nutritive value. For example, the cereal grains, plantains, and avocados possess a relatively high amount of proteins, fats, carbohydrates, vitamins, and minerals. Many wild birds and mammals depend almost entirely upon the fruits from a variety of angiosperms.

A few commercial products are also supplied by certain fruits. Dyes, oils, waxes, and drugs are some examples of economically valuable fruit products.

CHAPTER IN RETROSPECT

Reproduction is a highly specialized mechanism which provides plants with a means of species continuity and cell maintenance. The typical patterns by which plant organisms are continued from one generation to the next can be regarded as either asexual or sexual. The most common types of asexual reproduction, all of which do not involve seeds, include fission, budding, fragmentation, spore formation, and vegetative reproduction. In the process of sexual reproduction, the nuclei of specialized sex cells (gametes) fuse, making possible genetic recombinations. The cycle of sexual reproduction, unlike asexual reproduction, involves a unique type of nuclear division (meiosis) in which the chromosomes are reduced from the diploid ($2n$) to the haploid number (n). Thus, in the subsequent formation of male and female sex cells, each contributes one half the number of chromosomes, thereby providing the resulting zygote with the normal number of chromosomes for that species.

In flowering plants, the sexual process is carried out in six basic steps: (1) gamete production (eggs and pollen), (2) pollination, (3) fertilization, (4) fruit and seed production, (5) fruit and seed dispersal, and (6) seed germination. The flower is the site of various reproductive activities, such as the development of pollen and ovules and the occurrence of pollination and fertilization. Pollen grains are formed within the anther of a typical flower, whereas ovules are produced within an ovary. The process by which pollen grains are transferred to the stigma of the female reproductive organ is called pollination. Fertilization occurs when the nuclei of the sperm and egg, contained within the pollen grain and ovule, respectively, are united and a zygote is formed. The zygote later develops into a seed, which consists essentially of an immature plant surrounded by stored food and a protective coat. Once seeds have been dispersed, by whatever means, and germination occurs, the reproductive cycle of activities is terminated and a new generation is initiated.

The fruit of an angiosperm is composed of one or more ripened ovaries with or without seeds. Fruits are generally classified into three main groups: (1) simple, those derived from the ovary of one pistil (corn, walnut, bean, cranberry, apple, and cotton); (2) aggregate, those formed by a single flower containing several pistils (raspberry, blackberry, and strawberry); and (3) multiple, those formed by the flowers of an inflorescence (pineapple, fig, and mulberry). Flowers, fruits, and seeds rank high in the economic welfare of man. This is especially true in the case of seeds, which are represented by staple foods such as wheat,

corn, rice, barley, beans, soybeans, and peanuts. The various kinds of fruits eaten by man and other animals are also economically significant. In addition to their food value, nonvegetative plant structures provide the raw materials for various pharmaceutical and industrial products which include dyes, waxes, oils, ornaments, fibers, insulating materials, soap, castor oil, chaulmoogra oil, and strychnine.

Questions and Problems

1. Define the following terms:

asexual reproduction
sexual reproduction
fission
budding
fragmentation
vegetative propagation
micropyle
fertilization
germination
meiosis
homologous chromosomes
flower
fruit
seed
microsporangium
megasporangium
pollination
testa
parthenocarpy

2. Briefly describe the various means of artificial vegetative propagation. How do plants vegetatively propagate themselves by natural means?

3. Why do most organisms have even numbers of chromosomes? What problems would arise by an odd chromosomal number?

4. Using a diagram and appropriate labels, indicate the meiotic sequence of a hypothetical plant which possesses a diploid number of six. (Employ the following terms in your diagrammatic sketch: chromosomes, haploid, diploid, homologous chromosomes, chromatids, zygote, spore-mother cells, anthers, ovules, microspores, germination, pollen tube, sperms, megaspores, fertilization, and embryo plant.)

5. What are the similarities and differences between the processes of mitosis and meiosis? Where do these processes occur in multicellular plants?

6. Briefly describe the function of each structure found in a typical flower.

7. Contrast staminate and pistillate flowers. Explain the process by which these flowers reproduce sexually.
8. Compare the methods by which pollen and ovules are developed within flowers. Use a parallel word diagram which will show, side by side, the various structures involved in the production of these gametes.
9. What is the sequence of physiological events following pollination and resulting in the development of a new plant?
10. Describe the economic significance of fruits, flowers, and seeds. How do these compare with the vegetative structures of angiosperms?

Suggested Supplementary Readings

Biale, J. B., "The Ripening of Fruit," *Scientific American,* May 1954.

Eames, A. J., *Morphology of the Angiosperms.* New York: McGraw-Hill, Inc., 1961.

Esau, K., *Plant Anatomy.* New York: John Wiley & Sons, Inc., 1965.

Grant, V., "The Fertilization of Flowers," *Scientific American,* June 1951.

Johnson, M. A., "The Epiphyllous Flowers of *Turnea* and *Helwingia*," *Bulletin Torrey Botanical Club,* vol. 85, pp. 313–323, 1958.

Keller, D., "Germination," *Scientific American,* April 1959.

Pool. R., *Flowers and Flowering Plants.* New York: McGraw-Hill, Inc., 1941.

Rickett, H. W., "The Classification of Inflorescences," *Botanical Review,* vol. 10, pp. 187–321, 1944.

Torrey, J. G., *Development in Flowering Plants.* New York: The Macmillan Company, 1967.

Vasil, I. K., "Studies on Pollen Germination of Certain Cucurbitaceae," *American Journal of Botany,* vol. 47, pp. 239–247, 1960.

Wilson, C. L., and T. Just, "The Morphology of the Flower," *Botanical Review,* vol. 5, pp. 97–131, 1939.

Wodehouse, R. P., *Pollen Grains.* New York: McGraw-Hill, Inc., 1935.

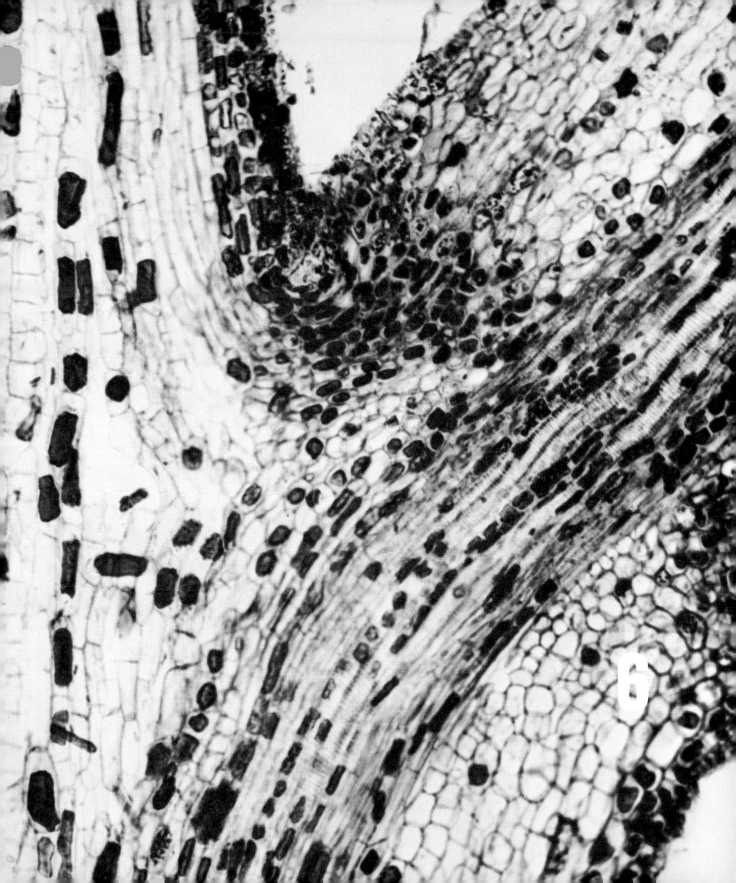

6
The Integrated Plant Body: Form and Function

INTRODUCTION

Thus far, emphasis has been placed on the structural organization of higher plants in terms of increasing levels of complexity. These levels are represented by atoms, molecules, cells, tissues, and organs. Each organizational level is prerequisite to the preceding one, and each new level features characteristic properties over and above those found at lower levels. For example, the cellular level of organization exhibits certain new structural characteristics not found at the molecular level, and the organ level represents certain structural departures from the tissue level.

The structural and functional integration of organs constitutes an **organism,** or complete living thing. It should not be thought, however, that all organisms consist of tissues and organs for a number of plants such as bacteria, some algae, and certain fungi consist of single cells, yet they, too, are complete organisms. All organisms, whether single-celled algae or complex flowering plants, share certain commonalities and distinguishing characteristics. In a very broad sense, these commonalities may be viewed as structural and functional characteristics, both of which contribute to the formation of a living, integrated plant. Structurally, all plants are composed of constituent atoms and molecules, and the degree of complexity of the plant is dependent upon the nature of the organizational level achieved. Functionally, all plants are related through various chemical processes collectively called **metabolism.**

THE NATURE OF PLANT METABOLISM

A living green plant cell is a complete self-sustaining chemical factory of incredible complexity. Within the cell wall boundaries are found thousands of substances, ranging from simple ions to exceedingly large and complex proteins. These compounds are continually producing hundreds of other substances that form a series of chemical reactions that keep the cell in operation. Even more amazing than the numbers of compounds and the multiplicity of chemical reactions is the orderly and systematic synchronization of these myriad reactions that maintain the organization and efficiency of the cell. In these reactions, some molecules supply, others transfer, and still others store energy: some molecules are broken down into simpler ones; others are built up into larger ones. In addition, cytoplasmic components wear out and are replaced while other molecules move about within the cell as well as into and out of the cell.

Phases of Plant Metabolism

The sum total of all these *biochemical reactions,* that is, reactions within living systems, is called **metabolism.** Some of these biochemical reactions are concerned with *synthesis,* or the buildup of molecules, in which case they are collectively termed **anabolism.** Food synthesis, protein synthesis, the formation of new cell walls, the replacement of worn-out cell parts, and the production of new cytoplasm are typical examples of the anabolic phase of metabolism. In all such synthetic reactions, energy is required and the ultimate source of this energy is the sun. By contrast, **catabolism** is a collective designation for those biochemical reactions that are concerned with the degradation, or breakdown, of molecules. Representative catabolic reactions are digestion, aerobic respiration, anaerobic respiration, and fermentation. In general, catabolic reactions are energy-yielding reactions; that is, they provide energy for various cellular activities, including anabolism. In addition to synthesis and degradation, metabolism also involves nutrition. All three aspects of metabolism are in operation simultaneously, although, for convenience of study, they will be analyzed separately. The dynamic nature of metabolism is substantiated by the fact that approximately 90 per cent of the cellular molecules of an actively growing organism are degraded and resynthesized during its lifetime.

Nutrition. **Nutrition** may be defined as that aspect of metabolism in which the raw materials for anabolism and catabolism are supplied. The raw materials, or *metabolites,* utilized by plants are obtained from the physical environment. Chief among the metabolites are water, inorganic salts, and gases such as carbon dioxide, oxygen, and nitrogen. The water is obtained from the soil by root hairs and the epidermal cells of the root tip; the gases are taken in from the atmosphere, primarily by the leaves.

The overall process of nutrition implies at least two fundamental activities: the absorption of metabolites and their subsequent distribution to sites of metabolic reactions. **Absorption** is the movement of raw materials from the environment into the plant and the process is facilitated primarily by roots. In general, metabolites that enter plants are dissolved in a water solution. In this regard, then, the main physiological activity of roots is the absorption of water containing essential minerals.

In the soil, root hairs are in direct contact with the small particles of which soil is composed.

Surrounding these soil particles is a thin film of water from which root hairs absorb moisture. Even in relatively dry soil types, such as sand, there is always a film of water covering individual particles of soil. Water passes with comparative ease through the cell walls and plasma membranes of the root hairs and epidermal cells of the root tip in accordance with the laws of certain physical phenomena.

Once raw materials are absorbed by roots, they must be **translocated,** or distributed, to other parts of the plant. In this context, a number of factors are in operation to conduct materials through plants. Only a relatively small portion of the water absorbed by the roots is retained in the plant. The greater part evaporates into the atmosphere, chiefly from the leaves. This process by which water is lost by an aerial portion of a plant is called **transpiration.** Because the leaf blade provides a broad surface area permeated with stomata, it is inevitable that transpiration will occur.

The absorption of soil water by the roots and transpiration from the leaves produces a current, termed *transpiration pull,* which passes through the axis of the plant. The liquid in this current is transported in the xylem and throughout the plant. Other physical and chemical factors are also involved in the translocation process.

Types of Plant Nutrition. According to the pattern by which nutrition is derived, plants are subdivided into two principal classes: autotrophic plants and heterotrophic plants. **Autotrophs** are self-nourishing plants that are capable of synthesizing all of their organic molecules from inorganic raw materials using energy provided by the environment. Inasmuch as the simplest organic compounds synthesized are sugars, which contain carbon, hydrogen, and oxygen, all autotrophs require an inorganic source for these elements. In addition, synthetic reactions require energy, so some source of primary energy must be supplied.

To satisfy these presynthetic conditions, all autotrophs utilize environmental carbon dioxide as a source of carbon and oxygen, and most derive hydrogen from water. The source of primary energy is solar radiation. Once a sugar is synthesized by an autotroph, it may be converted into other carbohydrates, lipids, proteins, or any other organic molecules needed by the cell. This unique ability of plants to synthesize organic molecules out of inorganic materials makes them the link between the inorganic and organic worlds.

The source of primary energy required by autotrophs for synthesis further divides them into two groups: photosynthetic autotrophs and chemosynthetic autotrophs. *Photosynthetic autotrophs* use solar radiation as a primary energy source, and all such plants contain chlorophyll. Thus all green plants are photosynthetic autotrophs and are capable of carrying on photosynthesis. All higher green plants, as well as the lower vascular plants, bryophytes, and algae, are photosynthesizers.

Chemosynthetic autotrophs, by contrast, obtain the energy for carbohydrate synthesis from a variety of inorganic chemicals in the environment. These autotrophs consist entirely of certain types of bacteria, none of which contain chlorophyll. Among the substances from which energy is derived by chemosynthetic autotrophs are inorganic compounds containing hydrogen, iron, sulfur, and nitrogen. In most cases, these compounds combine with oxygen in the cells, resulting in the production of energy. Depending upon which inorganic substances these bacteria oxidize as a source of energy, they are referred to as *hydrogen bacteria, iron bacteria, sulfur bacteria,* and *nitrogen bacteria,* respectively.

Heterotrophs also obtain inorganic materials from the environment but lack the ability to synthesize them into organic molecules. In view of this inability, they must take in organic molecules previously synthesized by autotrophs, and there-fore they may be considered as nutritionally dependent plants. Most bacteria and fungi, as well as slime molds, colorless algae, and some higher plants, exhibit heterotrophic characteristics. As a result, they subsist on other living plants or on non-living organic material of all kinds.

Synthesis: The Manufacture of Plant Materials. Many of the raw materials provided through the agency of nutrition are used by the plant for various types of synthetic reactions. Of these reactions, by far the most important single chemical process known to man is **photosynthesis.** Photosynthesis, a term that literally means putting together (*synthesis*) in the presence of light (*photo*), is an extremely complex series of anabolic reactions in which chlorophyll-bearing plants manufacture sugars. The necessary metabolites needed for photosynthesis are carbon dioxide and water. These inorganic raw materials are united through the chemical machinery of the green plant cell and sugars are produced as products of the reactions. The energy for photosynthesis is derived from sunlight and the most significant aspect of the process is that solar energy is trapped by chlorophyll and converted to chemical energy in the bonds of the organic molecules (foods) synthesized. Once light energy has been absorbed and converted into chemical energy, it may be used to perpetuate all other anabolic reactions.

Plants synthesize various organic molecules to maintain their existence. The principal compound manufactured through photosynthesis is glucose. Other carbohydrates, such as starch, cellulose, and sucrose, are made directly out of glucose by a series of biochemical reactions, or some of the glucose may be converted to fatty acids and glycerol from which fats are synthesized. Still other reactions in which glucose is degraded and minerals such as nitrogen, phosphorus, and sulfur are incorporated will produce amino acids. The amino acids are, in turn, joined together via peptide bonds to form

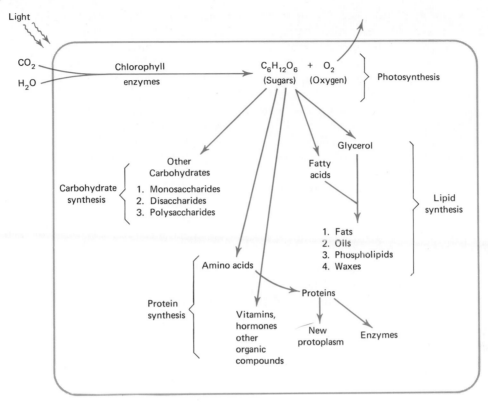

Figure 6-1. Summary of a few anabolic reactions of plant metabolism.

proteins. Further synthetic reactions may produce enzymes (functional proteins), vitamins, hormones, other complex organic compounds, or protoplasm itself. Figure 6-1 summarizes some of the important synthetic activities of plant cells.

Degradation: The Breakdown of Plant Materials. Just as complex organic molecules are continually being synthesized, many are also being broken down to provide energy and to participate in resynthesizing reactions. Such catabolic or degradation reactions may be categorized into two general types: digestion and respiration.

Before some organic molecules can take part in energy-yielding reactions, they must first be prepared for reaction. For example, storage products in plants, such as starch, fats, and proteins, which have been synthesized from the products of photosynthesis, can neither be used in cells for energy production nor as molecules for various synthetic reactions until they are broken down and prepared. This breakdown of complex organic molecules into soluble, diffusible substances is called **digestion.** Because digestion is a low-energy producing process, its primary function is the preparation of molecules for subsequent metabolic reactions.

Digestion occurs in plants primarily in areas where foods are stored. After digestion the products may be translocated to other parts of the plant, especially to regions of active growth (*meristems*). Here they are used in the production of new protoplasm, cells, and tissues, or they may be used to supply energy. Foods digested in one part of the plant are translocated to different regions and resynthesized again in new storage areas.

Some of the products of photosynthesis, especially simple sugars, contain large amounts of stored chemical energy and require no digestion prior to participating in catabolic reactions. As such, these molecules are capable of immediate degradation with the subsequent release of great quantities of energy for cellular use. The process by which organic molecules are degraded to furnish cellular energy is called **respiration** (biological oxidation). Many plants, including all the higher plants, require oxygen for respiration. This type of respiration in which oxygen is utilized is called *aerobic respiration,* and it represents the principal energy-supplying process of most plants. Other plants, however, do not require oxygen for respiration. In fact, in the presence of oxygen these plants may die. This alternative variation on respiration in which oxygen is not utilized is called *fermentation.* It is a common process carried on by many lower plants such as yeasts and bacteria, although in times of oxygen deficiency, the process may also be employed by higher plants. In either case, respiration makes available the potential chemical energy stored in the food materials elaborated by the plant through synthetic reactions. Figure 6-2 summarizes some aspects of the catabolic phase of metabolism.

Factors Affecting Plant Metabolism

The synchronization and methodical functioning of all plant metabolic activities is obviously not the result of random chance. Rather, it is the result of an orderly internal system of controls operating under the influence of various internal and external environmental factors. Ultimately, all phases of plant metabolism are controlled by DNA, a nuclear constituent. However, inasmuch as most metabolic reactions occur in the cytoplasm, the plant cell must have some mechanism for mediating the "messages" for metabolic control between DNA and specialized cytoplasmic organelles. The chemical mediator is RNA. Once the RNA receives a message from DNA, a second group of compounds, called **enzymes,** assumes the function of controlling metabolic reactions. Enzymes are compounds that are protein in nature and their role is to catalyze or accelerate the myriad metabolic reactions of the cell.

Both DNA and enzymes determine what types of metabolic reactions plant cells are capable of performing, but the extent to which reactions are actually performed is dependent upon the influences of various internal and external environmental conditions. For example, enzymes operate inside cells most efficiently within certain ranges of pH and temperature, and any significant deviations from normal operational ranges not only affects enzyme activity but also metabolic efficiency. Among the external factors that might alter metabolic rates are the availability of metabolites and moisture, light, and temperature.

GROWTH AND DEVELOPMENT: Extensions of Metabolic Activities

All interrelated aspects of metabolism contribute to two general activities of plants: (1) maintenance, and (2) growth and development. With respect to maintenance, various metabolic activities are concerned with synthesis and degradation reactions in which materials are constantly being built up and broken down. As a result of these reactions, new protoplasm is made, foods are

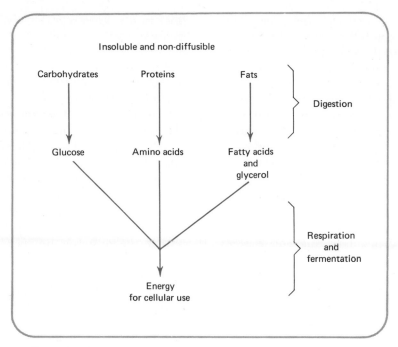

Insoluble and non-diffusible

Carbohydrates   Proteins   Fats

Digestion

Glucose   Amino acids   Fatty acids and glycerol

Respiration and fermentation

Energy for cellular use

Figure 6-2. Summary of a few catabolic reactions of plant metabolism.

stored, various molecules are synthesized, and energy is trapped, stored, and made available for cellular use. However, not all metabolic reactions are directed toward maintenance. Some reactions, especially synthetic ones, are directed toward plant growth and development.

The phenomena of growth and development are the results of directed metabolic activities and systems of controls operating within the confines of various influencing factors. Growth, in a very general sense, is an increase in protoplasmic constituents culminating in cell division. Once daughter cells are formed, developmental forces become operational. Development, consisting of cell enlargement and differentiation, involves progressive changes in form and complexity of undifferentiated cells. Essentially, growth is a quantitative concept, whereas development is qualitative in nature. Both growth and development are easily exemplified by considering a meristematic cell in the stem of a plant. During the very early stages of growth, cells in the undeveloped stem are basically similar in structure and function and are concerned with maintenance through metabolic activities. As the plant matures, some of the stem cells remain meristematic, but most undergo certain developmental changes. Through division, enlargement, and differentiation, meristematic cells are progressively changed into xylem, phloem, cortex, or pith cells. In each case, structural and functional modifications are brought about by various developmental changes.

Whereas the internal system of metabolic controls consists primarily of enzymes, growth and development are controlled principally by substances called **growth regulators.** Common examples of growth regulators are groups of compounds referred to as auxins, gibberellins, cytokinins, and vitamins; the first three groups are referred to as plant hormones. In general, these growth substances are produced in one part of the plant in minute quantities and are translocated to other parts, where they elicit certain growth and developmental responses. Among the plant activities under the control of growth substances are cell division, enlargement, and differentiation; leaf and fruit abscission; fruit development; root growth; stem elongation; flowering; and tropisms, or plant movements.

All internal and external environmental factors that influence metabolic activities concerned with maintenance must also have an influence on growth and development. In addition, the hereditary potentialities (genes) also play a prominent role in determining the patterns of plant growth and development.

REPRODUCTION:
The Continuity of Metabolism

Although the patterns of reproduction vary considerably among different species of plants, they do share one overriding commonality. This is the continuity or propagation of metabolism. All normally functioning metabolic cells are capable of reproduction, but prior to reproduction, cells must pass through stages of growth and development.

All plant reproduction may be viewed as cellular or species propagation. Cellular reproduction or mitosis in a multicellular plant is concerned primarily with various activities of the individual organism exclusive of other members of the species. For example, mitosis might be carried out

prior to the death of a cell, owing to an adverse environmental influence; or to replace worn-out, diseased, or damaged cells; or to form new cells and tissues as part of the growth process. In this regard, mitosis is directed toward maintenance of the individual plant.

In addition to cell division within individual plants, reproduction may also involve propagation of the species by the formation of new plants from existing parent plants. This phenomenon is called **heredity.** In species continuity, specialized cells are provided by parent plants, which, upon subsequent union, will form another generation of plants. The salient feature of species continuity is that the offspring produced are genetically similar to the parent plants and are thus provided with the potential for performing similar metabolic activities, and exhibiting similar growth and developmental patterns and reproductive mechanisms. In other words, the offspring inherit metabolic, growth and developmental, and reproductive traits from parent plants.

In Chapters 7 through 12, detailed discussions of plant metabolism, growth and developmental phenomena, and heredity will be presented. In all these discussions the importance of form and its relationship to function will be emphasized.

CHAPTER IN RETROSPECT

Plants may be viewed as sharing common characteristics in terms of both structural and functional attributes. At the structural level, all plants are composed of constituent atoms, molecules, and cells. Functionally, plants are related through numerous chemical processes collectively called *metabolism*. The raw materials for all plant metabolism (water, inorganic salts, and so on) are supplied by a process known as *nutrition*. However, not all plants make their own food; some are *heterotrophs* and are nutritionally dependent upon other living plants or animals (parasites) or on nonliving organic matter (saprophites). All green plants are autotrophs and can synthesize organic molecules from inorganic materials. Synthesized organic molecules are broken down (degraded) for further metabolic reactions (*digestion*), including biological oxidation (respiration), which is the chief energy-supplying process of most plants. All metabolic activities are directed toward the maintenance of plants as well as their growth and development. A number of complex organic substances, such as enzymes, auxins, gibberellins, cytokinins, vitamins, and genes, act together as an internal system of metabolic regulators. The transfer of metabolic activities from one cell to another (mitosis) or to genetically similar plants (heredity) can be regarded as the continuity or extension of metabolism.

Questions and Problems

1. Briefly describe the ways in which molecules may react during metabolic activities. Define metabolism.
2. What is the relationship between roots and plant nutrition?
3. Compare the nutritional pattern of autotrophic plants and heterotrophic plants.
4. Why is photosynthesis regarded as the most important chemical process known to man? How is photosynthesis related to the production of fatty acids, glycerol, and proteins?
5. Contrast digestion and respiration with regard to their similarities and differences.
6. What is the role of DNA and RNA in plant metabolism?
7. Describe the mechanism by which meristematic cells are converted to xylem, phloem, cortex, and other specialized cells. What is the role of auxins in this process?
8. What essential differences exist between cellular and species propagation?

Suggested Supplementary Readings

Biddulph, Susan, and O. Biddulph, "The Circulatory System of Plants," *Scientific American,* February 1959.

Bonner, J. T., "Volvox: A Colony of Cells," *Scientific American,* May 1950.

Cheldelin, V. H., and R. W. Newburgh, *The Chemistry of Some Life Processes.* New York: Reinhold Publishing Corporation, 1964.

Fischberg, M., and A. W. Blacker, "How Cells Specialize," *Scientific American,* September 1961.

Moscona, A. A., "How Cells Associate," *Scientific American,* September 1961.

7
The Raw Materials for Metabolism

INTRODUCTION

All phases of plant metabolism are dependent upon a constant supply of chemicals provided by the physical environment. The term *nutrient,* or metabolite, is used to designate any chemical substance that is utilized by plants as a raw material for metabolism. The environment not only supplies these raw materials but is also the external medium in which plants carry on metabolic activities. Plants function in an environment characterized by daily and seasonal fluctuations of light, temperature, precipitation, and mineral availability. Consequently, the procurement, fate, and utilization of metabolites provided by the environment are also affected by these daily and seasonal variations as well as other conditions.

It should not be thought that the movement of metabolites into the plant is in one direction only and that the plant maintains itself at the expense of the environment. In fact, the interaction of the plant with its environment is cyclic in nature. While the plant is living, some raw materials are returned to the environment as waste

Figure 7-1. Interaction of plants with the environment. Living plants procure metabolites from the physical environment for growth and development and return them as waste products of metabolic activities. Decay organisms eventually return all metabolites to the environment upon the death of an entire plant or plant part.

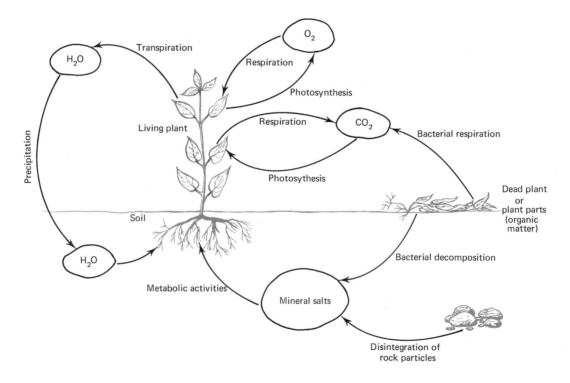

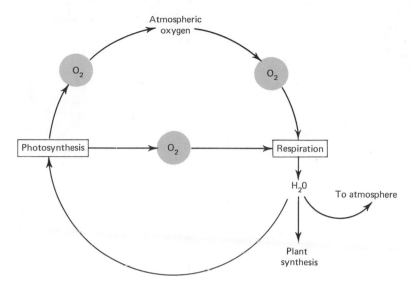

Figure 7-2. The oxygen cycle. Atmospheric oxygen, utilized by plants and animals during respiration, is returned to the atmosphere through the photosynthetic activities of green plants and certain chemosynthetic bacteria.

products of various metabolic reactions. When the plant dies, decomposition is brought about through the action of microorganisms such as bacteria and fungi, and all nutrients are eventually returned to the environment. The cyclic pattern of metabolite procurement and return is shown in Figure 7-1.

SOURCES FOR METABOLITES

The physical environment in which plants procure nutrients and perform metabolic processes is made up of three distinguishable, although interrelated, subdivisions. These are the atmosphere, the hydrosphere, and the geosphere.

The Atmosphere

The *atmosphere,* or gaseous envelope around the earth, consists principally of nitrogen, oxygen, carbon dioxide, and water vapor, and minute quantities of inert gases. Excluding the inert gases, all other atmospheric constituents serve as metabolites. These atmospheric nutrients dissolve in water, so they are readily accessible to water plants as well as land plants.

Oxygen Cycle. The *oxygen cycle* concerns two basic processes—respiration and photosynthesis (Figure 7-2). Land plants take in oxygen from the air, whereas aquatic plants utilize oxygen that is dissolved in water for the process of respiration. During respiration, molecular oxygen combines with hydrogen and forms water. This water becomes part of all the other water present in a plant cell and may serve three possible functions. Some of it is returned directly to the environment, where it is incorporated into the water cycle. Another portion may be used for synthesis of larger molecules, because it represents a source of hydrogen

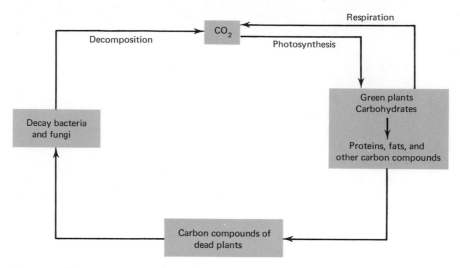

Figure 7-3. The carbon cycle. Carbon, obtained from the atmosphere as carbon dioxide, is used by green plants for the synthesis of organic compounds. It is also returned to the atmosphere as carbon dioxide, a product of respiration.

and oxygen atoms. Finally, some water may be used as a raw material for photosynthesis. In this process, water molecules are broken down into hydrogen and oxygen atoms, the hydrogen being used in food synthesis. The oxygen may be used again in respiration or it may be released to the atmosphere, completing the cycle.

Carbon Cycle. Carbon dioxide is the exclusive source of carbon for organic synthesis, and, together with water, is one of the two major sources of oxygen comprising the *carbon cycle.* Chlorophyll-containing autotrophs absorb CO_2 as a raw material for photosynthesis and incorporate it into organic compounds. Some of the synthesized organic nutrients are used in assimilation and others are used as energy-producing molecules in respiration. During respiration, CO_2 is given off as a by-product and may be used again in photosynthesis or returned to the atmosphere (Figure 7-3).

Nitrogen Cycle. Nitrogen is needed by plants for the synthesis of proteins as well as other nitro-gen-containing compounds. Atmospheric nitrogen (N_2), although in great abundance, exists in a form that is not usable by most plants. It must be combined with other elements, such as nitrates (NO_3^-) or ammonium salts (NH_4^+) before it becomes available to green plants. The most common usable nitrogen source is the *nitrate* ion, NO_3^-. The chemical and physical forces operating in the soil, together with the activities of certain bacteria, are important factors in rendering this nitrogen usable.

Roots of green plants absorb nitrates from the soil. In synthesizing proteins, the plant utilizes the nitrogen from nitrates and adds it to the carbon, hydrogen, and oxygen which have been organized during photosynthesis. Sulfur and phosphorus, which are constituents of some proteins, may be added from other soil minerals as sulfates (SO_4^{--}) and phosphates (PO_4^{3-}). Eventually the organic nitrogen compounds are returned to the soil by either plants or animals. For example, plants may die after a season or two, they may shed their leaves, or they may be eaten by animals. The animals, in turn, being heterotrophs, use the plant protein to assimilate their cytoplasm, and either during excretion or at death they also return the nitrogen to the soil. Plants, in association with man and other animals, compose a cycle in which nitrogen is continually being converted from one form to another. These conversions and close associations constitute the basis for the *nitrogen cycle* (Figure 7-4).

When an organism dies, the process of decay or decomposition results in the breakdown of its fairly complex components into a number of simple compounds which are returned to the environment. The agents of decay are certain bacteria and fungi, and these organisms are associated with the various stages in the complex

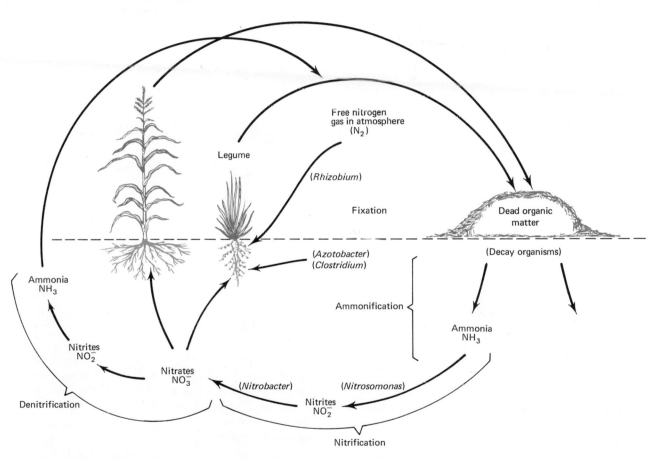

Figure 7-4. The nitrogen cycle. Through the continual circulation and transformations of nitrogen in the atmosphere and the soil by microorganisms, nitrogen is converted into a form that plants can utilize for the synthesis of proteins and other nitrogen-containing compounds. See text for details of the cycle.

process of decomposition. The end result of decay is the formation of ammonia (NH_3) from organic nitrogenous compounds and the process is called *ammonification.*

$$dead\ or\ waste\ products \xrightarrow{decay} \underset{ammonia}{NH_3}$$

Living in the soil are two genera of bacteria, *Nitrosomonas* and *Nitrobacter,* that are capable of oxidizing ammonia. In the oxidation process, the energy liberated is used by these bacteria to synthesize various compounds just as the photosynthetic autotrophs utilize the energy from sunlight. The oxidation of ammonia, or *nitrification,* involves two steps: (1) ammonia is converted to nitrites (NO_2^-), and (2) nitrites are changed to nitrates (NO_3^-).

$$\underset{ammonia}{NH_3}\ +\ \underset{oxygen}{O_2}\ \xrightarrow{Nitrosomonas}\ \underset{nitrite}{NO_2^-}\ +\ energy$$

$$\underset{nitrite}{NO_2^-}\ +\ \underset{oxygen}{O_2}\ \xrightarrow{Nitrobacter}\ \underset{nitrate}{NO_3^-}\ +\ energy$$

At various points in the cycle, atmospheric nitrogen is either added or removed. The loss of nitrogen from the cycle is called *denitrification,* and the addition of nitrogen is termed *nitrogen-fixation.* During denitrification many species of bacteria are capable of converting NO_3^- into N_2 by the following sequence of reactions:

$$\underset{nitrate}{NO_3^-} \longrightarrow \underset{nitrite}{NO_2^-} \longrightarrow \underset{ammonia}{NH_3} \longrightarrow \underset{\substack{atmospheric \\ nitrogen}}{N_2}$$

In this reaction, *denitrifying bacteria* return free or molecular nitrogen to the atmosphere and deplete nitrates from the soil. From the standpoint of soil fertility, this is an unfavorable process. However, denitrifying bacteria are anaerobic; that is, they do not function well in the presence of oxygen. Denitrification is most active in soils that are poorly aerated (oxygenated).

Nitrogen fixation involves the change of free atmospheric nitrogen into nitrates.

$$\underset{\substack{atmospheric \\ nitrogen}}{N_2} \longrightarrow \underset{nitrates}{NO_3^-}$$

Certain types of bacteria and blue-green algae are unique in their ability to use N_2 as a source of nitrogen for the synthesis of organic compounds. Essentially, there are two groups of *nitrogen-fixing* bacteria, distinguishable on the basis of where they

Figure 7-5. Nodules containing nitrogen-fixing bacteria on the roots of a soybean plant. (Courtesy of Nitragin Company, Inc.)

live. Some live in the roots of legumes (beans, peas, and clovers) and are called *symbiotic bacteria,* and others live free in the soil with no direct plant association and are termed *nonsymbiotic bacteria.*

The most common symbiotic genus is *Rhizobium.* These bacteria live in swellings (*nodules*) on the roots of legumes (Figure 7-5). **Symbiosis** is a nutritional relationship in which two organisms live together and both benefit. In this case, the bacteria supply nitrogen as NH_3 to the leguminous plants, while the plants supply the bacteria with an adequate supply of carbohydrates as a source of energy to produce NH_3. The free-living, nonsymbiotic bacteria, mainly *Azotobacter* and *Clostridium,* also supply the soil with considerable amounts of NH_3, which is converted by other bacteria to NO_3^-.

The Hydrosphere

The *hydrosphere,* or liquid portion of the physical environment, consists of oceans, rivers, lakes, streams, ponds, and other bodies of water on land. Water is the most abundant compound of all living matter, as well as the major constituent of the earth (73 per cent). As the principal metabolite required by plants, it is the exclusive source of hydrogen and one of several sources of oxygen. Like other inorganic materials provided by the atmosphere, water also follows a cyclic path.

Through evaporation, water is removed from the hydrosphere and geosphere and is deposited in the atmosphere. As the water cools and condenses, it is returned to the hydrosphere and land in the form of precipitation. Aquatic plants derive water principally from the hydrosphere, whereas terrestrial plants obtain it from the soil. Land plants transport water through roots, stems, and leaves, retaining what is needed for metabolic activities and returning the remainder through evaporation (transpiration) in the leaves. After death, any liquid in the bodies of plants returns to the hydrosphere as part of decomposition reactions (Figure 7-6).

The Geosphere

The *geosphere,* or solid component of the environment consists primarily of soil, the solid medium in which land plants live. The soil not only provides anchorage and support for plants, it also furnishes water (soil water) and contributes the minerals essential for maintenance. In this regard, the physical condition of the soil and its chemical nature and properties are important considerations with respect to plant growth and development.

Soils are relatively complex mixtures of rocks, minerals, water, air, living organisms, and products of their decay in which numerous physical and chemical changes are constantly occurring. A *soil profile,* or vertical cross section of soil, shows three general layers (Figure 7-7). The uppermost layer, called the *topsoil* is the most important as far as plants are concerned, because plants are in direct contact with this layer and it is the principal source of essential materials needed for growth. In the middle layer, the soil material consists principally of rock and represents the parent material from which the upper layers are eventually formed.

Soil Constituents. Topsoils, as well as soil profiles in general, exhibit considerable variations in physical texture, chemical constituents, origin, depth, and fertility. Despite these differences, the distinct components of the soil mixture may be separated into the following categories: (1) rock particles and mineral matter, (2) soil water, (3) soil air, (4) organic matter, and (5) organisms, both plants and animals.

Rock Particles and Mineral Matter. The principal portion of soil consists of a mixture of rock fragments and particles which have been formed primarily by the weathering of preexisting rock.

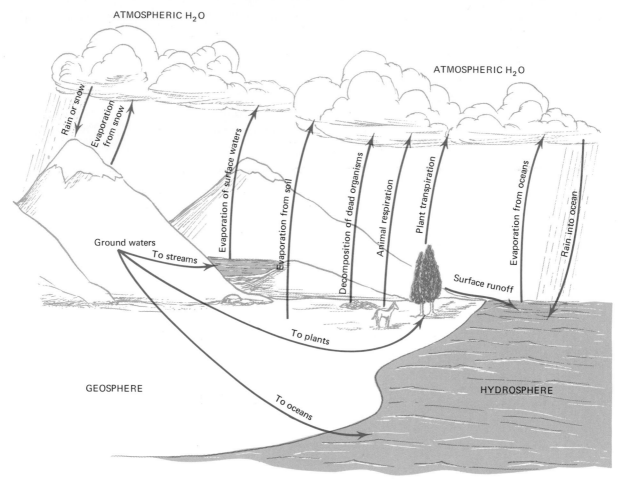

ATMOSPHERIC H₂O

ATMOSPHERIC H₂O

Rain or snow

Evaporation from snow

Evaporation of surface waters

Evaporation from soil

Decomposition of dead organisms

Animal respiration

Plant transpiration

Evaporation from oceans

Rain into ocean

Ground waters

To streams

Surface runoff

To plants

GEOSPHERE

To oceans

HYDROSPHERE

Figure 7-6. The water cycle. Water, the principal metabolite required by plants, is returned to the atmosphere by evaporation from bodies of water and the soil, by transpiration of plants, and as a product of decomposition reactions. It is returned to the geosphere and hydrosphere as various forms of precipitation.

Various weather phenomena such as precipitation, temperature fluctuations, and wind are the agents of mechanical breakdown of rock. Chief among the rock materials that are subjected to weathering to form soil particles are shale, feldspar, sandstone, limestone, and granite. The nature and size of the mineral particles thus formed are directly related to the degree of weathering, the chemical composition of the parent rock, and the resistance of the rock to breakdown. The sizes of the mineral particles in a soil provide a basis for distinguishing various soil textures. The *texture* of a soil (coarse, medium, or fine) is a relative measure of the size of the most abundant mineral particles in a given

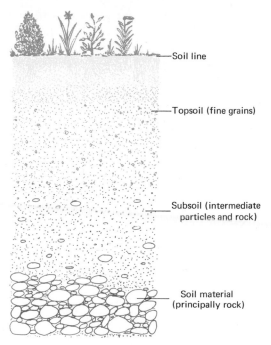

Soil line

Topsoil (fine grains)

Subsoil (intermediate particles and rock)

Soil material (principally rock)

Figure 7-7. Soil profile. The topsoil in which plants grow is formed by the chemical and physical breakdown of parent rock located in the lowest layer. The intermediate layer contains some parent rock and soil particles and is used to a lesser extent by plants as a growth medium.

soil mixture. Usually the sizes of these particles are expressed as a range in diameter per millimeter (Table 7-1).

In general, soils do not consist exclusively of sand or silt or clay but contain proportions of each

TABLE 7-1. Soil Texture Classification Based upon Size of the Dominant Mineral Particles

Particle	Range in Diameter (mm)
Coarse sand	2.0–0.2
Fine sand	0.2–0.02
Silt	0.02–0.002
Clay	Smaller than 0.002 (colloidal dimensions)

of these constituents in varying amounts. Soils are named on the basis of the percentage of sand, silt, or clay they contain. A *sandy soil* is a mixture in which the silt and clay particles comprise less than 20 per cent of the material by weight. A *clay soil* is one which contains at least 30 per cent clay particles. A *loam soil* contains about 50 per cent sand and 50 per cent silt. Clay is the most important fraction in determining the characteristics of soils.

Although the particles which comprise various soil types are cemented together, spaces called *pore spaces* are present. The pores, which make up between 30 and 60 per cent of the soil volume, contain air and water. The larger the particles, the larger the pore spaces between them. Accordingly, sandy soils have large pore spaces which permit excellent penetration of water and air. Such penetration in clay soils is somewhat more difficult because of the extremely small pore spaces that exist.

The ability of soil particles to retain water is called the *water-holding capacity*. In sandy soils, aeration is good and water is absorbed rapidly, but retention is poor and water drains away quickly. Clay soils, by contrast, have greater water-holding capacity than sandy soils, even though aeration and water absorption are less adequate. The reason for this is that most clays are of colloidal size, and this results in a tremendously large surface volume ratio. Once water has penetrated, a clay soil may hold three to six times as much water as a sandy soil. The addition of organic matter increases the water-holding capacity of soils because it contains relatively small pores and because it, like clay, is colloidal in nature.

Soil Water. Not all the water found in the soil is readily available to plants. Much of the water added to the soil by precipitation or irrigation seeps deeply into the subsoil. A smaller percentage, which fills the pore spaces and forms a film around the soil particles, is usable. The importance of soil water to plants is that it may be utilized as a metabolite and it is the medium in which all entering substances must be dissolved. Various inorganic salts such as nitrates, phosphates, and sulfates and other water-soluble substances are the principal materials found dissolved in soil water.

The soil solution, generally very dilute, contains a large number of soluble inorganic substances which are present as ions. Potassium nitrate (KNO_3) is found as K^+ and NO_3^- ions, calcium phosphate [$Ca_3(PO_4)_2$] exists as Ca^{++} and PO_4^{3-} ions, and magnesium sulfate ($MgSO_4$) exists as Mg^{++} and SO_4^{--} ions. These ions of salts furnish plants with the necessary mineral elements for growth and development.

Just as water is held to the surface of soil particles to form a film, so also are mineral ions held to these surfaces. This condition is possible because soil particles act as negatively charged ions and are able to attract and hold positively charged mineral ions such as K^+, Ca^{++}, and Mg^{++}. Clay soils and organic matter are especially effective in adsorbing positive ions because of their large surface areas, a characteristic of their colloidal nature. This is an important property of soils, because fine soil particles act as a reservoir of mineral nutrients and prevent the leaching of minerals out of the topsoil with the water that drains through. The finer the texture of the soil, the greater the fertility. Sandy soils in constant need of irrigation require a continual application of fertilizers because of the leaching of essential minerals. Negatively charged ions such as nitrate (NO_3^-) and sulfate (SO_4^{--}) are found as part of the soil solution. They are not adsorbed to the surfaces of soil particles.

The storehouse of mineral elements found in the soil is utilized by plants for a variety of metabolic activities. At least 15 of these elements have been found to be *essential* for plant growth and development. Other elements have been shown to be essential for the normal growth of certain plants, but evidence of their essential requirement for growth in the majority of plants is lacking.

The essential mineral elements may be subdivided into **macronutrient elements** and **micronutrient (trace) elements.** Macronutrients are required by plants in large amounts, and if a plant is deprived of one of these elements, deficiency symptoms develop within a few days. The macronutrient elements are carbon (C), hydrogen (H), oxygen (O), nitrogen (N), phosphorus, (P), potassium (K), calcium (Ca), sulfur (S), magnesium (Mg), and iron (Fe). The trace elements, by contrast, are required by plants in minute amounts. These include manganese (Mn), copper (Cu), zinc (Zn), boron (B), and molybdenum (Mo).

Table 7-2 lists the essential mineral elements, the forms in which they are utilized by plants, their roles, and some deficiency symptoms. Although carbon, hydrogen, and oxygen are not included in the table, they comprise approximately 94 per cent of the whole plant and are found as constituents of all cellular organic molecules.

Soil Air. For plants to grow, an adequate supply of *soil air* is essential. This air, found in pore spaces, contains the same gases as the air present above the soil. In general, however, the relative proportions of soil air gases may differ from those of the atmosphere. For example, the average percentage by volume of atmospheric carbon dioxide (0.03 per cent) is lower than that of the soil air, which may go as high as 5 per cent. This disproportionate condition exists because roots, together with the plant and animal life of the soil, produce carbon dioxide during respiration. Inasmuch as roots and soil organisms require oxygen for respiration, the oxygen concentration of the soil decreases while the carbon dioxide supply increases. The soil

TABLE 7-2. Mineral Elements Required by Higher Plants

Essential Element	Form Utilized by Plants	Function	Deficiency Symptoms
Macronutrients			
Nitrogen (N)	NO_3^- (nitrate) NH_4^+ (ammonium)	Structural component of proteins, nucleic acids, vitamins, hormones, coenzymes	Chlorosis (yellowing of leaves), purple coloring of leaf petioles and veins
Phosphorus (P)	$H_2PO_4^-$ (phosphate) HPO_4^{--}	Constituents of ATP, nucleic acids, phospholipids, coenzymes	Necrotic (dead) areas on leaves, petioles, fruits; dark to blue-green leaf coloration; stunted growth; anthocyanin formation
Calcium (Ca)	Ca^{++}	Constituent of cell walls; activator for several enzymes; influences membrane permeability; forms buffers	Disintegration of meristems; distorted leaves, especially hooking of tips; rigid cell walls; abnormal ion uptake
Magnesium (Mg)	Mg^{++}	Component of chlorophyll; cofactor for many enzymes	Chlorosis; anthocyanin formation; necrotic spotting
Potassium (K)	K^+	Enzyme, amino acid, and protein synthesis; membrane integrity	Chlorosis; necrotic areas at leaf tips; downward bending of leaves; stunted growth
Sulfur (S)	SO_4^{--} (sulfate)	Constituent of some amino acids, vitamins, and coenzymes	Similar to nitrogen deficiency symptoms
Iron (Fe)	Fe^{++} (ferrous) Fe^{3+} (ferric)	Constituent of hydrogen transport enzymes; chlorophyll synthesis; cytochromes	Chlorosis
Micronutrients			
Manganese (Mn)	Mn^{++}	Cofactor of respiratory enzymes; photosynthetic enzymes; nitrogen metabolism	Chlorosis; necrosis
Copper (Cu)	Cu^{++}	Component of many oxidation enzymes	Withering of tips of young leaves
Zinc (Zn)	Zn^{++}	Needed for synthesis of tryptophan and IAA; component of several amino acids	Chlorosis; white spotting of leaves; distorted leaves
Boron (B)	BO_3^{3-} (borate) B_4O^- (tetraborate)	Influences Ca^{++} uptake and utilization; differentiation; pollination; may assume a role in sugar translocation	Death of shoot tip; disintegration of internal tissues; copper-textured leaves
Molybdenum (Mo)	MoO_4^{--} $HMoO_4^-$ (molybdate)	Nitrogen metabolism	Chlorosis; marginal necrosis; infolding of leaves

air, however, is never saturated with carbon dioxide or depleted of oxygen, because carbon dioxide diffuses out of the soil and oxygen diffuses in. For most plants both the amount and movement of soil air depends not only upon adequate pore spaces but also upon the water content of the soil. Some plants, however, such as willows, several species of ash, bald cypress trees, rice, and mangroves, flourish when their root systems are submerged in unaerated water or water-soaked soil.

Organic Matter. All organic matter in the soil is derived from the decayed remains of plants and animals or the waste products of living organisms by the activities of various microorganisms such as bacteria and fungi. The organic matter contained, which consists of carbohydrates, proteins, lipids, lignins, mucins, and other materials, is ultimately reduced to simpler inorganic compounds such as ammonia (NH_3), water (H_2O), carbon dioxide (CO_2), nitrites (NO_2^-), nitrates (NO_3^-), sulfur compounds, phosphates (PO_4^{3-}), calcium compounds, and others. The greater portion of the organic matter of soils is of plant origin, principally from dead roots, decaying wood and bark, and abscised leaves. The breakdown of these organic substances through the action of soil organisms produces and maintains a continuous supply of inorganic substances which plants require for growth. In most agriculturally important soils, organic matter comprises about 2 to 10 per cent. Swamps and bogs, by contrast, contain a higher content of organic matter, reaching a proportion of 95 per cent in some peat bog soils.

A considerable part of the organic matter in soils occurs as *humus,* a colloidal complex. This dark-colored material is composed chiefly of organic materials which are more resistant to decay, such as cellulose and lignin. In this regard, humus is partially decomposed organic matter.

The addition of organic matter, either completely or partially decomposed, is essential to continued soil fertility. Most organic materials in soils are able to absorb and hold water to a greater degree than even clay particles. In addition, because of their spongy nature, organic materials loosen the soil to prevent the formation of heavy crusts and increase the proportion of pore spaces in the soil. A considerable part of organic materials occur as colloids, so they bear negative charges and thus attract positively charged ions on their surfaces, as do clay particles.

Soil Organisms. The soil contains not only innumerable bacteria and fungi which are important in decay but also many animals, ranging from microscopic forms to insects, millipedes, centipedes, spiders, slugs, snails, earthworms, roundworms, mice, moles, gophers, and reptiles. Most of these are beneficial in some way or another in that they promote some mechanical movement of soil, thus keeping the soil loose and open. In addition, all soil organisms contribute to the organic matter of soils as a result of their waste products and decomposition.

Without the presence of microorganisms, especially bacteria, the soil would soon become unfit to support plant life. Bacteria influence soils in many ways. Some decompose organic matter into simple products such as water, carbon dioxide, hydrogen sulfide, ammonia, sulfates, and phosphates. In these reactions, nutrients are made available for plant life. Other soil bacteria are associated with transformations of nitrogen and its compounds, so that an available supply of nitrogen is continually maintained.

PRINCIPLES GOVERNING THE ABSORPTION OF METABOLITES

In the performance of their metabolic activities, plants are continually taking in metabolites and allowing others to pass out. The process of intake through which metabolites of all kinds pass from the environment into plant cells is called **absorption.** The term environment in this context means both the external (physical) environment, in which materials enter the outermost plant cells, and the internal (biological) environment, in which the outermost cells pass metabolites to underlying cells. These nutrient materials ordinarily enter the plant through the root in water solution, although other parts of the plant may also serve as absorptive centers.

Diffusion

Most of the movement of dissolved inorganic and organic materials into and out of plant cells occurs by a process called diffusion. **Diffusion** may be defined as the movement of molecules, or ions, from a region of higher concentration to a region of lower concentration. Consider the following example. If a large crystal of copper sulfate (blue) is placed in the bottom of a beaker filled with water, it will be noted that initially a dark blue color is evidenced around the crystal. As the distance from the crystal increases, the color becomes less and less intense (Figure 7-8). As diffusion of molecules and ions of the crystal progresses, it will be seen, in a few days, that the entire water solution becomes uniform in color. This is due to the fact that the dissolved crystal particles have diffused and become evenly distributed among the water molecules. Similarly, water molecules have migrated from more concentrated to less concentrated regions. When the water and crystal particles have become evenly distributed, an *equilibrium* is attained and diffusion ceases, although molecular movements still continue. In this regard, diffusion may be viewed as a tendency toward reaching an equilibrium, because ultimately it results in the equal distribution of molecules or ions within a given space.

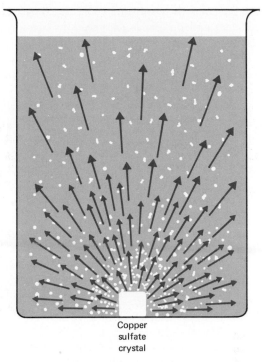

Copper
sulfate
crystal

Figure 7-8. Diffusion. Copper sulfate molecules and ions move from a more concentrated area to a less concentrated region. Similarly, water molecules move from more concentrated to less concentrated regions. Diffusion ceases at equilibrium when the crystal and water particles become evenly dispersed.

In the example just cited, diffusion was considered in terms of a solid (copper sulfate) being the diffusible material. The dissolved solid is the solute and the substance in which the solute is dissolved (water) is the solvent. Living plants also contain systems in which liquids such as water and gases diffuse. For example, diffusion occurs between gases in the atmosphere and those in leaf cells. During photosynthesis oxygen molecules diffuse out of mesophyll cells and exit from the leaf through the stomata, while molecules of carbon dioxide enter the stomata and diffuse into mesophyll cells.

As particles of a substance diffuse, they exert a pressure called *diffusion pressure*. This pressure is caused by the high molecular activity of the diffusing molecules or ions and is proportional to the concentration, or number of diffusing particles. In the copper sulfate system, the diffusion pressure is greater near the dissolving crystal, because the concentration of copper sulfate particles is greater at that point. Similarly, the diffusion pressure is least at a point farthest from the dissolving crystal, because there the concentration of particles is lowest. The behavior of diffusing molecules may be described in terms of diffusion pressures by stating that they diffuse from areas of higher diffusion pressure to areas of lower diffusion pressure.

The relative concentrations of a substance determine both the direction and the rate of diffusion. The direction of diffusion of copper sulfate is from the area of higher concentration to the area of lower concentration. Other diffusible substances migrate in a similar manner. In addition, the greater the difference in the relative amounts of a substance, the more rapid is the rate of diffusion. Temperature and pressure also affect the rate and direction of diffusion. The diffusion pressure of a substance increases with a rise in temperature because of the increased velocity of the diffusing molecules. Any pressure imposed upon the molecules of a diffusing substance will increase the diffusion rate in proportion to the amount of pressure applied.

The direction of diffusion of one substance is independent of the movement of other substances. If a cube of sugar is added to the beaker containing copper sulfate, the molecules or ions of each of these substances will diffuse until they are equally distributed in the solvent. The rate of diffusion may decrease slightly because of collisions between the copper sulfate and sugar particles, but the direction of movement depends upon factors concerning each type of particle separately. In living plants, as certain molecules or ions are diffusing into cells others are simultaneously diffusing out at varying rates.

Membranes and Diffusion

An analysis of the structure of plant cells reveals the presence of various membranes, each of which differs greatly in the ease with which gases, solutes, and solvents can pass through. From the standpoint of diffusion, membranes are of three types: permeable, impermeable, and differentially permeable. *Permeable membranes* permit the diffusion of all types of substances. The walls of cork cells produced by the cork cambium form *impermeable membranes* and prevent the diffusion of all molecules. Other cellular membranes, such as those delimiting the cytoplasm, vacuoles, nucleus, and mitochondria, are *differentially permeable*. They will permit the diffusion of some substances to the selective exclusion of others. A membrane of this sort allows water to pass through it but prevents the diffusion of certain solutes.

The plasma membrane of a living cell is differentially permeable. This type of membrane is extremely important in absorption of water and dissolved substances because it governs the exchange of materials diffusing into and out of the cell. The differentially permeable nature of plasma membranes is not fixed but may vary under certain conditions. The reason membranes show differential permeability is not completely understood; probably this characteristic is related to the physical and chemical properties of the membrane. These properties include the structure of the membrane, especially the orientation of lipid and protein molecules; the size and character of the diffusing substance; and the electrical charges on the membrane.

Most certainly, membranes decrease the rate of diffusion but not the direction, which is a function

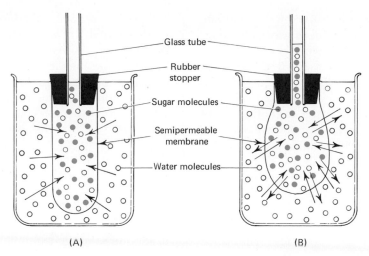

Figure 7-9. Demonstration of osmosis in an osmometer. **(A)** Osmometer at start of experiment. **(B)** Osmometer at equilibrium. Apparatus A consists of an internal sugar solution enclosed within a semipermeable membrane and immersed in a water solution. Arrows indicate the direction of movement of water molecules, which pass freely across the membrane whereas the sugar molecules cannot. As more water molecules move into the sac than leave it, the sugar solution is diluted and the volume of liquid in the sac increases. This flow of water is demonstrated in the sac on the right, where the increasing volume forces the sugar solution into the glass tube. The final height attained by the column of solution is directly related to the concentration of the sugar solution at the start of the experiment. When the final height is reached, an osmotic equilibrium is established and the same number of water molecules passes across the membrane in each direction.

of diffusion pressure on either side of the membrane. The general order of penetration of materials remains fairly constant, but individual rates may vary. For example, substance *A* may diffuse more rapidly than substance *B*, but under certain conditions, such as an increase in temperature, both substances may diffuse more rapidly. Despite the temperature increase, *A* will penetrate more rapidly than *B*. Thus the rate of diffusion is governed by the nature of the membrane as well as by all the other factors that influence diffusion.

Osmosis

The diffusion process of special importance to living plant cells is the diffusion of water and dissolved substances into and between cells. The term **osmosis** refers to the diffusion of water through a differentially permeable membrane from a region of higher diffusion pressure to a region of lower diffusion pressure of water.

Osmosis can be demonstrated quite readily by reference to an apparatus called an *osmometer* (Figure 7-9). The apparatus may consist of a sac made of parchment, cellophane, dialysis tubing, or other differentially permeable material filled with a concentrated sugar solution. For ease of observation the sugar solution may be colored with a dye such as Congo Red. The top portion of the sac is stoppered with a rubber stopper through which a glass tube is fitted, and the entire apparatus is placed in a distilled-water solution.

The relative concentrations of water are different on either side of the membrane. There is a lower concentration of water inside the sac than in the beaker. As a result of these unequal concentrations, there is a tendency toward equalization of the concentrations of water on both sides of the membrane. The sugar molecules cannot diffuse to reduce the concentration difference because of the presence of a differentially permeable membrane, so water must move from the beaker into the sac. Actually water diffuses into and out of the sac, but the net inward movement is much greater. The greatest difference in water concentration exists at the moment the sac is submerged in the beaker containing pure water. As a result, it is at this time that the rate of water movement into the sac is highest and the rate of water movement out of the sac is lowest. As the volume of water in the sac increases, the sugar solution becomes more and more diluted and moves up the thistle tube. After a period of time the rate of water moving into the sac slows down. As the water builds up in the sac and the tube, the weight of the column of solution begins to exert a pressure. This pressure continues to increase until an equilibrium is reached in which the pressure exerted by this water forces water out of the sac at the same rate as water molecules enter from the beaker. The maximum amount of pressure which can be developed in a solution separated from pure water by a differentially permeable membrane is called the **osmotic pressure** of the solution. Inasmuch as living plant cells in their environment are not immersed in pure water, the theoretical maximum pressure never develops. In this regard, the term osmotic pressure simply refers to a potential pressure, if a cell were placed under prescribed conditions.

A number of analogies exist between the osmometer just described and a living plant cell. The cell cytoplasm, vacuoles, and organelles are enclosed by differentially permeable membranes comparable to the membrane forming the sac. The membrane around the sac, however, is nonliving, with unchanging properties. The living membranes of a cell continually alter their permeabilities. The cell sap within vacuoles consists of a water solution containing various dissolved materials such as sugars and salts. In this instance, the tonoplast is impermeable to many materials dissolved in the cell sap, just as the membrane forming the sac is impermeable to sugar. The concentration of dissolved substances in the cell sap, however, as well as the concentration of dissolved materials in the soil solution, continually fluctuates.

Turgor Pressure

The actual pressure that develops as a result of osmosis and pushes the cell membrane against the cell wall is called **turgor pressure** (Figure 7-10A). This pressure is always less than osmotic pressure unless the cell is in pure water under laboratory conditions. The cellulose wall of a plant cell, being rigid, exerts an equal and opposite pressure called *wall pressure*. As a result of the equal and opposite forces of turgor pressure and wall pressure, the cell is said to be *turgid* (firm or stiff). Organs whose cells are free of turgor undergo a partial collapse, as externally exhibited by *wilting*. Turgor, as in the case of cut flowers, may be regained if the wilted plants are supplied with water. Turgor is extremely important for the normal growth and development of plants, as it is the basis of mechanical support in most flowers, leaves, and nonwoody stems. It also provides the force for cell elongation. One of the first signs of a water deficit in a plant is the loss of turgor by its leaf cells, which results in a wilting of the leaves.

When the turgor pressure is less than the osmotic pressure, water will tend to enter with a force equal to the difference between the two. The

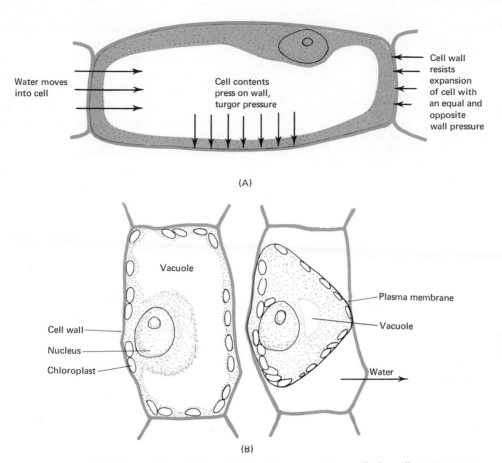

(A)

(B)

Figure 7-10. Turgidity and plasmolysis. **(A)** As water moves into a cell, the cell contents are pushed against the wall by a pressure called turgor pressure. The wall resists this pressure with an equal and opposite force called wall pressure. Once these equal and opposite forces are established, the turgidity (rigidity) of the cell is maintained. **(B)** If a cell is placed in a solution containing more solute than cell sap, water diffuses out of the cell and the cytoplasm shrinks away from the cell wall. *Left:* Before plasmolysis. *Right:* After plasmolysis.

entrance of any water into a living cell will compress the contents of the vacuole (cell sap) and raise the turgor pressure of the vacuolar contents to the value of the osmotic pressure, and the net transfer of water inward will stop. It can be seen that the net tendency of water to enter a cell depends on the difference between the osmotic pressure and the actual turgor pressure. This difference is known as the **diffusion pressure deficit (DPD),** and the relation may be summarized as follows:

$$DPD = OP - TP$$

Plasmolysis

If cells are placed in pure water, there is a net movement of water into cells whose contents consist of water diluted by certain solvents. If, instead of water, a highly concentrated sugar solution is placed around the cell, water will move out of the cell rather than in. As water diffuses out of the cell, the cytoplasm contracts and shrinks away from the cell wall. The shrinkage of the cytoplasm due to a water loss to an external solution with a high solute concentration is called **plasmolysis** (Figure 7-10B). Although continued plasmolysis results in the death of the cell, if a plasmolyzed cell is placed in water soon enough it will normally regain turgidity.

Imbibition

Still another process is available to the plant for the uptake of water. This process, a special type of diffusion, is known as imbibition. **Imbibition** may be defined as the swelling of an apparently solid, pore-free substance due to the uptake of a liquid (usually water) or vapor. The cell wall, although apparently solid when viewed through a light microscope, is actually quite porous, with spaces between the interwoven, cellulose microfibrils. Many of these spaces contain pectic substances and are filled with water, forming a continuous pathway through which solute particles can freely diffuse.

It appears that imbibition is related to the ability of certain colloidal particles, such as cellulose, pectin, and cytoplasmic proteins, to *adsorb* water molecules, that is, to attract and hold water molecules to their surfaces. In the process colloidal particles move apart, swell, and allow the passage of water and dissolved substances. One of the best examples of imbibition is the swelling of a seed or dry fruit when placed in water or in moist soil.

A dry piece of sponge or a piece of gelatin soaking up water also demonstrates imbibition.

Root hairs contain the necessary colloids to adsorb water and thus are important in the uptake of water into plants. In fact, practically all parts of a plant, except waterproofed layers, are capable of imbibition. This ability to hold water by adsorption is related to the capacity of the plant to withstand drying (desert plants) as well as freezing (evergreens). Imbibition also assumes a role in the transfer and distribution of water between cells.

The Absorption of Solutes

The movement of many substances into and out of plant cells in most cases is due simply to differences in diffusion pressure. Such movement of materials into a cell as a result of diffusion only is called *passive absorption*. In passive absorption, the molecules or ions of a dissolved substance diffuse from the region of greater concentration to a region of lesser concentration. The continuation of this type of absorption depends upon the maintenance of a concentration difference (gradient) of the particles of the diffusing substance. In other words, passive absorption is operational only when the diffusing material is present in a greater concentration outside the cell than inside. As substances enter cells, they become incorporated into various chemical compounds or they may be passed on to other cells. As a cell uses a certain substance, the cellular concentration of that material decreases and more of it diffuses inward as long as the external concentration is greater than the internal concentration.

Passive absorption, a result of simple diffusion, is not the only method by which materials enter cells. There is evidence that dissolved substances, especially mineral ions, continue to diffuse into plant cells, even though there is a greater concentration of them within the cells than outside. Such a movement of materials is directly opposite what would occur by diffusion alone, because the movement is from a region of lesser concentration to a region of greater concentration. The ability of plant cells to accumulate ions in vacuoles in concentrations higher than those of the surrounding medium was first recognized in certain types of algae. In the freshwater alga *Nitella*, potassium ions were found to be 1000 times more concentrated inside the cells than in the surrounding medium. Similarly, iodine may accumulate in kelp, at least 100 times more concentrated than in the sea. Such accumulation of ions also occurs commonly in root cells of higher plants.

The accumulation of ions against a concentration gradient is called *active transport,* or *active absorption.* Such a movement of materials requires an expenditure of energy by the absorbing cells. Many data suggest that this energy comes from respiration, specifically energy supplied by ATP.

Studies of ion intake and accumulation have been carried out largely with root systems of higher plants. In these studies it has been demonstrated that the rate of respiration in root cells is accompanied by an increase in ion accumulation and that if certain chemicals are used to block respiration, the ion intake by active absorption is stopped. The way in which energy from ATP is utilized is not completely understood, but various cellular membranes and enzymes probably assume an active role. It is believed that carriers, which are enzyme-like molecules located in or on cell membranes, combine with an ion and transport it from one side of the membrane to the other, thus bringing it into the cell. In addition, it is likely that the attraction of colloids in the cell wall and cytoplasm for certain ions also aids active absorption.

XYLEM TRANSPORT OF WATER AND DISSOLVED SUBSTANCES

The Root and Absorption

Before a plant can utilize metabolites provided by the environment, they must be absorbed and then distributed to sites of metabolic activities. The most important metabolites furnished by the environment consist of water, gases, and minerals in the form of ions of their salts. Among the processes that facilitate the absorption of materials are osmosis, passive absorption (simple diffusion), active absorption, and imbibition.

Roots of higher plants are the principal organs of absorption. They are in direct contact with the subsoil, a medium which provides a means of anchorage and a reservoir of water and mineral nutrients. Extensive root–soil contact is accomplished not only by the numerous branchings of the root but also by the presence of root hairs. These epidermal projections enormously increase the area of root–soil contact, and there is much evidence that root hairs provide a major portion of the actual absorbing area of the root system.

Water and solutes move independently of each other through the walls of the root hairs. Water absorption is probably accomplished by osmosis. The total concentration of dissolved materials is normally less in the soil solution than in the cell sap (water and dissolved materials in the vacuoles), so water will pass from the soil through the plasma membrane of the root hair and into the cell sap. This causes a dilution of the cell sap in the root hair. As a result, water passes from the root hair cell, by osmosis, and is absorbed by the cortex cell immediately adjacent to the root hair. This process of absorption continues as water is absorbed by successively deeper layers of cells. After being absorbed by the layers of cortical cells,

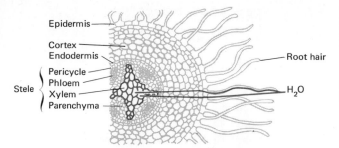

Epidermis
Cortex
Endodermis
Stele { Pericycle, Phloem, Xylem, Parenchyma
Root hair
H_2O

Figure 7-11. Pathway of the uptake of water and metabolites from the soil through a root. Most entering materials are absorbed by the root hairs, although some may be taken in by epidermal cells. The processes of osmosis, passive absorption (simple diffusion), and active absorption facilitate the transfer of materials from the soil solution into the xylem for distribution throughout the plant body.

it is moved through the cells of the endodermis, the pericycle, and into the xylem vessels (Figure 7-11).

In the soil solution are dissolved a variety of mineral salts. If the concentration of a particular mineral ion is greater in the soil solution than in the sap solution, it will diffuse into the root hair if the selectively permeable membranes permit its passage. Certain ions can enter the cells of the root even though there is a greater concentration of them inside the cells. This mechanism of ion accumulation against a concentration gradient is active transport, in which cellular energy is expended to transfer ions across plasma membranes.

Conduction

Once water and mineral substances are absorbed by cells, internal distribution is accomplished mainly by diffusion and cyclosis (Figure 7-12). Both processes enable every part of the cell to receive adequate supplies of all metabolites that the cell absorbs at specific points on its surface. Diffusion not only transports materials within cells but may also occur between plasmodesmata of adjacent cells or through the permeable cell walls. Inasmuch as diffusion is a relatively slow transport process, the movement of materials may be appreciably increased by protoplasmic streaming. The streaming of cytoplasm has been measured at rates of a few to several centimeters per hour. In considering a single cell, this rate of movement

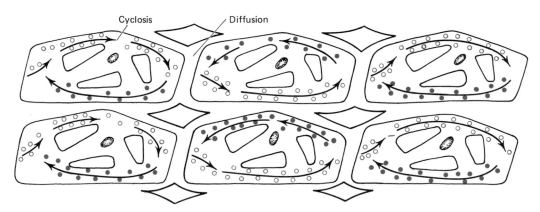

Cyclosis   Diffusion

Figure 7-12. Internal distribution of metabolites by cells. Materials gain entrance into cells by diffusion through the walls or the plasmodesmata. Once inside, they circulate in the streaming cytoplasm and diffuse along concentration gradients.

is fairly rapid, because a substance may travel the length of a cell in seconds. Lateral conduction of materials across roots, or other plant organs, probably involves the combined mechanisms of cyclosis and diffusion through vascular rays or other cell to cell transfers. The upward conduction of sap absorbed by roots, however, must take into account other processes than diffusion and cyclosis because, in considering the entire plant, both are relatively slow processes.

It has been known for many years that materials absorbed by roots move upward in the plant through the tracheids and vessels of the xylem. The mechanisms involved, however, are not completely understood. Most of the water absorbed by roots of plants is returned to the atmosphere through their leaves. The distances separating roots and leaves are quite commonly of considerable magnitude. Many trees of large forests are over 200 feet high, and the tops of the tallest trees may exceed 400 feet. Therefore, any explanation of the rise of sap in xylem tissue must take into account the forces required to move materials along such distances.

Two principal explanations offered to identify the forces capable of raising and conducting water in the xylem are (1) root-pressure forces, in which a pushing force is involved, and (2) the cohesion–tension theory, in which a pulling force is involved.

Root Pressure. The absorption of water by root cells creates a hydrostatic pressure in the root system known as *root pressure.* The manifestations of this pressure can be readily observed. It can be seen in some plants whose stems have been cut off a short distance above ground level, such as in the stump of a recently cut tree or a freshly cut herbaceous plant. Sap will continue to flow or *bleed* from the surface of the cut for some time. If a glass tube is attached to the cut stump, the sap can be seen to rise in it against the force of

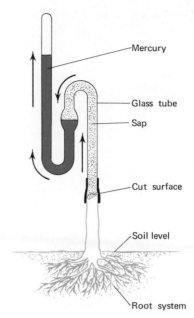

Figure 7-13. Measurement of root pressure. Sap exudes from the cut surface of a stem because of root pressure. The force of this pressure may be measured by placing a curved glass tube over the surface of the cut. The force of the root pressure causes the sap to rise in the glass tube, which in turn causes the level of the mercury column to become elevated. Root pressure is determined by calculating the distance through which the column of mercury is elevated. (From M. S. Gardiner and S. C. Flemister, *The Principles of General Biology,* New York: The Macmillan Company, © 1967.)

gravity (Figure 7-13). Root pressure can also be observed when conditions are excellent for water absorption by roots, but the humidity is so high that minimal amounts of water are lost to the atmosphere through evaporation. Given these conditions, root pressure may force water out of the veins of leaves, forming droplets along the leaf margins (Figure 7-14). This process is referred to as **guttation.** Both processes, bleeding and guttation, demonstrate the existence of root pressure.

Root pressure has been suggested as an impor-

Figure 7-14. Guttation in a strawberry (*Fragaria*) leaf. Under conditions of maximum absorption and minimum evaporation, root pressure may develop to such an extent that water is forced out of the leaf veins, forming droplets along the margin of the lamina. (Courtesy of Dr. J. Arthur Herrick, Kent State University.)

tant pushing force in the ascent of sap through xylem, but a number of serious objections make this explanation inadequate. First, not all plants bleed when they are detopped, and bleeding is negligible in summer when the evaporation rate is highest. Second, the root pressures which forces the sap upward have been measured, and in only rare instances have these pressures exceeded 1 to 2 atmospheres (1 *atmosphere* at sea level is equal to a force of 14.7 pounds per square inch). These low values suggest that root pressure is not the principal force involved in the ascent of sap because a pressure of 20 atmospheres is needed to lift sap from the roots to the tops of trees of ordinary heights. Third, the rates at which exuda-

tion (bleeding) occur are generally much slower than normal evaporation (transpiration) rates. Moreover, water movement through stems continues even though roots have been removed. This can be demonstrated by the failure of cut stems or flowers to wilt when placed in adequate amounts of water.

Cohesion–Tension Theory. This theory offers the most adequate explanation for the ascent of sap in plants. As contrasted to the root-pressure hypothesis, the cohesion–tension theory is based upon a pulling force rather than a pushing force. This theory assumes that water lost by transpiration through the leaves is pulled upward and that the pull is maintained by the attraction of water molecules for each other and to the walls of the xylem.

Transpiration is the loss of water vapor from the internal tissues of living plants. Although such water loss may occur from any exposed part of the plant, most of it actually takes place through the leaves. Less than 10 per cent of the total amount of transpiration from leaves occurs directly through the cuticle of the epidermis (*cuticular transpiration*). The greatest loss of water takes place through the stomata (*stomatal transpiration*).

The rate at which transpiration takes place varies greatly with changes in the plant itself, leaf structure, and various environmental conditions. As a result of these factors, which alter transpiration, any figures of isolated transpiration rates have little significance. Nevertheless, the quantity of water transpired by plants is very great. Of the total quantity absorbed by roots, as much as 98 per cent of it is lost by transpiration.

Under natural conditions, the air of the atmosphere usually contains water vapor in a concentration of 1 to 3 per cent (Figure 7-15). This concentration is considerably lower than the water concentration of living cells in the leaf. The parenchyma cells of the spongy mesophyll in leaves as well as the air spaces which permeate them are in contact with the atmosphere through the stomata. Water which has been imbibed by the cell walls of these cells evaporates and saturates the air in the air spaces. The diffusion (vapor) pressure of water is then higher than in the outside atmosphere, even if the relative humidity is high. Water-vapor molecules diffuse from the air spaces through the stomata into the atmosphere. Air currents, a dry breeze, or wind may facilitate the movement of this water vapor from the immediate vicinity of the plant. This water loss results in a slight drying of the air in the intercellular spaces. Such a water deficit is made up by the diffusion of water molecules from the wet cell walls into the air spaces. As water leaves the walls this deficit is compensated for by the diffusion of water from the cell sap in the vacuoles and the cytoplasm. Water enters the water-deficient cells of the mesophyll by osmosis from neighboring cells. Ultimately, the water losses resulting from transpiration are made up by absorption from the xylem. Essentially, a diffusion gradient exists from the air spaces, through the mesophyll cells, to the water-filled xylem elements. In this manner, all the mesophyll cells are supplied with water from the xylem as it is lost through transpiration.

The removal of water from the top of the plant pulls a column of water upward. This column represents a continuous path between the water on the evaporating surfaces of the leaf mesophyll cells, the water in the xylem of the leaf, and the xylem running from the top of the plant to the root. If this continuity of water from the leaf cells to the root is broken in any way, such as by the entrance of air bubbles, that particular xylem system could not facilitate the ascent of sap. Water molecules being lost by transpiration must pull other water molecules behind them in such a manner that there is no break in the water column. In this regard, it is assumed that there is some force in operation that holds water molecules together and that the tensile strength of the water is sufficient to form an unbroken column. The intermolecular force between water molecules is called *cohesion*. It is this force that affords the water column its tensile strength. If this force is greater than the force developed by transpiration pull, the water column will rise. Measurements of the tensile strength of sap in xylem tubes have shown that in some cases this exceeds 150 atmospheres. This is more than enough to withstand the osmotic pull developed by transpiration. It has been estimated that the tensile strength of water is not only sufficient to form an unbroken column against osmotic pull, but it is also adequate to overcome the frictional and gravitational forces encountered in its vertical rise in the plant. Furthermore, the attraction of water molecules for the walls of the xylem (*adhesion*) is very great.

PHLOEM TRANSPORT OF NUTRIENTS

Water and dissolved minerals absorbed by roots and transported by xylem represent the essential metabolites needed by plants for growth and development. These inorganic metabolites are the materials from which plant cells synthesize organic molecules for the maintenance of metabolic activities. The principal cells which assemble metabolites into nutrients are the photosynthetic cells of leaves. Anatomically, the distances separating photosynthetic cells from other living cells are relatively large. Considering this, the need for a rapid and efficient transportation system becomes

Figure 7-15. Transpiration. See text for amplification.

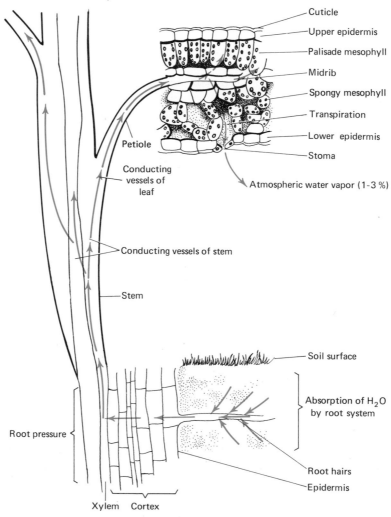

Cuticle

Upper epidermis

Palisade mesophyll

Midrib

Spongy mesophyll

Transpiration

Lower epidermis

Stoma

Atmospheric water vapor (1-3 %)

Petiole

Conducting vessels of leaf

Conducting vessels of stem

Stem

Soil surface

Absorption of H$_2$O by root system

Root pressure

Root hairs

Epidermis

Xylem   Cortex

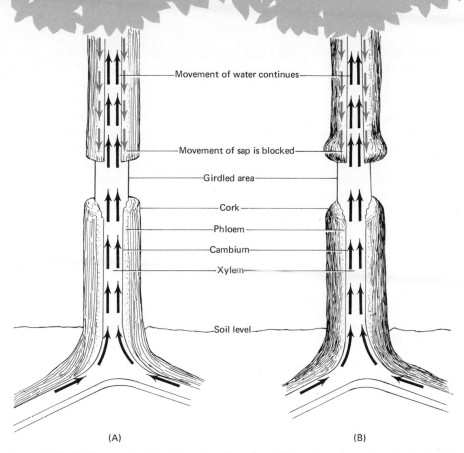

Movement of water continues

Movement of sap is blocked

Girdled area

Cork

Phloem

Cambium

Xylem

Soil level

(A)   (B)

Figure 7-16. Girdling and nutrient translocation. By girdling a tree (removing the bark, but leaving the wood intact), it can be demonstrated that phloem tissue transports nutrients downward. This is evidenced by the fact that the stem of a girdled tree bulges because of nutrient accumulation. (**A**) Immediately after girdling. (**B**) Several weeks later.

apparent. The movement of dissolved nutrients is accomplished by the specialized cells of the phloem called *sieve-tube elements*. These cells, like those of xylem, form a continuous conducting network that extends to every part of the plant.

A number of methods have been employed to ascertain that the phloem is the principal channel for the conduction of organic nutrients. Chemical analysis of phloem cell contents reveals that they consist largely of dissolved carbohydrates, usually sucrose, proteins, and other nitrogen-containing

compounds. If the bark (which includes the phloem) is removed, a process called *ringing* or *girdling,* and the wood (xylem) is left intact, it can be shown that after several weeks nutrients will accumulate above the ring. In addition, the stem will bulge after the damaged tissues have regenerated (Figure 7-16). Girdling prevents the movement of nutrients to all parts of the plant below the ringed section. Thus the phloem facilitates the downward movement of nutrients. The structure of sieve tubes also suggests a conducting function.

Although there is considerable evidence that organic substances are transported in the phloem, some evidence also suggests the bidirectional conduction of inorganic substances, such as salts. Studies using radioactive tracers have demonstrated the downward movement of salts in phloem tissue. Radioactive phosphorus (^{32}P), for example, has been shown to move in such a manner. It would appear, then, that there is a bidirectional movement of both organic solutes and certain mineral salts in the phloem.

Nature of Phloem Translocation

The term **translocation** is generally used to describe the movement of dissolved organic nutrients through phloem tissue. The directions of translocation are not only upward and downward but also laterally in vascular rays. The problem of translocation is a difficult one to resolve because of the interplay of a number of factors. For example, simple diffusion alone cannot account for the rapid movement of translocated materials. Furthermore, different materials move with different translocation rates through sieve cells that are filled with cytoplasm and contain sieve plates perforated by extremely small pores. In addition, translocation rates are affected by temperature, light, metabolic inhibitors, concentration gradients, hormones, and mineral deficiencies among other factors. Considering all the aspects of phloem translocation, no single theory has gained general acceptance in explaining the mechanism involved. Two hypotheses, however, have suggested possible explanations.

Mass (Pressure) Flow Hypothesis. According to this hypothesis, there is a mass flow of water and solutes through sieve tubes due to differences in turgor pressure. Cells of the leaf contain high sugar concentrations as a result of photosynthetic activity. As a result of a high solute concentration, water from the xylem diffuses into them, thereby

increasing their turgor pressure. Other cells of the plant such as those that are actively engaged in respiration, growth, or storage processes have a lower solute concentration because sugars are being consumed. This loss of sugar causes these cells to lose water, thereby decreasing their turgor pressure. At one end of the phloem, then, are cells with high turgor pressure, and at the other end are cells with low turgor pressure. The result of these vary-

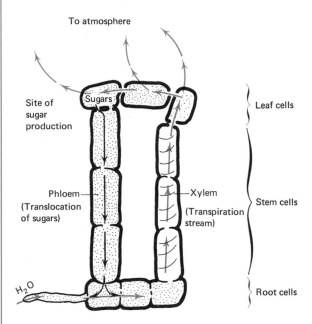

Figure 7-17. Diagram to illustrate mass (pressure) flow hypothesis of phloem transport. Due to differences in solute concentration between cells of the leaf (high) and cells in which nutrients are being utilized (low), a turgor pressure gradient develops. Leaf cells have a high turgor pressure whereas cells such as root cells have a lower turgor pressure. The result of these unequal pressures is a mass flow of the contents of the phloem cells from the regions under higher pressure to the regions under lower pressure. (Modified from A. S. Crafts, "Movement of Organic Materials in Plants," *Plant Physiology,* Vol. 6, p. 1, 1931.)

ing turgor pressures at opposite ends of the phloem is a mass flow of the contents of the sieve cells from regions under higher pressure to regions under lower pressure (Figure 7-17).

Protoplasmic Streaming Hypothesis. This theory assumes that solute particles diffusing into sieve cells are caught up in the circulating cytoplasm and carried from one end of the cell to another. The particles are then passed on to the next sieve cell through plasmodesmata and are again picked up by the streaming cytoplasm (Figure 7-18). According to this hypothesis, materials move through sieve tubes by alternating streaming within cells with diffusion between cells through plasmodesmata.

CHAPTER IN RETROSPECT

The chemicals needed for plant metabolism are procured from the physical environment, that is,

the atmosphere (air), the hydrosphere (water), and the geosphere (earth). Certain atmospheric nutrients are provided through various gaseous cycles of oxygen, carbon, and nitrogen. The liquid portion of the physical environment (hydrosphere) provides the chief source of water, a principal metabolite, and the solid portion of the environment (geosphere) furnishes land plants with the minerals essential for maintenance and growth. The means by which these various metabolites pass from the external environment into plant cells is called absorption. Various biophysical forces, such as diffusion, osmosis, turgor pressure, plasmolysis, and imbibition, appear to account for the movement of metabolites from the nonliving physical environment to the living environment found within plants.

Once inside the plants, the metabolites must be equitably distributed to various sites of metabolic

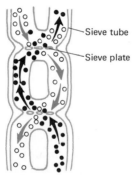

Sieve tube

Sieve plate

Figure 7-18. Protoplasmic Streaming Hypothesis. According to this hypothesis of the mechanism by which nutrients are transported by phloem tissue, it is assumed that solute particles diffusing into sieve cells are picked up by the streaming cytoplasm and carried from one end of the cell to another. It is further assumed that the particles are then passed on to the next sieve cell through the plasmodesmata of the sieve cell plates and are once again picked up by the circulating cytoplasm. Essentially, this hypothesis purports that solute particles move through sieve tubes by alternating streaming within cells with diffusion between cells through plasmodesmata. The proposed mechanism is shown diagrammatically on the left. On the right, a series of time lapse photographs illustrates the translocation of castor oil droplets by rotational cytoplasmic currents in living phloem cells of *Vicia faba* (horse bean). (Photo Courtesy of U.S.D.A.)

activity. This is accomplished mainly by the processes of diffusion and cyclosis through the tracheids and vessels of the xylem. The principal cells which assemble metabolites into plant nutrients are the photosynthetic cells (mesophyll) of leaves. Obviously, it is necessary to transport these nutrients around the entire plant. This movement of dissolved nutrients is achieved by the specialized cells of the phloem called sieve-tube elements. Thus green plants have developed a means for absorbing needed raw materials from their physical environment, transporting them to chemical factories within leaves, and distributing the needed products throughout the living system.

Questions and Problems

1. Define each of the following terms.

nutrient	absorption
nitrogen fixation	diffusion
hydrosphere	differentially permeable
geosphere	membrane
humus	osmometer
imbibition	turgor pressure
guttation	root pressure
translocation	girdling

2. What inorganic compounds are the exclusive sources of carbon and hydrogen in plant metabolism?
3. Briefly describe each of the five categories of soil.
4. What are the main inorganic substances found in soil water? How are the ions of these substances held to soil particles?
5. How would the appearance of a plant be changed as a result of the following mineral deficiencies: nitrogen, calcium, copper, and boron?
6. List ways in which plants are aided by the addition of organic materials to the soil.
7. Discuss the relationship between the presence of microorganisms in soil and soil fertility.
8. Devise an experiment which will indicate the effect of the relative concentration of a substance on the rate and direction of diffusion.
9. List examples of permeable, impermeable, and differentially permeable membranes. What factors are involved in differential permeability?
10. What are the factors which bring about equilibrium in an osmotic system?
11. What is the meaning of DPD? What are the conditions by which water enters or leaves a living cell?
12. What are the means by which water is absorbed by the root system? How is water transported to the various cells throughout the plant?
13. Contrast the relative importance of root pressure and the cohesion–tension theory with respect to the ascent of sap in plants.
14. Compare the nature of xylem and phloem transport. (Consider substances transported, types of cells involved, rate of movement, and various explanations of the mechanism.)

Suggested Supplementary Readings

Biddulph, Susan, and Orlin Biddulph, "The Circulatory System of Plants," *Scientific American*, February 1959.

Bollard, E. G., "Transport in the Xylem," *Annual Review of Plant Physiology*, vol. 2, 1960.

Bonner, James, "Water Transport," *Science*, vol. 129, pp. 447–450, 1959.

———, and A. W. Galston, *Principles of Plant Physiology*. San Francisco: W. H. Freeman and Company, 1952.

Crafts, A. S., *Translocation in Plants*. New York: Holt, Rinehart and Winston, Inc., 1961.

Ferry, J. F., and H. S. Ward, *Fundamentals of Plant Physiology*. New York: The Macmillan Company, 1959.

Greulach, V. A., "The Rise of Water in Plants," *Scientific American*, October 1952.

Holter, H., "How Things Get into Cells," *Scientific American*, September 1961.

Hudson, J. P., "Plants and Their Water Supplies," *Endeavour*, vol. 16, pp. 84–89, 1957.

Kellogg, C. E., "Soils," *Scientific American*, July 1950.

Kramer, P. J., *Plant and Soil Water Relationships.* New York: McGraw-Hill Book Company, Inc., 1949.

———, "Roots as Absorbing Organs," *Encyclopedia of Plant Physiology,* vol. 3, pp. 188–214, 1956.

Mitchell, J. W., and J. F. Worley, "Intracellular Transport Apparatus of Phloem Fibers," *Science,* vol. 145, pp. 409–410, 1964.

Revelle, Roger, "Water," *Scientific American,* September 1963.

Slatyer, R. O., "Absorption of Water by Plants," *Botanical Review,* vol. 26, pp. 331–392, 1960.

Solomon, A. K., "Pumps in the Living Cell," *Scientific American,* August 1962.

Steward, F. C., *Plants at Work.* Reading, Mass.: Addison-Wesley Publishing Company, Inc., 1964.

Sutcliffe, J. F., *Mineral Salts Absorption in Plants.* New York: Pergamon Press, Inc., 1962.

Thaine, R., "The Protoplasmic-Streaming Theory of Phloem Transport," *Journal of Experimental Botany,* vol. 15, pp. 470–484, 1964.

Waggoner, P. E., and Israel Zelitch, "Transpiration and the Stomata of Leaves," *Science,* vol. 150, pp. 1413–1420, 1965.

Zimmermann, M. H., "How Sap Moves in Trees," *Scientific American,* March 1963.

8
Enzymes: Catalysts of Plant Metabolism

INTRODUCTION

Wherever attention is turned in the plant world, there is evidence of chemical change. Chlorophyll-bearing plants synthesize organic compounds from inorganic raw materials; bacteria increase soil fertility, cause milk to sour, ripen cheeses, and cause disease; and fungi decompose dead organic matter and ferment sugar. Plants produce flowers and fruits, leaves fall from trees, and seeds germinate and develop into mature plants. In fact, close observation of the activities of any plant over a period of time would show evidence of chemical changes taking place.

The vast majority of chemical reactions outside living plant cells occur relatively slowly at the temperatures under which plant cells perform similar reactions. Yet, the rate of these reactions within the plant body is extremely high. In fact, similar chemical reactions carried out in the laboratory under substantially higher temperatures still do not proceed as rapidly as those in a cellular environment. The obvious conclusion to be drawn from these data is that plant cells possess some unique mechanisms that can accelerate metabolic reactions without an increased temperature.

Various lines of research and investigation have demonstrated the presence of a special group of organic substances associated with nearly all chemical reactions in living systems. These substances have been proved to be the agents of acceleration and are known as **enzymes.** Practically every step in the sequence of metabolic reactions in plants is activated and regulated by one of these biochemical accelerators. In essence, as mediators between the DNA of the nucleus and the rest of the cell, they control, direct, and regulate the metabolic cycle and enable the chemical reactions of the cell to proceed as orderly, integrated, and efficient systems.

THE MECHANISM OF ENZYME ACTION

Enzymes and Activation Energy

Biochemical reactions occur when molecules interact with each other through random collisions. Consider the following example. Assume that molecules of substance X, the reactants, are to be converted to molecules of substance Y, the products, and that X is spontaneously converted to Y, in the absence of an enzyme. In a given number of molecules of substance X, at a specific temperature, there will be some possessing relatively little energy; a large number of molecules, indeed the major fraction of the population, possessing average energy; and a small fraction of the population having quite a high energy. Only the energy-rich X molecules are able to react and thus be converted to Y molecules. Therefore, only a relatively few molecules at any one time, as a result of molecular collisions, can reach the level of energy necessary to react. This minimum energy of collision required for chemical reaction is termed *activation energy*. The rate of reaction, that is, the frequency of collision of sufficient energy, depends upon the proportionate number of reacting molecules at the activation energy level. One way to increase the frequency of collisions is to increase temperature so that more molecules can attain the necessary activation energy. Note that the nonenzymatic agent of acceleration is heat (Figure 8-1A).

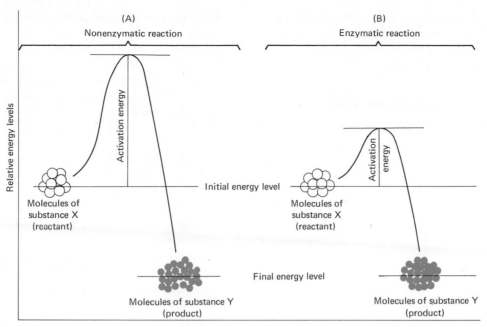

Figure 8-1. Diagrammatic representation of the energy requirements of a chemical reaction in the absence and in the presence of an enzyme. Before the chemical reaction can occur, molecules of substance X must collide with each other with sufficient energy (activation energy) to form molecules of Y, the product. In reaction **(A)** the frequency of collisions may be increased by raising the temperature. In reaction **(B)**, an enzyme-catalyzed reaction, the presence of the enzyme lowers the required activation energy. In this way more molecules of X are converted to Y (at a lower temperature) since a larger fraction of molecules of substance X possess the required activation energy for reaction. (From *Principles of Plant Physiology*, by James Bonner and Arthur W. Galston, W. H. Freeman and Company, copyright © 1952.)

If on the other hand, the same reaction is considered in which an enzyme is present, a significant difference is noted. The activation energy of the reaction is lowered. It is thought that an enzyme will react with the energy-rich and energy-poor molecules alike. In this way, many more molecules of X can participate in the reaction than if the enzyme were absent. The presence of an enzyme, therefore, speeds up the reaction because

more of the X molecules attain sufficient activation energy (Figure 8-1B).

An enzyme is capable of accelerating a reaction without an increase in temperature. This is of considerable importance, because all enzymes are protein in composition and proteins are susceptible to breakdown by heat. Consequently, a significant temperature increase would destroy the enzymes as well as other cellular proteins. The essential *function of enzymes,* therefore, is to accelerate biochemical reactions by lowering the activation energy level within living cells without an increase in temperature. In this way the structural integrity of proteins is maintained.

Lock and Key Theory of Enzyme Action

The manner in which an enzyme actually lowers activation energy is not completely understood. However, from what has been learned from the study of the behavior of enzymes under a variety of experimental circumstances, some approaches to the understanding of enzyme action have been made. When an enzyme and a substrate are mixed, the following sequence of events occurs: (1) the surface of the *substrate* molecule, that is, the molecule with which the enzyme acts, comes into close contact with the surface of the enzyme molecule; (2) a temporary intermediate compound called an *enzyme–substrate complex* forms; (3) the substrate molecule is transformed; (4) the transformed substrate molecules (products) move away from the surface of the enzyme molecule; and (5) the recovered enzyme, now freed, reacts with other substrate molecules.

These events, which purport to explain the mechanism of enzyme action, are part of the *Michaelis–Menton,* or lock and key, *theory,* of enzyme action (Figure 8-2). Essentially, an enzyme and substrate will combine to form the complex only if there is a specific molecular fit. There is an analogy here to the complementarity between lock and key. Only a particular type of key will fit a particular type of lock. Similarly, only a particular type of molecule will fit the substrate binding site of any enzyme. As a result, only those substrate molecules that possess specific molecular configurations may serve as potential substrates for a given enzyme. Consequently, a myriad of enzymes may operate in the same plant cell without confusing their different substrates.

The concept of an enzyme–substrate complex and the realization that enzymes are frequently much larger than their substrates suggests that only a relatively small portion of an enzyme is in contact with substrate. Furthermore, it is reasonable to suppose that only those chemical groupings on the enzyme which are in the very near neighborhood of the enzyme-bound substrate will play a central role in transforming the substrate. The relatively small portion of the enzyme that is directly responsible for binding and acting upon the substrate is called the *active site.* It is at the active site that substrate transformation occurs.

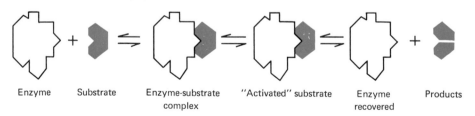

Enzyme   Substrate   Enzyme-substrate   "Activated" substrate   Enzyme   Products
complex   recovered

Figure 8-2. Lock and key theory of enzyme action. During a biochemical reaction, an enzyme and substrate molecule combine to form an unstable enzyme-substrate complex. In the process of combination, the substrate is activated and transformed into products. Following transformation, the enzyme is recovered and may be used again to catalyze a similar biochemical reaction.

It is believed that the formation of the enzyme–substrate complex lowers the activation energy so that substrate molecules may be activated for reaction without a significant temperature increase. Once the complex has been formed, the transformation of the substrate takes place. Some substrates become ionized whereas others are split or joined together, and still others lose atoms, take on atoms, or undergo changes in molecular configuration in enzymatic reactions.

After the substrate has been transformed, it leaves the surface of the enzyme and the enzyme is freed for reaction with similar substrate molecules. Although an enzyme molecule is recovered after a reaction, it eventually breaks down and a new one must be synthesized by the cell.

CHARACTERISTICS OF ENZYMES

All enzymes, including plant enzymes, have certain characteristics that contribute to their function as agents of biochemical acceleration. These include properties of catalysis, biotic origin, specificity, quantitative efficiency, and colloidal nature, among others.

Catalysis

Catalysts are chemical substances which increase the rates of chemical reactions without being altered or destroyed in the process. In general, catalytic agents accelerate chemical reactions, exhibit a certain degree of specificity, are effective in small quantities, and are recovered after the reaction is catalyzed. The available data support the conclusion that enzymes possess these properties of catalysts, and, in this context, enzymes are known as *biochemical catalysts.*

Although an enzyme accelerates the completion of a reaction, it does not affect the equilibrium of that reaction. Most biochemical reactions in living systems are reversible and would proceed toward equilibrium at a very slow rate in the absence of enzymes. However, an enzyme will accelerate a reaction in either direction. It will bring about the equilibrium of that reaction at a much faster rate.

Biotic Origin

All known enzymes, including plant enzymes, are synthesized inside living cells. However, a given chemical reaction involving an enzyme may be performed outside of plant tissues or cells after the enzyme has been extracted and prepared. For example, the enzyme amylase may be extracted and placed in a test tube with starch to start the "digestion" of the polysaccharide. Although the reaction may be performed in the laboratory, the enzyme was previously synthesized and extracted from the cells of a living plant. One such plant that regularly carries on the extracellular hydrolysis of starch is the common bread mold, *Rhizopus.*

Specificity

One of the most striking characteristics of plant enzymes is their relatively high degree of specificity. Each enzyme catalyzes the transformation of only one specific substrate, or at most the transformation of a group of closely related substrates. Enzyme specificity is conveniently subdivided into two types: (1) absolute specificity, and (2) group specificity. In the case of *absolute specificity,* an enzyme will act upon only one particular substrate and will not catalyze the transformation of a closely related or dissimilar compound. This type of specificity is illustrated by the action of the enzyme maltase on the substrate maltose. In the plant, maltose does not usually accumulate but is rapidly converted into glucose molecules. Maltase catalyzes the hydrolysis of maltose into two glucose molecules, and in the reaction, like other enzymatically controlled reactions, the enzyme is recovered.

In the second type of specificity, *group specificity,* a given enzyme will act upon a group of chemical substances which are structurally related to each other. That is, the substrates are related by some common chemical structure. The commonality is usually a chemical bond or *functional group* (a group of atoms that participates in the chemical reaction).

Proteins, for example, consist of a series of amino acids linked together to form a chainlike molecule. While living organisms contain only 20 amino acids, there can be anywhere from 100 to 3000 in various sequences and arrangements comprising a protein molecule. Chemically, the component amino acids are joined together by peptide bonds, and it is these bonds, regardless of the particular protein, that are susceptible to attack by certain group-specific enzymes. In germinating seeds, for example, a group of enzymes called *proteases* hydrolyze the peptide bonds of storage proteins into component amino acids, which are required for the growth of the organs of the seedling.

Quantitative Efficiency

The most outstanding quantitative property of plant enzymes is the extremely high proportion of substrate molecules that can react with relatively few enzyme molecules. It has been demonstrated, for example, that in certain plant cells the enzyme sucrase (invertase) can catalyze the breakdown of the substrate sucrose in the approximate ratio of 1:1,000,000 and that a single molecule of the enzyme catalase can bring about the decomposition of 5,000,000 molecules of the substrate hydrogen peroxide per minute at O°C.

A single enzyme molecule can, therefore, catalyze the transformation of a great many substrate molecules in rapid succession. In this way a small amount of an enzyme can bring about a chemical transformation which is disproportionate to the quantity of enzyme present.

Colloidal Nature

All known enzymes consist of molecules that fall into the dimensions of colloidal particles, which range in diameter from 0.001 to 0.1 μ. Given a constant volume, as the colloidal particles are subdivided into smaller and smaller sections, the surface area increases. For illustrative purposes, assume that a colloidal particle is a 1-inch cube. As the 1-inch cube is cut into smaller and smaller cubes, the total surface area increases while the volume remains constant. If the volume of the starting cube is 1 cubic inch, the total surface area is 6 square inches. Now, if the same 1-inch cube is divided into smaller cubes, each $\frac{1}{2}$ inch on a side, the surface area increases to 12 square inches. A subdivision of the 1-inch cube into 64 smaller cubes, each $\frac{1}{4}$ inch on a side, increases the surface area to 24 square inches and a subdivision into 512 cubes, each $\frac{1}{8}$ inch on a side, increases the surface area to 48 square inches.

It becomes apparent that as the particle size within the constant volume decreases, there is a proportionate increase in the total surface area. This large, exposed surface area is one of the factors responsible for the quantitative efficiency of enzymes, because there are a number of active "sites" available for various substrate molecules to react with enzyme molecules. The relationship of particle size to surface area is shown in Figure 8-3 and in Table 8-1.

THE COMPOSITION OF ENZYMES

Protein Structure

Since the first isolation and crystallization of the enzyme urease from jack bean seeds by Sumner in 1926, the nature of enzyme composition

TABLE 8-1. Relationship of Particle Size to Surface Area

	Length of Single Edge of Cube	Number of Cubes in 1 cubic cm	Total Surface Area Exposed
	1 cm	1	6 square cm
	1 mm	10^3	60 square cm
	0.1 mm	10^6	600 square cm
	0.01 mm	10^9	6000 square cm
	1.0 μ	10^{12}	6 square m
Colloidal	0.1 μ	10^{15}	60 square m
particle	0.01 μ	10^{18}	600 square m
sizes	0.001 μ	10^{21}	6000 square m

has been elucidated. By 1956, about 75 enzymes had been crystallized and many more had been purified. Based upon Sumner's investigations and the research of subsequent studies, enzymes have been identified as protein in composition. One of the definitive proofs that enzymes are proteins is the experiment in which a given enzyme (enzyme A) is acted upon by a protein-digesting enzyme such as papain. At the termination of the reaction the catalytic activity of the enzyme decreased and the protein disappeared, indicating the protein composition of the substrate (enzyme A). In crystalline form, enzymes, as well as other proteins, exhibit symmetric shapes and patterns. The crystalline structure of the enzyme pepsin is represented in Figure 8-4.

Because of their chemical nature, plant enzymes share with other plant proteins the property of alteration by heat, called *denaturation*. In the denatured state, enzymes lose their chemical config-

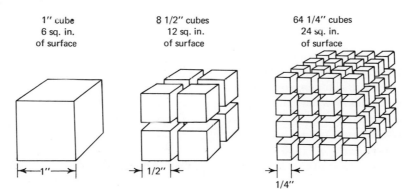

1″ cube
6 sq. in.
of surface

8 1/2″ cubes
12 sq. in.
of surface

64 1/4″ cubes
24 sq. in.
of surface

1″   1/2″   1/4″

Figure 8-3. Relationship of particle size to surface area. If the volume of the particle being considered is one cubic inch, as the particle size within the constant volume decreases, there is a proportionate increase in the total surface area. Similarly, note in Table 8-1 the effect of successive subdivisions on the total surface area of a given mass of material.

Figure 8-4. Crystalline structure of pepsin. (Courtesy of Dr. John N. Northrop, University of California.)

uration, become insoluble, and precipitate out of solution. As such, they are no longer capable of catalysis and are inactive, being deprived of their vital biological activity. In addition to heat, other agents which may denature plant enzymes include high concentrations of heavy metal ions (for example, copper, zinc, silver, arsenic, and mercury), alcohol, ultraviolet radiation, and concentrated acidic and basic solutions.

Some plant enzymes appear to consist solely of proteins. Many other enzymes, however, consist of two distinct portions. The first part, called an *apoenzyme,* is the protein portion composed of only amino acids. The other part is a nonamino acid constituent referred to as a *prosthetic group,* which assists the enzyme in catalysis. The apoenzyme and its related nonprotein portion are collectively designated the *holoenzyme.* The relationship between the apoenzyme and coenzyme is such that, if they are dissociated, the holoenzyme is rendered ineffective, but if they are recombined, the catalytic activity of the holoenzyme is reestablished (Figure 8-5).

Prosthetic Groups

Prosthetic groups associated with enzymes are generally of two types: activators and cofactors (coenzymes). **Activators** are metals, usually ions of salts such as potassium, iron, copper, magnesium, manganese, zinc, calcium, and cobalt. Definite correlations between the catalytic properties of some enzymes and their association with various metal components have been shown. Although the exact function of these activators is not completely understood, it is believed that they form a bridge between the enzyme and the substrate, thus binding them together to facilitate substrate transformation. In general, it may be stated that most trace elements are probably required as activators for cellular enzymes.

In contrast to enzymes requiring metals, some enzymes require organic prosthetic groups called **coenzymes.** Generally, coenzymes assist the enzyme in transforming the substrate by acting as an

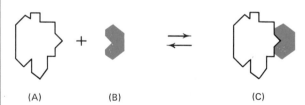

(A)   (B)   (C)

Figure 8-5. Components of an enzyme. Most enzymes are composed of (**A**) an apoenzyme (protein portion) and (**B**) a prosthetic group (nonprotein portion). Both portions together constitute (**C**) the holoenzyme, or complete enzyme. If the apoenzyme and prosthetic group are separated, the holoenzyme becomes inactive and will not catalyze a biochemical reaction.

acceptor of atoms being removed from the substrate or as an acceptor of atoms by contributing atoms in a synthetic reaction. Many coenzymes are vitamins.

Coenzymes in general have been found associated chiefly with enzymes concerned with oxidation-reduction reactions. Two of the most important coenzymes in plant cellular metabolism are *NAD* (*nicotinamide adenine dinucleotide*) and *NADP* (*nicotinamide adenine dinucleotide phosphate*). Both compounds are derivatives of the B vitamin *nicotinic acid* (*niacin*) and function with their respective enzymes in the removal and transfer of hydrogen atoms and electrons from substrate molecules. Enzymes that participate in reactions in which hydrogen atoms are removed (*dehydrogenation*) are collectively called **dehydrogenases.**

The flavin coenzymes, such as *FMN* (*flavin mononucleotide*) and *FAD* (*flavin adenine dinucleotide*), contain the B vitamin *riboflavin* and are important in some of the photosynthetic reactions. Like NAD and NADP, these coenzymes function in hydrogen transfer reactions with dehydrogenase enzymes.

Cytochromes are another group of coenzymes that function as electron carriers in photosynthesis and respiration. These compounds are metalloprotein pigments which contain iron and are structurally related to hemoglobin and chlorophyll. Cytochromes function in serial succession in reactions in which electrons are systematically passed from one cytochrome to another. This methodical, synchronized passage of electrons from one cytochrome to another froms the *cytochrome system*. This system plays an important role in both respiration and photosynthesis.

Another important coenzyme, called *coenzyme A (CoA)* contains a derivative of *pantothenic acid,* another B-complex vitamin. This coenzyme assumes important roles in the synthesis and breakdown of fats and in a series of oxidizing reactions of respiration called the *Krebs cycle.* Unlike the aforementioned coenzymes, CoA is not involved in transferring hydrogen or electrons but is a carrier of certain useful fragments of organic molecules.

Finally, mention should be made of the *adenosine phosphates,* specifically, *ATP* (*adenosine triphosphate*) and *ADP* (*adenosine diphosphate*). Although not coenzymes in a strict sense, these compounds act as agents of phosphate transfer between substrate molecules. Both compounds participate in energy-yielding and energy-requiring metabolic reactions.

TYPES OF ENZYMES

In 1878 Kuhne proposed the word enzyme from the Greek derivation meaning "in yeast." From the discussion so far it is probably apparent that enzymes typically end in *-ase,* although this is not always the case. Enzymes are usually named according to three criteria: (1) the specific substrate, (2) the nature of substrate transformation, and (3) the group of related substrates with which they react. In most cases, the prefix represents the criterion according to which the enzyme is named, and the suffix (*-ase*) designates the compound as an enzyme.

Many different classification schemes for enzymes have been devised and most have been based primarily on the nature of the substrate transformation involved. As a result, enzyme classification has lacked complete standardization. In an attempt to establish a universally accepted system for classifying enzymes, the Commission of Enzymes of the International Union of Biochemistry (1961) drafted and adopted a set of specific rules. In the discussion of the following section, this system of classification will be employed.

According to this scheme of classification, enzymes are grouped into six major classes. These are hydrolases, lyases, transferases, isomerases, oxidoreductases, and ligases (synthetases). These designations represent classes of enzymes and not the names of specific enzymes.

Class 1. Hydrolases

Plant *hydrolases* bring about *hydrolysis* of substrate molecules by adding water to the substrate and simultaneously decomposing it into smaller, simpler substances. Hydrolases usually lack prosthetic groups and are concerned with the process of digestion. Digestion may be viewed as an organized sequence of hydrolysis reactions bringing about the solubility and stepwise degradation of organic food molecules. In each reaction, the phases of digestion are catalyzed by specific enzymes which act upon specific substrates.

Some hydrolase enzymes operate inside (intracellular) plant cells, whereas others catalyze reactions outside (extracellular) plant cells. An example of the action of extracellular hydrolases is represented by the digestive process of the insectivorous or carnivorous plants. The pitcher plant, for example, has a funnel-shaped leaf into which insects crawl and are trapped in water at the base of the leaf and are further held by downward projecting bristles. Digestive enzymes, secreted by the leaf cells, hydrolyze the insect body. The products of digestion are then absorbed and utilized by the plant as a source of energy. Similar digestive processes are carried on by the Venus's-flytrap and the sundew plant.

Germinating seeds exemplify the action of intracellular hydrolases. Food contained in seeds is in an insoluble form. Before it can be utilized, it must first be made soluble and then digested. If conditions are favorable, water is absorbed (imbibition), the seed swells, and the intracellular hydrolysis of stored fats, proteins, and starch begins. The resultant fatty acids and glycerol, amino acids, and sugars are then available and are utilized to pro-

duce energy for the formation of roots, stems, and leaves for the developing embryo plant. Nutrients that are not immediately digested or utilized in some other way are stored in plants. For example, sugar is stored in fruits, starch in potatoes, fats and oils in nuts, and proteins in beans.

Class 2. Lyases

Enzymes of this class are concerned with the breakdown of substrate molecules and are therefore very similar in their action to hydrolases. The notable difference is that, whereas hydrolases add water to the substrate molecule to decompose it, *lyases* do not utilize water. In addition, lyases split or cleave the substrate molecules symmetrically. For example, a six-carbon sugar might be cleaved into 2 three-carbon sugars in the presence of a lyase. In addition to cleaving sugars, lyases also participate in various other catabolic reactions.

Class 3. Transferases

Transferase enzymes transfer atoms or groups of atoms from one substrate to another. One of the most common transfer reactions in plant cells is the transfer of high-energy phosphate groups from ATP to substrate molecules to activate them for reaction. For example, certain enzymes in the presence of ATP transfer high-energy phosphate groups to glucose during the initial reactions of respiration. This serves to prepare glucose for further reactions.

Class 4. Isomerases

As a class, these enzymes isomerize (rearrange) atoms or groups of atoms within substrate molecules. An example of the action of an *isomerase* is seen as part of fermentation in which a molecule of glucose phosphate is isomerized into a molecule of fructose phosphate. It should be recalled that glucose and fructose are isomers, each represented by the general formula $C_6H_{12}O_6$.

Class 5. Oxidoreductases

This class of enzymes is concerned with the process of biological oxidation. In all biological oxidations, electrons or hydrogen or both are transferred from one substrate to another and *oxidoreductases* together with certain coenzymes (NAD, NADP, cytochromes, and so on) assist in these electron and hydrogen transfers. Photosynthesis and cellular respiration involve sequences of transfer reactions in which electrons and hydrogen are systematically passed from one oxidoreductase to another.

Class 6. Ligases ((Synthetases)

As a class these enzymes catalyze reactions in which molecules are joined together. Usually ATP is closely associated with these reactions. *Ligases* are commonly found as participants of anabolic reactions in which molecules are synthesized.

The six classes of enzymes just discussed are involved in numerous metabolic reactions of plant cells. Some hydrolyze substrate molecules, others cleave substrate molecules, and still others transfer and rearrange atoms of substrate molecules. Some catalyze oxidation–reduction reactions and a few join substrate molecules together. It is the synchronized action of a galaxy of enzymes that maintains the metabolic plant as an efficient, organized molecular system.

FACTORS AFFECTING ENZYME ACTIVITY

Within a normal cellular environment, the metabolic activities of plants proceed in an efficient, organized, and sequential pattern. At times, however, the cellular environment may become adverse, and as a consequence enzyme behavior is affected. The result may be a change in the rate or direction of a chemical reaction resulting from an alteration or destruction of the enzyme. Chief among abnormal environmental factors, both intercellular and intracellular, which affect enzymes are the substrate concentration, enzyme concentration, pH of the medium, temperature, and the presence of inhibitors.

Substrate Concentration

In view of the fact that enzymes react with substrates, it seems reasonable that the availability of substrate should influence enzyme activity. Substrate availability may be viewed as low, adequate, or excess. Too low a concentration of substrate results in a low velocity of reaction (Figure 8-6A). This result is because all the available reactive sites of the enzymes are not being occupied by substrate molecules. In general, as the concentration of substrate molecules increases, the velocity of the reaction increases. If substrate molecules are added to the enzyme system in gradually increasing quantities, the rate of the reaction would increase in proportion to these additions (Figure 8-6B). However, a point is reached where any further additions of substrate molecules would not affect the speed of the reaction. The speed is maintained at a *maximum velocity,* and subsequent additions of the substrate in many enzyme systems would actually decrease the reaction rate (Figure 8-6C). This result may be explained on the basis that each enzyme molecule has only limited numbers of active sites available for combination with substrate molecules.

During photosynthesis, for example, if the sugars (substrates) are not produced rapidly enough, their utilization by the cell is low. If, on the other hand, sugar production is excessive, photosynthesis may decrease due to an accumulation of products. In general, however, the enzymatically controlled phases of photosynthesis proceed normally because the substrate concentration is usually adequate.

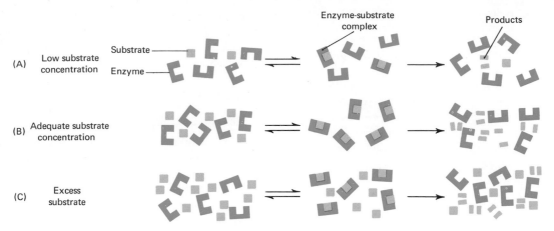

Figure 8-6. Substrate availability (concentration) and enzyme activity. (From Richard A. Goldsby, *Cells and Energy,* New York: The Macmillan Company, © 1967.)

Concentration of the Enzyme

Within fairly wide limits the velocity of an enzyme-catalyzed reaction is directly proportional to the concentration of the enzyme. This statement must be qualified by stating that the substrate concentration must remain relatively high to ensure continual saturation of active sites and that there are no substances competing with the substrate for an active site on the enzyme. Such substances are called *inhibitors.* Under normal cellular conditions, the enzyme concentration is fairly constant. Even though enzymes eventually break down, new ones are constantly being synthesized. In this way an adequate concentration of enzymes is maintained.

pH of the Medium

Enzymatic reactions within plant cells are markedly influenced by the pH of the medium in which they operate. Each enzyme has its particular *optimum pH,* that is, a specific pH at which it operates most efficiently. Most plant enzymes have optima in the pH range 5 to 8. The optimum pH is so critical that deviations of either increasing acidity or alkalinity result in a decrease in the reaction rate. In fact, if the changes in pH are extreme, the enzyme may become inactivated or denatured. It should be recalled (Chapter 2) that the two mechanisms by which plant cells maintain a constant pH are by neutralization reactions and by the formation of buffer systems.

The Factor of Temperature

At 0°C the rate of an enzyme-catalyzed reaction is practically zero. As the temperature increases, there is a proportionate increase in the rate of reaction up to the *optimum temperature,* that is, the temperature at which the maximum rate of reaction is achieved. For most plant enzymes this is about 25°C. At higher than optimum temperatures (35°C) the rate is decreased to the point where the enzyme may be destroyed.

In general, the enzymes in plants have an optimum temperature between 20 and 30°C. The combined factors of the formation of an enzyme–substrate complex, the opening and closing of stomata, and the high specific heat of water all tend to maintain fairly constant temperatures. Temperature is one of the most crucial factors determining enzymatic activity, as excessive temperatures denature the protein portion of enzymes as well as other cellular proteins.

Inhibitors

Enzymes are proteins consisting of a large number of amino acids, so there are a number of possible sites for reaction. As a result, a variety of substances, in addition to specific or group substrate molecules, are capable of combining with the active sites. Such substances are referred to as *enzyme inhibitors* and the process is called *enzyme inhibition.* Based upon the mechanism of the process, inhibitors may be classified as (1) competitive, and (2) noncompetitive.

Competitive Inhibitors. *Competitive inhibitors* are compounds that closely resemble the enzyme's natural substrate. Because of this close resemblance, they can compete with the normal substrate for the active site on the enzyme surface. Consider the following example. Suppose that substrate A is to be converted to products B and C and that substance I is a competitive inhibitor of A. Under normal cellular conditions, in the absence of an inhibitor, the reaction goes to completion. In the presence of I, however, the inhibitor

Figure 8-7. Enzyme inhibition. **(A)** Competitive inhibition. Competitive inhibitors compete with normal substrate molecules for an active site on the enzyme. **(B)** Noncompetitive inhibition. Noncompetitive inhibitors do not bind the active site on the enzyme. (From Thomas P. Bennett and Earl Frieden, *Modern Topics in Biochemistry: Structure and Function of Biological Molecules,* New York: The Macmillan Company, © 1966.)

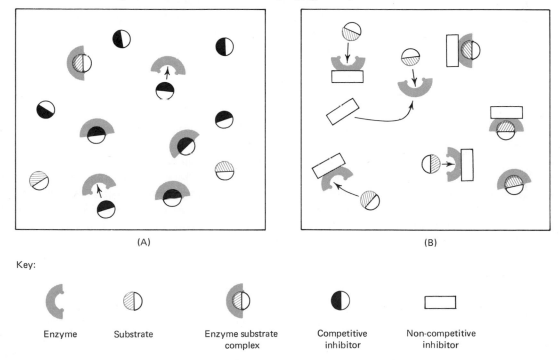

(A)   (B)

Key:

Enzyme   Substrate   Enzyme substrate complex   Competitive inhibitor   Non-competitive inhibitor

may combine with the enzyme and less B and C will be formed. Competitive inhibitors do not combine permanently with the enzyme, so the degree of inhibition will depend upon the relative concentration of I and the concentration of normal substrate molecules. The further addition of substrate molecules can restore the total catalytic capacity of the enzyme (Figure 8-7A).

Noncompetitive Inhibition. *Noncompetitive inhibitors* differ from competitive inhibitors in that the union of enzyme and noncompetitive inhibitor is permanent and the enzyme undergoes a structural alteration (Figure 8-7B). The amino acids of plant enzymes have a number of reactive groups present as side chains; $-CO_2H$, $-NH_2$, and $-SH$ are examples of such groups and any substances capable of combining with these is a potential inhibitor. Heavy metal ions in high concentrations, such as copper, zinc, silver, arsenic, and mercury, are potential inhibitors. Whereas competitive inhibition may be reversed by the addition of more normal substrate molecules, noncompetitive inhibition is an irreversible chemical process. This fact accounts, in part, for the poisonous nature of noncompetitive inhibitors.

The phenomenon of inhibition is of significant practical importance in terms of preventing the growth of one organism in relation to another by altering and even stopping an essential enzymatic reaction in the metabolic cycle. For example, the mechanism of inhibition could alter the growth of bacteria in man, or weeds and crabgrass on a lawn, or insects which damage vital crops. The principle involved is that whereas one organism requires a specific enzyme to complete a metabolic reaction, the other organism has no need for that enzyme or can accomplish the same metabolic process by an alternative reaction. Thus one organism carries on its normal metabolism in the presence of an inhibitor, whereas the other is affected.

Other Factors

The rate and efficiency of enzyme-catalyzed reactions are also directly related to and are affected by the physical nature of the enzyme, ultraviolet radiation, hormones, certain amino acids, and other biochemical agents.

CHAPTER IN RETROSPECT

Practically every step in the sequence of metabolic reactions in plants is activated and regulated by one or more protein catalysts called enzymes. In general, enzymes are colloidal substances which exhibit a certain degree of specificity (lock and key theory), are effective in small quantities, and are recovered after a reaction is catalyzed. Many different classification systems have been proposed for naming enzymes; however, in most cases, they are named according to (1) the specific substrate, (2) the nature of substrate transformation, and (3) the group of related substrates with which they react. Accordingly, enzymes are grouped into six major classes—hydrolases, lyases, transferases, isomerases, oxidoreductases, and ligases (synthetases). Each of these enzyme classes acts in specific ways. For example, hydrolases are concerned with the process of digestion and may be viewed as an organized sequence of hydrolysis reactions which eventually bring about the solubility and degradation of organic food molecules. Oxidoreductases engage in the process of biological oxidation and assist in electron and hydrogen transfers.

Certain factors may affect the behavior of enzymes, such as the pH of the medium, temperature, substrate concentration, enzyme concentration, and the presence of inhibitors. In the case of pH, for example, it has been found that each enzyme operates more efficiently at a particular pH (optimum pH). The same is true with respect to temperature. Most enzyme-catalyzed reactions function best within a specific temperature range, usually between 20 and 30°C. Other factors which relate to the rate, direction, and efficiency of enzymatic reactions in the metabolic cycle are ultraviolet radiation, hormones, and amino acids.

Questions and Problems

1. What is meant by each of the following?
   enzyme catalyst
   activation energy denaturation
   substrate coenzymes
   Michaelis–Menton theory enzyme inhibitor
   cytochromes
2. What is the role of NAD, NADP, FMN, and FAD?
3. Contrast cytochromes and coenzyme A with regard to the role each plays in oxidizing reactions of respiration.
4. List representative enzymes from the six major classes and indicate how each functions.
5. Briefly discuss the various factors which affect enzyme activity. Consider reactions of both high (maximum) and low velocity for each factor listed.
6. Consider the economic and practical possibilities of further research in the area of noncompetitive inhibitors, particularly as they apply to the fields of medicine and agriculture.

Suggested Supplementary Readings

Baldwin, Ernest, *The Nature of Biochemistry.* New York: Cambridge University Press, 1967.

Bennett, T. P., and Earl Frieden, *Modern Topics in Biochemistry.* New York: The Macmillan Company, 1967.

Goldsby, R. A., *Cells and Energy.* New York: The Macmillan Company, 1967.

Green, D. E., "Enzymes in Teams," *Scientific American,* September 1949.

Jellinck, P. H., *Biochemistry, an Introduction.* New York: Holt, Rinehart and Winston, Inc., 1963.

Phillips, D. C., "The Three Dimensional Structure of an Enzyme Molecule," *Scientific American,* November 1966.

9
Anabolism: The Synthesis of Plant Materials

INTRODUCTION

If a plant cell is to be metabolically active enough to grow, maintain itself, and reproduce, certain materials must be constantly synthesized. Beginning with the simple metabolites, water and carbon dioxide, green plants are able to synthesize a multiplicity of organic molecules (foods). The pivotal synthetic process upon which all others are initially dependent is photosynthesis, a process in which green plants synthesize sugars which are used in turn in the synthesis of fats, proteins, vitamins, hormones, and other organic compounds.

Although this chapter is concerned primarily with anabolic plant activities, the interdependence and simultaneity of these activities with catabolic processes should be kept in mind at all times. They are separated only for convenience of study and understanding; but in living plant cells they occur together in a constant cyclic pattern.

PHOTOSYNTHESIS—FROM SIMPLE METABOLITES TO SUGARS

Significance of Photosynthesis

Green plants, in the presence of light, are capable of transforming CO_2 and H_2O into carbohydrates and O_2 by a fundamental anabolic process called **photosynthesis.** This process, by which solar energy is converted into chemical energy, is the single most important chemical process in the world. The foods produced are not only utilized by plants for various activities but are also used by animals and man for their metabolic processes. In other words, all heterotrophs depend, either directly or indirectly, upon autotrophs for their food. It is this almost unlimited capacity which plants have for making their own foods that gives them so prominent a place in the physical and biological environments. This ability of plants to synthesize foods out of inorganic metabolites makes them the connecting link between the inorganic and organic worlds.

The Magnitude of Photosynthesis

Such an important physiological activity involving all components of the environment must have equally impressive quantitative dimensions. Each year, through the agency of photosynthesis, green plants produce 200 billion tons of sugars and 470 billion tons of net organic matter by a process that has not yet been duplicated in the test tube and by a process that is just beginning to be understood. The magnitude of photosynthesis can be further appreciated by considering the quantities of raw materials absorbed and the amount of material produced as a product of the reaction. Each year, 690 billion tons of CO_2 and 280 billion tons of water are taken in by plants to produce organic compounds with the release of 500 billion tons of oxygen to the atmosphere. Considering that green plants themselves use 80 billion tons of the food produced by photosynthesis, the actual synthetic production is about 550 billion tons. By comparison, each year, the world's steel mills produce only 350 million tons of steel, and the world's cement factories turn out a mere 325 million tons of cement.

Historical Development and General Nature

The discovery of photosynthesis goes back some 200 years, and since then, all data and knowledge of the process have been derived from a long series of tedious, piecemeal research. The first significant clue regarding photosynthesis came in 1772 when John Priestly observed that oxygen was given off by green plants. Seven years later, Jan Ingen-Housz, a Dutch physician and chemist, investigated the work of Priestly and observed that oxygen is evolved by green plants only if they are placed in light. He also noted that, in the process, plants used carbon dioxide as a raw material. In 1804 Nicholas Theodore de Saussure, a Swiss botanist, correlated the uptake of CO_2 and the evolution of O_2 to the light requirement and also suggested the role of water as a raw material. Thirty-three years later, Dutrochet, a French scientist, demonstrated the necessity for chlorophyll in the process, and in 1845 the German biochemist Liebig indicated that all organic compounds synthesized by green plants were derived from CO_2. The extensive research of these scientists, as well as the investigations of others from many fields of scientific disciplines, finally reduced the entire photosynthetic process to a generalized chemical formula:

$$6\,CO_2 + 6\,H_2O \xrightarrow[\text{chlorophyll}]{\text{light}} C_6H_{12}O_6 + 6\,O_2$$

carbon dioxide   water   sugar   oxygen

raw materials   necessary conditions   products

This equation is a mere representation of the raw materials, necessary conditions, and partial products of photosynthesis; there is no indication that a variety of other molecules are also synthesized. It provides some quantitative information in terms of the numbers of molecules involved, but it does not indicate the myriad of intermediate reactions that occur in the formation of sugars. It is known, for example, that carbohydrates are not formed by mixing CO_2 and H_2O; the product of such a reaction would be carbonic acid, H_2CO_3. Obviously, then, the equation does not indicate how sugars are actually synthesized. To understand the mechanisms that produce sugars photosynthetically, each aspect of the generalized equation will be considered.

The Fate of the Metabolites

The chemical composition of carbohydrates (carbon, hydrogen, and oxygen) indicates that there must be a supply of compounds containing these elements for sugar synthesis. Water and carbon dioxide are the simple inorganic forms in which these elements are found. Generally, land plants absorb water through the roots; the water is conducted through the stems to the leaves, where photosynthesis typically occurs in the chloroplasts of mesophyll cells. Carbon dioxide enters in gaseous form from the atmosphere through the stomata of the leaves by simple diffusion into the mesophyll cells. Considering that both H_2O and CO_2 are constantly being consumed during photosynthesis, their concentrations are less inside the plant than outside, a condition conducive to continuous diffusion into the photosynthetic sites.

The generalized equation for photosynthesis indicates that the carbon in the synthesized sugar comes from CO_2 and that the source of hydrogen is H_2O. The source of oxygen, however, is another consideration which is not self-evident.

The source and fate of oxygen during photosynthesis was determined by using an isotope of oxygen (^{18}O). Because isotopic oxygen is heavier than normal oxygen (^{16}O), it can be traced by various techniques through a series of reactions.

Water containing the tagged or labeled isotopic oxygen was given to a plant with the following results:

$$6\,CO_2 + 6\,H_2O^* \xrightarrow[\text{chlorophyll}]{\text{light}} C_6H_{12}O_6 + 6\,O_2^*$$

carbon dioxide   water containing isotopic oxygen   sugar   isotopic oxygen

It is obvious from the modified equation that the oxygen in water is not incorporated into the synthesized carbohydrate. Instead, the oxygen in water is eventually evolved as a gas. Logically, then, the oxygen source for sugar must be CO_2. It should also be noted in the above equation that if CO_2 is the source of oxygen for the synthesized sugar, there are twice as many oxygen atoms (12) in the CO_2 than in the sugar molecule (6). Similarly, note that the 12 O^* atoms produced as a product cannot result from only 6 O^* atoms supplied by water. In view of this unbalanced situation, the above equation should be modified as follows:

$$6\,CO_2 + 12\,H_2O^* \xrightarrow[\text{chlorophyll}]{\text{light}} C_6H_{12}O_6 + 6\,H_2O + 6\,O_2^*$$

carbon dioxide   water containing isotopic oxygen   sugar   water   isotopic oxygen

To summarize this relationship, the atoms in synthesized sugar ($C_6H_{12}O_6$) are derived from the 6 carbon atoms of carbon dioxide ($6\,CO_2$), 12 of the hydrogen atoms from water ($12\,H_2O^*$), and 6 of the oxygen atoms from carbon dioxide ($6\,CO_2$). The atoms in the 6 molecules of water as a product ($6\,H_2O$) are derived from 12 of the hydrogen atoms in ($12\,H_2O^*$) and 6 of the oxygen atoms from ($6\,CO_2$). The 6 molecules of oxygen ($6\,O_2^*$), another product of the reaction, are de-rived from water ($12\,H_2O^*$). This relationship is summarized in Figure 9-1.

The Necessary Conditions— Chlorophyll and Light

It has been known for some time that the process of photosynthesis occurs in the chloroplasts of the mesophyll cells in green leaves (see Figure 4-23). Other green structures, however, such as herbaceous stems, also have photosynthetic capacities. Seven different types of chlorophyll pigments are known to exist in plant cells. These are chlorophyll a, b, c, d, and e, bacteriochlorophyll, and bacterioviridin. Chlorophylls a and b are the most abundant and are found in all autotrophs except pigment-containing bacteria, which contain the bacteriochlorophyll and bacterioviridin. Chlorophyll b is also absent from certain forms of algae. Chlorophyll c, d, and e are found only in algae along with chlorophyll a.

The two principal chlorophylls found in chloroplasts of higher plants are blue-green chlorophyll a ($C_{55}H_{72}O_5N_4Mg$), and yellow-green chlorophyll b ($C_{55}H_{70}O_6N_4Mg$). It is believed that the lipid tail ($H_{39}C_{20}$) of these chlorophylls, built up of H—C—H units, is anchored to the lipid and protein layers of the grana of the chloroplast (see Figure 2-23). Although the relative proportions of chlorophyll a and chlorophyll b vary somewhat in higher plants, on the average, there are three molecules of chlorophyll a to every molecule of chlorophyll b.

Chloroplasts contain, in addition to chlorophyll, other pigments called accessory pigments. They are so designated because they assist chlorophyll molecules during the initial phase of photosynthesis. The two general types of accessory pigments are (1) carotenoids, pigments ranging in color from yellow to purple; and (2) phycobilins, red or blue pigments that are found in certain groups of algae.

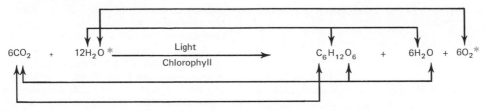

$$6CO_2 \quad + \quad 12H_2O\,^* \underrightarrow{\quad \overset{\text{Light}}{\underset{\text{Chlorophyll}}{}}\quad} \quad C_6H_{12}O_6 \quad + \quad 6H_2O \quad + \quad 6O_2\,^*$$

Figure 9-1. The fate of carbon dioxide and water during the process of photosynthesis.

Light, the second prerequisite photosynthetic condition, provides the energy to drive the photosynthetic process. Physicists have proposed two models to describe the behavior of light. The first model of the duality of light is the *wave theory*, which states that light is the result of minute, rapid vibrations or waves called *wavelengths* (Figure 9-2). These wavelengths vary from one type of light to another; the longer the wavelength, the smaller is the energy conveyed (Figure 9-2A). Conversely, the shorter the wavelength, the greater is the energy conveyed (Figure 9-2B). Sunlight, although appearing to be white, is a mixture of light of all colors, the colors being determined by different wavelengths. When white light is passed through a glass prism, the various wavelengths, seen as different colors, become apparent. This band of colors, the *visible spectrum*, includes those wavelengths perceived by the eye (Figure 9-3A). The wavelengths of the visible spectrum range from the relatively long waves of red light to the short waves of violet light with gradations in between. Violet light conveys more energy than red light.

The second model of the duality of light is the *quantum theory* which holds that light is composed of a stream of tiny light particles called *quanta* or *photons*. These photons represent discrete packets of energy given off by any light-emitting object which interact with a material object. This conception of the nature of light is particularly helpful, because although light seems to travel as a wave, it apparently interacts with matter, such as chlorophyll, as a particle.

Some solar radiation is absorbed by various gases and dust particles in the atmosphere before

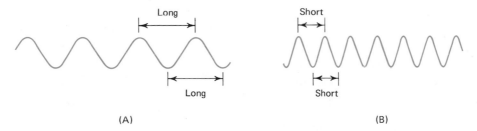

(A) (B)

Figure 9-2. Wave theory of light. According to one proposed model of the behavior of light, light energy travels in waves as a result of minute, rapid vibrations. **(A)** The longer the wavelength, as in the case of red light, the less energy is conveyed. **(B)** By contrast, the shorter the wavelength, as in the case of blue light, the more energy conveyed.

striking the surfaces of leaves. Of that portion of light actually striking leaves some is absorbed, a part is reflected, and another part is transmitted. In general, leaves absorb about 83 per cent of the light, while reflecting 12 per cent and transmitting 5 per cent. Of the 83 per cent absorbed, only 4 per cent is actually used by plants during photosynthesis (Figure 9-4). The remainder of the absorbed light energy, which is not used by the plant during photosynthesis, is dissipated as heat.

To determine what portion of the visible spectrum is absorbed by leaves during photosynthesis, chlorophyll may be extracted from leaves with organic solvents such as carbon tetrachloride or alcohol. The extract is then spread on a clean glass microscope slide and placed between a source of white light and a prism. Examination of the spectrum now shows that certain wavelengths are partially or completely absent, indicating that they have been more or less completely absorbed by the chlorophyll molecules on the slide (Figure 9-3B). By comparing the two spectra in Figure 9-3 it can be noted that the chlorophyll absorption spectrum contains dark bands, indicating that chlorophyll has absorbed these wavelengths. In regions where little absorption occurs, changes in the spectrum will be slight.

As implied previously, only those portions of the spectrum that are absorbed by chlorophyll can be used in photosynthesis. The chlorophyll spectrum indicates that much of the red, orange, blue, indigo, and violet regions are absorbed. Basically, chlorophyll can absorb all wavelengths of the visible spectrum, but the maximum absorption is in the red and blue regions and the minimum absorption is in the green region, which incidentally accounts for the green color of chlorophyll. The absorption spectrum of chlorophyll *a* is depicted graphically in Figure 9-5. An *absorption spectrum* is a plot of the efficiency with which a substance absorbs light as a function of wavelength.

Figure 9-3. (Opposite) Comparison of visible light spectrum and chlorophyll absorption spectrum. **(A)** Visible white light passing through a prism separates into a range of wavelengths from red (longest) to violet (shortest). **(B)** Sunlight passing through a chlorophyll extract in front of the prism reveals dark bands in the spectrum in the red, blue, and violet regions. These bands indicate that chlorophyll has absorbed light of these wavelengths.

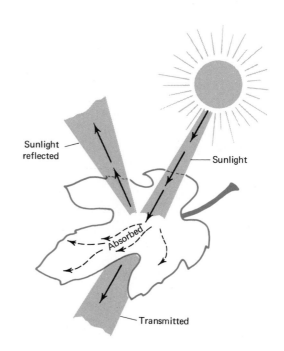

Figure 9-4. Fate of visible white light striking the surface of a leaf. The greatest portion of light is absorbed, although only about 0.5 to 2 per cent is actually used during photosynthesis. A somewhat lesser amount of light is reflected while the least amount is transmitted.

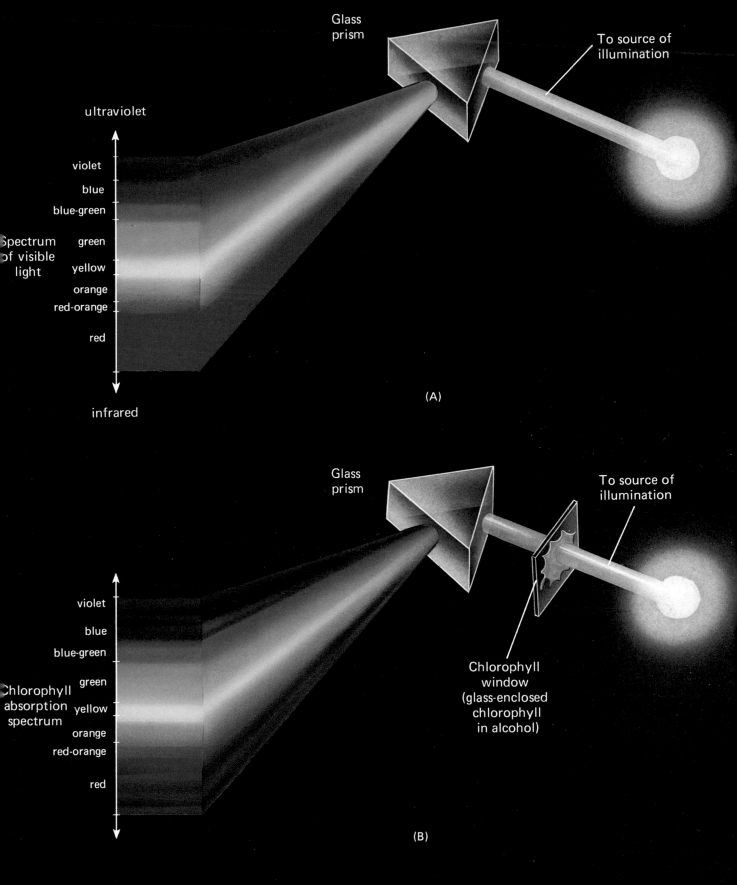

ultraviolet

violet
blue
blue-green

Spectrum of visible light

green
yellow
orange
red-orange

red

infrared

Glass prism

To source of illumination

(A)

violet
blue
blue-green

Chlorophyll absorption spectrum

green
yellow
orange
red-orange

red

Glass prism

To source of illumination

Chlorophyll window (glass-enclosed chlorophyll in alcohol)

(B)

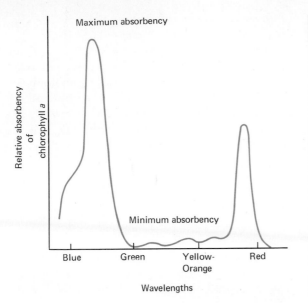

Maximum absorbency

Relative absorbency of chlorophyll *a*

Minimum absorbency

Blue   Green   Yellow-   Red
                 Orange

Wavelengths

Figure 9-5. Graphic representation of the absorption spectrum of chlorophyll *a*. The wavelengths of red and blue light are absorbed to a greater degree by chlorophyll *a* than other wavelengths of the visible light spectrum. (Modified from Arthur W. Galston, *The Life of the Green Plant*, 2nd edition, © 1964, Prentice-Hall, Inc., Englewood Cliffs, N.J. By permission.)

The Reactions of Photosynthesis

Many data indicate that photosynthesis is not a simple process but consists of at least two main steps. For example, if light is supplied to a green plant in flashes separated by dark periods, a higher photosynthetic yield for a given quantity of light is possible than if the light is given in one continuous period of illumination. Accordingly, the entire process of photosynthesis may be subdivided into two interrelated phases: light reactions and dark reactions. The *light reactions* refer to a group of reactions that depend upon light to take place. The *dark reactions* refer to reactions that do not require light energy to occur. They will proceed in the light or in the dark.

Light Reactions. Although many reactions of photosynthesis are incompletely understood, it is agreed that the initial reaction is the absorption of light by pigments within the chloroplasts. Chlorophyll *a* appears to be the principal light-absorbing pigment, although other chlorophylls, as well as other pigments, such as carotenes and xanthophylls, may also absorb light energy and transfer it to chlorophyll *a*. All atoms normally possess a given number of electrons which orbit about the nucleus at a certain distance. The outermost electrons are the most weakly bound to the nucleus and are the most chemically reactive. Atoms can frequently absorb energy from an outside source and the added energy raises the outermost electrons into orbits more distant from the nucleus. Such "activated" or "excited" electrons tend to return to their normal orbits and, in doing so, release the added energy that has been absorbed.

When a photon of light strikes a molecule of chlorophyll, and is absorbed, its energy is trans-

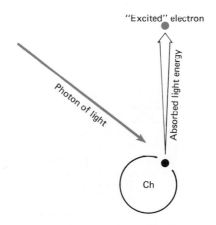

Figure 9-6. Effect of light on chlorophyll. When a photon of light strikes a molecule of chlorophyll, energy is transmitted from the light to an outer electron of a chlorophyll molecule. As a result, the chlorophyll electron is "excited" to a higher energy level.

ferred to an outer electron of the chlorophyll molecule (Figure 9-6). This "excited" electron is then raised to a higher energy level (moves farther from the nucleus), but because the molecule is now unstable, the electron tends to fall back to its normal energy level, giving up the absorbed energy. It has been estimated that the shift of an "excited" electron to a higher energy level and back again occurs in a millionth of a second. The energy-rich "excited" electrons do not fly off freely into space before falling back to their normal energy levels; they are trapped by various electron acceptors. These acceptors function in serial succession; that is, the electrons are passed from one acceptor to another in a series of chain reactions. The essential role of chlorophyll, therefore, is to absorb photons of light energy and transfer this energy to electron acceptors.

Through various pathways of electron transfer, chloroplasts are able to convert light energy into chemical energy. In other words, light energy (photons) absorbed by chlorophyll electrons is converted into chemical energy (ATP) by the passage of excited electrons from chlorophyll to specific electron acceptors. This conversion of light energy to chemical energy is called **photophosphorylation.** There are two plausible, but rather different pathways, to explain the routes which electrons given up by chlorophyll may take. These are cyclic photophosphorylation and noncyclic photophosphorylation.

In *cyclic photophosphorylation* (Figure 9-7) a molecule of chlorophyll loses an excited electron to some other substances and then takes it back, thus setting up a cycle. When chlorophyll donates an electron to an appropriate acceptor molecule, it becomes a powerful electron acceptor. Inasmuch as the chlorophyll molecule has lost an electron, it is left as a positive ion ($+1$). Thus chlorophyll serves both as an electron donor and an electron acceptor.

A number of cofactors (coenzymes) have been

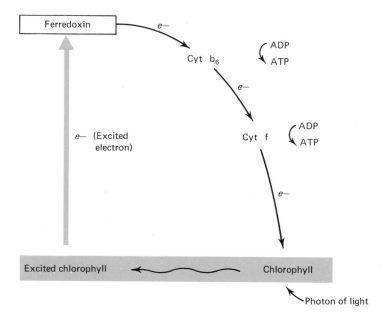

Figure 9-7. Cyclic photophosphorylation. Light energy is absorbed by an outer electron of chlorophyll and, in the process, the electron becomes energized and is raised to a higher energy level. The energized electron is picked up by cofactors such as ferredoxin and passed in serial succession through the cytochrome system of electron carriers. With each electron transfer, the electron is systematically de-energized as light energy is converted into chemical energy in the form of ATP. Ultimately, the de-energized electron returns to the chlorophyll molecule. (Adapted from D. I. Arnon, H. Y. Tsujimoto, and B. D. McSwain, "Photosynthetic Phosphorylation and Electron Transport," *Nature,* vol. 207, pp. 1367–1372, 1965. By permission of *Nature* and Dr. D. I. Arnon.)

identified as initial acceptors of excited electrons from chlorophyll. One of the most important of these is water-soluble iron protein related to chloroplasts called *ferredoxin*. Next the electron is passed to a group of acceptors collectively called *cytochromes*, iron-containing pigments structurally related to hemoglobin. Each cytochrome passes the electron to another cytochrome acceptor molecule, and so on. As the electron is cycled in this manner, it progressively loses energy. When the electron left the chlorophyll molecule it was energy-rich, but upon its return it is energy-poor. Thus the electron is eased down an energy gradient from the "excited" state to the normal state and each electron transfer results in a gradual loss of energy. At several points along the carrier chain of electron acceptors, the released light energy is converted into chemical energy by being incorporated into molecules of ATP.

It will be recalled that if an inorganic (low-energy) phosphate group is added to ADP with the addition of energy, a molecule of ATP is formed.

$$\text{ADP} \quad + \quad \text{P} \quad + \text{ energy} \longrightarrow \text{ATP}$$

| adenosine | inorganic | adenosine |
| diphosphate | phosphate | triphosphate |

After the cytochromes have accepted the electron and ATP has been formed, the electron is transferred to ionized chlorophyll. Chlorophyll, now an electron acceptor, takes the electron from the final cytochrome. Thus it again becomes a neutral molecule capable of absorbing more light energy. The ATP thus produced during cyclic photophosphorylation is now available to the cell as potential chemical energy. Some of this energy, as will be discussed later, is used during the dark reactions for the synthesis of sugar.

At this point it might be well to examine, in somewhat more detail, the mechanism of electron transport, because energy transformations in cells result from electron transfers. In living plants, as well as in other living organisms, the term *oxidation* refers to the removal of electrons (e^-) from atoms or molecules and the process whereby electrons (e^-) are added to atoms or molecules is called *reduction*. With rare exceptions, the removal or addition of electrons may be considered synonymous with the removal or addition of hydrogen atoms, because hydrogen is usually involved in cellular oxidation–reduction reactions. An electron is added to one molecule and so must have been removed from another; with every oxidation there is a coupled reduction. Another way of stating this is that whenever one substance is reduced, another is oxidized. Atoms or molecules that are oxidized lose energy, whereas atoms or molecules that are reduced gain energy.

Inasmuch as electron transfers in oxidation–reduction reactions are rather complex, they may simply be visualized as additions and removals of hydrogen and oxygen. Two of the commonest processes by which a compound may be reduced in plant cells are by removal of oxygen or by the addition of hydrogen (Figure 9-8A). Similarly, a compound may be oxidized by the addition of oxygen or by the removal of hydrogen (Figure 9-8B). Reaction A is an oxidation–reduction reaction involving the addition or removal of oxygen, whereas reaction B, which is also an oxidation-reduction reaction, involves the addition or removal of hydrogen. In plant systems, reaction B is the basic mechanism by which oxidation–reduction takes place. Note, once again, that each reduced compound stores energy and each oxidized compound releases energy.

If cyclic photophosphorylation is considered in these terms it can be seen that the stored energy in a chlorophyll electron, which is light energy, is passed from one compound to another, each compound being oxidized and reduced in the

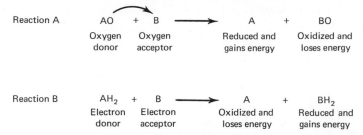

Reaction A   AO + B → A + BO
              Oxygen Oxygen   Reduced and   Oxidized and
              donor   acceptor   gains energy loses energy

Reaction B AH_2 + B → A + BH_2
              Electron Electron   Oxidized and Reduced and
              donor   acceptor   loses energy gains energy

Figure 9-8. Types of oxidation-reduction reactions that occur in plant cells. Reaction A: Removal and addition of oxygen. Reaction B: Removal and addition of hydrogen (and electrons). Reaction B is the common type of oxidation-reduction that occurs in plant systems.

process. Ultimately the energy is stored in ATP. Chlorophyll, by losing the electron temporarily, is oxidized and therefore releases energy. Because of the lost electron, it assumes a positive charge. Next the electron is passed to cofactors, which in turn are reduced and therefore store energy. The cofactors then transfer the electron to cytochromes, being oxidized in the process. Each cytochrome passes the electron to another cytochrome, each being reduced first and then oxidized. The final cytochrome passes the deenergized electron to chlorophyll and the chlorophyll molecule is reduced. As a result of regaining an electron, the chlorophyll loses its positive charge. Thus cyclic photophosphorylation is completed.

The dark reactions of photosynthesis, which will be treated later, are concerned with combining the carbon and oxygen of CO_2 with hydrogen to synthesize sugars. The source of hydrogen is water, and this implies a splitting of water to make the hydrogen available. Green plants possess such a mechanism whereby the hydrogen of water is made available for incorporation into sugars. This process is called *noncyclic photophosphorylation* (Figure 9-9). It is so named because the excited electron lost by chlorophyll is passed to an acceptor, which

passes it to another acceptor and not back to chlorophyll.

In noncyclic photophosphorylation, the electron lost by chlorophyll is picked up by ferredoxin and eventually passed to a confactor called *NADP* (*nicotinamide adenine dinucleotide*). Unlike the electron acceptors in cyclic photophosphorylation, NADP does not immediately pass the electron to an acceptor but retains it. Because an electron has been added to NADP, this acceptor is unstable. The electron takeover by NADP is stabilized by the addition of protons (H^+) resulting from the decomposition of water. The resulting compound is the reduced form, *NADPH*. It is this compound that furnishes hydrogen atoms for combination with carbon and oxygen during the dark reactions.

The OH^- ions, a second decomposition product of water, are used as a source of electrons to replace those lost from chlorophyll. These electrons are transferred through a series of cytochromes until they are finally accepted by chlorophyll, thus reducing it. The transference of electrons from OH^- to chlorophyll is coupled with ATP formation as in cyclic photophosphorylation. The (OH) groups left after removal of electrons react

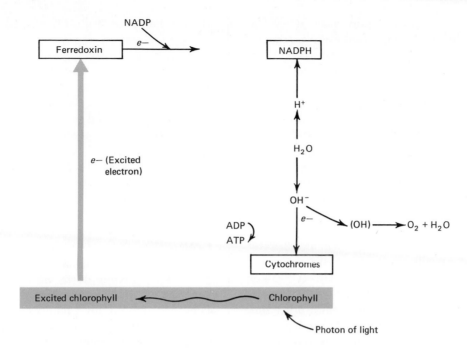

Figure 9-9. Noncyclic photophosphorylation. Like cyclic photophosphorylation, noncyclic photophosphorylation consists of a series of oxidation-reduction reactions in which electrons are transferred from one coenzyme to another in serial succession. The electrons do not return to chlorophyll as in cyclic photophosphorylation. They are passed to NADP. The electrons required to restore chlorophyll to the ground state come from the splitting of water to form oxygen. The two principal outcomes of noncyclic photophosphorylation are the formation of NADPH and the production of ATP. (Adapted from D. I. Arnon, H. Y. Tsujimoto, and B. D. McSwain, "Photosynthetic Phosphorylation and Electron Transport," *Nature*, vol. 207, pp. 1367–1372, 1965. By permission of *Nature* and Dr. D. I. Arnon.)

with each other to produce water and oxygen, products of photosynthesis. The oxygen diffuses through the mesophyll cells of the leaf into the air spaces and through the stomata into the atmosphere. The important outcomes of noncyclic photophosphorylation are the production of reduced NADP (NADPH), the formation of ATP, and the release of oxygen as a waste product.

In all probability both cyclic and noncyclic photophosphorylation occur in living green cells when illuminated. Thus, in a chloroplast as a whole, some chlorophyll molecules are involved in cyclic photophosphorylation and the production of ATP only, whereas others are concerned with noncyclic photophosphorylation and the production of both ATP and NADPH. The formation of NADPH signals the end of the light phase of photosynthesis. This compound, together with ATP, now enters the dark phase of photosynthesis.

Dark Reactions. For many years it was believed that a hexose sugar (glucose) was the first stable product of photosynthesis. However, extensive research carried on by Melvin Calvin and his associates at the University of California demonstrated that a three-carbon compound, *phosphoglyceraldehyde* (*PGAL*), is one of the first stable substances produced during the dark reactions.

Essentially, Calvin sought to trace the path of carbon through the dark reactions by labeling carbon dioxide with ^{14}C. His experimental organism, the green alga *Chlorella*, was placed in a culture medium containing labeled carbon dioxide and permitted to carry on photosynthesis. To determine which compounds had incorporated the labeled carbon, the reaction was stopped at various intervals. Analysis of the earliest products of photosynthesis revealed that PGAL contained radioactive carbon. Samples taken at other time intervals showed that other compounds had incorporated the labeled carbon into its structure. In this manner, a number of intermediate products were identified as compounds of the dark reaction.

The dark reactions are concerned with combining the hydrogen atoms from NADPH with CO_2 to form sugars. In this series of synthetic reactions, energy is needed to reduce carbon dioxide. ATP and NADPH function as energy donor and reducing agent, respectively.

Sugars contain a great deal of stored chemical energy, whereas CO_2 possesses relatively little. Just as energy is slowly released in a stepwise series of oxidation-reduction reactions during photophosphorylation, so also is energy slowly built up during *CO_2 fixation*, that is, during the reduction of CO_2 by NADPH to form sugars. In effect, CO_2 is slowly pushed up an energy gradient through a series of intermediate compounds, some of them unstable, until a stable end-product sugar is formed. However, CO_2 cannot be directly reduced as a free molecule. It must first be combined with a larger molecule, a step which makes possible its subsequent reduction (Figure 9-10).

The first step of the dark reactions, preparatory to combining with CO_2, consists of the conversion of a five-carbon sugar called *ribulose phosphate* (*RuP*) into a phosphorylated sugar called *ribulose diphosphate* (*RuDP*). RuP is continually being produced in photosynthetic cells and is always present in chloroplasts as a reactant. This conversion is necessary because CO_2 is not able to combine with RuP as such, whereas RuDP represents a form that will facilitate the combination readily.

Within a fraction of a second after CO_2 reaches a chloroplast, it combines with the RuDP and forms an unstable six-carbon molecule which immediately breaks down into 2 three-carbon compounds called *phosphoglyceric acid* (*PGA*). Each molecule of PGA is then phosphorylated to a triose sugar by ATP. It is at this juncture that energy from the light reactions participates in the dak reactions. In the process of phosphorylation molecules of ATP are broken down into ADP + P. In addition, hydrogen is transferred to the PGA molecules from NADPH. In this process NADPH is oxidized back to NADP. The product of the phosphorylation and the reduction of PGA is the formation of energy-rich triose sugars called phosphoglyceraldehyde (PGAL). The ADP + P and the NADP return to the light reactions to assume their functions as cited earlier.

For every three molecules of CO_2 that are used, six molecules of PGAL are produced. One molecule of PGAL represents the stable end product of photosynthesis, which is converted into other sugars, usually hexoses, such as glucose and fructose. For this reason, it is customary to consider a hexose ($C_6H_{12}O_6$) as the end product of photosynthesis and not PGAL. The remaining five PGAL molecules are converted into three molecules of RuDP in an intricate series of reactions

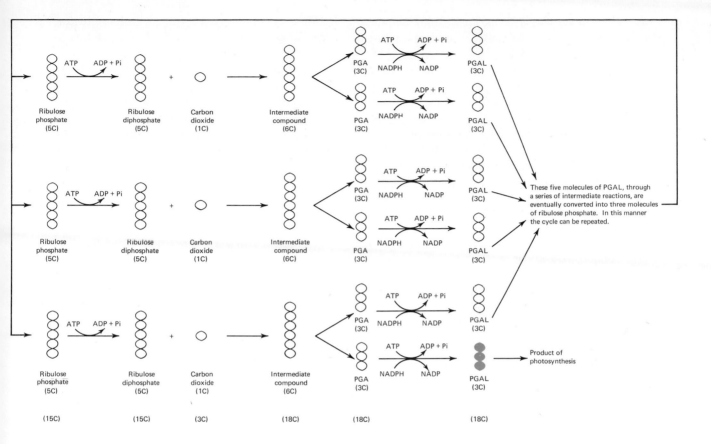

Ribulose phosphate (5C) — ATP → ADP + Pi — Ribulose diphosphate (5C) + Carbon dioxide (1C) → Intermediate compound (6C) → PGA (3C) — ATP → ADP + Pi / NADPH → NADP — PGAL (3C)

PGA (3C) — ATP → ADP + Pi / NADPH → NADP — PGAL (3C)

Ribulose phosphate (5C) — ATP → ADP + Pi — Ribulose diphosphate (5C) + Carbon dioxide (1C) → Intermediate compound (6C) → PGA (3C) — ATP → ADP + Pi / NADPH → NADP — PGAL (3C)

PGA (3C) — ATP → ADP + Pi / NADPH → NADP — PGAL (3C)

These five molecules of PGAL, through a series of intermediate reactions, are eventually converted into three molecules of ribulose phosphate. In this manner the cycle can be repeated.

Ribulose phosphate (5C) — ATP → ADP + Pi — Ribulose diphosphate (5C) + Carbon dioxide (1C) → Intermediate compound (6C) → PGA (3C) — ATP → ADP + Pi / NADPH → NADP — PGAL (3C)

PGA (3C) — ATP → ADP + Pi / NADPH → NADP — PGAL (3C) → Product of photosynthesis

(15C)   (15C)   (3C)   (18C)   (18C)   (18C)

in which their atoms undergo extensive reshuffling and phosphorylation. The three RuDP molecules thus formed can now react with more CO_2 to initiate another cycle of dark reactions.

SYNTHESIS OF OTHER CARBOHYDRATES

The synthesis of PGAL will be considered the termination of photosynthesis. Although other carbohydrates are synthesized in chloroplasts, these reactions are not considered as part of photosynthesis, because they can occur elsewhere in the

Figure 9-10. Dark reactions of photosynthesis. These reactions, which proceed independent of light, are concerned with combining the hydrogen atoms of NADPH with carbon dioxide to form PGAL. ATP and NADPH function as energy donor and reducing agent, respectively. (Modified from J. A. Bassham and M. Calvin, *The Path of Carbon in Photosynthesis,* © 1957, Prentice-Hall, Inc., Englewood Cliffs, N.J. By permission.)

cells independently of photosynthesis as well as in other nonphotosynthetic cells.

As PGAL forms in chloroplasts, only one molecule represents the product of photosynthesis; the other five are converted to RuDP to initiate another cycle. The single PGAL molecule, as well as other single PGAL molecules from similar photosynthetic reactions, may be used by the plant in a variety of ways. Before being utilized, PGAL is usually rapidly converted into a hexose sugar such as fructose or glucose, although it may be utilized directly. The hexose sugars thus produced are utilized by the plant in various ways (Figure 9-11). Some are used by the cell which produced them to provide energy (via respiration) or to serve as building materials. Others are translocated to nonphotosynthetic cells of the plant for these same purposes. Still others may be stored in certain parts of the plant. In addition, the sugars produced during photosynthesis also serve as the raw materials for the synthesis of fats, proteins, and other compounds.

The function of sugars as energy-providing molecules will be considered in Chapter 10. As building materials, various sugars such as glucose are synthesized into wall material (cellulose), various cytoplasmic constituents, and components of certain organelles. Proteins and fats, formed from synthesized sugars, also assume respiratory and construction roles. Chief among these construction processes involving proteins are the formation of new cytoplasm (assimilation), the repair of organelles, and the synthesis of enzymes. Fats, in conjunction with proteins, are important structural constituents of cellular membranes and organelles as well as cytoplasmic components.

A photosynthetic cell produces many more sugar molecules than are needed for its own maintenance; the extra sugars are translocated to other parts of the plant. This translocation starts in the leaf mesophyll cells, proceeds through the phloem, and continues into stems, roots, and nonphotosyn-

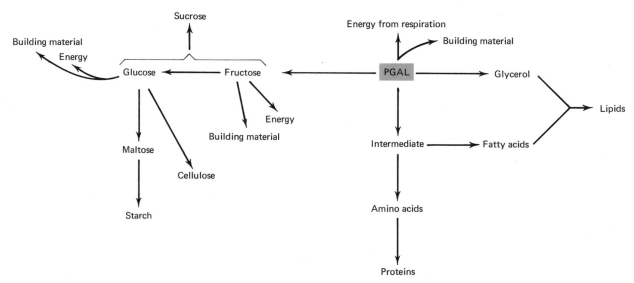

Figure 9-11. Some of the possible fates of PGAL, the first stable product of photosynthesis. Once PGAL is formed, it may be used in the synthesis of other carbohydrates, lipids, and proteins. These larger synthesized molecules may then be utilized to provide energy through respiration, to serve as storage materials, or to serve as building materials for cellular organelles and protoplasm.

thetic portions of the plant. The conversion of PGAL into glucose or fructose is prerequisite to translocation becuase PGAL as such is very reactive and in the course of translocation might combine with other materials and never reach its final destination.

If the photosynthetic sugars are not directly used for energy production, or as a building material, or as a translocated substance, they are stored by the plant. Such storage generally occurs in roots and stems and to a lesser extent in leaves. As in translocation, storage also implies inactivation of PGAL by conversion into glucose or fructose. In addition, any storage product must be compacted to occupy a minimum of space within plant storage areas. Compaction of storage products is facilitated by dehydration synthesis. By this mechanism a number of glucose molecules may be joined together to form a molecule of starch. By the same mechanism, sucrose may be formed by the combination of glucose and fructose molecules. The essential point here is that, each time two molecules undergo dehydration synthesis for storage, a molecule of water is removed. This water removal is an important factor in the complete utilization of the available space in various plant storage centers. Once a molecule, such as starch, is stored, it must be made both soluble and diffusible again if it is to be made available for cellular use. This is accomplished by digestion.

FACTORS AFFECTING PHOTOSYNTHESIS

The rate of metabolic reactions, including photosynthesis, depends upon the relative intensity or amount of the combined influences of both external and internal factors in the environment. For example, photosynthesis will proceed very slowly at low light intensities, considerably faster at progressively higher ones, and then slower under the influence of very bright light until the process stops. In other words, depending upon light intensity, photosynthesis may proceed at a minimum rate (low intensity), an optimum rate (the intensity at which the process is most rapid), and a maximum rate (the highest intensity at which photosynthesis will occur). In this ragard, light may be viewed as a *limiting factor,* that is, any factor, which, if deficient, decreases the rate of a reaction. Even though light is most frequently the limiting factor under natural conditions of day and night, other factors operate in a similar manner. The amount of carbon dioxide may affect the rate of photosynthesis even though there is an otherwise optimum light intensity. In addition, temperature may be a limiting factor in evergreens during the winter months. Essentially, most activities of a plant are affected by more than one factor, and these activities cannot operate at their maximum if any one of these factors is below a certain critical level.

Chief among the external factors that affect photosynthesis are temperature, light, CO_2 concentration, water supply, and the availability of mineral elements in the soil. Among the internal factors that affect the process are anatomical features of leaves, the presence of products of photosynthesis and enzymes.

External Factors

Temperature. The temperature requirement for photosynthesis varies according to the climate in which plants live. For example, evergreens of cold climates can carry on photosynthesis at temperatures below 0°C, whereas some plants, such as algae in the water of hot springs, operate effectively at temperatures as high as 75°C. The vast majority of temperate climate plants function best at temperatures between 10 and 35°C, assuming that other variables are constant. Within this range, for each 10°C rise, the rate of photosynthesis increases about 2.5 times up to 35°C, at which temperature the rate starts to decline. Although temperature is being considered an external factor, it should be recalled that temperature also affects the activity of enzymes, which are considered to be internal factors.

Light. Light is one of the most important external factors varying in intensity, quality, and duration, each affecting the rate of photosynthesis. As noted earlier, all parts of the visible spectrum do not equally affect photosynthesis. The intensity of light decreases from the leaf surface to the chloroplasts, owing to certain anatomical features such as the thickness of the cuticle, the orientation of chloroplasts in the mesophyll cells, the presence of trichomes, and the thickness of the epidermis. In general, prolonged intense light tends to decrease the photosynthetic rate.

The duration of light also affects the quantity of carbohydrate production. Quite obviously, during the short days of autumn and winter, for those plants that remain green, the rate is much slower than during the long days of spring and summer.

Water Supply. Water is one of the necessary metabolites, so it may affect the rate of photosynthesis but to a somewhat lesser extent than the other factors. Only when the water supply becomes so low that wilting occurs, however, does this factor assume any significance. During prolonged periods of drought the stomata may close, thereby affecting water conservation but at the same time reducing diffusion of oxygen and carbon dioxide.

Mineral Availability. From earlier discussions of the structure of chlorophyll it becomes obvious that certain minerals such as Mg, N, and Fe are necessary for chlorophyll formation as well as for the synthesis of certain electron acceptors. Any deficiency of these minerals causes a distinct yellowing of leaves, called **chlorosis.** This mineral deficiency may be caused by other contributing factors, but when it occurs, the rate of photosynthesis is significantly decreased.

CO_2 Concentration. Perhaps no other single factor has a greater influence on the rate of photosynthesis than the CO_2 supply from the atmosphere. The content of CO_2 in the atmosphere is rather constant at about 0.03 per cent (3 parts per 10,000 parts of air), a seemingly small amount but totaling approximately 2,000,000,000,000 tons in the atmosphere. The amount of atmospheric CO_2, however, is limited and, if the present supply were not constantly being renewed, it would be depleted in about 22 years. Fortunately, several natural processes are continually releasing CO_2 to the atmosphere. These include the respiration of plants and animals, the decay of organic materials, combustion of fuels, weathering of rock, volcanic activity, and diffusion from the oceans.

Internal Factors

Leaf Structure. Various anatomical features of leaves affect the amount of CO_2 and the intensity and quality of light reaching chloroplasts. For example, the size, position, number and physiology of stomata, together with the amount of intercellular space between mesophyll cells, affect the diffusion of gasses. Also, the quality and intensity of light are affected by the thickness of the cuticle and epidermal cells, the presence of trichomes, the orientation of mesophyll cells, and the position of chloroplasts.

Photosynthetic Products. Assuming the normal operation of other factors, the rate of any biochemical reaction is determined by the accumulation of products of the reaction. An accumulation of products tends to decrease the rate of a biochemical reaction. PGAL, for example, may build up into sucrose in the chloroplasts during the day when the photosynthetic rate is fastest and it may not be translocated at a comparable rate. The result would be a decrease in sugar production.

Enzymes. In view of the fact that enzymes catalyze energy transfers and synthetic reactions, they assume an important role in photosynthesis. The same factors that control enzymatic reactions (discussed in Chapter 8) correspondingly regulate the rate of photosynthesis. These include enzyme concentration, substrate concentration, products of the reaction, pH, temperature, and inhibitors.

FAT SYNTHESIS

Fats are built up from molecules of glycerol and fatty acids by a dehydration synthesis reaction. Prerequisite to synthesizing a fat, a plant cell must first be provided with the essential building units and, in addition, must contain the necessary enzymes. Both glycerol and fatty acids are supplied by carbohydrates by a series of intermediate reactions. In this context, glycerol originates from PGAL or glucose, whereas fatty acid synthesis begins with the fermentation of sugars. Assuming that the raw materials are present, a plant cell can combine a molecule of glycerol with three molecules of fatty acids in a dehydration synthesis reaction to form a fat, in the presence of the appropriate enzymes. The complex multienzyme system, *fatty acids synthetase*, is one of the principal catalysts that is involved in the production of fatty acids.

PROTEIN SYNTHESIS

It will be recalled that proteins consist of atoms of carbon, hydrogen, oxygen, and nitrogen, and, in addition, some contain sulfur and phosphorus. Previously synthesized carbohydrates serve as the source of carbon, hydrogen, and oxygen, and nitrates, sulfates, and phosphates absorbed by roots from the soil supply the necessary minerals. The proteins ultimately synthesized by plant cells serve as structural molecules in organelles, membranes,

and cytoplasm; as functional molecules (enzymes); as storage products (gluten and aleurone grains); and as energy molecules during respiration, although most cellular energy is derived from the oxidation of carbohydrates and fats.

To understand the mechanism of protein synthesis it will be necessary to review the structures of DNA and RNA and the amino acids and proteins in Chapter 2. Each of these molecules represents an important aspect of the total synthetic process.

Various experimental data have supported the conclusion that DNA possesses the proper chemical code to control all metabolic reactions of the cell. The DNA code is based upon information stored in a base sequence composed of varying arrangements of the four nucleotides, and all information contained in this code is passed from DNA to other parts of the cell. Within the chemical storehouse of the cell there are 20 different amino acids which may be built up into proteins of various sizes and levels of complexity. Inasmuch as there are only four different nucleotides in DNA, and if each nucleotide coded one amino acid, only four amino acids would have the potential for protein synthesis. Similarly, if two nucleotides specified one amino acid and if all possible combinations were considered, then still only 16 (4^2) amino acids could be built up into proteins. According to mathematical reasoning, then, at least three nucleotides must specify each amino acid. In this manner 64 (4^3) combinations are possible. This concept, whereby at least three nucleotides specify an amino acid, is called the *triplet code*. These triplets are commonly referred to as **codons.**

Most of the information concerning DNA coding was accumulated from various experiments with RNA, a very similar nucleic acid. The initial breakthrough came with the discovery of an enzyme, *polynucleotide phosphorylase,* that could catalyze the synthesis of RNA nucleotides into chain-

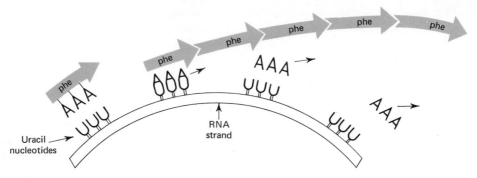

Figure 9-12. Synthesis of phenylalanine. A single RNA strand containing only uracil nucleotides will synthesize a protein molecule consisting only of the amino acid phenylalanine.

like nucleic acid molecules. By placing a mixture of amino acids in a test tube with certain RNA sequences it was found that specific proteins were formed. For example, it was determined that an RNA chain consisting of only uracil nucleotides would synthesize a protein molecule consisting only of the amino acid *phenylalanine*. For every three uracil nucleotides (codon) in sequence in the RNA chain, a molecule of phenylalanine was formed (Figure 9-12).

It was stated earlier that the code for protein synthesis is contained in DNA. By considering the nucleotide sequence of RNA, the "message" can be translated into a DNA sequence, because both nucleic acids are very similar. Remember that uracil replaces thymine in RNA; the other nucleotides remain the same. In a series of related investigations, the codons for all but three of the 64 trinucleotides were elucidated. Perhaps it would appear as though there are more codons than there are amino acids. In fact, however, experimental data have demonstrated that most of the codons are used during protein synthesis. Of the 20 amino acids, all but one (*tryptophan*) is specified by at least two different codons and several are specified by as many as six different codons. Thus,

among the codons, there are a number of synonyms.

In the process of protein synthesis the proper kinds of amino acids must be assembled in a specific linear sequence. To accomplish this, the cell must expend energy for the polymerization of amino acids, the proper amino acids have to be selected from those available, and these amino acids have to be arranged according to the sequence prescribed by DNA. In addition, most metabolic reactions are carried out in the cytoplasm in various organelles, and because DNA is restricted to the nucleus, DNA must logically exert indirect metabolic control. It appears that the metabolic mediator between nuclear DNA and cytoplasmic biochemical reactions is RNA, which carries the various coded messages to the sites of these reactions.

Having considered some of the necessary background information, let us now examine the mechanism of protein synthesis. The first step involves the synthesis of an RNA strand from a DNA strand (Figure 9-13A). In the cell there is an enzyme, *RNA polymerase*, that catalyzes the build up of RNA nucleotides into RNA strands. Both the kind and sequence of nucleotides in an RNA strand are determined by DNA. In accordance with the concept of complimentary nucleotide pairings, the adenine of DNA pairs off with the uracil of RNA, the thymine of DNA with the adenine of RNA, and cytosine of either nucleic acid with guanine of the other. When the RNA strand is assembled, DNA acts as a *template;* that is, the RNA molecule is copied on the DNA code. This RNA molecule, called *messenger RNA (mRNA),* now has the DNA information to synthesize a protein according to a specific amino acid sequence.

Messenger RNA separates from DNA and leaves the nucleus through the porous nuclear membrane. It attaches itself by one end around a ribosome in the cytoplasm and there it forms a template or a pattern for the synthesis of a specific protein with had been predetermined by DNA (Figure 9-13B). In the cytoplasm are found all the necessary amino acids to synthesize any protein. These amino acids are activated by ATP and united with smaller RNA molecules called *transfer RNA (tRNA).* Each transfer RNA is specific for each of the 20 amino acids. After the amino acid is picked up, the transfer RNA carries it to the messenger RNA, where it is fitted into place (Figure 9-13C). This "fit" is not a random one, as a particular portion of the messenger RNA will accept only a transfer RNA that contains the proper complimentary nucleotide sequence.

As the amino acids become assembled at the ribosome, they are joined together via peptide bonds until a whole chain of them results in the formation of a protein. Each protein molecule thus formed may consist of hundreds of amino acids in a precise sequence as orginally determined by DNA. After synthesis, the protein separates from the ribosome and is transported to any part of the cell, where it assumes a particular function (Figure 9-13D).

Proteins, like other molecules of the cell, are constantly being broken down or worn down so that the process of synthesis is repeated over and over again to form identical protein molecules from the same template. Perhaps the salient feature concerning protein synthesis is that it is the process by which enzymes (functional proteins, are produced. These biochemical catalysts, as discussed earlier, regulate all phases of metabolic activity, including protein synthesis. In short, DNA makes RNA, RNA makes proteins, and proteins control the metabolism of cells.

ASSIMILATION—THE CLIMAX OF ANABOLISM

A great deal of energy expended in the plant cell is concerned with **assimilation,** the processes by which food is converted into cytoplasm (living material) and cell walls (nonliving material). Assimilation represents the climax, or end result, of all anabolic processes. It has been shown that the sugars produced during photosynthesis may be converted into other carbohydrates, or to fats, or, with the addition of certain chemical elements from the soil, proteins may be formed.

The transformation of inorganic raw materials into these foods requires energy. The plant cell then takes foods (nonliving) and with further energy expenditure converts them into cytoplasm by assimilation. Little is known about the mechanism of assimilation by which nonliving substances become living except that it can occur only in a living system. A great deal of information has been accumulated concerning the synthesis of raw materials into foods, but the conversion of foods into cytoplasm remains a poorly understood process.

Before any substance can be assimilated, it must be soluble and it must be translocated to actively growing regions of the plant, such as the meristems. Once there, in the presence of auxins (plant

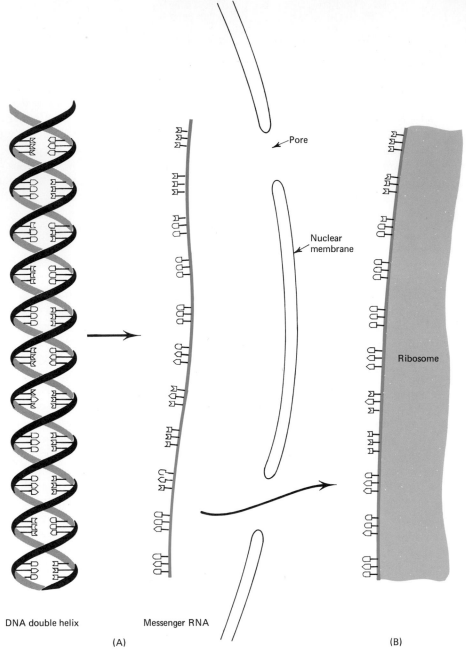

DNA double helix   Messenger RNA

(A)   (B)

Pore

Nuclear membrane

Ribosome

Figure 9-13. Diagrammatic representation of a possible mechanism of protein synthesis. **(A)** In the nucleus of the cell, DNA temporarily uncoils, exposing a sequence of nitrogenous bases. In this condition, DNA acts as a template for the formation of messenger, RNA. The sequence of bases on the messenger RNA is assembled in a strand in an order complementary to that of the nucleotides of the DNA strand. **(B)** The messenger RNA strand leaves the nucleus and moves to the ribosomes. **(C)** In the cytoplasm, transfer RNA molecules become bonded to free amino acids. The type and sequence of the unpaired triplet of bases at the end of each

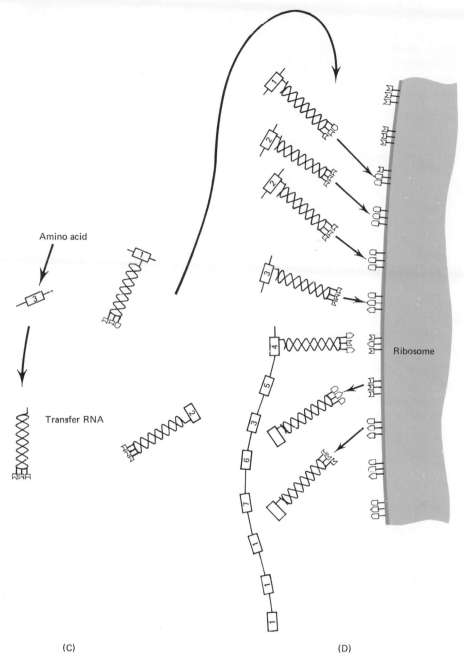

Amino acid

Transfer RNA

Ribosome

(C)   (D)

transfer RNA molecule determines the kind of amino acid to which it becomes bonded. (**D**) Transfer RNA brings the activated amino acids to a ribosome. The type and arrangement of the unpaired bases determines where each molecule of transfer RNA fits onto the messenger RNA. The side-by-side attachment of the transfer RNA's on the messenger RNA brings the amino acids that they carry into close contact, and the adjacent amino acids polymerize to form a polypeptide chain (protein).

hormones), these substances are assimilated as part of cellular growth, enlargement, and differentiation. Some data suggest that assimilation is probably more a matter of physical organization of the various cell structures than of chemical change.

CHAPTER IN RETROSPECT

The most significant anabolic process upon which all others are dependent is photosynthesis, a process by which green plants synthesize sugars which are then used to form fats, proteins, vitamins, hormones, and other essential organic compounds. Chlorophyll and light are the prerequisite photosynthetic conditions. The essential role of chlorophyll is to absorb photons of light energy and transfer this energy, through various pathways, to chemical energy in the form of ATP (photophosphorylation). Like other metabolic plant activities, photosynthesis is affected by various conditions in and out of cells. To name a few, light, temperature, water availability, mineral supply, and carbon dioxide concentration can individually or collectively influence the rate of photosynthetic activity. For instance, photosynthesis will proceed slowly at either extremely low or extremely high light intensity. The same is true with temperature. An intermediate range of temperature ($10°$ to $35°C$) is ideal for a vast majority of temperate-climate plants, assuming that other variables are constant. It should also be emphasized that the same factors that control enzymatic reactions (discussed in Chapter 8) correspondingly govern the rate of photosynthesis.

The culmination or end result of all anabolic processes, including the synthesis of carbohydrates, fats, and proteins, is the conversion of nutrients into the living substance of cytoplasm (assimilation). The process by which this is accomplished is still relatively unknown.

Questions and Problems

1. Compare the magnitude of photosynthesis with that of the products of mankind. Approximately what percentage of all foods produced by plants in consumed by man? Explain your answer in terms of other competitors for the world's food supply.
2. Briefly describe the intermediate reactions which occur prior to sugar formation in the generalized photosynthetic reaction

$$6\,CO_2 + 6\,H_2O \xrightarrow[\text{chlorophyll}]{\text{light}} C_6H_{12}O_6 + 6\,O_2$$

3. Show how the use of isotopic oxygen in a photosynthetic reaction indicated that the oxygen in glucose originally came from carbon dioxide rather than water. How must the equation in problem 2 be modified to reflect this information?
4. What is the effect of different wavelengths of light on the photosynthetic process? Describe a method by which the absorption of different wavelengths of light by leaves can be measured.
5. What is meant by cyclic photophosphorylation? Is this a dark reaction or light reaction? Compare with noncyclic photophosphorylation.
6. What was the significance of Calvin's work in the field of photosynthesis? Why did he label carbon dioxide with ^{14}C?
7. Describe in detail the synthesis of PGAL. Why must PGAL be converted to either glucose or fructose prior to translocation?
8. Briefly outline the various factors in the internal and external environment which affect the rate of photosynthesis. Which of these factors can be controlled by man under greenhouse conditions?
9. In what specific way does DNA control protein synthesis? What is the related role of RNA?
10. Why is assimilation regarded as the culmination of anabolism? Compare our knowledge of the synthesis of raw materials into foods with that of the conversion of foods into cytoplasm.

Suggested Supplementary Readings

Ackley, M. E., and P. B. Whitford, *The Chemistry of Photosynthesis*. New York: Appleton-Century-Crofts, 1965.

Arnon, D. I., "Ferredoxin and Photosynthesis," *Science,* vol. 149, pp. 1460–1470, 1965.

————, "The Role of Light in Photosynthesis," *Scientific American,* November 1960.

Baker, J. J. W., and G. E. Allen, *Matter, Energy, and Life*. Reading, Mass.: Addison-Wesley Publishing Company, Inc., 1965.

Bassham, J. A., "The Path of Carbon in Photosynthesis," *Scientific American,* June 1962.

Borograd, L., "Photosynthesis." In W. A. Jensen and L. G. Kavaljian (eds.), *Plant Biology Today,* 2nd ed. Belmont, Calif.: Wadsworth Publishing Company, 1966.

Goldsby, R. A., *Cells and Energy*. New York: The Macmillan Company, 1967.

Hill, R., and C. P. Whittingham, *Photosynthesis*. New York: John Wiley & Sons, Inc., 1955.

Kamen, M. D., *Primary Processes in Photosynthesis*. New York: Academic Press, Inc., 1963.

Lehninger, A. L., "How Cells Transform Energy," *Scientific American,* September 1961.

Park, R. B., "The Chloroplast." In J. Bonner and J. Varner (eds.), *Plant Biochemistry*. New York: Academic Press, Inc., 1965.

Rabinowitch, E. I., "Photosynthesis," *Scientific American,* August 1948.

————, and Govindjee, "The Role of Chlorophyll in Photosynthesis," *Scientific American,* July 1965.

Rosenberg, J. L., *Photosynthesis*. New York: Holt, Rinehart and Winston, Inc., 1965.

Stumpf, P. K., "ATP," *Scientific American,* April 1953.

Wald, G., "Life and Light," *Scientific American,* October 1959.

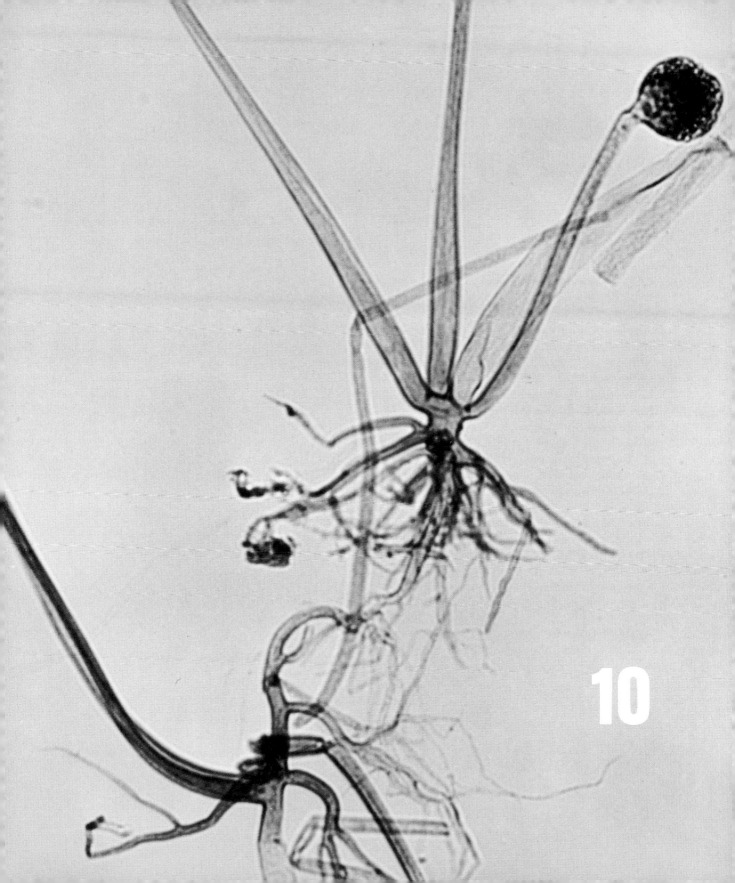

10
Catabolism: The Breakdown of Plant Materials

INTRODUCTION

In Chapter 9 the fates of organic molecules produced during photosynthesis were considered. In summary, these were (1) the production of more complex carbohydrates from simple sugars; (2) the synthesis of fats, proteins, and other organic compounds from the products of photosynthesis; (3) the immediate utilization of PGAL or glucose as a source of energy; and (4) the conversion of synthesized materials into storage products. The purpose of this chapter is to consider the breakdown and energy liberation of the materials produced during photosynthesis. Some of these materials, such as glucose, exist in a relatively simple form and require little preparation for breakdown and energy liberation. Others, such as storage products, which are larger, more complex molecules, must first be converted into simpler forms for translocation from storage centers prior to breakdown and energy liberation. This conversion into simpler form is accomplished by digestion.

DIGESTION

Digestion is a series of *hydrolysis* reactions, that is, a series of reactions by which complex compounds are broken down into smaller, simpler ones by the enzymatic addition of water. Chemically, digestion is the reverse of dehydration synthesis,

both processes being catalyzed by specific enzymes. The essential features concerning digestion are that the products of digestion become soluble and diffusible, and they are prepared for respiration and other syntheses.

Consider the fate of glucose produced during photosynthesis. If it is not used immediately as a source of energy or as a synthetic molecule glucose is converted to starch via dehydration synthesis reactions and then stored in the plant. When starch, stored fats, or proteins are needed by the plant at night when photosynthesis ceases or when the rate of photosynthesis is altered by some environmental factor, these materials must be converted into simpler forms before utilization by the plant. Starch, as well as other storage products, is insoluble and nondiffusible and unavailable for energy production and synthesis. These storage materials are prepared for cellular use through digestion.

In higher animals, including man, the process of digestion occurs in a series of structurally and functionally related organs collectively designated as the digestive system from which the products of digestion are transported to an elaborate circulatory system extending to all parts of the body. In plants, however, no such system exists and digestion occurs primarily where foods are stored, although it may take place in any living cell. There is some evidence to suggest that within the plant cell at least some digestive enzymes are produced by specialized organelles called *lysosomes*. These structures, similar to mitochondria in external form and size, possess a single outer membrane and lack the intricate, internal convoluting cristae of mitochondria. The current concept of lysosome function is that within these organelles are present assorted digestive enzymes, which, when released into the cell, become activated and catalyze the hydrolysis of certain substrates. The digestion of substrates such as carbohydrates, proteins, and fats into glucose, amino acids, and fatty acids and glycerol, respectively, does not release large amounts of energy. The end products of digestion are the molecules that provide vast quantities of energy through the agency of respiration.

Carbohydrate Digestion

The carbohydrates found in the greatest abundance in flowering plants are glucose, fructose, sucrose, starch, cellulose, and hemicelluloses. *Glucose* and *fructose,* being monosaccharides, are relatively simple and require no further digestion. In fact, in a great many instances, they represent the end products of digestion of other carbohydrates.

Sucrose, a disaccharide, requires digestion before being utilized in the plant cell. It is the most abundant and most commonly translocated disaccharide in plants. It is found as a storage product in roots, stems, leaves, and fruits. Although sucrose is soluble in water and generally readily diffusible, some plant cell membranes are impermeable to it. In the presence of the enzyme *sucrase,* sucrose is hydrolyzed into glucose and fructose. Sucrase, especially abundant in yeast cells, is also produced in green leaves, stems, grains, fruits, and some roots.

Starch, unlike sucrose, is an insoluble storage product, and must be digested prior to respiration, translocation, and synthesis into building materials. Starch, as the principal storage product of plants, is usually found in the form of grains in leucoplasts and consists of two distinguishable chemical components, a straight-chain glucose polymer, *amylose* and *amylopectin,* a branched polymer. Although synthesized from glucose units, starch is not converted directly into glucose but is hydrolyzed into a number of intermediates by a stepwise series of reactions catalyzed by specific enzymes.

First, it is believed that starch is made soluble by an enzyme called *amylophosphatase.* Another enzyme, called *β-amylase,* attacks the amylose portion of starch and hydrolyzes it into maltose units (Figure 10-1A). At this point, *maltase,* usually present, can catalyze the hydrolysis of maltose into glucose units. *β*-Amylase is much less effective, however, in hydrolyzing amylopectin, the branched component of starch. *β*-Amylase converts only the short branched chains of amylopectin to maltose and is unable to hydrolyze molecules beyond the point of branching (Figure 10-1B). Only about 50 per cent of amylopectin is converted to maltose by *β*-amylase, the remaining portions, called *dextrins,* which are intermediates of starch hydrolysis, must be acted upon by a different enzyme.

The enzyme *α-amylase* acts upon the dextrins converting them into maltose and some glucose, the resultant maltose, in turn, being hydrolyzed into glucose by maltase. The combined action of amylophosphatase, *β*-amylase, *α*-amylase, and maltase eventually hydrolyzes starch into its component glucose units. It is for this reason that dextrins and maltose are not found in plants in any appreciable quantities. In the plant the glucose resulting from starch digestion may immediately be converted into fructose and then into sucrose.

The hydrolysis of starch can be studied quite easily in the laboratory based upon certain color reactions of the products of digestion. When starch is stained with iodine, a characteristic blue-black color results. This color indicates that the starch consists of amylose and amylopectin molecules. As digestion starts and progresses, color changes can be noted in the iodine-stained intermediates ranging from blue-violet to red-brown to colorless. These colors are the physical expression of the breakdown of starch into dextrins, maltose, and finally glucose. Larger dextrins (12 glucose units

Figure 10-1. Hydrolysis of starch by amylases. **(A)** β-Amylase completely hydrolyzes amylose into maltose units. **(B)** β-Amylase hydrolyzes about 50 per cent of the amylopectin portion of starch into maltose units. The remaining dextrins are broken down by α-amylase. Eventually, all maltose units are hydrolyzed into glucose by the action of maltase. (Modified from W. Z. Hassid and R. M. McCready, *Journal of the American Chemical Society*, vol. 65, p. 1159, 1943.)

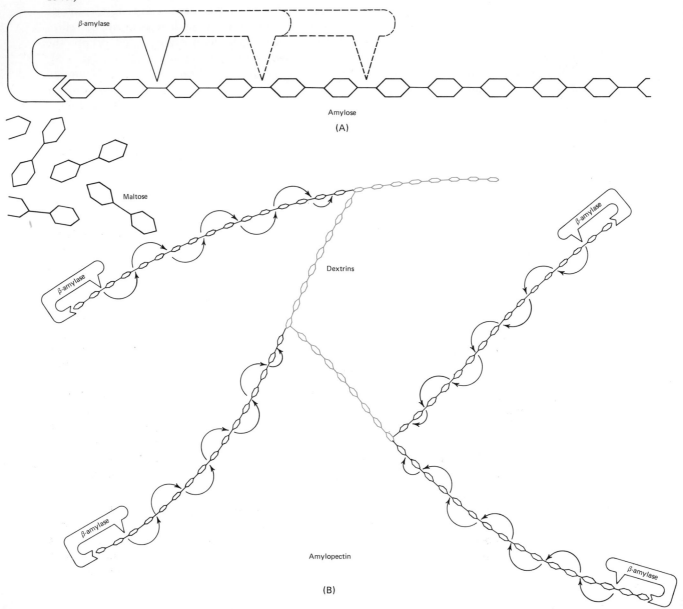

or more) produce the blue-violet color with iodine; smaller dextrins (8 to 12 glucose units) form a reddish-brown color; maltose (2 glucose units) and glucose give no color reaction with iodine.

The various enzymes associated with starch digestion are collectively designated as *diastase,* which is really an enzyme complex and not a single enzyme. It is found widely in plants being distributed in most seeds, leaves, roots, and fruits. The seeds of cereals are especially rich in diastase and extracts of the enzyme complex from barley seeds have been used extensively in starch hydrolysis studies.

Cellulose and *hemicellulose,* two common building materials of plant cell walls, like starch, are insoluble polysaccharides. Most animals, including man, lack enzymes for cellulose and hemicellulose digestion. Certain bacteria and fungus plants, however, which act as organisms of decay of plant and animal remains, are good sources of these enzymes.

Cellulose is hydrolyzed first into a disaccharide called *cellobiose* by the enzyme *cellulase.* The disaccharide is then converted into glucose molecules by the action of *cellobiase* according to the following reactions:

$$(C_6H_{10}O_5)_n + H_2O \xrightarrow{\text{cellulase}} C_{12}H_{22}O_{11}$$

cellulose   water   cellobiose

$$C_{12}H_{22}O_{11} + H_2O \xrightarrow{\text{cellobiase}} 2\,C_6H_{12}O_6$$

cellobiose   water   glucose

It should be noted that cellulose, being a complex polysaccharide like starch, probably involves a more complex digestive pattern than is suggested in the above reactions.

Hemicellulose, a polysaccharide closely related to cellulose, containing sugars other than glucose, may sometimes be utilized as a food during respiration and is digested by a variety of enzymes collectively designated as *cytases.* The end products thus formed consist of glucose, galactose, and pentoses, among other sugars. The nature of the end product is dependent upon the chemical constitution of the hemicellulose and the particular cytase that catalyzes the reaction.

Fat Digestion

Fats are insoluble in water and are therefore found in the plant cell in the form of small droplets or globules dispersed in the cytoplasm. These globules are frequently large enough to be seen microscopically, particularly if they are stained with special dyes, such as Sudan III. The digestion of fats is catalyzed by the enzyme *lipase,* an enzyme isolated readily from bean. Whether lipase acts as a synthetic or hydrolytic enzyme appears to be a function of the water content of the cells in which it occurs. For example, during periods when water content decreases, as when seeds are being formed on the plant, fats are synthesized. During periods of increasing water content, as when seeds are germinating, lipase hydrolyzes fats.

The glycerol formed as a result of hydrolysis, although diffusible, is probably converted to carbohydrates prior to utilization.

Protein Digestion

The digestion of proteins into amino acids, like polysaccharide hydrolysis, involves a series of specific enzymes at various reaction sites. Essentially, plant proteins are broken down into proteoses, peptones, and polypeptides, and finally into amino acids. Each of these subdivisions of protein digestion, from peptones to amino acids, represents a progressively smaller hydrolysis product of the starting protein molecule.

TABLE 10-1. Summary of Digestion Showing Substrates, Enzymes Involved, and Products

Substrate	Enzyme	Result
I. Carbohydrates		
Sucrose	Sucrase	Glucose and fructose
Starch	Amylophosphatase	Makes starch soluble
1. Amylose	β-Amylase	Maltose
2. Amylopectin	β-Amylase	Maltose and dextrins
3. Dextrins	α-Amylase	Maltose and some glucose
4. Maltose	Maltase	Glucose
Cellulose	Cellulase	Cellobiose
Cellobiose	Cellobiase	Glucose
Hemicellulose	Cytase	Glucose, galactose, and pentoses
II. Fats		
Fat	Lipase	Fatty acids and glycerol
III. Proteins		
Starting protein	Proteinases 1. Papain 2. Pepsin 3. Ficin 4. Bromelin 5. Solanain	Proteoses, peptones, and polypeptides
Proteoses, peptones, and polypeptides	Exopeptidases and endopeptidases	Amino acids

The general term *protease* is commonly used for all enzymes that catalyze protein digestion. In terms of the protein or specific hydrolytic product acted upon, proteases are classified into several types. For example, proteases that act on the starting protein molecule are called *proteinases.* Representative proteinases are *papain,* from the fruit of papaya trees; *pepsin,* from insectivorous plants; *ficin,* from the milky sap of fig trees; *bromelin,* from pineapples; and *solanain,* from the horse nettle.

Once the protein has been hydrolyzed into proteoses, peptones, or polypeptides by the action of various proteinases, the smaller hydrolytic products are further split into amino acids by endopeptidases and exopeptidases. The *endopeptidases* are proteases that act on the internal peptide bonds of proteoses, peptones, and polypeptides, while *exopeptidases* act on the terminal peptide bonds of these hydrolysis products. In either case both exopeptidases and endopeptidases produce amino acids. The resultant amino acids may be translocated and utilized in respiration of for resynthesis into structural proteins, enzymes, or other compounds.

Table 10-1 summarizes the substrates, enzymes, and digestive products discussed in this chapter.

RESPIRATION

The essential feature of photosynthesis is the conversion of solar energy into potential chemical energy. As a synthetic phase of metabolism, it is an energy-requiring process, the energy being supplied by the sun. PGAL, the photosynthetic product, along with the sugars and other compounds synthesized from it, are the recipients of stored chemical energy. Before any of these compounds can release stored energy, they must be broken down chemically through the agency of respiration.

Although photosynthesis and respiration are reverse metabolic processes, one storing energy and the other releasing it, and, although the pattern of energy-storing reactions differs somewhat from the energy-releasing reactions, certain commonalities do exist. For example, in respiration, the chemical breakdown of organic molecules is not a single gross reaction but a series of smaller step by step reactions, each enzymatically catalyzed. In addition, the energy released in respiration does not dissipate as heat or light but is conserved and stored in ATP for cellular utilization. These similarities shared by photosynthesis and respiration, as well as others, are but a few examples of the interrelationships that exist between anabolic and catabolic processes in plant cells, all contributing to the total functioning plant. Some aspects of photosynthesis and respiration are compared in Table 10-2.

Respiration, or **biological oxidation,** may be defined as the oxidation of organic compounds (principally glucose) in cells, resulting in the release of chemical energy. The overall process may be stated according to the following general equation:

$$C_6H_{12}O_6 \xrightarrow{\text{enzymes}} 6\,CO_2 + 6\,H_2O + \text{energy}$$

glucose carbon water (ATP)
dioxide

As in the case of photosynthesis, this equation is a mere representation of the raw materials, necessary conditions, and products of the reaction. It in no way accounts for the myriad intermediate chemical reactions that actually occur.

The energy released by the oxidation of organic molecules is actually transferred to the high-energy terminal bonds of ATP. In other words, the chemical energy in the bonds of food molecules is transformed into the energy in ATP, a form that can be readily utilized by the cell to do work. If such a transfer and storage of energy from food molecules to ATP did not occur, the energy contained in foods would dissipate as heat in the cell, becoming unavailable as cellular energy. The energy stored in ATP is used for reactions which require energy; thus energy from one reaction (respiration) is transferred to drive another reaction (synthesis). In addition to providing energy for synthetic reactions, such as those which bring about the production of amino acids, proteins, and lipids, the energy of ATP may also be used in growth processes, active absorption, cyclosis, and translocation to mention only a few.

Once ATP is formed, its energy may be utilized at various places in the cell to drive energy-requiring reactions. In the process, one of the three phosphate groups in removed from the ATP molecule. This leaves a compound, adenosine diphosphate (ADP), which is less rich in energy than ATP, and an inorganic phosphate (P) group. Molecules of ADP and P may be synthesized back

TABLE 10-2. Comparison of Some Aspects of Photosynthesis
and Respiration

Photosynthesis	Respiration
1. Occurs only in chlorophyll-containing cells of plants	1. Occurs in all plant and animal cells
2. Takes place only in the presence of light	2. Takes place continually both in the light and in the dark
3. Sugars, water, and oxygen are products	3. CO_2 and H_2O are products
4. CO_2 and H_2O are raw materials	4. O_2 and food molecules are raw materials
5. Synthesizes foods	5. Oxidizes foods
6. Stores energy	6. Releases energy
7. Results in an increase in weight	7. Results in a decrease in weight

again into ATP by trapping energy released from the oxidation of glucose or other organic molecules. The role of ATP as an intermediate energy-transferring compound between energy-releasing and energy-consuming reactions is shown in Figure 10-2.

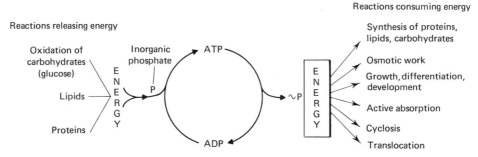

Figure 10-2. The ATP cycle. ATP is an intermediate energy-transfer compound between energy-releasing and energy-consuming reactions. When ATP is broken down to release energy for cellular activities, it is built up by the addition of a phosphate group. The energy for this build-up is supplied by the breakdown of food molecules during respiration.

The Reactions of Respiration

The first stage of respiration, called **glycolysis** ("sugar splitting"), consists of a sequence of reactions occurring in the cytoplasm in which glucose, the starting substrate, is converted into 2 three-carbon organic acids called *pyruvic acid* (Figure 10-3). The major events of glycolysis are the preparation of glucose for reaction, splitting it into 2 three-carbon compounds, partial oxidation of the three-carbon compounds, and the formation of ATP. Let us now consider these events in more detail.

Glucose, a product of photosynthesis, is a fairly stable molecule in that it resists breakdown. If the energy contained in its molecular configuration is to be released, it must first be made more reactive. This "activation" of glucose is accomplished by transferring some of the energy of ATP, derived from previous respiration, to the glucose molecule. This reaction is called **oxidative phosphorylation,** in which ATP is converted to ADP and some of the stored energy is transferred to glucose. The product of this reaction is *glucose-6-phosphate* (activated glucose), so named because the phosphate group of ATP ($\sim \textcircled{P}$) is attached to the sixth carbon atom of the glucose molecule. Catalysis is facilitated by the enzyme *hexokinase* and Mg^{++} as an activator (Figure 10-3A).

The glucose-6-phosphate now undergoes a series of reactions in which atoms are rearranged, another phosphorylation occurs, and cleavage (splitting) of a hexose molecule takes place. In the presence of an *isomerase* enzyme (*phosphoglucoisomerase*), the atoms of glucose-6-phosphate are rearranged to produce a molecule of *fructose-6-phosphate* (Figure 10-3B). Following this, another oxidative phosphorylation reaction, catalyzed by a *kinase* (*phosphofructokinase*), converts fructose-6-phosphate into *fructose-1,6-diphosphate*. In this reaction a phosphate group is added to the first carbon atom of the fructose molecule and the enzyme is activated by Mg^{++} (Figure 10-3C).

The fructose-1,6-diphosphate molecule is then split into 2 three-carbon sugar phosphate molecules in the presence of an *aldolase* enzyme. One of these sugars in **PGAL,** the other is a compound called *dihydroxyacetonephosphate*. In the presence of an appropriate *isomerase* enzyme (*phosphotriose isomerase*), dihydroxyacetone phosphate is usually converted into PGAL (Figure 10-3D). All these reactions are preparatory to the oxidation of glucose and the release of energy. In effect, at this juncture, respiration is an energy-consuming process, because two molecules of ATP have been used to phosphorylate the starting glucose molecule.

The next step in glycolysis involves the oxidation of PGAL (and dihydroxyacetone phosphate, because it is converted to PGAL) to a compound called *1,3-diphosphoglyceric acid* (Figure 10-3E). In this oxidation reaction, a pair of electrons is removed from PGAL by a *dehydrogenase* enzyme (*phosphoglyceraldehyde dehydrogenase*). It should be noted that the removal of electrons presents somewhat of a problem to the cellular machinery because electrons ($-$) are attracted to the protons ($+$) of the molecule to which they are attached. Inasmuch as a great deal of energy would have to be expended to remove only the electrons, a proton is removed along with an electron, a process requiring less energy. In effect, then, oxidation of the PGAL molecule involves the removal of hydrogen atoms (an electron and a proton together), a process called *dehydrogenation*. It should be kept in mind, however, that the energy released during oxidation reactions resides in the electrons and not the protons. The hydrogen atoms are picked up by the coenzyme NAD^+ (*nicotinamide adenine dinucleotide*), reducing it to $NADH + H^+$. In addition to the oxidation of PGAL, it also picks up an

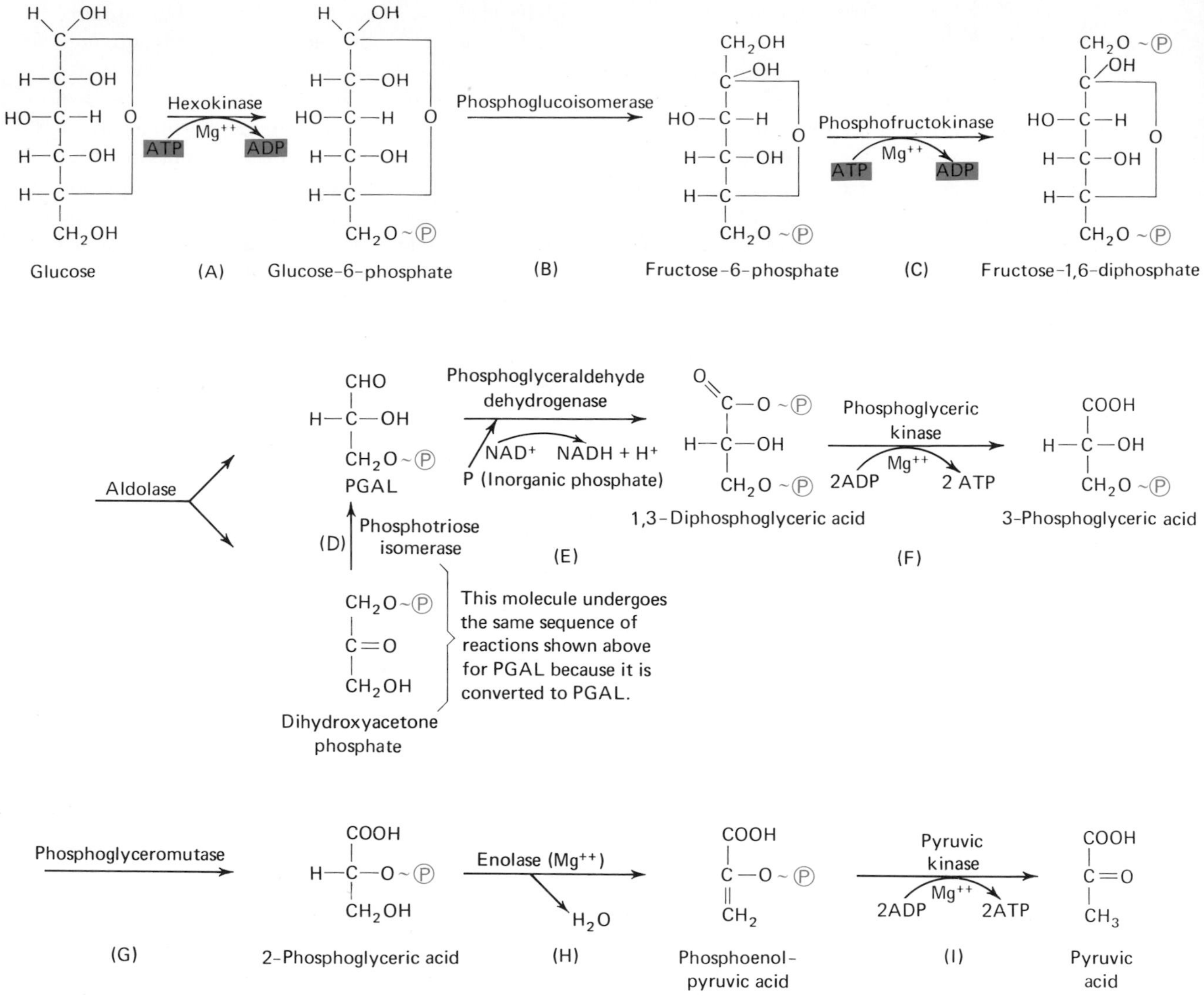

H—C—OH (Glucose)

Hexokinase
Mg++
ATP → ADP

(A) Glucose-6-phosphate

Phosphoglucoisomerase

(B) Fructose-6-phosphate

Phosphofructokinase
Mg++
ATP → ADP

(C) Fructose-1,6-diphosphate

Aldolase

CHO
H—C—OH
$CH_2O \sim P$
PGAL

Phosphoglyceraldehyde dehydrogenase
NAD^+ → $NADH + H^+$
P (Inorganic phosphate)

(D)

Phosphotriose isomerase

$CH_2O \sim P$
C=O
CH_2OH

Dihydroxyacetone phosphate

This molecule undergoes the same sequence of reactions shown above for PGAL because it is converted to PGAL.

(E)

$C—O \sim P$
H—C—OH
$CH_2O \sim P$

1,3-Diphosphoglyceric acid

Phosphoglyceric kinase
Mg++
2ADP → 2 ATP

COOH
H—C—OH
$CH_2O \sim P$

3-Phosphoglyceric acid

(F)

Phosphoglyceromutase

(G)

COOH
H—C—O $\sim P$
CH_2OH

2-Phosphoglyceric acid

Enolase (Mg++)
H_2O

(H)

COOH
C—O $\sim P$
CH_2

Phosphoenol-pyruvic acid

Pyruvic kinase
Mg++
2ADP → 2ATP

(I)

COOH
C=O
CH_3

Pyruvic acid

Figure 10-3. Glycolysis, the conversion of glucose into pyruvic acid. A single glucose molecule undergoes activation, splitting, oxidation, and phosphorylation. The result of these reactions is the production of two molecules of pyruvic acid. Considering the net input and the net output, glycolysis yields two ATP molecules.

inorganic (low-energy) phosphate, raising the energy level of the inorganic (low-energy) phosphate, raising the energy level of the inorganic phosphate to the high-energy level of ATP.

In the presence of ADP and a *kinase* enzyme (*phosphoglyceric kinase*) activated by Mg^{++}, 1,3-diphosphoglyceric acid is converted to *3-phosphoglyceric acid* and ATP is formed (Figure 10-3F). This acid, in turn, is transformed to *2-phosphoglyceric acid* by the activity of the *mutase* enzyme, *phosphoglyceromutase* (Figure 10-3G). This enzyme transfers the phosphate from the third carbon atom of glyceric acid to the second carbon atom. As glycolysis nears completion, a molecule of water is removed from 2-phosphoglyceric acid by the enzyme *enolase*, which is activated by Mg^{++}. The elimination of water from 2-phosphoglyceric acid results in the formation of *phosphoenolpyruvic acid* (Figure 10-3H). Finally, in the presence of ADP and the enzyme *pyruvic kinase*, phosphoenolpyruvic acid is converted to *pyruvic acid*. In this reaction, the phosphate group of phosphoenolpyruvic acid is transferred to ADP to form ATP (Figure 10-3I).

The formation of pyruvic acid signals the end of the glycolysis phase of respiration. It should be noted that whereas two molecules of ATP are converted to ADP to activate glucose, for each molecule of glucose that undergoes glycolysis, four are produced for a net gain of two. Remember that since dihydroxyacetone phosphate is converted to PGAL, it also undergoes the same sequence of reactions as PGAL. In addition, note that glycolysis requires no molecular oxygen for its operation. Of the total energy built into PGAL, about 5 per cent is transferred and stored in the ATP, 15 to 20 per cent is absorbed by NADH + H^+, and the remainder is left in pyruvic acid. In view of this unequal energy distribution, it would appear that the fate of pyruvic acid, especially, is of great importance to the plant cell.

Types of Respiration

During glycolysis, hydrogen atoms are removed from glucose and, in the process, NAD^+ is reduced to NADH + H^+. Organisms possess various mechanisms for disposing of the hydrogen held as NADH + H^+, so that NAD^+ can be freed to pick up more hydrogen. The removal of hydrogen from NADH + H^+ is extremely important to organisms. If NADH + H^+ could not be freed of hydrogen and reused as NAD^+, glycolysis would stop and the organism would die. Based upon the manner in which organisms oxidize NADH + H^+ to NAD^+, three principal types of respiration exist. If the hydrogen is passed on to pyruvic acid, or an organic compound derived from pyruvic acid, the type of respiration is called **fermentation.** If the hydrogen is released, ultimately to combine with molecular oxygen (O_2) from the air, the process is known as *aerobic respiration.* Some bacteria carry on *anaerobic respiration,* in which the hydrogen is passed on to oxygen found as part of an inorganic compound such as sulfate ($SO_4^=$) and nitrate (NO_3^-). Inasmuch as anaerobic respiration is carried on by only a relatively few bacteria, the following discussion of respiration will be concerned with fermentation and aerobic respiration, the more common types. It should be noted that glycolysis is a common feature of all types of respiration and that the fate of pyruvic acid and the method of hydrogen disposal determine the particular respiratory pathway followed by an organism.

Fermentation. The most common type of fermentation is *alcoholic fermentation,* a process carried on by yeasts, some bacteria, and even green plants when deprived of oxygen. The first sequence of reactions in alcoholic fermentation is the same as those occurring during glycolysis, in which a glucose molecule is split and oxidized into two pyruvic acid molecules. In this part of the process there

Figure 10-4. Fermentation. **(A)** In alcoholic fermentation ethyl alcohol acts as the final hydrogen acceptor. **(B)** In lactic acid fermentation, lactic acid is the final hydrogen acceptor. Inasmuch as these fermentations are only partial oxidations of glucose molecules, only two ATP molecules are produced. The remaining energy is still stored in the chemical bonds of ethyl alcohol and lactic acid.

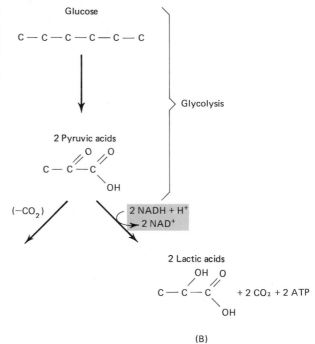

is a net gain of two ATP molecules. In alcoholic fermentation, the pyruvic acid resulting from glycolysis is then converted in two steps into alcohol and CO_2. First, CO_2 is removed from pyruvic acid, leaving *acetaldehyde* (CH_3CHO). Second, the acetaldehyde is converted into *ethyl alcohol* (C_2H_5OH). These reactions are shown in Figure 10-4A.

During alcoholic fermentation, there is a net production of only two ATP molecules for each molecule of glucose oxidized. The bulk of the energy originally present in the sugar is now in the alcohol. In view of the fact that the alcohol is an end product of the reactions and undergoes no further oxidation, alcoholic fermentation is an inefficient energy-releasing process. Also note that the final hydrogen acceptor in alcoholic fermentation is the alcohol.

A second type of fermentation, called *lactic acid*

fermentation, is characteristic of animals and some forms of bacteria and is very similar to alcoholic fermentation. The main difference is that pyruvic acid, formed by glycolysis, acts as the hydrogen acceptor and *lactic acid* is an end product (Figure 10-4B). Like alcoholic fermentation, there is a net production of two ATP molecules during lactic acid fermentation.

Different fermentations produce different byproducts, many of which are of economic importance. The two principal commercial uses of yeast, baking and brewing, are dependent on the two products of alcoholic fermentation, carbon dioxide and alcohol. The fermentation of fruit juices, malted grains, and molasses by yeasts produces wine, beer, and rum. Different types of yeast are also used to leaven bread. The CO_2 produced during fermentation makes bubbles in the dough

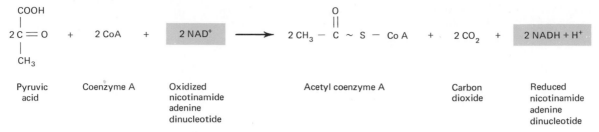

$$2\,C\!=\!O \ \ + \ \ 2\,CoA \ \ + \ \ \boxed{2\,NAD^+} \longrightarrow \ \ 2\,CH_3-\overset{\overset{\displaystyle O}{\|}}{C}\sim S-CoA \ \ + \ \ 2\,CO_2 \ \ + \ \ \boxed{2\,NADH + H^+}$$

| Pyruvic acid | Coenzyme A | Oxidized nicotinamide adenine dinucleotide | Acetyl coenzyme A | Carbon dioxide | Reduced nicotinamide adenine dinucleotide |

Figure 10-5. Formation of acetyl coenzyme A. Pyruvic acid, which represents the last product formed during glycolysis, is converted to acetyl coenzyme A if sufficient oxygen is present. The principal mechanisms involved in this conversion are decarboxylation (loss of CO_2) and dehydrogenation (removal of hydrogen atoms). Although acetyl coenzyme A formation is shown here as a single reaction, it should be noted that four intermediate reactions are involved.

and causes the bread to rise, giving it lightness and a desirable texture when baked. Other bacterial fermentations produce a wide variety of metabolic by-products. Lactic acid, for example, is used commercially in the manufacture of many dairy products, including cheese, buttermilk and yogurt.

Aerobic Respiration. Both glycolysis and fermentation, as well as anaerobic respiration, occur in the absence of free molecular oxygen. In the presence of free oxygen, however, an alternative respiratory pathway called *aerobic respiration* is followed. During aerobic respiration, the complete oxidation of glucose into CO_2 and H_2O is brought about by a complex series of reactions in which oxygen serves as the final hydrogen acceptor. Essentially two mechanisms are involved in these complete oxidation reactions. These are the Krebs cycle and the cytochrome system.

Krebs Cycle. The first phase of the complete breakdown or oxidation of pyruvic acid, an energy-rich compound, into carbon dioxide and water is achieved on the outer mitochondrial membrane by a complex series of reactions collectively called the *Krebs cycle.* In a fashion characteristic of many biochemical reactions, the pyruvic acid is oxidized

in a cyclic stepwise pattern. Pyruvic acid, like glucose during glycolysis, must first undergo preparation prior to continuing its course through the molecular pathway of aerobic respiration. Before entering the Krebs cycle, pyruvic acid is converted to a two-carbon compound by the loss of CO_2 (*decarboxylation*) and two hydrogen atoms (*dehydrogenation*). This compound, bonded to a coenzyme called *coenzyme A,* is an activated substance known as *acetyl coenzyme A* (Figure 10-5). When a molecule of pyruvic acid is oxidized to acetyl coenzyme A and carbon dioxide, hydrogen is removed and is picked up by NAD^+, forming $NADH + H^+$. Two pyruvic acid molecules are formed from each glucose molecule during glycolysis, to two molecules of $NADH + H^+$ are formed.

Acetyl CoA, which contains only two carbons of the original pyruvic acid, is now activated and prepared to enter the Krebs cycle. The cycle starts in the mitochondria when a molecule of acetyl CoA combines with a four-carbon organic acid, *oxaloacetic acid,* already present. The result of this reaction is the formation of a six-carbon organic acid called *citric acid* and CoA (Figure 10-6A). The next step in the cycle involves the removal of H_2O from citric acid, which results in the formation of

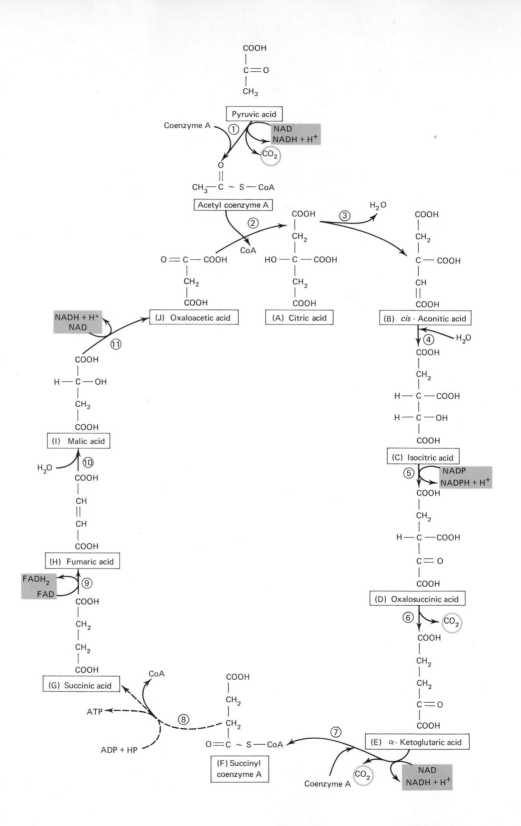

cis-aconitic acid, also a six-carbon acid (Figure 10-6B). In another reaction, H_2O is added to *cis*-aconitic acid. This addition produces a six-carbon acid, *isocitric acid* (Figure 10-6C).

At this juncture in the cycle the first oxidation takes place. In the presence of $NADP^+$ (*nicotinamide adenine dinucleotide phosphate*), a cofactor closely related to NAD, two hydrogen atoms are removed (*dehydrogenation*) from isocitric acid. The products of this reaction are reduced nicotinamide adenine dinucleotide phosphate ($NADPH + H^+$) and the six-carbon acid *oxalosuccinic acid* (Figure 10-6D). Next, oxalosuccinic acid is decarboxylated (CO_2 is removed) and a five-carbon acid, *α-ketoglutaric acid,* is formed (Figure 10-6E).

The oxidation of α-ketoglutaric acid, the second oxidation step of the cycle, is very similar to that of pyruvic acid. The α-ketoglutaric acid is decarboxylated and dehydrogenated. In the dehydrogenation reaction, NAD^+ is reduced to $NADH + H^+$. In addition, CoA enters the reaction and the product is a four-carbon compound called *succinyl coenzyme A* (Figure 10-6F). The energy locked within succinyl CoA may be released in the next reaction to form ATP, CoA, and *succinic acid,* a four-carbon organic acid (Figure 10-6G).

A third oxidation occurs when succinic acid is dehydrogenated to form the four-carbon acid *fumaric acid* (Figure 10-6H). It is interesting to note that this reaction is the only oxidation in the cycle in which a pyridine nucleotide (NAD^+ or $NADP^+$) does not serve as the hydrogen acceptor coenzyme. Instead, *FAD* (*flavin adenine dinucleotide*), a deriva-tive of riboflavin, is reduced to $FADH_2$. Fumaric acid, by the addition of water, is then converted to the four-carbon acid *malic acid* (Figure 10-6I).

In the fourth oxidation step of the Krebs cycle, malic acid is converted to oxaloacetic acid and, in the process, NAD^+ is reduced to $NADH + H^+$. This reaction, which results in the regeneration of oxaloacetic acid, completes the cycle (Figure 10-6J). This is the same acid to which acetyl CoA was attached at the start of the cycle. The oxaloacetic acid can now combine with more acetyl CoA to form citric acid, and another cycle is initiated. Remember that because one molecule of glucose produces two pyruvic acid molecules during glycolysis, two turns of the cycle occur for each molecule of glucose oxidized. The carbon dioxide produced as a result of decarboxylation of the organic acids in the Krebs cycle is the source of the CO_2 given off as a product of respiration.

Cytochrome System. Approximately 95 per cent of the ATP produced during aerobic respiration is formed by the transfer of electrons from hydrogen acceptor molecules ($NADH + H^+$, $NADPH + H^+$, and $FADH_2$) formed during glycolysis and the Krebs cycle. These high-energy electrons are passed along, in serial succession, to electron acceptor molecules and the energy of the electrons is converted into ATP. Because the high-energy electrons are united with protons as the hydrogen atoms, the electrons are separated from the hydrogen, which is released into the cytoplasm as H^+ ions. The entire sequence of electron transfers resulting in the production of ATP involves an

Figure 10-6. (Opposite) The Krebs cycle. In this complex series of reactions, activated pyruvic acid (acetyl CoA) enters the cycle and undergoes dehydrogenations and decarboxylations. The oxidations (dehydrogenations) of the organic acids produce reduced coenzymes, whereas the decarboxylations form CO_2, a gaseous waste product. The enzymes as numbered in the reactions are: ① pyruvic dehydrogenase, ② condensing enzyme, ③ aconitase, ④ aconitase, ⑤ isocitric acid dehydrogenase, ⑥ carboxylase, ⑦ α-ketoglutaric dehydrogenase, ⑧ succinic thiokinase, ⑨ succinic dehydrogenase, ⑩ fumarase, and ⑪ malic dehydrogenase.

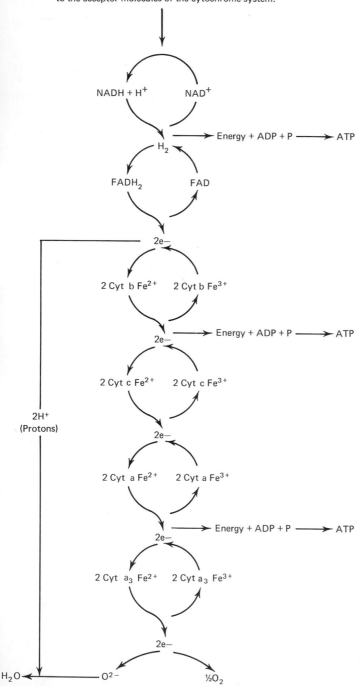

Reduced coenzymes (NADH + H⁺, NADPH + H⁺
and FADH₂) from glycolysis and the Krebs cycle
pass their hydrogen atoms (protons and electrons)
to the acceptor molecules of the cytochrome system.

NADH + H⁺   NAD⁺

→ Energy + ADP + P ⟶ ATP

H₂

FADH₂   FAD

2e−

2 Cyt b Fe²⁺   2 Cyt b Fe³⁺

2e−   → Energy + ADP + P ⟶ ATP

2 Cyt c Fe²⁺   2 Cyt c Fe³⁺

2H⁺
(Protons)

2e−

2 Cyt a Fe²⁺   2 Cyt a Fe³⁺

2e−   → Energy + ADP + P ⟶ ATP

2 Cyt a₃ Fe²⁺   2 Cyt a₃ Fe³⁺

2e−

H₂O ◄—   O²⁻   ½O₂

organized group of electron acceptor molecules on
the inner mitochondrial membrane referred to as
the *cytochrome system* (Figure 10-7).

The hydrogen removed from the original glucose molecules are picked up by NAD⁺, reducing
it to NADH + H⁺. This compound is the initial
hydrogen acceptor in the cytochrome system. The
other acceptor molecules are FAD, *cytochrome b,
cytochrome c, cytochrome a,* and *cytochrome a₃* (*cytochrome
oxidase*), the terminal acceptor molecule in the
chain. *Cytochromes* are proteins that have a reversible valency change of their iron-containing group.
As components of the cytochrome system, they are
alternately oxidized and reduced as they transfer
electrons. The iron portions of the molecules can
reverse their valency by going from Fe⁺⁺ (reduced
form) to Fe³⁺ (oxidized form). Entire hydrogen
atoms are passed from NADH + H⁺ to FAD.
The ionization of hydrogen occurs with the oxidation of FADH₂, so from this point on in the system
only electrons are transferred through the various
cytochromes. When these electrons reach the end
of the system, they are passed by cytochrome
oxidase to molecular oxygen. The oxygen, now
negatively charged, picks up the H⁺ ions from the
ionization of hydrogen to form water, a product
of respiration. In this regard, oxygen serves as the
final hydrogen acceptor during aerobic respiration. As electrons are passed along the system, they

Figure 10-7. The cytochrome system. In this series of reactions, the main line of oxidation-reduction in cells that use
oxygen, electrons are passed to various electron acceptors
and, in the process, energy is released and stored in ATP.
Dehydrogenase enzymes remove hydrogen atoms (protons
and electrons) from reduced coenzymes formed during
glucolysis and in the Krebs cycle. These hydrogen atoms
are passed to NAD⁺ and then to FAD. FAD liberates the
protons into the mitochondrial matrix and passes the electrons to the cytochromes. The last cytochrome in the series,
cytochrome oxidase (cytochrome a₃), passes the electrons
to oxygen, and protons are picked up to form water. Note
the three junctures at which ATP is formed.

pass down an energy gradient, and at specific points their energy is transferred into ATP. Three ATP molecules are believed to be formed from ADP and inorganic phosphate molecules for each pair of electrons that passes through the system.

Fermentation and Aerobic Respiration Compared. Fermentation produces a total of four ATP molecules for each molecule of glucose that is oxidized. Inasmuch as two ATP molecules are utilized to activate glucose for reaction during glycolysis, only two of these four molecules of ATP are made available as stored chemical energy. The greater portion of the energy of glucose is transferred to pyruvic acid or a compound derived from pyruvic acid, and these compounds undergo no

further oxidation, so fermentation is an inefficient energy-producing process.

In contrast, aerobic respiration, through the agency of the Krebs cycle and the cytochrome system, completely oxidizes glucose into CO_2 and H_2O with the net production of 38 ATP molecules. These 38 ATP molecules represent 40 to 60 per cent of the total energy present in a glucose molecule. The remaining energy not recovered by ATP is lost in the cell as heat.

Both fermentation and aerobic respiration involve a series of oxidation reactions in which the potential energy of glucose is systematically passed to ATP (Figure 10-8). Two molecules of ATP are initially expended to activate glucose for reaction.

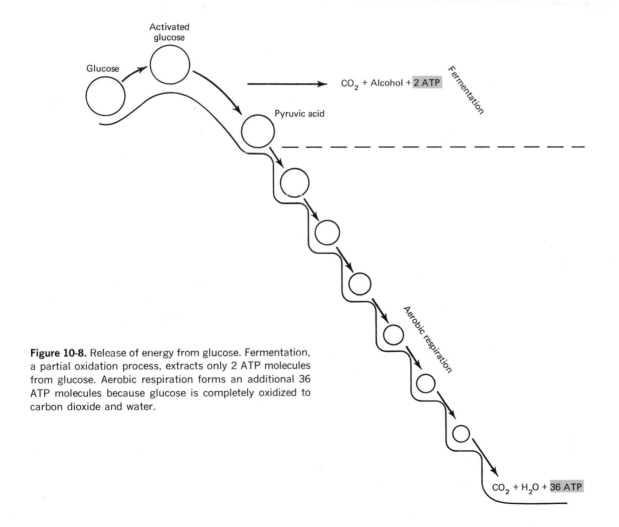

Figure 10-8. Release of energy from glucose. Fermentation, a partial oxidation process, extracts only 2 ATP molecules from glucose. Aerobic respiration forms an additional 36 ATP molecules because glucose is completely oxidized to carbon dioxide and water.

257

Once activated, glucose may be visualized as passing down an energy hill, releasing energy at various steps in the formation of ATP. Fermentation is a partial oxidation of glucose, so only 2 ATP molecules are formed. Aerobic respiration releases an additional 36 ATP molecules from glucose for a net production of 38 ATPs. Whereas CO_2 and ethyl alcohol are the end products of fermentation (alcoholic), CO_2 and H_2O are the end products of aerobic respiration.

FACTORS AFFECTING RESPIRATION

In general, the same factors which affect the rate of photosynthesis also have a marked effect on the respiratory rate. These factors may be either internal or external.

Internal Factors

Among the internal factors that govern the rate of respiration are the amount of food available for respiration and the activity of the respiratory enzymes.

It has been found that photosynthesizing leaf cells kept in the dark contain small amounts of sugar and do not respire vigorously. But when the plant is placed in light or given sugar in the dark, the rate of respiration immediately increases. Logically, any factor which limits photosynthesis, thus food availability, must affect respiration.

One of the most important internal factors affecting respiration is the activity of *enzymes*. It was stated earlier that digestive enzymes assume a prominent role in preparing storage products for respiration. In addition, the myriad reactions of glycolysis, the Krebs cycle, and the cytochrome system are all catalyzed by various classes of enzymes.

External Factors

Chief among the external factors that affect the rate of respiration are water supply, temperature, oxygen supply, and carbon dioxide concentration.

Water Supply. The effect of water supply on respiration has been investigated by studying seeds and wilted plants. Many seeds remain viable when stored in dry air and because of the low water content, respiration is very slow. When seeds imbibe water, however, the rate increases. The respiration of a growing plant is also affected by the water content of its tissues. If excessive water is lost by transpiration, and wilting occurs, there is a decline in respiratory activity.

Temperature. Temperature regulates the rate of respiration by affecting the activity of the respiratory enzymes. From about 0 to 30°C, for each 10°C rise in temperature, the rate of respiration doubles. As the temperature increases, the rate of increase becomes less prominent. The loss of catalytic properties in typical cases begins at about 35°C and is complete as 60°C is approached.

Oxygen Supply. Oxygen is generally not a limiting factor unless its concentration is less than 20 per cent. Although most plants can continue for a time to oxidize foods anaerobically if the concentration of oxygen is very low, normally oxygen is necessary for other plant activities. Obviously, the amount of available oxygen affects the type of respiration (aerobic or anaerobic) and consequently the end products produced.

Carbon Dioxide Concentration. If carbon dioxide, a product of respiration, is allowed to accumulate, the rate of respiration decreases. This is easily demonstrated in germinating seeds in a sealed container into which oxygen is pumped. As the carbon dioxide accumulates, respiration decreases. Upon removal of the lid, germination continues and the rate of respiration increases.

THE METABOLIC SIGNIFICANCE OF THE KREBS CYCLE

The discussion of respiration thus far has centered about the oxidation of glucose and its importance as an energy-supplying molecule. Typically, glucose is the principal energy-supplying substrate; however, other molecules, as well as other metabolic pathways, also assume a role in cellular energetics. For example, once polysaccharides and disaccharides are hydrolyzed to monosaccharides, the hexoses are readily converted to glucose. In addition to degradation by glycolysis and the Krebs cycle, glucose may be oxidized by an alternative pathway called the *hexose monophosphate shunt*. This cyclic pathway produces the five-carbon sugar intermediate ribulose phosphate (RuP), which is important in the dark reactions of photosynthesis and is a source of five-carbon sugars for the synthesis of nucleic acids. If a molecule of glucose is oxidized via the hexose monophosphate shunt, there is a net production of 36 ATP molecules. Thus capture of energy released in the oxidation of glucose in this pathway is almost as efficient as that of the glycolytic Krebs cycle pathway.

The Krebs cycle, in addition to representing the major pathway for carbohydrate metabolism, also serves as a metabolic route for fats and proteins. Once a fat is degraded by hydrolytic enzymes, the products of digestion may, after a series of conversions, enter the Krebs cycle. Glycerol, one of the products of hydrolysis, is converted to dihydroxyacetone phosphate and may be metabolized in glycolysis and then in the Krebs cycle. Fatty acids are metabolized by a series of reactions that remove two-carbon fragments from the long-chained fatty acid molecules. Before it can be metabolized, a fatty acid must be activated by ATP and then with coenzyme A. The product of these reactions is a two-carbon unit acetyl coenzyme A which then enters the Krebs cycle. The term *beta oxidation* is applied to the metabolism of fatty acids by the removal of two-carbon fragments in this manner.

Amino acids, the products of protein hydrolysis, are oxidized by reactions in which the amino group is first removed, a process called *deamination.* The amino acid alanine, for example, yields pyruvic acid when deaminated, glutamic acid yields α-ketoglutaric acid, and aspartic acid yields oxaloacetic acid. Once deaminated, these amino acids can enter the Krebs cycle directly. Other amino acids may require, in addition to deamination, several other reactions to produce substances that can enter the cycle. Ultimately the carbon chains of all amino acids are metabolized in this fashion.

The few examples given above show that the Krebs cycle is an important metabolic hub in which many biochemical reactions of cells are interrelated in one way or another (Figure 10-9). It is more than just a metabolic pathway for the oxidation of glucose; it is a central route for most of the biochemical activity of the cell.

CHAPTER IN RETROSPECT

Catabolism consists of a chemical breakdown and energy liberation of the materials produced during photosynthesis. When viewed in broad terms and as a cycle of anabolic-catabolic activities, three significant metabolic events should be noted: (1) carbon dioxide and water are converted to foods (photosynthesis), (2) foods are broken down to simpler substances by hydrolytic enzymes (digestion), and (3) organic molecules are degraded back to inorganic molecules of carbon dioxide and water (respiration). One very significant end result of this complex series of chemical activities is the conversion of solar energy into

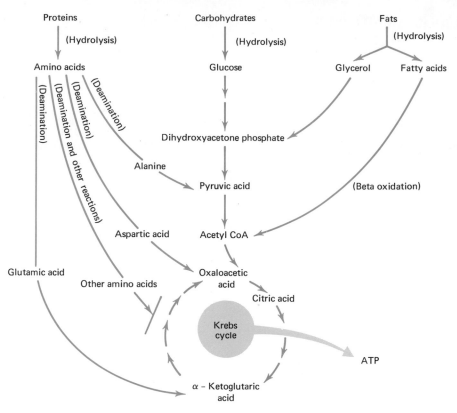

Figure 10-9. The interrelations of metabolic (oxidation) reactions via the Krebs cycle.

a usable form of chemical energy as found in molecules of ATP. For example, respiration, in addition to carbon dioxide and water production, yields energy-rich ATP.

Glycolysis and fermentation occur in the absence of free molecular oxygen and are examples of anaerobic respiration. In the presence of free oxygen, an alternative respiratory pathway is followed (aerobic respiration) which produces higher yields of ATP and can be illustrated by the Krebs cycle or citric acid cycle. This process involves the chemical breakdown of pyruvic acid in the mitochondria through a series of complex biochemical reactions in which ATP, carbon dioxide, and water are formed. In a very general sense, the Krebs cycle may be viewed as a chemical pathway for glucose oxidation and energy production as well as an important metabolic hub in which numerous biochemical activities are interrelated. The rate of respiration, like photosynthesis, is affected by a variety of internal and external factors, which include food availability, activity of respiratory enzymes, water and oxygen supply, temperature, and carbon dioxide concentration.

Questions and Problems

1. Compare photosynthesis and aerobic respiration with respect to energy transfer, influencing factors, intermediate steps, raw materials and products, and biological significance.
2. Briefly describe the process by which carbohydrates, fats, and proteins are digested. How does the digestion of these substances compare with their photosynthetic production?
3. Indicate, using chemical shorthand, the various steps in glycolysis which lead to the formation of pyruvic acid.
4. Contrast aerobic and anaerobic respiration in terms of ATP production.
5. Describe that portion of the Krebs cycle which occurs in the mitochondria.
6. What is the mechanism by which electron acceptor molecules within the cytochrome system produce molecules of ATP?
7. Compare lactic acid fermentation with alcoholic fermentation. Consider ATP production, chemical reactions, and economic importance.
8. Describe the biological significance of the Krebs cycle.
9. What is the effect of temperature, carbon dioxide concentration, and water supply on the rate of respiration?
10. The physical environment of plants contains the same chemical elements as are found within plants. How would you compare the chemical organization of the nonliving environment with that of the living plant?

Suggested Supplementary Readings

Baker, J. J. W., and G. E. Allen, *Matter, Energy, and Life*. Reading, Mass.: Addison-Wesley Publishing Company, Inc., 1965.

Baldwin, Ernest, *The Nature of Biochemistry*. New York: Cambridge University Press, 1962.

Baserga, R., and W. E. Kisieleski, "Autobiographies of Cells," *Scientific American*, August 1963.

Beevers, H., *Respiratory Metabolism in Plants*. New York: Harper & Row, Publishers, 1961.

Bennett, T. P., and Earl Frieden, *Modern Topics in Biochemistry*. New York: The Macmillan Company, 1966.

Goldsby, R. A., *Cells and Energy*. New York: The Macmillan Company, 1967.

Green, D. E., "Biological Oxidation," *Scientific American*, July 1958.

————, "The Mitochondria," *Scientific American*, January 1964.

Lehninger, A. L., "How Cells Transform Energy," *Scientific American*, September 1961.

Siekevitz, P., "The Powerhouse of the Cell," *Scientific American*, July 1957.

Stiles, Walter, "Respiration III," *Botanical Review*, vol. 26, pp. 209–260, 1960.

Stumpf, P. K., "ATP," *Scientific American*, April 1953.

11
Plant Growth and Development

INTRODUCTION

Growth is one of the unique attributes of all living things, yet, as a phenomenon, it is quite difficult to define. Normally, increases in weight and size are associated with growth, although these are not necessarily indicative of growth taking place. A germinating seed, for example, may swell to many times its original size by the imbibition of water before any growth occurs, or increased turgidity may cause plant cells to increase in size without actually growing. A seed developing on a plant may increase in weight through the deposition of food without really growing. In other words, increase in size or weight are not absolute criteria for growth. For our purposes, however, **growth** may be defined as an increase in the amount of protoplasm in an organism generally accompanied by an irreversible increase in the size of the plant. In this respect, then, growth may be viewed as a quantitative phenomenon.

Most logically, quantitative growth cannot completely account for the total maturation of plants. If plants grew only quantitatively they would consist of aggregates of similar cells lacking any variation in form, size, and function. It is quite obvious, however, that a fully matured angiosperm consists of highly differentiated and specialized cells capable of absorption, translocation, transpiration, division, photosynthesis, respiration, and storage among other highly specialized activities. Growth, therefore, must involve qualitative as well as quantitative aspects.

The qualitative aspects of growth are cellular enlargement and differentation. Both involve a progressive change in form and complexity and are collectively referred to as **development.** Like growth, development is an irreversible process. Once meristematic cells of the root tip develop into primary tissues, the differentiated tissues cannot revert exactly to the undifferentiated form from which they originated. Although growth and development are quantitative and qualitative processes, respectively, no sharp line of distinction exists between them. Both are in simultaneous operation in the same organs or organisms as part of the total maturation process.

The growth and development of an angiosperm from a seed into a mature plant affords an excellent example of the integrated and simultaneous operation of growth and developmental phenomena. Starting as a single cell, a fertilized egg, the units of its structure at first grow quantitatively, bringing about an increase in size and weight. As the plant matures, the units do not remain alike; most of them develop qualitatively into specialized cells. Groups of cells form tissues, tissues give rise to organs, and organs form the integrated plant body (see Figure 4-1).

NATURE OF GROWTH AND DEVELOPMENT

The Pattern of Growth

There are a number of ways in which plant growth can be measured and, in most of these, quantitative data are recorded as a function of time. For example, growth may be analyzed by measuring size in terms of height, length, or width of the intact plant or any of its parts, or by measuring the fresh or dry weight of the whole

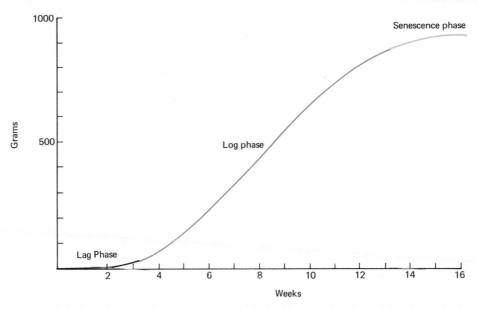

Figure 11-1. Growth curve of an entire corn plant. In this curve the dry weight of the plant in grams is plotted against time in weeks. Note the typical S-shaped curve. (Redrawn from D'Arcy W. Thompson, *On Growth and Form,* Cambridge University Press, New York, © 1942 [from G. Bachmann, after Stefanowska].)

plant or its component parts. If the analysis of growth is to have any significance, it must be measured over an interval of time. A typical growth rate might be stated as centimeters per day for size measurements or dry weight per day for weight measurements. In either case, the growth rate is expressed as a quantitative aspect per unit of time.

Assume that the growth of an intact corn plant is measured by increase in dry weight plotted against time (Figure 11-1). In general, it will be found that from the time of germination to maturity, three distinct phases of growth can be noted. The initial phase, or **lag phase,** represents a period in which the seedling plant grows rather slowly at first, then gradually grows more rapidly. During this period, internal changes preparatory to growth are taking place. The second phase, the

log phase, is characterized by a constant maximum rate of growth. Some plants reach maximum growth for only a day or two, whereas in other plants the maximum rate may be sustained for days or even weeks. The final growth phase, the **senescence phase,** is characterized by a slow or even decreasing growth rate at which point the plant has reached maturity.

It should be noted that the growth curve is an S-shaped or *sigmoid curve,* and this type of curve is typical of the growth pattern of the plant as a whole or its individual organs. If the growth rate for size instead of dry weight had been determined, a similar curve would have resulted. In fact, the sigmoid curve is characteristic of growth patterns for individual cells, tissues, organs, organisms, and populations whether plants or animals are being studied. The general nature of the sigmoid curve is fairly constant for all plants that have been investigated, although it may be modified in response to certain variations in the environment.

Stages of Plant Growth

Fundamentally, all plant growth is cellular growth, because tissue, organ, and organismic growth are all products of cell growth. The growth of organs of higher plants involves not only the enlargement and differentiation of cells already present but also the formation of new cells from previously existing ones. The formation of new cells by the division of those already present and their subsequent enlargement and differentiation into permanent forms provides the principal means of growth of any multicellular organism. In essence, then, the stages or phases through which cells pass to reach maturity are division, enlargement, and differentiation.

Division. In general, the process of growth is initiated in regions of higher plants in which cell division takes place. These regions are the meristems (see Figures 4-9A and B and 4-14B), and the degree and length of their mitotic activity depends upon seasonal as well as other factors. Apical meristems, which are characteristic of shoot, root, and bud tips, engage in primary growth and the product of this activity is primary tissue. By contrast, the vascular cambium and cork cambium are involved with secondary growth, and the product of their activity is secondary tissue. Although the bulk of plant growth is vegetative growth, a small but extremely significant part of the growth is reproductive growth in which reproductive structures reach maturity.

Mitotic cells of meristematic zones are structurally characterized by small size, thin walls, dense cytoplasm, small vacuoles, and prominent nuclei (see Figure 3-25). Functionally, these mitotic cells are engaged in active synthesis and water and mineral absorption which culminate in the assimilation of new protoplasm.

Enlargement. After a cell has divided, the resulting two daughter cells together are at first only as large as the parent cell. Before further growth can take place, the daughter cells must undergo *enlargement,* the dominant and most obvious phase in the growth of the plant. Enlargement is essentially an increase in volume caused primarily by the absorption of water (see Figure 4-12). As a cell elongates, the thickness of the wall is maintained. In addition, during elongation, the vacuoles grow in size from the inward diffusion of water in response to a decreased diffusion pressure caused by elongation. The protoplasm appears as a narrow layer between the cell wall and the vacuole (Figure 11-2). Even though an elongating cell is also assimilating protoplasm, the protoplasmic constituents appear to be sparse, owing to the increase in size.

Differentiation. As cells enlarge, they gradually assume their permanent shapes and forms. Both cell division and cell enlargement adequately account for an increase in size as the plant grows but not for the specialization of cells. The differentiation of cells, involving a multiplicity of structural and functional specializations, terminates in a series of modifications of the original cell. As part of these modifications, cells may differentiate and become specialized with respect to size, shape, nature, and extent of secondary walls, or protoplasmic contents. Structurally these modifications may be represented as changes in shape (tracheids), loss of end walls (vessel segments), perforation of end walls (sieve cells), impregnation with lignin (stone cells), and deposition of suberin (cork cells) or cutin (epidermal cells). Some functional specializations brought about by chemical changes involve the decomposition and death of protoplasm (tracheary elements), the formation of specialized organelles (chloroplasts), or the storage

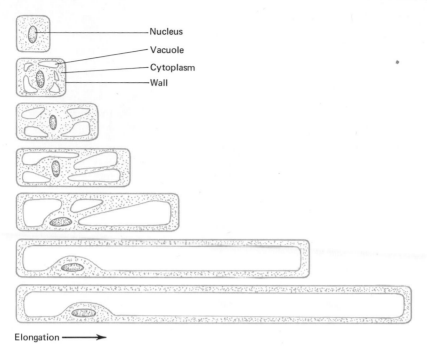

Nucleus
Vacuole
Cytoplasm
Wall

Elongation ⟶

Figure 11-2. Cellular elongation. Note the progressive structural changes in the cell wall, the cytoplasm, the vacuoles, and the nucleus as a result of elongation.

of certain types of foods or ergastic substances. **Differentiation** is, then, the transformation of apparently identical cells of the meristem into a number of highly specialized ones.

Distribution of Plant Growth

Plant growth is not uniformly distributed but is concentrated in specific growth regions. The distribution of plant growth was first demonstrated by Julius von Sachs in the middle of the last century. He marked young roots, stems, and leaves with inked lines and then observed the subsequent growth pattern for each organ. After the particular organ had grown for a period of time, it could be noted that the marks were no longer equidistant. Using this procedure, the growth rate for part of the organ can be determined from the distance between each pair of marks (Figure 11-3).

In roots, elongation occurs most rapidly in a region just behind the root tip (Figure 11-3A). Similarly, stem elongation occurs in a region just below the shoot tip, but the length of the stem is determined not only by this elongation but also by growth of the internodes below. The expanding leaves of such broad-leaved plants as tobacco have their own pattern of growth. In this case, expansion occurs in a relatively uniform manner throughout the entire surface of the leaf (Figure 11-3B). Elongating leaves of narrow-leaved plants

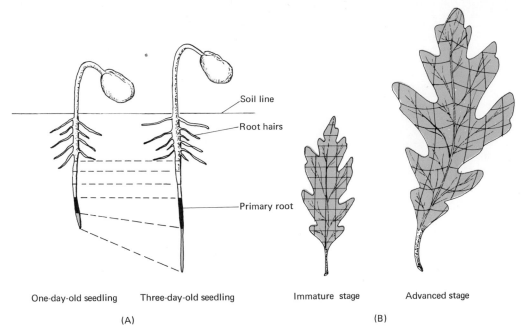

Soil line

Root hairs

Primary root

One-day-old seedling   Three-day-old seedling

(A)

Immature stage   Advanced stage

(B)

Figure 11-3. The distribution of growth in roots and leaves. **(A)** The progressive change in length of the zones between markings, initially equidistant, demonstrated that the elongation of the root occurs near the tip. **(B)** By comparing a young leaf marked with equally distant parallel lines and the same leaf in an advanced stage of growth, it can be seen that the pattern of growth is relatively uniform throughout the lamina.

such as grasses grow primarily at the base, so the mature leaf tissue is being continuously expanded and pushed up by the growing region.

FACTORS INFLUENCING PLANT GROWTH

The rate at which a plant or organ grows, the pattern according to which it develops, and the final form that it assumes are determined by the combined influences of a number of complex internal and external factors. Among the internal factors affecting plant growth are the influences of growth substances produced by the plant, the balance of metabolic activities, and the structural and functional integration of plant parts. Chief among the external factors that influence plant growth are light, temperature, moisture, and certain physical, chemical, and biotic factors in the environment. The hereditary potential of the plant is also a significant factor in determining growth patterns. The interrelationship of hereditary, environmental, and internal factors as they relate to plant growth and development may be represented as follows:

Heredity

Heredity is an important internal factor in regulating plant growth. Hereditary potentials (genes) exert control over all internal processes and conditions of plant cells. It should be recalled that genes (DNA) contain the genetic code for the regulation of all life processes and that this code is passed on to RNA, the intermediate between nuclear DNA and cytoplasmic organelles. The RNA then migrates from the nucleus to the ribosomes where specific enzymes (proteins) are synthesized. The kinds of enzymes that a plant cell can synthesize are thus controlled by its genes, and the enzymes in turn determine the nature of the biochemical reactions that the cell is capable of performing. Included in these biochemical reactions are the regulation of plant growth, development, and behavior.

In essence, it is the **genotype,** or gene complement, which the plant inherits that gives the organism its potentialities for development according to certain structural and functional patterns. Each plant species contains a particular group of inherited factors which is capable of influencing the development of definite characteristics, providing the proper conditions are supplied to express these traits. In other words, heredity determines what a plant *can* do, and the environment, both internal and external, determines which of these things the plant *will* do.

Many of the hereditary potentialities that plants receive from previous generations are so constant that even extreme variations in the environment cannot change them. For example, morning glories typically have twining vines and weak stems and, although their rate of growth may be influenced by environmental variations, they still remain as twiners. As another example, consider the buds of black locust trees, which give rise to pinnately compound leaves. Although the rate of growth of these leaves is subject to the influences of environmental factors, the compound nature of the leaves cannot be altered by moisture, temperature, or light variations. Thus the environment can induce certain changes in form and function, but the degree and quality of the changes are limited by the genotype of the plant.

Internal Factors

Growth Substances. The action of **growth substances,** or **hormones,** constitutes one of the prominent internal mechanisms that regulates the growth of plants. A *plant hormone* is an organic compound produced by the plant, which, in extremely low concentrations, is capable of influencing physiological activities including the promotion, inhibition, and modification of growth and development. Plant hormones are synthesized in one part of the plant and translocated to another part where a particular physiological activity is influenced and a response is elicited. The influence of hormones in extremely minute quantities is substantiated by the fact that auxin, a hormone that has been shown to promote the growth of roots, is effective in concentrations of about 0.0000001 gram per liter (1000 grams of water).

Balance of Metabolic Activities. Inasmuch as green plants are photosynthetic autotrophs, the quantity and balance of nutrients produced are relative to the supply of inorganic metabolites as well as to the factors that govern the rate of photosynthesis. It has been demonstrated that the growth and development of certain plants are related to a proper balance between sugars produced during photosynthesis and nitrogen-containing compounds. For example, tomato plants when provided with excess nitrate tend to become more vegetative and less reproductive. In other words, fewer fruits are produced. By contrast, if nitrates are made less available, protein synthesis is minimal, and the plant appears tough, stunted, nonvegetative, and nonreproductive. Tomato plants also exhibit similar abnormal growth and developmental characteristics if abundant nitrates are provided and if sufficient light for photosynthesis is removed. What happens in this case is that sugar production is so minimal that the synthesis of structural carbohydrates and other organic compounds necessary for growth is retarded. In all probability other relationships involving a metabolic balance of nutrients exist, but relatively little is known about them.

Integration of Plant Parts. The influence of one part of a plant on the pattern of growth and development of other parts is called *correlation*. It has long been known that the removal of one part of a plant exerts profound effects on the growth of other parts. If the terminal portion of a stem is removed by pruning, lateral buds, which would normally remain dormant, start to grow. Another example of correlation is the relationship existing between vegetative and reproductive structures. If the flower buds of tomato plants are removed, the plant undergoes a rejuvenated vegetative growth. Conversely, if flowering and fruiting are afforded continued growth, vegetative growth is terminated. Such a correlation between reproductive and vegetative structures is typical of most annual plants.

The vast majority of correlations in plants are far more subtle and less well understood than the examples just cited. However, the effect on organs that remain when others are removed affords ample evidence of correlation. The development of the shoot influences that of the root, the growth of roots influences the development of the shoot, and the development of reproductive structures affects the developmental patterns of vegetative organs.

External Factors

Light. Perhaps no other environmental factor assumes a more prominent role in the growth of plants than does light. From the time a seed germinates until it develops into a fully matured and reproductive plant, solar radiation is a continuing and significant factor. Light obviously affects various metabolic activities of plants; thus it must also influence growth and development, which are results of metabolism. In this regard, then, light influences the following plant activities: (1) synthesis of chlorophyll and other photosynthetically involved processes; (2) synthesis of the products of photosynthesis into various organic compounds; (3) temperature of aerial portions of the plant, thus the rates of synthesis, degration, and transpiration; (4) regulation of stomatal openings; (5) reaction or movement of plants or plant organs; (6) geographic distribution of plants; (7) anatomical features of leaves; and (8) germination of seeds of certain species.

Temperature. In general, plant growth is promoted as the temperature increases and is retarded as the temperature decreases. Every metabolic activity is influenced by temperature and each contributes to the total pattern of growth and development. Consider the effect of a significant temperature change on metabolic activities such as photosynthesis, respiration, digestion, assimilation, enzyme synthesis, permeability, diffusion, osmosis, translocation, transpiration, and reproduction. Obviously, the alteration of the rate of any of these activities would markedly affect plant growth.

The temperature ranges within which plants live and grow vary with the species of plant, different organs of the same plant, and with the nature of the internal and external conditions to which the plant is subjected. Although temperature ranges vary considerably among different plants, the *minimum temperature,* or temperature below which growth ceases for most plants, is about 0°C (freezing). The *maximum temperature,* or temperature above which no growth takes place, is about 45°C for most plants. The *optimum temperature,* or temperature at which the growth rate is highest, usually averages about 20 to 25°C for most plants.

Moisture. The importance of water to the plant in general, and therefore to growth, has already been considered, so further elaboration is not necessary at this point.

Other Factors. In addition to light, temperature, and moisture, other physical factors in the environment also affect plant growth. For example, the orientation of rocks limit, and to an extent determine, the nature of root systems; wind and rain alter the form and structure of branches, stems, and leaves; and contact with various objects such as fences, poles, and walls cause plants to twine about them, altering growth patterns.

In a similar manner chemical substances in the soil, water, and air, other than necessary metabolites, influence plant growth and development. The addition of fertilizers and similar beneficial chemical compounds to plants greatly facilitate growth. Unfortunately, however, the rapid destruction and disappearance of plants in heavily industrialized areas due to air and water pollution are proof of the adverse effects of certain chemicals on plant growth.

Finally, in considering some of the factors that affect plant growth, mention should be made of biological influences resulting from the activities of plants, animals, and man. Some representative biological influences are the competition of plants with each other for available space, light, and metabolites; the dependence of angiosperms on insects and other animals for pollination; the activities of soil bacteria; the grazing of animals; plant and animal parasites; and the destruction of vegetation by man, among others.

THE MECHANISMS OF GROWTH CONTROL

Hormones and Plant Growth

Most plant physiologists recognize three principal groups of hormones: auxins, gibberellins, and cytokinins. Prominent among these plant growth substances are the auxins, the first group of plant hormones to be discovered (1928) and the most intensively investigated group. The term *auxin,* meaning "to increase," is applied to a group of compounds characterized by their capacity to induce elongation in shoot cells. Although the major physiological activity of auxins is to promote growth, if present in excessive quantities, they tend to inhibit growth. Various experimental data have demonstrated that the required auxin concentration for growth promotion varies with different plant organs, being much lower for roots than for stems (Figure 11-4).

Most data concerning the effects of auxins on plant growth have been obtained from studies with seedlings of grasses, particularly oats (*Avena sativa*). In very young oat seedlings, the growing shoot is surrounded by a cylindrical sheath (covering) called a *coleoptile,* the first structure to rise above ground after germination (Figure 11-5). As the growth of the seedling progresses, the coleoptile ceases elongation and is broken through by the first foliage leaf of the enclosed shoot. In experimental work using *Avena,* the seedlings are allowed to grow in the dark until the coleoptiles are about 1 centimeter long and just prior to the first foliage protrusion.

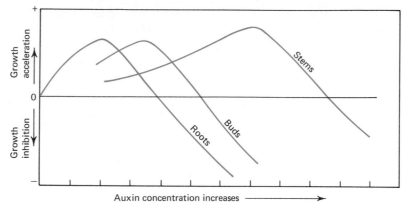

Figure 11-4. The effect of auxin concentration on the growth of different plant organs. Note that a similar auxin concentration may promote the growth of one organ (stem) but inhibit the growth of another (root). (After K. V. Thimann, *The Action of Hormones in Plants and Invertebrates,* New York: Academic Press, Inc., ©️ 1952.)

It has long been known that the extreme tip of the coleoptile is necessary for elongation of the coleoptile itself and the enclosed stem below it. As early as 1880, Charles Darwin noted that grass

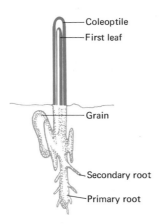

Figure 11-5. Structure of an *Avena* seedling. Much experimental work related to plant hormones and their physiological effects has been done with the coleoptiles of grass seedlings, especially oats.

coleoptiles exhibit a characteristic curvature toward light (Figure 11-6A). When he shielded the tips of the coleoptiles, however, no bending occurred (Figure 11-6B) even though the lower portion of the coleoptile was still illuminated from one side. When the upper portion of the coleoptile was illuminated and the lower portion shielded, normal curvature took place (Figure 11-6C). Carrying the experimental procedure one step further, Darwin removed the tip of the coleoptile and found that there was no curvature or elongation (Figure 11-6D). The conclusion drawn from this series of experiments was that some influence (hormone) was translocated from the tip of the coleoptile downward, causing curvature and elongation.

Further experimentation on coleoptiles was carried out in 1910–1913 by Boysen-Jensen. He demonstrated that if a transverse incision was made on the nonilluminated side of the coleoptile and a plate of mica placed in the incision, no curvature took place (Figure 11-6E). When a similar incision and placement of the mica plate was

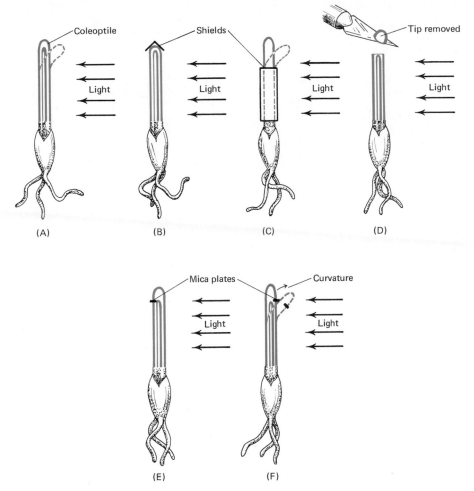

Figure 11-6. Experiments with *Avena* coleoptiles. **(A–D)** In the series of experiments conducted by Darwin it was demonstrated that the stimulus for growth is transmitted from the tip of the coleoptile downward. **(E, F)** Boysen Jensen demonstrated that the growth stimulus is transmitted down the shaded side of the coleoptile.

made on the illuminated side, however, curvature did occur (Figure 11-6F). These responses indicated that the influence translocated from the coleoptile tip downward moved down the shaded side of the coleoptile.

Boysen-Jensen also demonstrated the site of production and diffusible nature of the growth substance in the coleoptile tip. To do this he removed the tip, placed gelatin on the stump, replaced the tip on the gelatin, and found that

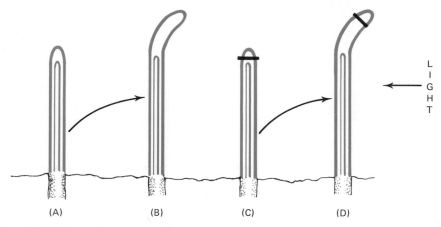

Figure 11-7. Boysen-Jensen's experiment. **(A, B)** The control (untreated) coleoptiles exhibit bending in the presence of light. **(C, D)** The coleoptiles are decapitated, gelatin is placed on the stump, and the tips are replaced on the gelatin. Even after treatment, the coleoptiles continue to bend toward the light. This experiment illustrates the diffusible nature of the growth substance.

curvature and elongation took place (Figure 11-7). The growth substance diffused from the tip through the gelatin to the stump, thus maintaining the normal growth pattern. If the tip was not replaced, no curvature or elongation took place.

These early experiments in conjunction with others eventually led Fritz Went in 1928 to the discovery of the hormonal nature of this substance in coleoptiles. Went showed that the growth substance found in coleoptile tips could diffuse into *agar,* a gelatin-like polysaccharide, if excised tips are placed on the agar for about 2 hours (Figure 11-8A). If the excised tips are then discarded and small blocks of agar are placed on the decapitated coleoptile stumps, growth and curvature take place (Figure 11-8B). Controls in the form of pure agar show no effect (Figure 11-8C). Went noted that if a treated agar block was placed on one side of a coleoptile stump a curvature resulted, indicating that growth was more rapid along the side containing the agar block. This test, called the

Avena test, was the basis for determining the amount of growth substance in the block of agar because the degree of curvature was found to be proportional, within limits, to the concentration of auxin in the agar blocks.

It was not until the mid-1930s that the growth substance in coleoptile tips was finally characterized as an auxin. The most commonly found naturally occurring auxin is *indole-3-acetic acid* (*IAA*) (see Figure 4-24A). This compound, as well as other auxins, has been found to be present in growing organs such as foliage leaves, stems, flower stalks, germinating seeds, roots, fruits, and pollen grains.

Influences of Auxins. *Leaf Abscission.* One of the important functions of auxins is believed to be the control of leaf fall. Leaves separate from most plants due to the formation of a weakened region of cells at the base of the petiole called the *abscission layer* (see Figure 4-25). The relationship between abscission and auxins is shown by certain experimental data. If a leaf blade is excised, the petiole stump soon falls off because of the formation of an abscission layer. If however, an auxin is applied to the cut end of the stump, abscission is delayed or prevented. Inasmuch as the leaf blade is one of the sites of auxin production, it can be concluded that auxin is one controlling factor in abscission.

It has been found that auxin content is high in young leaf blades as compared to the petiole, a condition conducive to leaf retention. As the leaf ages, however, the auxin content of the blade decreases to a point comparable to that found in the petiole and leaves abscise. There is evidence to suggest that the most important factor controlling abscission is the nature of the auxin gradient across the abscission layer, that is, the relative concentrations of auxin on the stem side and leaf side of the layer. If the gradient is steep, as when the auxin concentration is high on the leaf side of the layer and low on the stem side, no abscission occurs (Figure 11-9A). When the gradient disappears, there is leaf abscission (Figure 11-9B). When the gradient is reversed, that is, the auxin concentration is high on the stem side of the abscission layer and low on the leaf side, abscission is accelerated (Figure 11-9C).

Fruit Drop. Fruits also fall from plants at an abscission layer as a result of an auxin gradient on either side of the layer. The fruit-growing industry makes practical application of these relationships. For example, in the spring, at the beginning of the growing season, the stems of fruit-bearing plants are sprayed with auxins, increasing

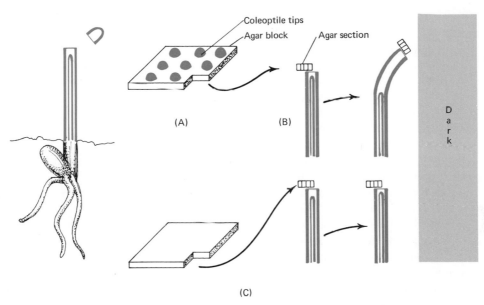

Figure 11-8. Went's experiment. **(A)** Coleoptile tips are placed on agar. **(B)** Excised tips are discarded and agar blocks are placed on the decapitated coleoptiles. Growth and curvature occur. **(C)** Plain (untreated) agar blocks show no effect on either the growth or bending of the coleoptiles.

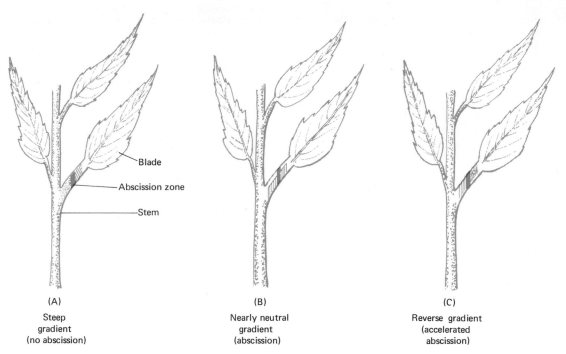

(A)
Steep
gradient
(no abscission)

(B)
Nearly neutral
gradient
(abscission)

(C)
Reverse gradient
(accelerated
abscission)

Blade

Abscission zone

Stem

Figure 11-9. Relationship between auxin concentration and leaf abscission. (Data from F. T. Addicott and R. S. Lynch, "Physiology of Abscission," *Annual Review of Plant Physiology*, vol. 6, p. 211, 1955.)

the auxin concentration in the stems. This reversed gradient causes premature fruit abscission, resulting in fruit "thinning." The remaining fruits that do not drop are usually larger and better because of the availability of more metabolites. At some point later in the season, auxin may again be supplied to the remaining fruits, setting up a steep gradient and delaying abscission and permitting longer tree ripening of the fruits (Figure 11-10).

Fruit Development. The development, as well as the drop of fruits, is also related to auxin control. The growth and development of the ovary of a flower into a fruit is controlled by an adequate supply of auxins produced after a flower has been pollinated. The initial enlargement of the ovary

is brought about by auxins supplied by the pollen grains and pollen tubes. Other auxins are produced by the embryo within the seed after fertilization has occurred. Auxins supplied through pollination and fertilization are necessary for fruits to develop fully.

Fruits fail to develop on most plants unless flowers are pollinated. In some plants, however, this is not the case. It appears that the ovaries and other floral structures of such plants contain adequate supplies of auxins to promote fruit development. Inasmuch as pollination and fertilization do not occur in such plants, seedless fruits are developed. This phenomenon, called *natural parthenocarpy,* is characteristic of certain varieties of

Figure 11-10. The effects of auxins on fruit drop. **(A)** Untreated Jonathan apple tree showing extensive fruit drop. **(B)** A Jonathan apple tree sprayed with an auxin (α-naphthalene-acetic acid, or NAA) showing almost no fruit drop. (Photos courtesy of U.S.D.A.)

grapes, bananas, pineapples, figs, cucumbers, and oranges. In addition to natural parthenocarpy, man has also developed a commercially important process of induced, or *artificial, parthenocarpy*. In this process, flowers are prevented from being pollinated and sprayed with auxin solutions or treated with auxin pastes. As a result, seedless fruits develop. Common products of artificial parthenocarpy are certain varieties of watermelons, peppers, strawberries, and tomatoes.

Apical Bud Dominance. In most stems, the terminal bud is the dominant bud; that is, it grows more rapidly and is physiologically more active than lateral buds. The inhibition of lateral bud growth by terminal buds is called *apical dominance,* and the phenomenon is controlled by auxins. Such a correlation phenomenon may be noted by removing the terminal bud of a sunflower plant (Figure 11-11A). After removal, the plant loses its apical dominance and lateral buds undergo exten-

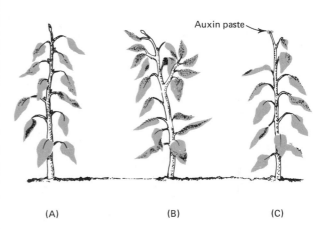

Figure 11-11. Apical dominance. **(A)** Normal sunflower plant exhibiting apical dominance. **(B)** Removal of terminal bud resulting in loss of apical dominance and extensive growth of two lateral buds into branches. **(C)** Apical dominance restored after application of an auxin paste on the cut end from which the apical bud was removed.

Plant Growth and Development 277

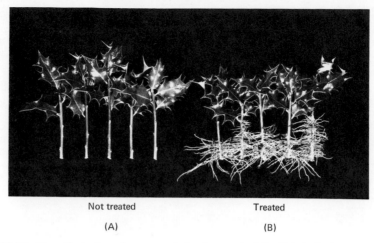

Not treated   Treated

(A)   (B)

Figure 11-12. The effect of auxins on root development of English holly (*Ilex aquifolium*). **(A)** Untreated plants fail to root. **(B)** Plants treated with auxin produce numerous adventitious roots. (Photo courtesy of Dr. A. E. Hitchcock, Boyce Thompson Institute for Plant Research, Inc., Yonkers, N.Y.)

sive growth into branches (Figure 11-11B). If, however, an auxin preparation is applied to the excised stem tip, lateral bud growth is inhibited and apical dominance is restored (Figure 11-11C).

Root Development. The control of the development of adventitious roots is another important function of auxins. If a stem is removed from a plant and placed into moist soil or water, development of roots is caused by the translocation of IAA from the tip of the stem to the base (Figure 11-12). Such cuttings, commonly used in horticulture, are used as a means of plant propagation. Cuttings of many species of hard to root plants form few or no adventitious roots unless supplied with adequate concentrations of auxins. In general, cuttings with leaves react more favorably to rooting than do cuttings void of leaves.

Figure 11-13 summarizes some of the influences of auxins on plant growth and development.

Influences of Gibberellins. *Gibberellins* represent a second group of plant hormones that were first discovered in 1941. The hormones are produced by a fungus plant, *Gibberella fujikuroi,* that causes a disease in rice grown in the Orient known as "foolish seedling disease." When infected with this fungus, rice seedlings grow rapidly according to abnormal, erratic patterns. It was determined that the substance responsible for this abnormal growth produced by the plants was *gibberellic acid* (see Figure 2-24B). Gibberellins, chemicals related to gibberellic acid, appear to promote cellular elongation, particularly in stems. In some shoots, auxins and gibberellins interact in their effects on cell enlargement, but details of such interactions are still vague. Both hormones affect aspects of cell elongation, perhaps at different points in the overall process.

Genetic Dwarfism. *Genetic dwarfism,* a condition usually caused by the mutation of a single gene,

Figure 11-13. Summary of some of the influences of auxins on various plant growth and development phenomena.

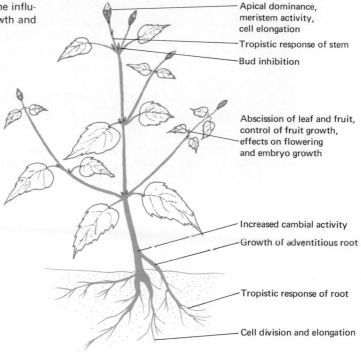

Apical dominance, meristem activity, cell elongation

Tropistic response of stem

Bud inhibition

Abscission of leaf and fruit, control of fruit growth, effects on flowering and embryo growth

Increased cambial activity

Growth of adventitious root

Tropistic response of root

Cell division and elongation

is characterized by a shortening of the internodes of a plant. Among the plants that exhibit this type of dwarfism are pea plants (*Pisum sativum*), horse bean plants (*Vicia faba*), and scarlet runner bean plants (*Phaseolus multiflorus*). When gibberellin is applied to such mutant plants, genetic dwarfism is overcome. Although the mechanism of the influence of gibberellin in genetic dwarfism is not known, two hypotheses purport to account for its role. According to one of these, the application of gibberellin alleviates the shortage of *endogenous* (produced by the organism itself) gibberellin. Such a shortage could be caused by the lack of an enzyme (as a result of the mutation) involved in the synthesis of the hormone. The second hypothesis holds that an excess of a natural inhibitor is present in dwarf plants and that the effects of this inhibitor, which retard growth, are counteracted by the activity of gibberellin.

Other Influences of Gibberellin. Aside from their influence on internode elongation, gibberellins are related to other physiological activities such as plant movements, flowering, parthenocarpy, and increased cambial activity. Many of the responses elicited by gibberellins are probably due to an interaction with auxin as well as other plant hormones. Others appear to be the result of gibberellins acting independently. Whether gibberellins and auxins act independently or together is a function of the species of plant under investigation, the conditions under which the plant is grown, and the particular response being studied.

Much work remains to be done on this aspect of plant growth regulation before any conclusions can be reached.

Influences of Cytokinins. In general, the principal role of auxins and gibberellins is to promote cell elongation and therefore effect certain physiological activities. *Cytokinins,* a third group of plant hormones structurally related to kinetin (see Figure 2-24C) appear to influence not only cell enlargement but also to induce cell division. Shortly after the discovery of kinetin, its effects on cell division and enlargement were studied in a number of plant growth systems.

Cell Division. It is believed that cytokinins assume a role in the initiation of cell division. It has been demonstrated that isolated plant cells grown in the laboratory and supplied with the necessary minerals, foods, and auxins enlarge enormously but do not divide. If a cytokinin is supplied to the cells, not only do the cells continue to enlarge, but they also divide. The mechanism by which cytokinins induce cell division is still unknown. Its interaction with auxin, however, is fairly well established.

Apical Dominance. In an earlier discussion, the influence of auxin on apical dominance was analyzed. Removal of the apical bud stimulates the growth of lateral buds; leaving the apical bud inhibits lateral bud growth. In a series of investigations it has been shown that if an intact shoot is soaked in a cytokinin solution, inhibition of lateral buds is overcome by the apical bud to a large extent. Other studies have shown that cytokinins have a stimulatory effect on lateral bud growth. It appears that the phenomenon of apical dominance may be under the influence of a balance between endogenous cytokinins and IAA.

Other Influences of Cytokinins. In addition to their effect on cell division and cell elongation, cytokinins assume a role, either directly or indirectly, in root initiation and growth, stem initiation and growth, and germination of certain types of seeds (for example, white clover and carpet grass). As in the case of gibberellins, cytokinins probably interact with endogenous IAA to effect and influence growth and development of the plant.

Vitamins as Plant Growth Substances

Vitamins are usually associated with various nutritional deficiency diseases and not with the control of growth and development. They are, however, like hormones, effective in minute quantities and, owing to their physiological effects, may be looked upon as growth substances. The physiological effects of vitamin deficiencies are studied quite readily in animals by depriving the animal of a certain vitamin and then noting the effect. In plants, however, the chief difficulty in studying vitamin deficiencies is that many autotrophic plants synthesize their own vitamins. Because certain heterotrophs such as fungi, as well as certain plant parts such as roots and other tissues of autotrophs, are unable to synthesize vitamins, much knowledge of the importance of vitamins as growth regulators has been obtained by studies with fungi or with tissue cultures of higher plants grown on synthetic media.

Among the vitamins that act as growth substances are *thiamin, pyridoxine, nicotinic acid, biotin,* and *ascorbic acid.* Using the culture of isolated root tips on a synthetic media, it has been shown that thiamin is necessary for the growth of plant roots. In addition, nicotinic acid and pyridoxine are also required by roots of certain plants along with thiamin for normal growth. Embryos of certain plants fail to develop if they lack vitamins such as biotin, thiamin, and ascorbic acid.

Light and Plant Growth

Considering that the nutritional dependency of green plants is directly linked to light, it is not surprising that solar radiation also has profound

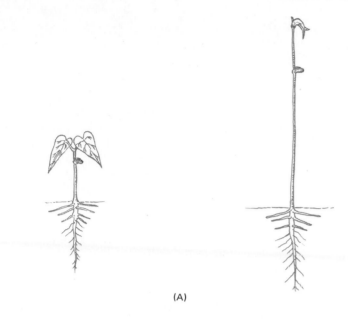

(A)

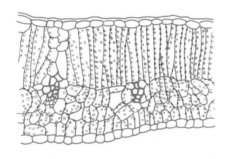 

(B)

Figure 11-14. Light and plant growth. **(A)** The effect of darkness on the growth of bean plants. *Left:* In the light, stem growth is inhibited while the growth of leaves is promoted. *Right:* In the darkness, bean plants develop long stems, smaller leaves, and other etiolated characteristics. **(B)** Sun and shade leaves of sugar maple (*Acer saccharum*). *Left:* Sun leaf showing well-defined internal tissues. *Right:* Shade leaf exhibiting a loose mesophyll, less vascular tissue, and the absence of heavy cutinization.

effects on plant growth and development. In considering the effects of light on plant growth, three aspects of radiation must be taken into account: intensity, quality, and duration.

Light Intensity. Prior to any known physiological cause and effect relationship existing between light and growth, certain peculiar developmental patterns of plants grown in darkness had been observed. The most striking influence of light on plant growth can readily be seen by comparing plants grown in complete darkness with ones grown under ordinary conditions of day and night (Figure 11-14). The complete absence of light affects not only the rate of growth but also certain anatomical features of plants. In the dark, leaves tend to be yellow because of their inability to synthesize chlorophyll; stems elongate abnormally and become spindly with poorly developed vascular tissue; and leaves fail to enlarge and remain folded as though still embryonic. A plant exhibiting these traits is said to be *etiolated* and the phenomenon is called **etiolation.**

Plants growing in shade (not darkness) also exhibit some characteristics of etiolation such as excessive stem elongation and, in addition, they show marked variations in leaf anatomy. For example, leaves grown in light have more sugar and less water than leaves grown in shade and, as a result, tend to be small in area, thick, and compact with a heavier cuticle and more conducting and supporting tissue (Figure 11-14B, left). Note that the structure of sun leaves is such as to prevent excessive transpiration and to facilitate translocation of photosynthetic products. Shade leaves, by contrast, have a loose palisade mesophyll consisting of single layers of cells and a very loose spongy mesophyll filled with air spaces. In addition, the epidermal cells lack heavy cutinization (Figure 11-14B, right).

All plants do not respond similarly to variations in light intensity. Some require more light than others and different species of plants grow best at different levels of light intensity. Some plants, such as sunflowers, tomatoes, and grasses, require direct bright light for maximum growth and development. Other plants, such as violets and ferns, have a relatively low light requirement and develop normally in shaded areas.

Light Quality. Light varies in quality in different regions of the spectrum. The effects of different wavelengths of the visible spectrum have been studied by growing plants under glasses that transmit limited regions of the spectrum. At any given light intensity, various wavelengths of light influence plant growth and development in different ways. For example, it has been found that plants grown under the red-orange part of the spectrum exhibit certain aspects of etiolation. Plants grown under the blue-violet end of the spectrum exhibit

growth patterns similar to those grown under the full visible spectrum.

Many striking results are noted when plants are grown in light from which the blue-violet portion of the spectrum has been eliminated. Plants grown under this condition elongate very rapidly during the first few weeks of growth. Some plants, such as *Coleus,* tomato, and four-o'clock, attain greater final height under these conditions, whereas others, such as sunflower and Sudan grass, do not. In either case, regardless of the final height, plants grown in the absence of blue-violet light have thinner stems, less well-developed vascular tissue, poorly developed flowers and fruits, and a higher degree of hydration than plants grown in the full sunlight spectrum.

Light Duration. The duration of daily exposure to light has significant effects on various developmental processes in plants, especially flowering. The response of plants to length of the daily period of illumination is referred to as **photoperiodism.** Normally, an angiosperm will flower only when the length of day falls within certain limits, a function of the season of the year. Based upon a 24-hour cycle of light and darkness, angiosperms are classified into three main groups:

1. Plants which flower when the day length is less than a certain critical length are called *short-day plants* (Figure 11-15A). Day lengths in excess of this critical value will keep a short-day plant vegetative. Common examples of short-day plants are ragweed, chrysanthemum, violet, aster, cocklebur, and strawberry. The critical day length for these, as well as other short-day plants, varies with the species.

2. *Long-day plants* flower after a critical day length is exceeded (Figure 11-15B). This critical day length also differs from species to species. Some representative long-day plants are radish, spinach, beet, iris, lettuce, wheat, and clover.

3. *Day-neutral plants* (Figure 11-15C) flower after a period of vegetative growth, regardless of the photoperiod. Some examples of day-neutral plants are tomato, four-o'clock, cotton, dandelion, sunflower, and certain varieties of pea.

Attempts to elucidate the mechanism of photoperiodism have been only partially successful. However, several contributory factors are known. First, it has been determined that the leaves of flowering plants are the receptor organs for the transmitted light stimulus (Figure 11-16A). Plants in which leaves have been removed fail to flower, even though the proper photoperiod has been provided (Figure 11-16B).

Second, it has been suggested that hormones might be involved in the flowering process. One line of evidence to suggest the presence of a floral hormone was presented by Chailakhian in 1936. Using chrysanthemums (short-day plants) as his experimental organism, he excised leaves from the upper part of the plants, leaving those on the lower half intact (Figure 11-16C). He then subjected the lower half to short-day photoperiods while simultaneously exposing the upper half to long-day photoperiods. Under these conditions, the plants flowered. Next, he reversed the experimental conditions (Figure 11-16D). This time the lower half was exposed to long-day treatments and the upper portion received short-day treatments. The plants remained vegetative and did not flower. From these data he concluded that day length causes leaves to produce a hormone which moves from the leaves to the buds to induce flowering.

Evidence favoring the concept of a diffusible hormone has also been presented by various graft-

Figure 11-15. Photoperiodism. Based upon a twenty-four-hour cycle of light and darkness, angiosperms may be classified as (**A**) short-day plants, (**B**) long-day plants, and (**C**) day-neutral plants.

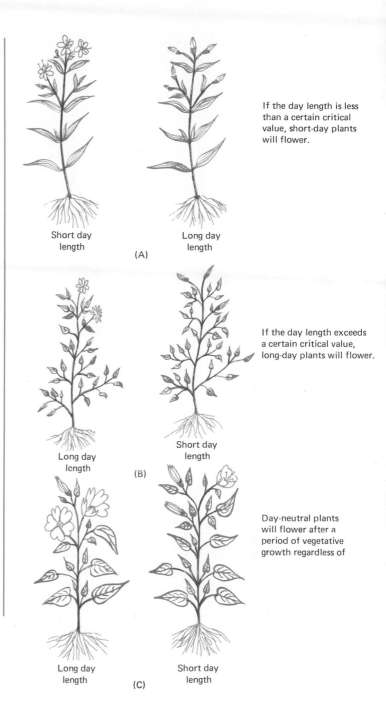

If the day length is less than a certain critical value, short-day plants will flower.

Short day length   Long day length
(A)

If the day length exceeds a certain critical value, long-day plants will flower.

Long day length   Short day length
(B)

Day-neutral plants will flower after a period of vegetative growth regardless of

Long day length   Short day length
(C)

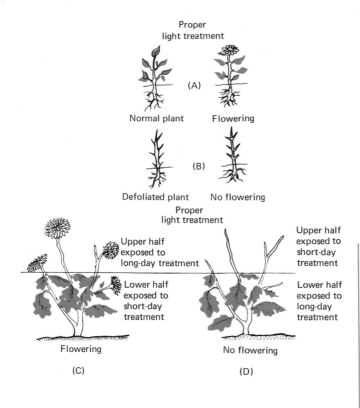

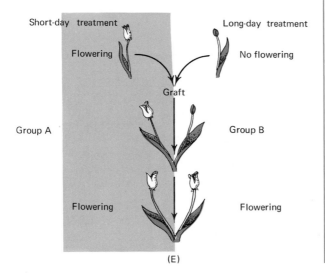

Figure 11-16. Various experiments concerning floral initiation. *Above:* Effects of defoliation on flowering. **(A)** Plants with normal foliage and proper light treatment produce flowers. **(B)** Plant with excised leaves and proper light treatment fails to flower. *Center:* Chailakyan's experiment. **(C)** Chrysanthemums (short-day plants) from which the upper leaves were removed were exposed to long days while the bottom half was exposed to short days. This treatment induced flowering. **(D)** When the experimental procedure was reversed, no flowering took place. *Below:* Experiment showing that florigen may diffuse from one plant to a diffused partner. **(E)** Both plants are short-day plants. The plant on the left has received a short-day photoperiod while the one on the right has received a long-day photoperiod. When grafted together, the plant on the right flowers.

ing techniques. One group of plants (short-day), group A, was grown under short-day photoperiods and they produced flowers. A second group of plants (also short-day), group B, was subjected to long photoperiods and flowering was inhibited. If plants from both groups are grafted together, and if a short-day photoperiod is maintained for group A and a long-day photoperiod is maintained for group B, the plants exposed to short days (group A) will flower, and soon thereafter the plants exposed to long days (group B) will also flower (Figure 11-16E). The nonflowering group will produce flowers. These data indicate that a diffusible substance has moved down from its point of origin, the leaf, to another part of the plant, the meristems, where the flowering response is initiated. This flowering hormone, tentatively named *florigen,* has not yet been isolated. Ringing experiments indicate that the flowering hormone is transported in the phloem, as are most organic substances.

Intensive investigation of photoperiodism has revealed that the photoperiodic response of short-day plants is more of a response to the dark period than to the light regime. For example, if a long-day plant such as henbane (*Hyoscyamus*) is deprived of light for about 1 hour during the day, it still

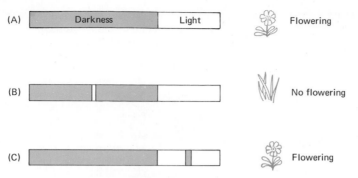

Figure 11-17. The importance of the dark period in short-day plants. **(A)** If a short-day plant is given a long dark period and a short light period, it flowers. **(B)** If the long dark period is interrupted in the middle by a brief light period, the short-day plant does not flower. **(C)** If the plant is given a long night period and a short light period and if the light period is interrupted in the middle by a brief period of darkness, it flowers. Long-day plants show no requirement for a dark period; they will flower in continuous light.

flowers. A short-day plant, by contrast, such as cocklebur (*Xanthium*), if illuminated for a short period of time during the night will not flower. Considering these relationships, a short-day plant is more properly termed a *long-night plant* that requires an uninterrupted dark period of certain minimal length to flower (Figure 11-17). Long-day plants, or *short-night plants,* differ from short-day plants in showing no requirement for a dark period. They will flower in continuous light. In fact, darkness tends to inhibit flowering in short-night plants.

Data concerning the importance of the dark period to floral initiation and the fact that flowering can be suppressed by a light break suggest that certain chemical reactions, sensitive to minute quantities of light, are in operation within the leaf. One of the initial problems was to determine what wavelengths of light are absorbed. This type of investigation was undertaken by a group of scientists at the U.S. Department of Agriculture, Beltsville, Maryland, in 1944. They were interested in determining the action spectrum for the inhibitory action of light breaks during the dark period. After exposing Biloxi soybeans (short-day plants) to light of different wavelengths for about 30 seconds in the middle of a long night, it was determined that red light (660 mμ) is most effective in inhibition of the flowering response in these short-day plants. Apparently, whatever happened during the long dark period was reversed by the flash of red light (Figure 11-18A).

In long-day plants, a brief treatment of red light acts in just the opposite way (Figure 11-18A). Under these conditions the dark interruption stimulated flowering. In this regard, short- and long-day plants show similar sensitivity to light regimes, but their responses are opposite to each other.

Further experimentation revealed that far-red light (730 mμ) had the same effect as darkness on short-day plants—that is, it had little or no effect alone. When short-day plants were exposed to brief periods of far-red irradiation during the long night, flowering was initiated (Figure 11-18B). If, however, brief far-red light followed red light, the effect of red light was negated and the short-day

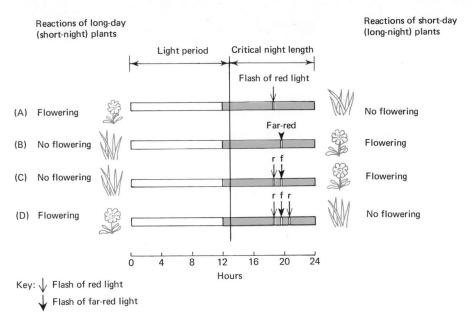

Light period   Critical night length

Flash of red light

(A) Flowering   No flowering

Far-red

(B) No flowering   Flowering

r f

(C) No flowering   Flowering

r f r

(D) Flowering   No flowering

0   4   8   12   16   20   24
Hours

Key: ↓ Flash of red light

↓ Flash of far-red light

Figure 11-18. Some responses of long-day and short-day plants under the influences of varying light regimes. See text for amplification.

plants flowered (Figure 11-18C). Almost any number of successive flashes were used (red, far-red, red, far-red, and so on), and the final effect depended solely on whether the last flash was red or far-red (Figure 11-18D). Not only did red and far-red light have opposite effects, but each reversed the effect of prior exposure to the other. Thus the short-day plants did not flower if the last flash given was red light, but they did flower if the last flash given was far-red light.

The reactions of long-day plants to red and far-red irradiation were determined in the same manner. In this case, red light promoted flowering, and far-red light alone acts essentially as darkness (Figure 11-18A-D). Here again, red and far-red reversibility was found.

These experiments and many others on light effects have led investigators to conclude that plants contain a receptor pigment capable of exist-

ing in two forms. One form of the pigment absorbs red light, the other absorbs far-red light. This photoreceptive protein-like pigment is called *phytochrome*. When the red absorbing form of phytochrome (P_R) absorbs red light, it is converted into the far-red absorbing form (P_{FR}). Absorption of far-red light transforms P_{FR} into P_R. These relationships may be symbolized as follows:

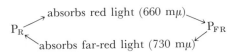

$$P_R \underset{\text{absorbs far-red light (730 m}\mu)}{\overset{\text{absorbs red light (660 m}\mu)}{\rightleftharpoons}} P_{FR}$$

In an earlier discussion it was noted that if a brief red flash is given to a short-day plant in the middle of a long night, the plant will not flower. If the red light is followed immediately by a brief flash of far-red light, the plant will flower, because the red-light effect is negated. However, if after a red-light treatment is given, the far-red flash is delayed for a period of time—about $\frac{1}{2}$ hour or more—the far-red treatment will not reverse the effect of red light. This observation suggests that when the pigment is in the P_{FR} form, it is coupled to biochemical reactions which proceed immediately. P_{FR}, therefore, appears to be the active form of the pigment in the photoperiodic responses. The mechanism of these coupled biochemical reactions is currently under investigation.

It has been shown that if phytochrome is exposed to both red and far-red irradiation simultaneously, the red light dominates and the pigment is converted from the P_R form into the P_{FR} form. Sunlight contains both red and far-red wavelengths, so that during the day phytochrome exists in the plant principally in the P_{FR} form. There is also some evidence to suggest that there is a spontaneous dark conversion of P_{FR} to the P_R form. Based upon half-life reconversion studies, it is postulated that after a period of light, dark conversion of P_{FR} will proceed during the darkness at a rate such that approximately 3 per cent of its initial activity as P_{FR} still remains after 10 hours of darkness. This time period closely approximates the range of critical night lengths for floral initiation in many plants.

Temperature and Plant Growth

The major influence of temperature on plants is on metabolic processes and, because growth is a result of such activities, temperature also influences growth. In general, within limits, growth is promoted as the temperature increases and is retarded with decreasing temperatures. Excessively high temperatures tend to dehydrate tissues through excessive transpiration as well as to denature enzymes, thus inhibiting vital metabolic activities.

In addition to influencing growth, temperature also affects the size and form of plants. These effects are most readily apparent when the temperature is near the minimum or maximum for growth. Constant low temperatures produce comparatively small plants, whereas high temperatures give rise to taller plants.

The most striking developmental consequences of temperature are produced by inducing growth and/or flowering by low-temperature treatments of seeds or plants. This temperature-treatment process is called **vernalization.** The temperature under which seeds are germinated may affect later growth and flowering of the plant. For example, seeds of plants such as winter varieties of wheat must be exposed to low temperatures if flowering is to occur when the plants mature. Such seeds are generally sown in the fall, at which time the low-temperature requirement is met. If wheat seeds are not vernalized in this manner, flowers do not develop. Seeds of plants grown in tropical zones must be vernalized at higher temperatures if flowering is to occur.

Whereas the leaf is the receptive organ for photoperiodism, the shoot tip is the structure that receives the vernalization stimulus. Once again, this may be demonstrated by grafting techniques. If plant A has been vernalized and is therefore capable of flowering, is grafted to plant B, which is a nonvernalized, nonflowering plant, plant A may induce flowering in plant B. In seeds, the receptor of the vernalization stimulus is the embryonic apex. It has been suggested that vernalization of seeds produces a hormone called *vernalin,* which, like florigen, has not yet been isolated. The vernalin, which then becomes incorporated into the mature plant, stimulates the production of florigen and flowering takes place.

PLANT MOVEMENTS

Reactions and movements of plants usually go unnoticed because the rate of movement is too slow for observation. For this reason it is often thought that plants lack movement. If, however, time-lapse photography is employed, many plant reactions and movements may be noted. With special photographic techniques, observation may be made of flowers opening, buds unfolding, organs growing, seeds germinating, stomates opening and closing, stems curving and turning, and organs reacting to various stimuli.

Plant movements may be categorized into two principal types: autonomic and paratonic. *Autonomic movements* are spontaneous movements of a plant brought about by internal activities of the plant which are relatively independent of the environment. *Paratonic movements* are induced movements brought about as a result of stimuli

in the environment. Examples of environmental stimuli are gravity, contact, light, heat, shock, and chemicals.

Both autonomic and paratonic movements are the result of two mechanisms: growth movements and turgor movements. *Growth movements* are irreversible movements in which the movement or reaction results from differences in growth rates of cells in different parts of an organ. These growth-rate differences, in turn, arise from unequal distribution of growth substances. *Turgor movements* are temporary and reversible movements which result from changes in the turgor pressures of certain cells. Such movements are usually more rapid than growth movements and may occur over and over again in the same organ.

Autonomic Movements

Aside from growth, which is the principal type of autonomic movement, several other types of autonomic movement may be distinguished. The most common is *nutation*. This is a back and forth motion of the tip of a growing stem or other organ caused by alternately changing growth rates on opposite sides of the organ (Figure 11-19A). Although nutation is most obvious in the stems and tendrils of climbing plants, it has also been observed in leaves, roots, flower stalks, and runners. Other autonomic growth movements include *circumnutation*, a rotational growth of an elongating shoot around its long axis (Figure 11-19B), and *twining*, the spiral orientation of a stem (Figure 11-19C).

Paratonic Movements

Paratonic movements are brought about as a result of external stimuli. Some paratonic movements are such that the direction of movement is independent of the direction from which the stimulus is received. These movements are called *nasties*. Other paratonic movements are such that the

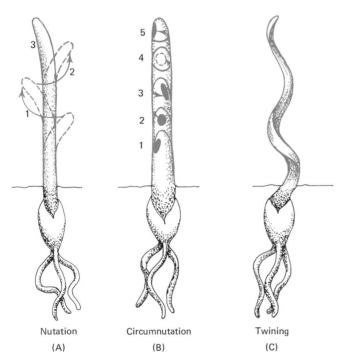

Nutation (A)   Circumnutation (B)   Twining (C)

Figure 11-19. Autonomic plant movements. These spontaneous movements are brought about by internal activities of the plants and they occur so slowly that they are not directly observable over a short period of time.

direction of movement is determined by the direction from which the stimulus is received. Such paratonic movements are called *tropisms*.

Nasties. Nasties are responses of organs such as leaves, flower petals, bud scales, and other flattened structures of plants. The principal nastic movements are those caused by changes in light intensity (*photonasties*), temperature changes (*thermonasties*), touch (*thigmonasties*), and alternation of day and night (*nyctinasties*).

The fluctuations of light and temperature as well as the alternation of day and night produce various nastic movements, especially in leaves and flowers. Many flowers, such as morning glory, tulip, and dandelion, are either completely or partially opened during the day but close at night. By contrast, flowers of other plants, such as four-o'clock and tobacco, close during the day under the influence of intense illumination and open in the light of low intensity. The rapid opening of certain flowers when brought into a warm room from a cold place is an example of a thermonastic response.

One of the most common examples of nyctinastic movements are the "sleep" movements of leaves in which the leaves of certain plants are horizontally oriented during the day and vertically oriented at night (Figure 11-20A). Another prominent and fairly well-known nastic movement is the thigmonastic response furnished by the sensitive plant, *Mimosa pudica*. This plant, indigenous to tropical and subtropical regions, is extremely responsive to shock and touch. When stimulated, the leaflets fold upward in pairs and the main petiole drops (Figure 11-20B). If a very strong stimulus is applied to one leaflet, it may be transmitted throughout the plant, the leaflets folding and dropping one after another.

Tropisms. Whereas nastic movements may be due to either growth or changes in turgor, tropisms are typically growth movements. Tropisms are plant responses in which the direction of movement is determined by the direction from which the stimulus originates. Depending upon the nature of the stimulus, a tropism may be called *phototropism* (light), *geotropism* (gravity), or *thigmotropism* (contact), and each tropism may be either positive or negative.

The response of a plant organ to light is called phototropism. Leaves and stems exhibit a positive tropism, that is, they bend toward light, whereas roots show no response to light or exhibit a negative response by bending away. Bending occurs because light reduces the auxin concentration on the lighted side of a stem. Because the shaded side has a higher auxin concentration it grows faster, and this unequal distribution of growth is responsible for the bending toward light (Figure 11-21A). Owing to phototrophic movements, the petioles of leaves bend in such a way that leaf blades orient themselves at right angles to the source of illumination. In such an arrangement, called a *leaf mosaic* (Figure 11-21B), few leaves shade others and few spaces between leaf blades are left unfilled.

Geotropism is the response of a plant organ to gravity. Roots exhibit a positive geotropism, whereas stems react negatively to gravity. This can readily be seen if germinating seeds are placed in vertical, horizontal, and inverted positions (Figure 11-21C). In each case, regardless of the orientation of the seeds, stems grow upward and roots grow downward. This growth movement also involves unequal auxin distribution. Auxin accumulates in greater concentration on the lower sides of both stems and roots grown horizontally due to the pull of gravity. The increased auxin concentration on the lower part of the stem causes a greater rate of growth on that side and the stem bends upward. According to this reasoning, roots in the horizontal position should also grow upward. However, this is obviously not the case. It has been found that a high auxin concentration

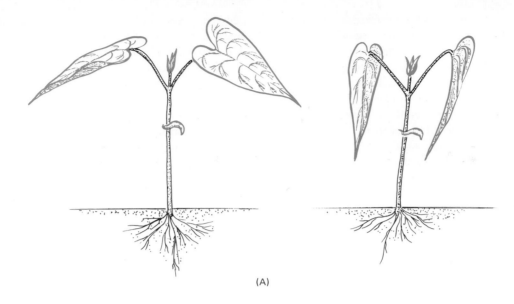

(B)

Figure 11-20. Paratonic movements. **(A)** Nyctinastic "sleep" movements of bean (*Phaseolus vulgaris*) leaves. *Left:* Horizontally oriented position of the leaves during the day. *Right:* Vertically oriented leaves at night. **(B)** Thigmonastic response of the sensitive plant, *Mimosa pudica. Left:* Normal position of the leaflets and the petioles. *Right:* Response of the leaflets and petioles to touch. (Photos courtesy of U.S.D.A.)

promotes growth on the lower side of stems but inhibits growth on the lower side of roots (see Figure 11-4). The lower concentration on the upper side of roots is more favorable for root growth and the root bends downward. Such a reaction of plant tissues to growth substances is not atypical, for in many instances high hormone concentrations may inhibit growth.

Thigmotropism is a growth movement in response to contact with a solid object. It is most readily apparent in tendrils or climbing plants such as members of the pea, grape, and passion-flower families. Once a tendril has made contact with a solid object, it bends or spirals because of increased growth on the side opposite to the point of contact (see Figure 11-21D).

CHAPTER IN RETROSPECT

Plant growth is a quantitative process resulting in a permanent increase in size caused by the assimilation of new protoplasm. Development, a qualitative process, involves progressive changes in form and complexity. Both processes are irreversible and occur at the cellular level. The formation of new cells by mitotic division and their subsequent enlargement and differentiation into permanent forms provides the chief means of growth of any cellular organism. Among the factors affecting plant growth are heredity, auxins, gibberellins, cytokinins, metabolic efficiency (internal influences), light, temperature, moisture, and biotic factors (external influences). Auxins influence such plant activities as cellular elongation in shoots, leaf abscission, fruit development, apical dominance, and root initiation. Stem elongation, tropisms, cambial tissue increase, and flowering are affected by gibberellins. Each factor of the internal and external plant environment influences growth in some specific way.

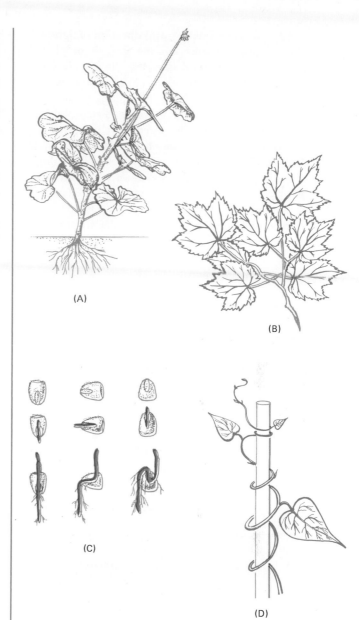

Figure 11-21. Plant tropisms. **(A)** Phototropism in a geranium plant. **(B)** Leaf mosaic formed in response to light. **(C)** Geotropism in germinating corn seeds. **(D)** Tendrils exhibiting thigmotropism.

The intensity, quality, and duration of light has a profound influence on the growth of plants. Plants grown in total darkness display etiolated characteristics. Those grown under different regions of the spectrum exhibit striking results. For example, plants grown in the absence of blue-violet light are characterized by thinner stems, poorly developed vascular tissue, retarded flower and fruit development, and a relatively high amount of hydration. Based upon a 24-hour cycle of light and darkness, angiosperms are classified as long-day plants, short-day plants, or day-neutral plants (photoperiodism). The stimulus for photoperiodic responses is received by leaves which transmit the stimulus to floral organs, where flowering is initiated.

Plant movements may be either autonomic (internally caused) or paratonic (externally caused); both are the result of growth and turgor movements. The principal autonomic movements are nutation, circumnation, and twining; paratonic movements include nasties and tropisms.

Questions and Problems

1. What are the various phases of plant growth? Identify those regions of plants which are characterized by rapid growth.
2. Prepare a chart which indicates the various internal and external factors which affect plant growth. In what specific way does each factor influence the growth of plants?
3. Distinguish between growth and development. Which is a qualitative process?
4. What is the relationship of phytochrome to red light and far-red light stimulus?
5. What is the relative importance of the dark period to floral initiation?
6. Give evidence to indicate the involvement of hormones in the flowering process.
7. Briefly describe the effect of low-temperature treatments on the seeds of temperate-climate plants.
8. Outline the various autonomic and paratonic plant movements and indicate cause and effect relationships.
9. What do the following terms mean?

differentiation	fruit drop
plant hormones	parthenocarpy (natural and
etiolation	artificial)
genotype	photoperiodism
abscission layer	long-day plant
phytochrome	short-day plant
vernalization	nasties
tropism	day-neutral plants

Suggested Supplementary Readings

Bonner, James, "On the Mechanics of Auxin-Induced Growth," in R. M. Klein (ed.), *Plant Growth Regulation.* Ames, Iowa: Iowa State University Press, 1961.

Borthwick, H. A., and S. B. Hendricks, "Photoperiodism in Plants," *Science,* vol. 132, 1960.

Butler, W. L., and R. J. Downs, "Light and Plant Development," *Scientific American,* December 1960.

Fogg, G. E., *The Growth of Plants.* Baltimore: Penguin Books, Inc., 1963.

Hillman, W. S., *The Physiology of Flowering.* New York: Holt, Rinehart and Winston, Inc., 1962.

Salisbury, F. B., "Plant Growth Substances," *Scientific American,* April 1957.

Siegelman, H. W., and W. L. Butler, "Properties of Phytochrome," *Annual Review of Plant Physiology,* vol. 16, pp. 383–392, 1965.

Sinnott, W., *Plant Morphogenesis.* New York: McGraw-Hill Book Company, Inc., 1960.

Steward, F. C., "The Control of Growth in Plant Cells," *Scientific American,* October 1963.

Sweeney, B. M., "Biological Clocks in Plants," *Annual Review of Plant Physiology,* vol. 14, pp. 411–440, 1963.

Thimann, K. V., "Growth and Growth Hormones in Plants," *American Journal of Botany,* vol. 44, pp. 49–55, 1957.

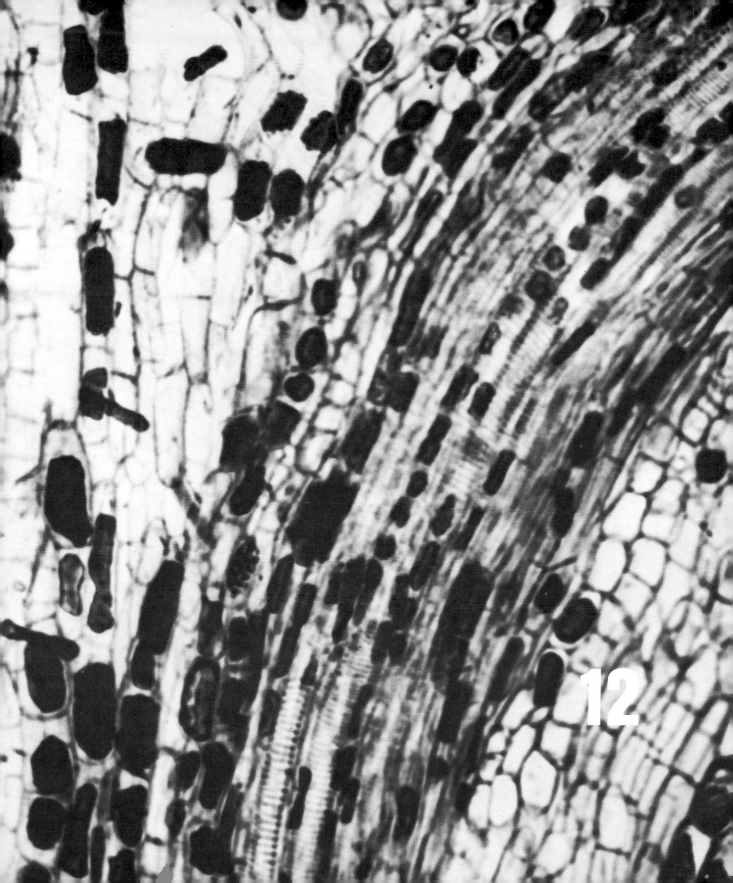

12
Plant Heredity

INTRODUCTION

Although the methods and patterns of reproduction, whether sexual or asexual, vary among different types of plants, the salient feature is that a new generation of living organisms is produced (see Chapter 5). Generally, newly produced generations undergo periods of development in which growth and differentiation are the dominant vegetative activities. After this period of development these organisms also become able to reproduce and to give rise to future generations.

Among the many unique qualities of protoplasm is its ability to produce more protoplasm of the same kind (self-propagation). The external manifestation of this protoplasmic property is the continual tendency of offspring to resemble their parents in all defining features of form and function. The phenomenon by which offspring resemble parents is called *heredity*. Through the mechanism of heredity, algae beget algae, mosses give rise to mosses, and oak trees produce oak trees. In other words, heredity implies a basic plan of inheritance, distinctive for each type of plant, in which future generations will resemble both parents and past generations.

All plants inherit certain characteristics from their parents, so it is not very difficult to distinguish between fern plants and pine trees, for example. Yet, subtle, but recognizable differences exist even between parents and their offspring. These differences are collectively termed *variation*.

The specialized branch of biology concerned with heredity and variation is called *genetics*. Today most people casually accept the fact that offspring inherit certain characteristics from their parents. However, it was not until 1866 that any data were systematically recorded and published concerning the basic concepts of inheritance. This important contribution to biology, formulated by Gregor Mendel, went unrecognized until 1900. In his document *Experiments in Plant Hybridization*, the fundamentals of genetics were established.

THE WORK OF MENDEL AND ITS MODERN INTERPRETATION

Gregor Johann Mendel (1822–1884), an Austrian monk, devised a series of experiments which were designed to demonstrate the fundamental principles of genetics. In his experiments, the garden pea plant (*Pisum sativum*) was used. This proved to be an excellent plant for experimentation for a number of reasons. First, pea plants exhibit a number of distinctive contrasting traits. This fact made it easy for Mendel to study the inheritance of one trait at a time and to maintain careful records of his findings. Moreover, pea plants although normally self-pollinating, can be cross-pollinated with relative ease. Another reason for the selection of pea plants was that they produce relatively large numbers of seeds.

Monohybrid Cross

One of the first steps in Mendel's experimental procedure was to study the inheritance of a single trait. As an example, height was considered. He crossed (pollinated) different pea plants from a variety averaging about 6 feet tall among themselves for many generations, and only tall plants were produced. On this basis he concluded that these plants contained some factor for tallness. In a similar manner, he crossed pea plants averaging 1 foot in height and once again concluded that these plants contained a factor for shortness. Each of these two groups of plants was considered to be pure for their respective traits. That is, tall plants contained no factors for shortness and short plants contained no factors for tallness.

Having developed a pure line of tall and short plants, Mendel next crossed these with each other. This cross, consisting of parents differing in a single trait, is called a **monohybrid cross.** To keep accurate records of the experimental crosses, Mendel designated the generations of peas by specific names. For example, the cross made in the starting generation was called the *P* **(parent) generation.** The first generation of offspring to be produced from the *P* generation was called the F_1, or **first filial** (daughter), **generation;** the second generation was referred to as the F_2, or **second filial, generation;** and so on.

In the *P* generation, then, Mendel crossed pure tall with pure short pea plants. He found that the F_1 generation contained all tall plants; none were either short or intermediate in height. Next he pollinated the tall pea plants of the F_1 generation and observed that approximately three fourths of the F_2 generation plants were tall and one fourth were short (Figure 12-1).

Based upon these observations, Mendel deduced certain conclusions. It was obvious that some factors were present which determined the inheritance of height and that these factors occurred in pairs because mating within the F_1 generation produced both tall and short plants, and, therefore, the F_1 plants must have contained both tall and short factors. Mendel concluded further that one factor (referred to as dominant) may mask the expression of the other because when tall and short pea plants of the *P* generation were crossed, all the F_1 plants were tall, but both tall and short plants appeared in the F_2 generation. Thus the F_1 plants must have carried the factor for shortness, but it was obscured by (was recessive to) the factor for tallness. A final conclusion reached by Mendel was that when gametes are formed, the

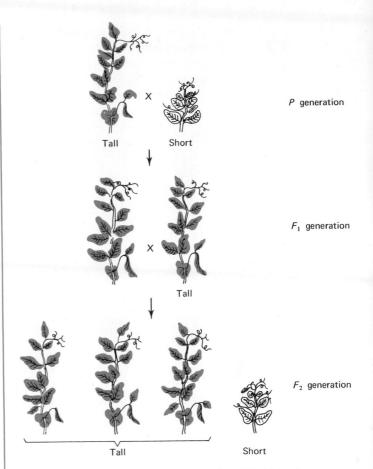

Figure 12-1. Simple monohybrid cross. Mendel obtained the results shown by pollinating tall and short pea plants.

factors separate from each other. Each of these conclusions later became known as Mendel's laws, but, before stating these laws, the results of Mendel's experiments will be analyzed in terms of modern principles of genetics.

In modern genetics, the factors to which Mendel made reference are called **genes** and the genes are usually symbolized by letters. Capital letters are generally used for dominant traits and the same lowercase letters are usually used to designate the recessive traits. Capital letter T is the symbol that represents the gene for tallness and lowercase t represents the gene for shortness. To interpret Mendel's results, the tall pea plants for the P generation would be represented by TT, because genes come in pairs on homologous chromosomes and because the tall P generation pea plants are pure or **homozygous** for tallness; there are no genes for shortness. Similar reasoning indicates that the short pea plants of the P generation should be represented by tt (Figure 12-2A). These data provide the basis for Mendel's first law of genetics, called the *law of unit characters,* which states that various hereditary characteristics are controlled by genes and that in diploid organisms these genes occur in pairs.

When homozygous tall pea plants are crossed with homozygous short pea plants in the P generation, the F_1 generation consists entirely of tall pea plants. Remember that the genetic makeup, or **genotype,** for the tall pea plants of the P generation is TT and that this represents the diploid state. Similarly, the diploid genotype for the short pea plants of the P generation would be tt. For sexual reproduction to occur both TT and tt must undergo meiosis, so that haploid gametes will be produced (Figure 12-2B). This separation of genes during meiosis is the basis for Mendel's second law of genetics, the *law of segregation,* which states that a pair of genes located on homologous chromosomes are separated from each other during the formation of the haploid products (microspores and megaspores in flowering plants—Chapter 5) of meiosis. The term **allele** is used to describe individual members of a gene pair. In other words, a particular gamete produced subsequent to meiosis contains only one gene of a pair, the other having been passed to another gamete. Applying this to the cross under discussion, the diploid cell TT would eventually produce, after meiosis, hap-

loid gametes containing only T and the diploid cell tt would give rise to haploid gametes containing only t.

Having analyzed the genotypes for the P generation plants as well as the constitution of the gametes they produce just prior to fertilization, the next obvious step is to determine how the gametes combine with each other to produce fertilized eggs (zygotes) which will eventually develop into F_1 generation plants. Special charts resembling checkerboards are used to predict possible gamete combinations. Assume that the homozygous tall pea plants of the P generation are male plants (pollen-producing) and that the short pea plants of the P generation are female (egg-producing). Typically, the male gametes are placed on the side of the chart and the female gametes are placed on the top. The four spaces in the chart represent the various possible combinations of male and female gametes; these possible combinations are the genotypes of fertilized eggs in the next generation—in this case, the F_1 generation. The possible gamete combinations are determined by "dropping" each of the female gametes on the left into the two lower spaces and by "dropping" the female gamete on the right into the two lower spaces; the upper male gamete is then moved across to the two spaces in line with it and the lower male gamete is moved across to the two spaces in line with it (Figure 12-2C).

It should be noted that all F_1 plants are tall, similar in appearance to the tall plants of the P generation. This observed characteristic, that is, the external appearance, is called the **phenotype.** In this regard, then, the phenotypic characteristic of tallness is common to the pea plants of the P generation used as pollen plants and to all the plants in the F_1 generation. Note also that the genotype for the P generation tall plants is homozygous (TT), whereas that of the F_1 plants is **heterozygous** (hybrid). In other words, a hetero-

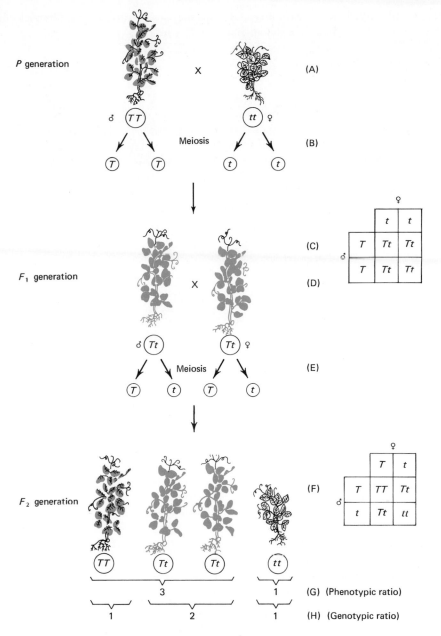

P generation   X   (A)

♂ TT   tt ♀

Meiosis   (B)

T   T   t   t

	♀	
	t	t
T	Tt	Tt
T	Tt	Tt

(C)

F₁ generation   X   (D)

♂ Tt   Tt ♀

Meiosis   (E)

T   t   T   t

	♀	
	T	t
T	TT	Tt
t	Tt	tt

(F)

F₂ generation

TT   Tt   Tt   tt

3   1   (G) (Phenotypic ratio)

1   2   1   (H) (Genotypic ratio)

Figure 12-2. Monohybrid cross showing products of meiosis and procedure for employing Punnett squares. See text for amplification.

Traits Selected for Parent-Generation Cross	Phenotype of F_1 Plants	Self-Pollination	F_2-Generation Plants (numbers and actual ratio)
1. Round × wrinkled seeds	All round	Round × round	5474 round, 1850 wrinkled; 2.96:1
2. Yellow × green seeds	All yellow	Yellow × yellow	6022 yellow, 2001 green; 3.01:1
3. Colored × white seed coats	All colored	Colored × colored	705 colored, 224 white; 3.15:1
4. Inflated × wrinkled pods	All inflated	Inflated × inflated	882 inflated, 299 wrinkled; 2.95:1
5. Green × yellow pods	All green	Green × green	428 green, 152 yellow; 2.82:1
6. Axial × terminal flowers	All axial	Axial × axial	651 axial, 207 terminal; 3.14:1
7. Long × short stems	All long	Long × long	787 long, 277 short; 2.84:1

zygous organism contains contrasting genes, or alleles (Tt), for the same trait.

Based upon the results of the F_1 generation, it is clear that even though the gene for shortness is inherited, none of the plants express the short trait. This observation is the basis for Mendel's third law, the *law of dominance,* which states that one gene of a pair may mask or inhibit the expression of the other. In this particular instance, the gene for tallness is **dominant** and that for shortness is **recessive.** This explains the appearance of all tall pea plants in the F_1 generation (Figure 12-2D).

When these tall plants of the F_1 generation are crossed with each other, the F_2 generation consists of a ratio of three tall to one short plant. This 3:1 ratio is a phenotypic ratio (Figure 12-2G). It does not mean that two plants were crossed in the F_1 generation and three tall and one short plant appeared in the F_2. It implies that many F_1 plants were crossed and they produced, in F_2, a ratio of 3:1. Actually, after many trials, Mendel recorded 787 tall pea plants and 277 short ones in the F_2 generation. These numbers closely approximate a

3:1 ratio. It should also be noted in the F_2 plants that a genotypic ratio also exists (Figure 12-2H). This ratio consists of a 1:2:1 distribution of homozygous tall to heterozygous tall to homozygous short plants. Both the phenotypic and genotypic ratios may be deduced by tracing meiosis in the F_1 plants (Figure 12-2E) and determining the possible gamete combinations by using the checkerboard (Figure 12-2F).

As indicated previously, Mendel studied the inheritance of a number of genetic traits. Thus far, the inheritance of only height has been traced. Table 12-1 summarizes the results Mendel obtained by following the inheritance of seven pairs of contrasting traits through two generations. The basic principles involved in the inheritance of height also apply for the other six traits, when each cross is considered as a monohybrid cross.

Probability and Genetics

Calculating or predicting genetic ratios is not unlike calculating the expected results of tossing two coins. Nothing is really changed except that instead of using gametic characters, coins are used. If two coins are flipped simultaneously, one of three possible combinations may result. These are head-head (HH), or the homozygous dominant trait; head-tail (HT), or the heterozygous trait; or tail-tail (TT), the homozygous recessive trait. The HT combination will occur with twice the frequency of either HH or TT combinations. Assume that the two coins are flipped simultaneously only 4 times. Although it is possible to get a 1:2:1 ratio, it is not probable based on only 4 trials. If the coins are flipped 40 times, however, the probability of getting a 1:2:1 ratio increases. Similarly, 400 or 4000 trials would further increase the probability. The point to be made here is that either actual or expected genetic ratios are based on large numbers of trials.

Test Cross

Phenotypically, the tall plants of the P generation are similar to those of the F_1 generation. Genotypically, however, the P generation plants are homozygous and the F_1 tall plants are heterozygous. These are known facts because the heredity of height in pea plants has just been traced. Suppose that a geneticist or a florist wanted to determine the genotype of the tall plants. To do this a **test cross** is employed. Such a cross consists of mating the plant with the unknown genotype with a plant exhibiting the recessive trait. In this case tall plants (TT or Tt) are crossed with short plants (tt). Note that the genotype of the recessive plant is always known.

If the test-cross plants produce a ratio of about half tall plants and half short plants, the genotype of the tall plant is Tt, since a tall plant with the genotype TT would produce all tall plants (Figure 12-3A). Even if only one or a few test-cross offspring plants are short, it can be concluded that the tall parent is heterozygous. If, on the other hand, all the test-cross offspring are tall, after a

dozen or so offspring are produced, it can be concluded that the parental genotype is TT (Figure 12-3B).

Dihybrid Cross

In addition to studying the inheritance of single traits, Mendel also performed a series of experiments involving the inheritance of two or more characteristics. A cross between individuals differing in two distinct characteristics is called a **dihybrid cross.** One of Mendel's dihybrid crosses consisted of mating plants having round yellow seeds with plants having wrinkled green seeds (Figure 12-4). These parents produced F_1 offspring all of which contained round yellow seeds. As in the monohybrid cross, the F_1 plants are heterozygous for the two dominant traits. When these plants were crossed among themselves, the F_2 generation consisted of four different phenotypes; 315 had round yellow seeds, 101 had wrinkled green seeds, 108 had round green seeds, and 32 had wrinkled green seeds. In simple terms, the four different phenotypes occurred in an approximate 9:3:3:1 ratio.

Mendel's dihybrid experiments provided important data not available from monohybrid crosses. For one thing, the results indicated that a dihybrid cross could produce two new types of plants phenotypically unlike either the P or F_1 generation plants—ones which developed either wrinkled yellow or round green seeds. The implication to be drawn from the appearance of new phenotypes is that the genes for seed color and the genes for seed texture do not necessarily stay together in the combinations in which they were present in the P generation. By further implication it can be reasoned that the genes for seed color are located on one pair of homologous chromosomes and that the genes for seed texture are on a different pair. Thus genes for these two traits segregate independently during meiosis (Chapter 5). These ob-

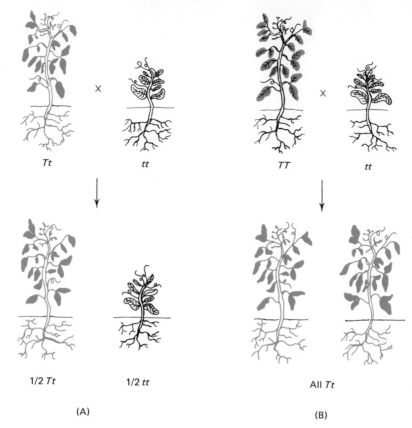

Tt $\times$ tt TT $\times$ tt

1/2 Tt 1/2 tt All Tt

(A)   (B)

Figure 12-3. Test cross. The purpose of a test cross is to determine the genotype controlling a dominant trait by crossing the unknown organism with its recessive. In pea plants, where tall dominates short, (**A**) if the unknown genotype is heterozygous, about 50 per cent of the plants will show the recessive trait. (**B**) If, however, the unknown genotype is homozygous, no recessive traits will appear.

servations provide the basis for Mendel's fourth law of heredity, the *law of independent assortment,* which states that the separation (segregation) of allelic genes of gene pairs on one pair of homologous chromosomes during meiosis is entirely independent of the segregation of allelic genes on other pairs of homologous chromosomes.

Figure 12-4 represents the mechanism of a dihybrid cross. R represents the allele for round seeds, r for wrinkled seeds, Y for yellow seeds, and y for green seeds. The round yellow parent can produce only gametes RY. The wrinkled green parents can produce only gametes ry. When RY gametes of one parent combine with ry gametes of the other, all the F_1 plants are heterozygous for both dominant traits ($RrYy$) and show the round yellow phenotype of the dominant parent. Each of these F_1 plants can produce four different types of gametes: $RY, Ry, rY,$ and ry. When these four possible gametes for one F_1 plant are combined with the same four possible gametes from another F_1 plant, 16 zygotic combinations are

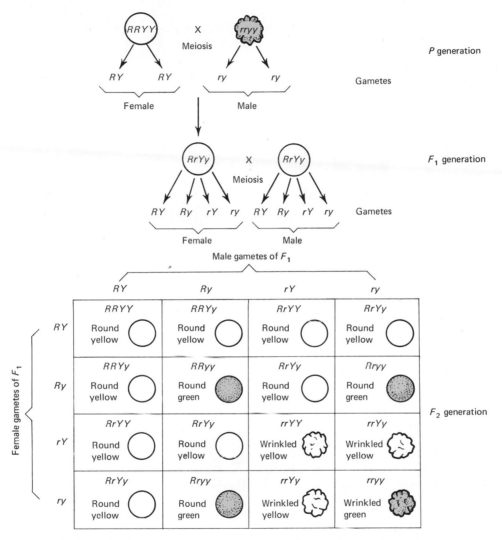

Figure 12-4. Dihybrid cross. Pea plants with round, yellow seeds are pollinated with ones containing wrinkled, green seeds. All F_1 plants contain yellow, round seeds. Upon pollination of the F_1 plants, a $9:3:3:1$ phenotypic ratio results in the F_2 generation.

possible. These possible combinations include nine different genotypes, which determine four different phenotypes in the ratio 9:3:3:1.

GENETICS SINCE MENDEL

Mendel's observations on inheritance in the garden pea were subsequently confirmed by numerous investigators and his general conclusions were found to apply in other plants and in animals. However, subsequent studies also showed

that certain modifications of Mendel's laws were necessary.

Incomplete Dominance

Genetic studies made after Mendel's work have clearly demonstrated that while dominance and recessiveness occur in many cases, they are rarely if ever absolute. Two common instances among plants in which the F_1 offspring exhibit an intermediate characteristic as compared with parental types are found in flower color in four-o'clock and

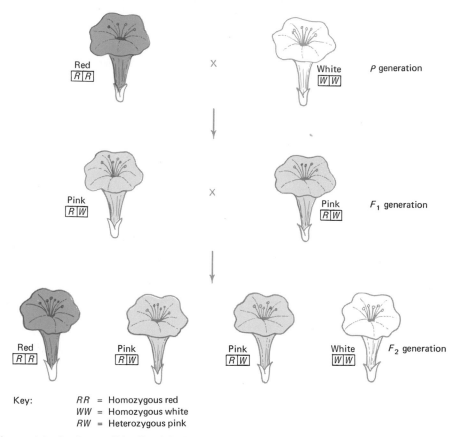

Key:
RR = Homozygous red
WW = Homozygous white
RW = Heterozygous pink

Figure 12-5. Incomplete dominance (blending inheritance) in four-o'clock flowers.

snapdragon. This type of inheritance in which offspring are phenotypically intermediate between parental phenotypes is called **incomplete dominance.** When homozygous red-flowered (RR) four-o'clocks are crossed with homozygous white-flowered (WW) four-o'clocks, all the F_1 flowers are pink (RW) (Figure 12-5). The pink phenotype is intermediate between red and white and indicates a lack of dominance for both red and white alleles. Subsequent crossing of the pink-flowered four-o'clocks produces the typical monohybrid genotypic ratio of 1 homozygous red to 2 heterozygous pink to 1 homozygous white.

Multiple Genes

It is now recognized by geneticists that many inherited traits are controlled by more than one gene. Indeed, it seems reasonable to assume that many genes of a plant exert some influence on the development of every hereditary trait, with one, two, or a few genes having a predominant effect. In this context, these minor genes probably assist the major genes to express a trait through various catalytic or inhibitory biochemical reactions. Generally, when several genes operate together to determine a trait, the trait will not appear if any one of the genes is missing. If several gene pairs are involved in the expression of a single characteristic, and if no dominance exists between alleles, their effect will be cumulative. Such heredity is due to the cumulative action of *multiple genes.*

In multiple-gene heredity the typical expected phenotypic ratios do not appear. Consider a cross between two varieties of wheat, each of which has two pairs of chromosomes that influences the coloration of the kernels (Figure 12-6). A plant homozygous for dark red kernels with the genotype $R_1R_1R_2R_2$ is crossed with a plant homozygous for white kernels with the genotype $r_1r_1r_2r_2$. All the F_1 offspring contain kernels of a medium red color ($R_1r_1R_2r_2$). Up to this point it is impossible

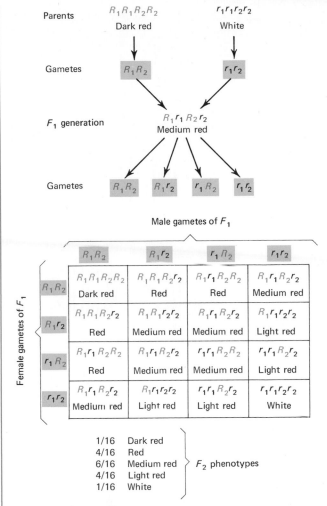

Figure 12-6. Multiple gene heredity in kernels of wheat. See text for amplification.

to determine whether kernel color is the result of incomplete dominance or multiple-gene heredity. But when the F_2 generation is analyzed, it is found that most of the kernels exhibit various shades of red and that the phenotypic ratio is 1 dark red : 4 red : 6 medium red : 4 light red : 1 white. This is far different from the 9 : 3 : 3 : 1 phenotypic ratio of the

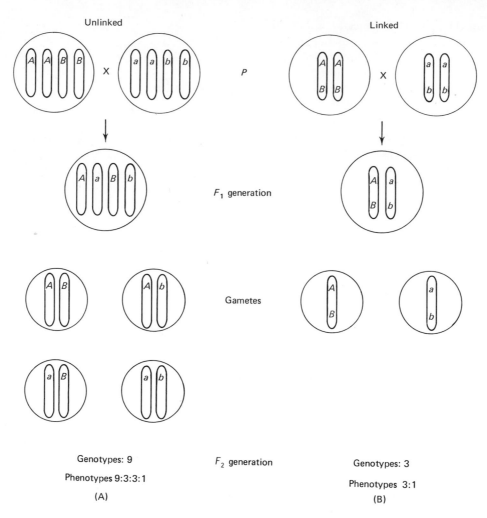

Figure 12-7. Comparison between **(A)** independent assortment and **(B)** gene linkage.

dihybrid cross analyzed previously. It should be noted that although the F_2 phenotypic ratios do differ, because of a different kind of gene interaction, Mendel's basic laws still apply. The assortment of genes during meiosis, gene recombinations during fertilization, and F_2 genotypic ratios are just the same as in any dihybrid cross involving independent assortment.

Many other hereditary variations in plants, such as height or the size and weight of fruits and other organs, vary quantitatively in that there are graded series of phenotypes between extremes. Such characters thus typically exhibit *continuous* rather than *discontinuous* hereditary variation. It is now known that numerous genes control the inheritance of characters exhibiting continuous variation even though each of these genes is presumably exerting a single effect at some level in the metabolism of the plant.

Gene Linkage

Dihybrid Mendelian ratios (9:3:3:1) depend upon independent assortment of genes during meiosis, and this assortment operates when genes are located on different chromosomes. With the increased knowledge of chromosomes and their behavior, it has been shown that individual chromosomes of plants contain hundreds and even thousands of genes. If two or more genes are located on the same chromosome they are said to be *linked*, and they do not separate independently but are inherited together.

As an example, consider the pollination of two plants, one having the genotype *AABB*, the other having the genotype *aabb* (Figure 12-7). If independent assortment does occur, the F_1 offspring will have the genotype *AaBb* and the gametes formed will be *AB, Ab, aB,* and *ab* (Figure 12-7A). In addition, nine genotypes expressing the 9:3:3:1 F_2 phenotypic ratio will result. By contrast, if both dominant genes are located on the same chromosome and linkage is complete, only two types of gametes will be formed: *AB* and *ab* (Figure 12-7B). As a result, instead of a 9:3:3:1 phenotypic ratio in the F_2 generation, a 3:1 ratio will result, because the two genes are inherited as though they were a single gene similar to monohybrid inheritance.

Gene linkage tends to modify phenotypic ratios as originally stated by Mendel because with linkage the number of chromosomes involved is reduced, and therefore the variety of gametes is also reduced. In summary, then, gene linkage represents an exception to independent assortment based upon the relationship of linked genes to their chromosomes and the segregation of chromosomes during meiosis.

Crossing Over

Although complete linkage of two or more genes appears to occur in certain hereditary patterns, it is comparatively uncommon for every gene on a chromosome to be passed down to the next generation as a linkage group. It has been found that some genes originally linked on the same chromosome do not always remain linked. Careful examination of meiosis shows that genes on one chromosome may be exchanged for corresponding genes of the homologous chromosome by a process referred to as *crossing over*. The frequency with which crossing over occurs is dependent upon the relative distances between pairs of allelic genes on homologs.

To understand the mechanism of crossing over during meiosis, consider a cell consisting of only one pair of homologous chromosomes (Figure 12-8). During the initial stages of meiosis, the chromosomes pair (Figure 12-8A) and each chromosome consists of two chromatids (Figure 12-8B). During this pairing or synapsis, the chromatids of homologous chromosomes become interwined and twist about each other (Figure 12-8C). As a result of breakage and rejoining of chromatids, there is an exchange of parts between chromatids (crossing over) and the broken ends then become joined in new sequences (Figure 12-8D). The net effect of this crossing over is that there is an exchange of hereditary material between chromatids of homologous chromosomes and the resulting haploid gametes contain four different types of chromsomes (Figure 12-8E). If crossing over did not occur the gametes would be only of two types: *AB* and *ab;* however, with crossing over four types of gametes are produced: *AB, Ab, aB,* and *ab*.

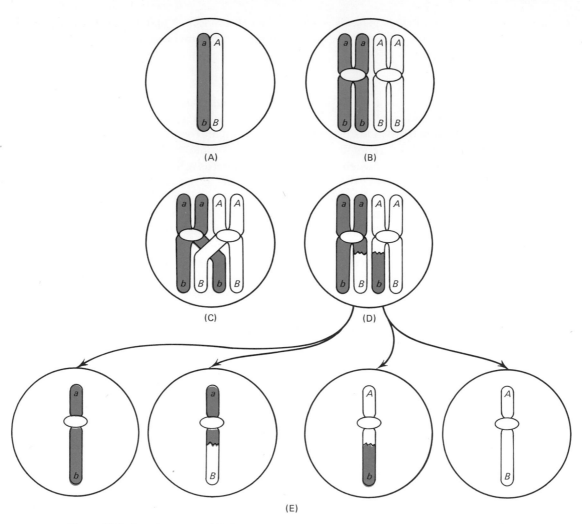

Figure 12-8. Crossing over. **(A)** Pairing of homologous chromosomes during meiosis. **(B)** Chromatid formation. **(C)** Chromatids twisting about each other (crossing over) during synapsis. **(D)** Exchange of parts between chromatids. **(E)** Formation of haploid cells.

HEREDITARY VARIATION: Causes and Effects

Thus far in this chapter, some of the mechanisms by which genes are transmitted from one generation to another have been considered. It was stated earlier that through the agency of heredity, plants resemble each other due to the inheritance of certain species characteristics. However, offspring are never exactly like the parents; they differ in some quality or degree. Among the principal factors that may cause variation are the environment, mutations, and recombination.

Variations Through the Environment

The form and function of a particular plant is dependent upon the interaction of two simultaneously operating factors: heredity and environment. Whereas the potential for growth and development is determined by heredity the actual degree to which the potential is realized is governed by the environment. In other words, genes are inherited and are not themselves expressed characteristics. The expression of a gene or group of genes is a function of the hereditary environment. If a certain plant is homozygous for red flowers, under favorable environmental conditions red flowers will be produced, but if the same plant is grown where light is deficient and pigment formation is blocked, the resulting flowers may be white. The expression of a particular genotype is influenced by such factors as light, temperature, moisture, and nutrient availability.

It has already been shown that plants grown under varying conditions of light exhibit markedly different characteristics. One distinct variation in most plants is the conspicuous absence of chlorophyll in plants grown in the dark. Although the genes for chlorophyll formation are inherited by plants maintained in the dark, such plants can express the chlorophyllous trait only in the presence of light. Similarly, plants with the same genotype may differ phenotypically if grown under varying conditions of temperature. In general, cold temperatures produce dwarf plants, whereas optimum temperatures give rise to normal-sized plants. Temperature, as well as light, may also influence flower color. Chinese primroses, for example, if grown at temperatures of 86°F will produce white flowers, but if genetically identical plants are grown at 68°F, floral pigmentation is red. As a final example of variation produced by the environment, consider the effect of soil nutrients on plant growth and development. Plants grown in soils deficient in minerals, especially in nitrates, exhibit many characteristics of etiolation, whereas soil fertility promotes normal patterns of growth and development.

Variations Through Mutations

The environmental variations just described, as well as all others, are not inherited by offspring. Mutations, or changes in the hereditary material, on the other hand, are heritable. **Gene mutations** result from a change in the chemical structure of a gene and **chromosome mutations** may result from either changes in the structure or number of chromosomes.

Gene mutations may occur at any time in any cell of the plant. Relatively few mutations are expressed phenotypically, and only those mutations that appear in gametes are transmitted from one generation to another. Many gene mutations frequently are not seen because only one gene of an allelic pair undergoes change and quite often the mutation is recessive. Even if the recessive mutation is passed down through many generations it may be some time before it appears, because it must be present in both alleles derived from male and female gametes. In addition, some gene mutations are *lethal* and may markedly alter an essential metabolic process so that in the homozygous condition neither the plant nor the mutation survive. In some instances gene mutations occur in only one part of a plant, usually vegetative cells. Such types, when dominant, are often called **bud mutations.** One such mutation appeared on the branch of a peach tree which produced atypical smooth-skinned "peaches" instead of the usual fuzzy peaches. Upon subsequent planting, the smooth-skinned "peach" seeds produced trees bearing only all smooth-skinned fruits, which are

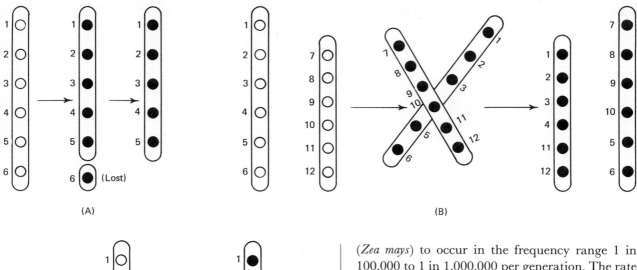

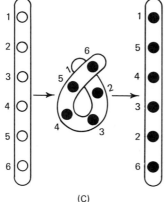

Figure 12-9. Three types of chromosome mutations. **(A)** Deletion, **(B)** Translocation, **(C)** Inversion.

now called nectarines. Other bud mutations considered desirable by man are those that produce new or more desirable flowers (for example, sweet pea), more desirable trees (for example, Lombardy poplar), and new varieties of fruits (for example, navel oranges and golden delicious apples).

Gene mutations occur under natural conditions at a very low frequency. Such natural, or *spontaneous, mutations* have been shown in the case of corn

(*Zea mays*) to occur in the frequency range 1 in 100,000 to 1 in 1,000,000 per generation. The rate of mutation may be artifically increased by certain agents called *mutagens*. Chief among mutagenic agents are X rays and ultraviolet radiation; certain chemicals such as nitrous acid and mustard gas; and temperature extremes.

In addition to gene mutations, mutations involving alterations of chromosome structure also occur. For example, during meiosis or under the influence of mutagens, a portion of a chromosome may break off. This phenomenon, called **deletion** (Figure 12-9A), results in the loss of a group of genes. In another type of chromosome mutation a portion of one chromosome may attach to another non-homologous chromosome by a **translocation** process (Figure 12-9B). Also, by a phenomenon called **inversion** (Figure 12-9C), a chromosome may twist about itself and where the ends cross, break, and exchange positions.

Another type of genetic change resulting in an altered chromosome number, occurs by a process called *polyploidization*. Typically, flowering plants have a paired set of chromosomes and are said to be diploid (2*n*). However, some plants contain more than two haploid sets of chromosomes, and these are referred to as *polyploids*. There are several kinds of polyploids: *triploids* (3*n*), *tetraploids* (4*n*), *pentaploids* (5*n*), and so on. Of these polyploid conditions, tetraploids are the most common in nature. Polyploidy may occur naturally because of a failure of a cell to undergo meiosis followed by gametic fusion or because of a division of the nucleus without subsequent cytokinesis. Tetraploids, especially, can be made to occur with greater frequency if plants are treated with a poisonous drug called *colchicine*. Cells treated with this drug, enter mitosis and the chromatids separate, but because there is no spindle mechanism they do not go to opposite poles. Rather, a single nucleus reorganizes and some multiple of the 2*n* chromosome number results. Tetraploids may be more vigorous plants than their diploid counterparts, producing taller stems, larger leaves, and larger flowers. However, excluding these desirable ornamental characteristics, tetraploids are not superior to the diploids because of increased sterility. Thus, although seeds and fruits are larger in tetraploid plants, fewer are produced than in diploid plants. Despite this reproductive disadvantage, such polyploids as primroses, tulips, petunias, watermelons, apples, and pears have been propagated artificially.

Variations Through Plant Breeding

The application of genetics to plant breeding has produced numerous variations of great economic importance to man. One technique of plant breeding, called *selection,* is perhaps the oldest procedure for the improvement of domesticated plants. Because of certain desirable variations which plants pass on to offspring, man has been able to select those plants which best serve his needs. For example, through selection, man has singled out such desirable plant traits as resistance to disease, high crop yield, and large fruits and

has propagated these traits either by seeds or by vegetative means. Seeds from selected plants are planted and selections again are made. This selection process is generally continued for many generations until a superior type of plant is isolated. Plants such as cotton, flax, and cereals are products of selection.

Despite the success of selection in plant breeding, some difficulties are encountered. Occasionally selected varieties gradually revert back to less desirable ancestral forms, so selection must be carried on continuously. In addition, selection toward a desired type is a very slow process. In view of these shortcomings, selection is usually supplemented by another plant breeding technique called hybridization.

Hybridization is a technique in which an overt attempt is made to bring the most desirable genes of two or more plants together. Crosses are made between two pure lines of selected parents that differ in one or more genes and the offspring are propagated for several generations. In the process, new gene combinations are developed naturally by chromosome assortment and crossing over or they may be artificially induced by the production of polyploidy through colchicine application, the increased rate of mutation by radiation treatments, or by other techniques. Eventually, the breeder selects the most desirable phenotypes and genotypes for propagation on a commercial scale.

To mention even a few of the improved varieties produced by plant breeders is far beyond the scope of this book. However, every farm crop, garden vegetable, orchard fruit, and ornamental species of importance to man has been developed through various scientific plant breeding techniques. Neither hybridization nor selection creates new genes, but they do combine the most desirable genes of parent plants. Generally, the most desirable traits plant breeders attempt to combine are vigor, disease resistance, and high crop yields.

THE CHEMICAL BASIS OF HEREDITY

The work of Mendel, important as it was, did not reveal much about the nature of genes or the mechanism by which genes exert control over the growth, development, characteristics, and functional aspects of plants. His work did, however, demonstrate the presence of a structural entity related to the reappearance of certain characteristics in successive generations. Much of the first half of the twentieth century was concerned with genetic research involving such phenomena as mutations, gene recombinations and interactions, and the construction of chromosome maps. All these investigations supplied conclusive proof that genes are physical units located on chromosomes and that chromosomes are composed principally of DNA and protein. The chemical nature of genes, however, had still not been elucidated.

Investigators postulated that whatever substance was in fact the genetic material must satisfy at least two basic requirements. First, it had to have some mechanism for duplicating itself exactly and, second, it must exert control over cellular activities. In the last two decades, intensive research into the chemical nature of the genetic material has demonstrated that DNA is the primary genetic material.

DNA as the Genetic Material

The experiments which ultimately led to the conclusion that DNA is the hereditary material were performed by Fredrick Griffith in 1928. He was working with two strains of bacteria. One of these strains, *Diplococcus pneumoniae,* causes pneumonia; another strain, an avirulent one, does not cause disease. The virulent strains are characterized by an enveloping polysaccharide capsule which is missing in the avirulent strains.

Griffith was interested in determining whether injections of heat-killed (60°C) virulent pneumocci could be used to vaccinate mice against pneumonia. He found that the injected mice were not infected by the heat-killed virulent strain. However, when these dead cells were injected simultaneously with live cells of the avirulent strain, the mice frequently died. When the blood of the dead mice was analyzed, it was found that it contained living virulent bacteria with capsules. It was concluded that something passed from the dead bacteria (virulent) to the live ones (avirulent) which caused the offspring of the harmless bacteria to become transformed into capsulated forms capable of causing pneumonia.

Subsequent investigations based upon Griffith's research revealed that bacterial transformation could also be duplicated using standard microbiological culturing techniques in broth and on agar petri dishes. Dead bacteria with capsules were added to a broth (liquid culture medium), which was inoculated with live bacteria without capsules and incubated for some hours. A drop of the culture was then smeared over agar (solid culture medium) in petri dishes. After a period of incubation it was found that some of the bacteria had capsules and had been transformed. The bacteria had acquired a new hereditary trait.

The next logical step was to extract various chemical components from the dead cells and to determine whether any of these chemicals would cause transformation. These crucial experiments were performed by O. T. Avery and his associates in 1944. After some years of work, they announced that the chemical substance responsible for transforming harmless pneumococci into virulent strains was DNA. Subsequent experiments showed that other genetic traits could also be passed from one bacterial cell to another and that the substance involved in the transformations was DNA. It was concluded that genes are composed of DNA.

There is also direct evidence that DNA is the primary genetic material of bacteriophages (viruses that attack bacterial cells). The bacterial viruses that have been studied extensively are those that infect the bacterium *Escherichia coli* and are referred to as T_2 *bacteriophages*. Electron-microscope studies reveal that T_2 viruses consist of hexagonal-shaped bodies and a protruding tail (see Figure 14-7A). When suspensions of *E. coli* and T_2 viruses are mixed, the viruses attach themselves by their tails to the surfaces of the bacteria, where they remain (see Figure 14-7B). After about 20 minutes the bacteria burst and each releases many complete new viruses, while the original infecting viruses still appear in outline attached to the surface of the bacterium (see Figure 14-7D). Based upon these observations it is clear that some substance passed from the infecting viruses into the bacteria, where it caused the synthesis of the new T_2 viruses. This substance must have contained the genes of the virus.

Chemical analyses of bacteriophages have shown that the principal constituents are protein and DNA, and that the DNA of the viruses is liberated when they burst, whereas most of the protein remains with the coats. These data suggest that the viral genes consist of DNA. The actual proof that the genes of T_2 viruses are made of DNA came in 1952.

In that year, A. D. Hershey and M. C. Chase prepared two kinds of bacterial viruses. One was grown in bacteria containing radioactive sulfur (^{35}S); the other was grown in bacteria containing radioactive phosphorus (^{32}P). After a cycle of multiplication, the newly synthesized viruses contained radioactive sulfur and radioactive phosphorus. Inasmuch as proteins contain sulfur but no phosphorus, all the radioactive sulfur was confined to the protein coat of the viruses. Similarly, because the DNA contained phosphorus and not the protein, only the DNA contained the radioactive phosphorus. A suspension of the radioactive viruses was then mixed with a suspension of ordinary *E. coli* cells. The viruses attached themselves to the bacteria, but after a few minutes, a brief treatment of the suspension with an electric blender broke the loose attachment between the viruses and the bacterial cells. This treatment had little effect on the number of infected bacteria; when a portion of the suspension was incubated, the bacteria burst, liberating the usual number of bacterial viruses. Thus, during the few minutes of attachment some substance must have passed from the virus particles into the bacteria, and this substance must have contained the genetic material of the viruses.

Hershey and Chase then measured the quantities of radioactive sulfur and radioactive phosphorus to determine whether DNA or protein (or both) was injected into the bacteria. They found that 85 per cent of the radioactive phosphorus of the viruses had passed into the bacteria and that 80 per cent of the radioactive sulfur was still with the remains of the viruses. The bulk of the DNA entered the bacteria but most of the protein did not. It was clear that DNA entered the host cell almost exclusively and that it was sufficient to code for the synthesis of complete virus particles.

The role of DNA in bacterial transformation and viral infection, as well as other investigations and lines of research, has conclusively demonstrated that DNA is the primary genetic material of most organisms. Exceptions have been found among viruses. All viruses are composed of a nucleic acid core and a protein coat, and many of those that infect higher plants contain a core of RNA and not DNA. *Tobacco mosaic virus* (*TMV*) is perhaps the best-studied example of this type (see Figure 14-6B and C). Experiments conducted with TMV have shown that the RNA portion of the virus contains all hereditary information required for the construction of new viruses.

A considerable amount of indirect evidence suggests that DNA is also the primary genetic material of higher organisms, including man. It has been shown, for example, that all the cells of a given diploid organism contain an equal amount of DNA; the only exceptions are gametes, which typically contain only half as much DNA. Also, DNA from any cell from any member of a given species of plant or animal contains the same relative amounts of the four nucleotides (see Chapter 2). Thus all cells of similar hereditary origin have DNA of similar composition. Furthermore, there are differences in the composition of DNA and the proportion of nucleotides among different species, so the greater the differences, the greater the hereditary variations.

Additional supportive data come from studies of mutations which appear in all organisms from viruses to man. Once a mutant form appears (if it is not lethal), it is continuously inherited according to Mendel's principles. The frequency with which mutants appear can be increased by certain treatments which alter the chemical structure of DNA.

DNA Replication

The next step in the elucidation of the genetic material was to demonstrate the phenomenon of duplication. Such an explanation is possible based upon the Watson–Crick model of DNA, which was constructed in 1953. At the time Watson and Crick started their investigations, there were a number of clues as to the manner in which DNA molecules are arranged in relation to one another. Among the data utilized by Watson and Crick in formulating their model were the following:

1. Many experiments had suggested that DNA was not a highly coiled molecule (for example, a tertiary or quaternary protein) but a long, thin molecule.

2. Some research had suggested that DNA was helical in configuration and that this structure was maintained by hydrogen bonds between turns in the helix.

3. The ratio of nucleotides containing adenine to those containing thymine is 1:1; the ratio of nucleotides containing guanine to those containing cytosine is also 1:1.

Based upon these data, Watson and Crick began to construct a model of DNA. After some months of hypothesizing and model building, they postulated the double-stranded helix structure of DNA (see Figure 2-21). In their model they proposed a mechanism by which replication of DNA might occur based upon the complementary nature of the two chains and base-pairing specificity. They suggested that the two strands separate, making available free hydrogen bonds for pairing with complementary bases. As the hydrogen bonds are progressively broken, free nucleotides become associated with unpaired complementary ones until each strand of the molecule is duplicated exactly. Each of the old strands serves as a pattern by which a new complementary strand is produced. The net effect of this duplication process is that each new strand contains the same sequence of nucleotides and thus the same genetic information as the old ones.

If such a model is correct, it would require that each daughter molecule be half old and half new. This type of replication is referred to as **semiconservative;** that is, half of each parent molecule (one strand) is conserved in each daughter molecule, so that the daughter molecule consists of one old strand from the parent and one new strand formed from free nucleotides in the cell. Support for the semiconservative model of replication was immediate and was derived from several lines of investigation.

In 1957 A. Kornberg and his associates isolated an enzyme from living cells that catalyzed the replication of DNA in a test tube. The enzyme, *DNA polymerase,* when placed in a test tube with a sample of DNA, a sufficient supply of nucleotides, and the required energy sources, catalyzed the synthesis of a DNA-like molecule having physical and chemical properties almost identical to those of the DNA sample. Approximately 20 times the amount of the DNA sample originally present was produced until one of the nucleotides was exhausted. Chemical analyses showed that the structures of the sample DNA and the product DNA were always identical.

An experiment performed by M. S. Meselson and F. W. Stahl also supported the belief that DNA replicates according to Watson and Crick's model. Using *E. coli* as the experimental organism, they employed a technique called equilibrium density-gradient centrifugation, a procedure which makes it possible to separate DNA molecules that differ only in their isotopic composition. They grew *E. coli* for several generations and labeled the DNA by culturing the bacteria in a solution containing glucose, mineral salts without nitrogen, and ammonium chloride in which almost all the nitrogen atoms were the heavy isotope ^{15}N. This treatment increased the density of the DNA in these bacteria over that in bacteria containing only the normally occurring isotope ^{14}N. The bacteria containing ^{15}N were incubated until many new generations had been produced. A portion of these bacteria was separated by centrifugation, and their DNA was isolated. A culture solution containing normal ammonium chloride (^{14}N) was added in large excess to the remainder and the incubation was continued. The bacteria whose DNA contained only ^{15}N continued to multiply and to form new DNA from the ammonium chloride, which contained only ^{14}N. Samples of the bacteria were withdrawn when they had doubled, and further increased in numbers, and DNA was isolated from each generation.

To compare the densities of the paired DNA molecules in the samples and their relative contents of ^{15}N, Meselson and Stahl mixed DNA with a solution of cesium chloride and centrifuged it at high speeds (140,000 times gravity) for long periods of time. Molecules of varying densities eventually separate into bands and the DNA comes to rest in a band in a region of the tube where its density equals that of the cesium chloride. The position of this band can be determined by photographing the tube in ultraviolet light. It was found that the DNA containing ^{15}N formed a band lower down the tube than the DNA containing ^{14}N.

Successive samples represented various stages in the replication of DNA and the transition from ^{15}N- to ^{14}N-containing DNA molecules (Figure 12-10). Meselson and Stahl centrifuged a number of samples of DNA, which they isolated after the bacteria containing ^{15}N had been growing for various periods of time in the solution containing ^{14}N. One of these samples was isolated after the bacteria had doubled their numbers (first generation), and therefore had doubled their quantity of DNA, in the solution. This generation of DNA molecules formed a band in the centrifuge tube which represented a density intermediate between those from ^{15}N and ^{14}N. The DNA consisted of bonded pairs of one heavy (^{15}N) and one light (^{14}N) strand (Figure 12-10B). A sample isolated after the bacteria had quadrupled in number (second generation) showed that their DNA contained particles of two densities; one half

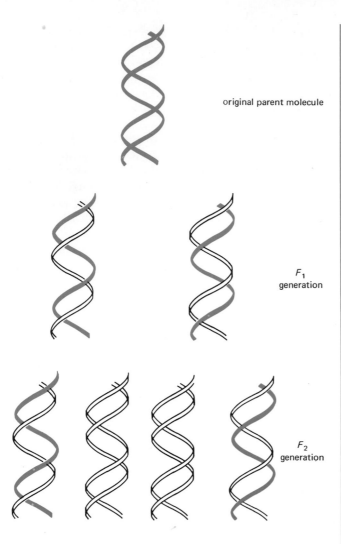

original parent molecule

F_1 generation

F_2 generation

Figure 12-10. The Meselson-Stahl experiment: a demonstration of semiconservative replication of DNA in *E. coli*. (**A**) In the parent DNA molecule, all the nitrogen is ^{15}N, or "heavy" nitrogen. (**B**) The F_1 generation molecules contained half ^{14}N, or "light" nitrogen, and half ^{15}N. (**C**) Half of the F_2 generation DNA molecules contained ^{14}N and ^{15}N; the other half contained only ^{14}N. The ^{15}N strands are shown in green. (From M. S. Meselson and F. W. Stahl, *Proceedings of the National Academy of Sciences*, vol. 44, p. 675, 1958.)

had the intermediate density, and the remainder had the density of DNA containing only ^{14}N (Figure 12-10C).

The Mechanism of Gene Function

The manner in which DNA exerts its control by specifying the synthesis of particular protein molecules has already been described in Chapter 9. To review this process briefly, genes direct the synthesis of a protein as follows:

1. Each triplet of nucleotides along a DNA molecule represents a codon for a particular amino acid.

2. The information in the codon is *transcribed* (transferred) from the DNA to a long single strand of mRNA (see Figure 9-13A).

3. This transcription occurs according to the concept of complementary nucleotide pairings as first suggested by Watson and Crick.

4. The mRNA strand attaches itself by one end around a ribosome and there it acts as a template for the synthesis of a particular protein (see Figure 9-13B).

5. Amino acids in the cytoplasm are activated and united with smaller RNA molecules (tRNA), each of which is specific for one of the 20 amino acids. A tRNA, carrying a particular amino acid, moves to the mRNA, where it is fitted into place based upon the same base-pairing principle (see Figure 9-13C).

6. Thus, as specific amino acids are brought into line and then linked via peptide bonds, the polypeptide chain (protein) is formed (see Figure 9-13D).

The DNA to RNA to protein pattern of information flow may be represented as follows:

$$\xrightarrow{\text{replication}} \text{DNA} \xrightarrow{\text{transcription}} \text{RNA} \xrightarrow{\text{translation}} \text{protein}$$

A. E. Garrod, an English physician, wrote a book in 1909 titled *Inborn Errors of Metabolism,* in which he discussed genetic abnormalities in man that were the result of certain chemical reactions of the body. One of the diseases he discussed was alcaptonuria, in which urine turns black in air because it contains an abnormal compound called homogentistic acid. Homogentistic acid is ultimately converted to CO_2 and H_2O and this conversion is catalyzed by an enzyme present in normal individuals and lacking in those with alcaptonuria. The importance of Garrod's work was that he was the first to postulate that certain diseases were caused by the lack of a specific enzyme and were hereditary in nature. In essence, he suggested over 50 years ago that a change in the genetic material could cause a change in an enzyme and that there is a direct relationship between genes and enzymes.

In 1935 G. Beadle began a series of experiments that were designed to discover the chemical reactions by which genes control inherited characteristics. One of his initial experiments was concerned with the inheritance of eye color in *Drosophila melanogaster* (fruit flies), and he determined that each gene involved controls one chemical reaction by which eye colors are formed. Tremendous progress was made a few years later (1941) when he collaborated with E. L. Tatum and they reversed the experimental procedure. Instead of selecting a genetic trait and then determining the sequence of chemical reactions which cause it, they worked with known sequences of reactions and determined how genetic changes affected these reactions.

For their experiments they selected the pink bread mold *Neurospora crassa.* Among the advantages of this organism are that: (1) it has a brief life cycle; (2) it exists in the haploid stage through most of its life cycle, so that when a mutation occurs its effects are immediately apparent, and (3) it has a simple nutritional requirement consisting only of sugar, a few inorganic salts, and the vitamin biotin. All other vitamins, amino acids, polysaccharides, and other essential substances are synthesized within the mold from the simple components of the growth medium, and the mold inherits the genes which enable it to do this.

Beadle and Tatum induced mutations in some of these genes with X rays and ultraviolet light and then studied the effect of these mutations on the reactions by which the chemical compounds were synthesized. As a result of the mutations they discovered that some of the *Neurospora* strains were unable to grow on a normal, minimal medium unless an extra vitamin or a certain amino acid was added to the growth medium. These molds had lost the ability to perform one or more of the synthetic reactions by which these compounds are formed, owing to the mutation of one or more genes. Careful analyses revealed that a number of distinct strains could be isolated each of which required the same vitamin or amino acid but which had mutations in different genes. This strongly suggested that more than one gene was involved in the synthesis of any one of these compounds and lead to the conclusion that each gene controlled the formation of a single enzyme that catalyzed one step in a sequence of reactions.

The evidence for this conclusion was derived in part from the following typical observations. Three strains of *Neurospora* were isolated which required the amino acid arginine and therefore could not synthesize it from the components of the minimal medium. Each strain had a mutation in a different gene. One of the strains would grow if arginine or the amino acids citrulline or ornithine were added to the minimal medium. Another mutant form would grow if either argi-

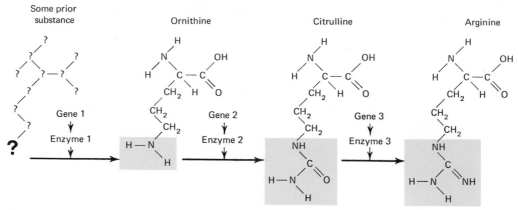

Figure 12-11. Beadle-Tatum hypothesis. Based upon their studies of *Neurospora,* Beadle and Tatum reasoned that genes synthesize enzymes that catalyze sequences of biochemical reactions. The action of enzyme 1 is to convert some prior substance into ornithine. Enzyme 2 adds four atoms (carbon, hydrogen, oxygen, and nitrogen) to ornithine, thereby converting it to citrulline. The catalytic activity of enzyme 3 is to replace the oxygen atom of citrulline with a nitrogen and a hydrogen atom. This conversion transforms citrulline into arginine. The basic differences among the amino acids are shown in the green boxes.

nine or citrulline was added, and the third mutant type would grow only if arginine was added to the minimal medium. Beadle and Tatum hypothesized that each mutant type was detective for a different enzyme involved in the synthesis of arginine. From these data it was reasoned that a given gene (gene 1) produces an enzyme (enzyme 1) that converts a prior substance into ornithine, another gene (gene 2) produces an enzyme (enzyme 2) that converts ornithine to citrulline, and still another gene (gene 3) produces an enzyme (enzyme 3) that converts citrulline into arginine (Figure 12-11). If gene 1 was mutated, the mold could no longer grow and would die unless ornithine (or citrulline or arginine) was supplied to the minimal medium. If gene 2 was mutated, the mold would die unless citrulline (or arginine, but not ornithine) was supplied. Finally, if gene 3 had mutated, the mold would survive only if arginine (but not ornithine

or citrulline) was added. The point here is that by altering the growth medium in various ways, the sequence of biochemical reactions for the synthesis of a specific compound necessary for growth can be arranged in the proper order.

The Beadle–Tatum hypothesis, called the *one-gene, one-enzyme hypothesis,* provided evidence that genes exert their influence through specific enzymes whose synthesis they control. Genes exert their control over cellular activities by specifying the enzymes that catalyze the numerous biochemical reactions of the cell. The sum of these reactions, in turn, determines the specific phenotypic traits.

Although the one-gene, one-enzyme hypothesis is compatible with most known data, at least two important exceptions should be noted. First, some genes do not appear to be primarily concerned with the expression of phenotypic characteristics

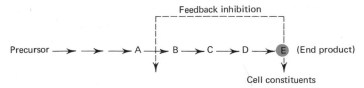

Feedback inhibition

Precursor → → → A —|→ B → C → D → E (End product)

Cell constituents

Figure 12-12. The mechanism of feedback inhibition. Accumulation of an end product in the cell results in the combination of the end product with an enzyme catalyzing an earlier step in the metabolic sequence (A → B). This feedback prevents the further synthesis of E from precursor A. Removal of the end product enables the metabolic sequence to proceed normally. (From A. Gib DeBusk, *Molecular Genetics,* New York: The Macmillan Company, 1968, p. 70.)

through the formation of specific enzymes; instead, they regulate other genes. Recent evidence suggests that they do so by forming specific kinds of nonenzymatic proteins. Second, a few genes in each cell appear to direct the formation of the RNA of the ribosomes and of the cell solution but not the formation of proteins.

Regulation of Gene Action

In recent years a great deal of research has demonstrated that genes do not function by forming their products in a continuous and unchanging manner, but that the genes themselves, as well as their products, are subject to some type of regulation. In addition, many other lines of investigation support the view that the environment can directly affect gene activity. Gene regulation appears to involve two principal mechanisms: (1) the control of enzymatic activity, and (2) the control of enzymatic synthesis.

Control of Enzymatic Activity. A number of investigations performed in the late 1950s were concerned with the control of enzymatic activity. In these investigations it was noted that when the end product of an enzymatic reaction is added to a culture of bacteria, the enzymatic activity of an earlier step is frequently inhibited. The effect of this addition is immediate, and there is no further synthesis of the end product as long as this

end product exists in high concentrations in the cell. This phenomenon, by which the product of an enzymatic reaction may inhibit the activity of one or more enzymes of a metabolic pathway, is called **feedback inhibition** (Figure 12-12).

An example of the phenomenon was demonstrated by H. E. Umbarger in strains of *E. coli.* He studied the biosynthetic pathway in which the bacterial cells produced the amino acid isoleucine. In this metabolic pathway, an early precursor, threonine, is enzymatically converted into isoleucine in five separate steps (Figure 12-13). It was found that if isoleucine was added to the *E. coli* culture medium, the bacteria no longer synthesized isoleucine; instead, they preferentially used the added isoleucine. This condition is maintained until the supply of isoleucine is depleted.

Feedback inhibition is characterized by its action on enzymes that have already been synthesized by the cell; it does not affect the synthesis of these enzymes. The data suggest that a single enzyme, usually the first one in a metabolic pathway, is affected by feedback inhibition. The chemical interaction between the enzyme and inhibitor (end product) appears to be unique, because the inhibitor is structurally different from the normal substrate. The normal substrate and the inhibitor do not compete for the same binding site on the enzyme; it seems probable that the enzyme has

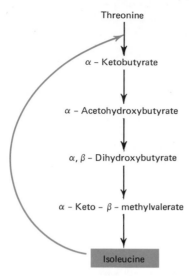

Threonine

↓

α – Ketobutyrate

↓

α – Acetohydroxybutyrate

↓

α, β – Dihydroxybutyrate

↓

α – Keto – β – methylvalerate

↓

Isoleucine

Figure 12-13. Feedback inhibition in *E. coli.* In the sequence of reactions that converts threonine to isoleucine, Umbarger observed that if isoleucine is added to the culture medium, its synthesis is blocked. The isoleucine combines with the enzyme that catalyzes the conversion of threonine into α-ketobutyrate. Thus, further synthesis is blocked.

two sites—one specific for the substrate and one specific for the inhibitor and that binding of the inhibitor results in a change at the catalytic site. This change in the activity of an enzyme that is brought about by the selective binding at a second site on the enzyme (not overlapping the substrate binding site) is called **allosteric interaction** (Figure 12-14). In this regard, the enzyme permits the interaction between inhibitor and substrate.

Allosteric interactions, then, represent one type of system in which regulation may be facilitated by controlling enzyme activity. Another regulatory mechanism involves a direct control of enzyme synthesis.

Control of Enzyme Synthesis. It has been shown from studies of microbial systems, especially those of the bacterium *Escherichia coli,* that certain organisms normally synthesize enzymes only in the

presence of specific substrates. For example, *E. coli* cells growing on a glucose medium contain very little of the enzyme β-galactosidase, an enzyme that catalyzes the hydrolysis of lactose into glucose and galactose. If, however, lactose is added to the culture medium, β-galactosidase molecules appear very shortly thereafter and can be easily detected. It has been estimated that the introduction of lactose to the culture medium causes a thousand-fold increase of the enzyme over the amount present in bacteria growing on a glucose medium. A substrate such as lactose which brings about an increased amount of an enzyme is called an *inducer* and enzymes that are synthesized in the presence of inducers are referred to as *inducible enzymes.* The phenomenon, which is under genetic control, is termed *enzyme induction.*

In addition to enzyme induction, there is a related regulatory phenomenon that results in the alteration of the synthesis of one or often several related enzymes. According to this mechanism, if cells are exposed to a particular end product there is a relative decrease in the rate of synthesis of enzymes involved in the formation of that end product. This is called *enzyme repression.* For example, cells of *E. coli* grown in a medium lacking any amino acids contain the enzymes necessary for the synthesis of all the amino acids contained in the protein molecules of the bacteria. The introduction of a particular amino acid to the culture medium greatly decreases the synthesis of the biosynthetic enzymes necessary for the production of that amino acid. Such enzymes that are reduced in amount by the presence of the end product of a metabolic pathway are called *repressible enzymes* and the molecule (end product) that brings about repression is termed the *corepressor.*

Both induction and repression of enzymes represent types of differential gene activity in which the synthesis of gene products is sensitive to a given set of environmental conditions. In this

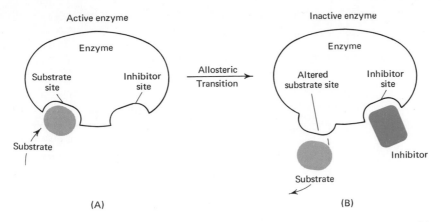

Active enzyme

Inactive enzyme

Enzyme

Enzyme

Substrate site

Inhibitor site

Altered substrate site

Inhibitor site

Allosteric Transition →

Substrate

Inhibitor

Substrate

(A)

(B)

Figure 12-14. Allosteric transition. **(A)** In a normal catalytic reaction, the substrate combines with the enzyme in such a way that the interaction facilitates transformation of the substrate. **(B)** However, when an inhibitor (end product) is attached to its specific site on the enzyme, a structural transition occurs so that the normal substrate can no longer interact with the enzyme. Such a transition inhibits the catalytic activity of an enzyme for a particular biochemical pathway.

regard, they are adaptive mechanisms that appear to be of survival value to the organism. When these mechanisms are in operation, cells do not expend large amounts of energy in synthesizing enzymes not immediately required; rather, such mechanisms allow for the synthesis of a limited number of enzymes in response to conditions in the environment. It should be noted that although enzyme repression is similar to feedback inhibition in that the end product acts to prevent its own synthesis, repression differs from feedback inhibition in that the end product in enzyme repression regulates enzyme *synthesis,* whereas in feedback inhibition the end product regulates enzyme *activity.*

In 1961, F. Jacob and J. Monod, as a result of their studies on induction and repression, formulated a general model to account for the regulation of enzyme synthesis. This model, called the *operon model,* was based upon studies of enzyme systems in *E. coli.* These studies, as well as other investigations of biochemical and genetical systems of gene regulation, have implicated a number of key components that are involved in the control of gene function. These are: (1) regulator genes, (2) repressor substances, (3) operator genes, (4) structural genes, and (5) cellular metabolites (inducers and corepressors). Figure 12-15 summarizes these components and their postulated relationships. Their proposed modes of action will not be discussed.

The cells of *E. coli* contain all the genetic information required for metabolic reactions, growth, and reproduction. They can synthesize from glucose and a number of inorganic ions all the complex molecules they need. Considering this, it can be assumed that a large number of enzymes is needed to accomplish so many syntheses. Some of these enzymes are present at all times. Among these are the enzyme systems involved in cellular respiration. Others however, are produced only when they are needed by the cell. Among these are β-galactosidase, β-galactoside permease, which is involved in the transport of lactose into the cell,

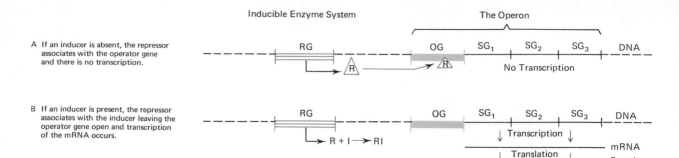

Inducible Enzyme System

A If an inducer is absent, the repressor associates with the operator gene and there is no transcription.

RG ... OG ... SG₁ ... SG₂ ... SG₃ ... DNA
No Transcription

B If an inducer is present, the repressor associates with the inducer leaving the operator gene open and transcription of the mRNA occurs.

R + I → RI

$\downarrow$ Transcription $\downarrow$ — mRNA
$\downarrow$ Translation $\downarrow$ — Protein

Repressible Enzyme System

C If a corepressor is absent the repressor does not combine with the operator gene and transcription of mRNA takes place.

R

$\downarrow$ Transcription $\downarrow$ — mRNA
$\downarrow$ Translation $\downarrow$ — Protein

D If a corepressor is present, the repressor combines with it, the complex associates with the operator gene, and there is no transcription.

R + C → CR

No Transcription

Key to Symbols

RG — Regulator gene

OG — Operator gene

SG₁, SG₂, SG₃ — Structural genes

I — Inducer

C — Corepressor

R — Repressor

Figure 12-15. Operon model for gene induction and gene repression according to Jacob and Monod. See text for amplification. (From F. Jacob and J. Monod, *Journal of Molecular Biology*, vol. 3, p. 318, 1961.)

and galactoside transacetylase, which is related to lactose utilization. The formation of each of these enzymes is controlled by a particular *structural gene*. These genes are so designated because they specify the amino sequence responsible for the structure of a particular enzyme. These three genes, which are closely linked on the bacterial chromosome, produce their respective enzymes rapidly and simultaneously when lactose is introduced into the culture medium.

Data suggesting that all genes are not structural genes have been derived from studies of mutant strains of *E. coli* which continue to synthesize high levels of an enzyme, regardless of whether its substrate is present. In other words, such mutants are not subject to the usual inductive responses.

Variations in the amount of a specific enzyme in bacteria are generally due to variations in the rate of synthesis of that enzyme. Similarly, the rate of synthesis of an enzyme is governed by the quantity of mRMA present in the cell. According to the operon model, the amount of mRNA template available for a given enzyme is controlled by special kinds of molecules, now known to be protein in nature, called *repressors*. Repressors, which block the synthesis of mRNA, are coded for by special genes termed *regulator genes*. Repressors control the rate of mRNA synthesis by combining with specific sites (called operator genes, as discussed subsequently) on DNA and blocking DNA transcription to form mRNA. If regulator genes synthesized repressors all the time, the production of mRNA would always be inhibited. It is therefore necessary to postulate further that repressors exist in two alternative and specific conformations—one active and the other inactive depending upon whether the repressors are combined with specific small molecules (inducers or corepressors).

The attachment of a specific inducer inactivates a specific repressor. For example, the combination of lactose with a repressor inactivates the repressor and permits the synthesis of β-galactosidase and related enzymes. The attachment of a corepressor (an amino acid, for example) to a repressor activates the repressor, permitting active repressor molecules to attach to a portion of the DNA molecule and inhibit the transcription of DNA, and therefore the number of RNA molecules that code for specific enzymes. The genes whose codes are transcribed on a single mRNA molecule and which are under the control of a single repressor are collectively termed an *operon*.

One further entity has been postulated in order to account for the control of operons. This structure, called the *operator gene,* is believed to be adjacent to the structural genes in an operon. Operator genes are described as sites in the DNA molecule to which active repressor substances are bound. This association inhibits the synthesis of mRNA by the adjacent structural genes. In the absence of active repressors, the structural genes in the operon are free to be transcribed, mRNA is formed, and enzymes are produced. When repressors are combined with the operator site, the operon is "switched off" and no mRNA is transcribed. This can be the case when either the operator site or the regulator gene is defective through mutation, or when sufficient end product (corepressor) is present. Presumably, combination of a corepressor with a repressor facilitates attachment of the repressor to the operator gene turning off the operon "switch." In the absence of the corepressor, the free repressor fails to combine with the operator site and the "switch" is open. Inducers also combine with repressor molecules, preventing their attachment to operator sites.

Inducible and repressible systems differ in the form of the repressor substance that will associate with the operator gene. In an inducible system, the repressor substance normally associates with the operator gene and there is no transcription; the presence of a cellular metabolite, which serves as an inducer, inhibits the association and permits transcription. In a repressible system, the repressor substance typically does not associate with the operator gene and transcription takes place; the presence of a metabolite, which acts as a corepressor, permits the association and prevents transcription.

The overall control mechanisms for the regulation of gene activity may be summarized as follows: (1) Structural genes function by specifying the structure (amino acid sequence) of an enzyme. (2) In addition to structural genes, there are regulator genes that, under certain unique conditions, code for the synthesis of repressors. (3) Repressors may be active or inactive depending on whether they are combined with inducers or corepressors. (4) Active repressors bind to operator sites and prevent transcription of adjacent structural genes. (5) Either repression or induction often occurs as a coordinate event involving a multigene region, the operon.

It should be noted that the evidence for the existence of control mechanisms involving operons has come almost entirely from bacterial systems. It is not only possible but quite probable that other types of genetic regulatory systems exist in more complex organisms, especially in higher plants and animals. Studies of regulation in these more complex organisms are currently being conducted in many laboratories and constitute an important contemporary field of study in genetics.

CHAPTER IN RETROSPECT

Through the mechanism of heredity there is a continual tendency of offspring to resemble their parents in all defining characteristics of form and function. The science dealing with this specialized branch of biology concerned with heredity and variation is called genetics. For the most part, this science was formulated about 100 years ago by the work of Gregor Mendel in his plant-hybridization experiments. Mendel developed the concept of monohybrid and dihybrid crosses and determined that when gametes are formed, the factors (later known as genes) separate from each other. On the basis of his experiments, published after his death in 1884, Mendel developed four important laws of genetics: (1) law of unit characters, which states that various hereditary characteristics are controlled by genes, and that genes occur in pairs; (2) law of segregation, which states that a pair of genes located on homologous chromosomes are separated from each other during the formation of gametes in the process of meiosis; (3) law of dominance, which states that one gene of a pair may mask or inhibit the expression of the other, and (4) law of independent assortment, which states that the separation of gene pairs on homologous chromosomes and the distribution of the genes to gametes during meiosis is entirely independent of the distribution of other gene pairs on other homologous chromosomes.

The study of genetics since Mendel has focused in such areas as incomplete dominance, multiple genes, gene linkage, crossing over, the chemical basis of heredity, the mechanism of gene function, and the regulation of gene action. Of particular interest is the so-called Beadle–Tatum hypothesis, or one-gene, one-enzyme hypothesis, which provides evidence that genes probably exert their influence through specific enzymes which they produce and the operon model which postulates a mechanism for the regulation of gene activity.

Questions and Problems

1. What do the following terms signify?

variation	monohybrid cross
genotype	allele
test cross	dihybrid cross
incomplete dominance	multiple genes
gene linkage	crossing over
mutation	mutagens
hybridization	enzyme repression
regulator gene	operator gene

2. How have the four basic Mendelian laws contributed to the development of the field of genetics?

3. Using the checkerboard technique, indicate the law of dominance by crossing a tall (dominant) and short (recessive) plant of the F_1 generation. Explain what happens in the F_2 generation. What is the phenotypic ratio?

4. What information was provided by dihybrid crosses which was not available from monohybrid crosses?

5. Why do the expected phenotypic ratios not appear in multiple gene heredity?

6. Briefly describe hereditary variations which can be attributed to the environment. Did Mendel consider these variations in his plant-hybridization experiments?

7. What are the chief mutagenic agents? Distinguish between a gene mutation and a chromosome mutation.

8. Summarize Griffith's research in terms of his contribution to the concept that genes are made of DNA.

9. Describe the Beadle–Tatum hypothesis. How would you evaluate this research in the light of subsequent discoveries?

10. What is feedback inhibition? How does this phenomenon help to explain the control of enzymatic activity?

11. Briefly describe the Umbarger experiment. What is an allosteric interaction?

12. Compare enzyme induction and enzyme repression.

13. List the principal components and their postulated relationships in an operon model.

Suggested Supplementary Readings

Ames, B. N., and R. G. Martin, "Biochemical Aspects of Genetics: The Operon," *Annual Review of Biochemistry,* vol. 33, p. 235, 1964.

Arnstein, H. R. V., "Mechanisms of Protein Synthesis," *British Medical Bulletin,* vol. 21, p. 217, 1965.

Atkinson, D. E., "Biological Feedback Control at the Molecular Level," *Science,* vol. 150, p. 851, 1965.

Barry, J. M., *Genes and the Chemical Control of Living Cells.* Englewood Cliffs, N.J.: Prentice-Hall, Inc., 1964.

Bennett, T. P., and Earl Frieden, *Modern Topics in Biochemistry.* New York: The Macmillan Company, 1966.

Borek, Ernest, *The Code of Life.* New York: Columbia University Press, 1965.

Brenner, S., "Colinearity and the Genetic Code," *Proceedings of the Royal Society (London),* vol. B164, p. 170, 1966.

Crick, F. H. C., "The Genetic Code," *Scientific American,* October 1962.

Crick, F. H. C., "The Structure of the Hereditary Material," *Scientific American,* October 1954.

Edgar, R. S., and R. H. Epstein, "The Genetics of a Bacterial Virus," *Scientific American,* February 1965.

Eigsti, O. J., "Induced Polyploidy," *American Journal of Botany,* vol. 44, pp. 272–279, 1957.

Fraenkel-Conrat, H., "The Genetic Code of a Virus," *Scientific American,* October 1964.

Hayes, William, *The Genetics of Bacteria and Their Viruses.* New York: John Wiley & Sons, Inc., 1964.

Ingram, V. M., *The Biosynthesis of Macromolecules.* New York: W. A. Benjamin, Inc., 1965.

Jacob, F., and J. Monod, "Genetic Regulatory Mechanisms in the Synthesis of Proteins," *Journal of Molecular Biology,* vol. 3, p. 318, 1961.

Jukes, T. H., "The Genetic Code II," *American Scientist,* vol. 53, pp. 477–487, 1965.

Kornberg, A., *Enzymatic Synthesis of DNA.* New York: John Wiley & Sons, Inc., 1962.

Lanni, Frank, "The Biological Coding Problem," *Advances in Genetics,* vol. 12, p. 1, 1964.

Monod, J., J. Wyman, and J. P. Changeaux, "On the Nature of Allosteric Transitions: A Plausible Model," *Journal of Molecular Biology,* vol. 12, p. 88, 1965.

Nishimura, S., D. S. Jones, and H. G. Khorana, "Studies on Polynucleotides," *Journal of Molecular Biology,* vol. 13, p. 302, 1965.

Sager, R., "Genes Outside the Chromosomes," *Scientific American,* January 1965.

Sinsheimer, R. L., "Single-Stranded DNA," *Scientific American,* July 1962.

Stahl, F. W., *The Mechanics of Inheritance,* Englewood Cliffs, N.J.: Prentice-Hall, Inc., 1965.

Stent, G. S., "The Operon: On Its Third Anniversary," *Science,* vol. 144, p. 816, 1964.

Strauss, B. S., *Chemical Genetics.* Philadelphia: W. B. Saunders Company, 1960.

Umbarger, H. E., "Intercellular Regulatory Mechanisms," *Science,* vol. 145, p. 674, 1964.

Watson, J. D., *Molecular Biology of the Gene.* New York: W. A. Benjamin, Inc., 1965.

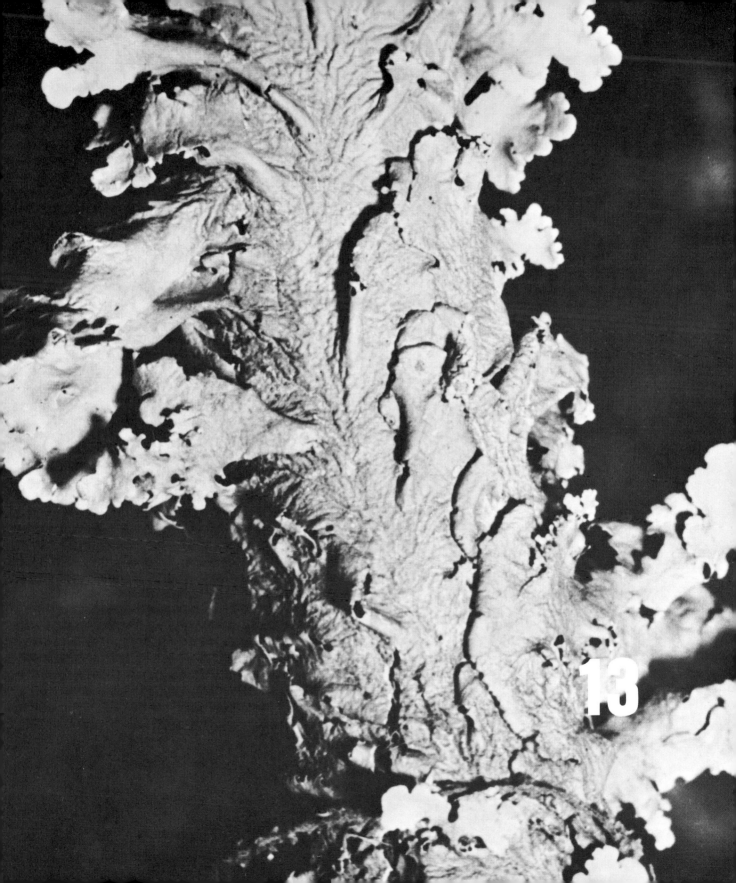

13
Plant Classification and Diversity

INTRODUCTION

At present, there are probably more than 375,000 known plant species. Considering that many more have not yet been discovered, identified, and named, it becomes apparent that a logical and meaningful system of classification is necessary. Such a system, if it is to be useful, must provide a set of criteria by which plants may be distinguished from each other on the basis of certain structures and relationships.

TYPES OF CLASSIFICATION SYSTEMS

The science of classification, called *taxonomy*, was one of the first specialized disciplines to evolve from the study of plants. In the course of its evolution, many contributions have been made by various men. Some of these ideas have been abandoned; others have been modified in terms of newly acquired knowledge. Despite the numerous systems devised for classifying plants, two principal types may be distinguished: artificial and natural.

An *artificial* system of classification is one that is based upon certain structural characteristics without regard for the ancestry and evolutionary development of plants. The system of Linnaeus, for example, was an artificial one. His system was based essentially on the number of stamens and styles in each flower. The difficulty with his system was that plants having the same number of stamens are not necessarily related to each other. Conversely, plants related to each other may differ in the number of stamens present. For example, mints and sages both possess strong odors, oppositely arranged leaves, and square-shaped stems and are related to each other on the basis of these characteristics as well as others. Sages, however, have only two stamens whereas mints have four, so the system of Linnaeus was more artificial than natural. For the most part, artificial classification systems have been greatly revised or replaced by natural ones.

A *natural* classification system is one which attempts to classify plants on the basis of their genetic relationships to each other. Such a system is based upon the similarities and differences of a great many external characteristics and the relationship of these structures as plants have evolved from relatively simple to more complex forms. In other words, it is a system of descent, in which plants are grouped according to the presence of primitive or advanced characteristics.

Whereas artificial systems are inadequate because they do not demonstrate evolutionary relationships, natural systems also contain inherent weaknesses. First, the number and diversity of plants in terms of form and function is very great. Second, new plant forms have appeared at various times and others have become extinct. In addition, inasmuch as the fossil record is discontinuous, many data concerning the evolutionary origin and development of various plant groups are either fragmentary, vague, or not available. In spite of these difficulties, as well as others, botanical taxonomists have generally attempted to devise classification systems that are more natural than artificial, because, even though subject to change, natural schemes do provide fairly accurate systems based upon genetic relationships and evolutionary patterns.

Classification Units (Levels)

To be useful a natural classification scheme must place a particular plant in a given position among all other plants. A specific plant is given two names, usually derived from either Latin or Greek, a convention designated as *binomial nomenclature.* The obvious advantage of employing scientific names for plants, and not common names, is that these names are universally accepted, and they convey specific meanings which show natural relationships. A scientific name consists of the *genus* and the *species* designation; both terms are underlined or italicized and the genus is capitalized. The scientific name for the black locust, for example, is *Robinia pseudoacacia.* The generic, or group, name is always a noun, whereas the specific, or individual, name is usually an adjective.

Scientific names for plants may be derived according to a number of criteria. Some are chosen on the basis of habitat. *Ranunculus aquaticus,* the water buttercup, is so designated because it inhabits a water environment. *Pyrus americana,* American mountain ash, is named on the basis of location. Still other plants such as *Rosa alba,* the white rose, are designated on the basis of a principal characteristic of the plant. Finally, some scientific names are chosen from the name of the discoverer of the plant. *Aster drummondii,* Drummond's aster, is such an example.

The species name of a plant represents the lowest category of taxonomic grouping. Occasionally, lower levels of classification, such as the variety, subspecies, and others, are used, but, for the purposes of this book, the species will be considered the lowest category. Despite the significance of the species as a taxonomic grouping, much debate still exists as to adequately defining it, although some definitive aspects of the species concept have been identified. It may be said that a *species* is a particular kind of living thing, all members of which are closely related through a common ancestry and persistently different from other groups. Further, species characteristics are maintained, under natural conditions, through many successive generations. Logically some species of plants show more variation than others. When these variations are quite conspicuous, some species are subdivided into *subspecies* and/or *varieties.* Taxonomists generally agree, however, that the species is the fundamental natural unit of classification.

A species, then, is a group of closely related individuals. A *genus* consists of a number of species different from each other according to certain variations but related to each other by descent. For example, *Quercus,* the oak genus, consists of all types of oak trees (white, red, bur, velvet, and so on); although each species of oak differs from all others, they are all related genetically and therefore constitute the genus *Quercus.*

Just as a number of species are combined into a larger unit, the genus, related genera constitute a *family.* A group of similar families constitutes an *order,* and a group of similar orders make up a *class.* Related classes in turn comprise a *division,* the highest category normally used in the scheme of plant classification. Through convention, and for purposes of consistency, division names usually end in *-phyta;* the ending *-ae* or *-eae* indicates class, the ending *-ales* is used for orders; and the ending *-aceae* indicates a family. The names of genera and species are represented by a multiplicity of endings. In general, the higher the unit, the more individuals it contains, and the fewer the number of units at the same level. By contrast, the lower the unit, the fewer the number of individuals in it, and the larger the number of units at that level. Consider the classification of the white oak in Figure 13-1.

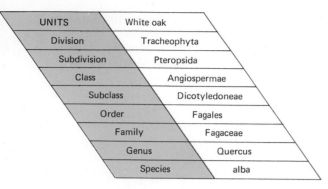

UNITS	White oak
Division	Tracheophyta
Subdivision	Pteropsida
Class	Angiospermae
Subclass	Dicotyledoneae
Order	Fagales
Family	Fagaceae
Genus	Quercus
Species	alba

Figure 13-1. Classification of the white oak (*Quercus alba*). Note the endings for each taxonomic unit.

It should be noted that in the foregoing discussion of classification, no mention was made of subunits, although two were used in the classification of the white oak. These were the subdivision and subclass. A *subdivision* is a unit between a division and a class; a *subclass* is a unit between a class and an order; and so on. Such subtaxonomic categories may be inserted, if needed, to show a natural relationship.

CLASSIFICATION OF THE PLANT KINGDOM: TRADITIONAL AND MODERN

The time-honored system of classification widely used in the later part of the nineteenth century and the early part of the twentieth century was based upon certain structural characteristics in which all plants were grouped into four divisions: Thallophyta, Bryophyta, Pteridophyta, and Spermatophyta (Table 13-1). Such a scheme of classification is not generally accepted today. The principal objection to this traditional classification scheme is that it does not take into account, and actually misrepresents, some of the phylogenetic (natural) relationships of plants. For example, the terms "Thallophyta," "Pteridophyta," and

"Spermatophyta" have no phylogenetic meaning and are used only for convenience. According to the traditional classification, "Thallophyta" represents a group of plants lacking specialized conducting tissues and not differentiated into true roots, stems, and leaves (nonvascular plants). Further, they are subdivided into algae, or photosynthetic thallophytes, and fungi, nonphotosynthetic thallophytes. The belief of many botanists is that fungi have not evolved from algae and that both "groups" of plants are so diverse and unrelated that they should be placed into a number of separate divisions. Moreover, it appears that no

TABLE 13-1. Traditional Classification of the Plant Kingdom

I. Division Thallophyta
  A. Subdivision Algae
    1. Class Euglenineae
    2. Class Cyanophyceae (blue-green algae)
    3. Class Chlorophyceae (green algae)
    4. Class Chrysophyceae (yellow-green algae)
    5. Class Bacillariophyceae (diatoms)
    6. Class Phaeophyceae (brown algae)
    7. Class Rhodophyceae (red algae)
  B. Subdivision Fungi
    1. Class Schizomycetes (bacteria)
    2. Class Myxomycetes (slime molds)
    3. Class Phycomycetes (algalike fungi)
    4. Class Ascomycetes (sac fungi)
    5. Class Basidiomycetes (club fungi)
    6. Class Fungi Imperfecti (imperfect fungi)
II. Division Bryophyta
    1. Class Hepaticae (liverworts)
    2. Class Musci (mosses)
III. Division Pteridophyta
    1. Class Lycopodinae (club mosses)
    2. Class Equisetinae (horsetails)
    3. Class Filicinae (ferns)
IV. Division Spermatophyta
    1. Class Gymnospermae (gymnosperms)
    2. Class Angiospermae (flowering plants)
      a. Subclass Dicotyledonae (dicots)
      b. Subclass Monocotyledonae (monocots)

close relationship exists either among the various algal divisions or the fungal divisions. Therefore, "Division Thallophyta" is a term of convenience used to designate a group of plants at a comparable level of structural complexity. It is a term used to describe an unnatural grouping of unrelated forms. Even the terms "algae" and "fungi" are unnatural taxonomic designations.

As another example of the artificial character of the traditional system, consider the division Pteridophyta. For many years, the ferns, lycopods, and horsetails were thought to be sufficiently related to each other to be considered as a single division. Evidence has shown that, although these plants do contain vascular tissues, they are, in fact, divergent (separate) lines from the psilophytes (primitive vascular plants). Further evidence suggests that the ferns are more nearly allied to the gymnosperms and angiosperms than they are to the horsetails and club mosses, and therefore the distinction between the "Pteridophyta" and "Spermatophyta" is more artificial than natural.

There is no single taxonomic scheme that can represent the phylogenetic relationships of all plants. In recent years botanists have attempted to synthesize all available data to formulate a natural classification system. No doubt, as more evidence of natural relationships is discovered, the system will be modified significantly. The classification scheme used in this book attempts to incorporate the available evidence about plant relationships into a logical, phylogenetic scheme. It is beyond the scope of the book, however, to include all plant groups in a single classification scheme. In subsequent chapters, only representative members from selected plant groups will be considered to emphasize form and function, a product of the evolutionary advances of plants.

Table 13-2 presents an outline of the classification scheme used in this book. Although the terms "Thallophyta" and "Embryophyta" are still re-

TABLE 13-2. Classification of the Plant Kingdom Used in This Book*

Subkingdom I. Thallophyta (thallus plants)
  Division 1. Schizophyta (bacteria and blue-green algae)
  Division 2. Euglenophyta (euglenoids)
  Division 3. Chlorophyta (green algae and stoneworts)
  Division 4. Xanthophyta (yellow-green algae and chloromonads)
  Division 5. Chrysophyta (golden algae and diatoms)
  Division 6. Pyrrophyta (dinoflagellates and cryptomonads)
  Division 7. Phaeophyta (brown algae)
  Division 8. Rhodophyta (red algae)
  Division 9. Myxomycophyta (slime molds)
  Division 10. Eumycophyta (true fungi)
    Class 1. Phycomycetes (algal fungi)
    Class 2. Ascomycetes (sac fungi)
    Class 3. Basidiomycetes (club fungi)
    Class 4. Deuteromycetes (imperfect fungi)
    Class 5. Lichenes (lichens)

Subkingdom II. Embryophyta
  Division 11. Bryophyta (mosses and allies)
    Class 1. Hepaticae (liverworts)
    Class 2. Anthocerotae (horned liverworts)
    Class 3. Musci (mosses)

  Division 12. Tracheophyta (vascular plants)
    Subdivision 1. Psilopsida (mostly fossil forms)
    Subdivision 2. Lycopsida (club mosses)
    Subdivision 3. Sphenopsida (horsetails)
    Subdivision 4. Pteropsida (ferns and seed plants)
      Class 1. Filicineae (ferns)
      Class 2. Gymnospermae (gymnosperms)
      Class 3. Angiospermae (flowering plants)
        Subclass 1. Dicotyledonae (dicots)
        Subclass 2. Monocotyledonae (monocots)

*The above classification scheme is a very general one, the lowest taxonomic grouping being the subclass. In subsequent chapters, as major groups of plants are studied, a more detailed classification is presented for the plants under consideration. The taxonomic schemes that appear in later chapters are designed to show possible phylogenetic relationships and to emphasize the group of plants being studied.

tained to designate the two principal subkingdoms, it should be remembered that they are artificial and not natural taxonomic units. Both imply that lower groups of plants are at similar levels of structural complexity.

The Principle of Classification

Modern systems of classification are based largely upon reproductive characters, because these features are the most stable morphological ones. Inasmuch as vegetative structures (roots, stems, and leaves) are subject to modification through the influence of various environmental factors and the reproductive structures remain unchanged, it is the reproductive characters that form the basis for classification.

It was stated in Chapter 12 that although plants exhibit a wide variety of reproductive patterns, two principal types may be distinguished. These are asexual and sexual reproduction. The common types of asexual reproduction are shown in Figure 12-1. In all forms of asexual reproduction there is no fusion of nuclei, and therefore no mixing of genes. In many plants, certain cells, called **spores,** are specialized for asexual reproduction. These cells are capable of producing new plants by cell division and growth. Not only are these cells adapted to resist adverse environmental conditions, but they are also capable of widespread dispersal. In lower aquatic plants such as algae and fungi, spores are provided with cilia or flagella so that new areas may be populated if existing conditions are unsuitable for growth. Such motile spores are referred to as **zoospores.** Spores are produced in a structure called a **sporangium,** although such a structure may be given a more specific name, depending upon the type of plant in which it appears.

Sexual reproduction involves the fusion, in pairs, of the nuclei of specialized sex cells called **gametes** (eggs and sperms). In the algae, the single cell that produces the egg is called an **oogonium,** whereas in higher plants the egg-producing chamber is called an **archegonium** and is multicellular. The male gamete-producing chamber, whether unicellular or multicellular, is called an **antheridium.** The collective designation for gamete-producing cells or chambers is the **gametangium.**

In a few primitive plants, the motile gametes are morphologically alike and they are referred to as **isogametes.** Their fusion is termed **isogamy.** The term **heterogamy** is used to designate the fusion of morphologically dissimilar gametes called **heterogametes.** In heterogamy both gametes are motile. In more advanced plant forms, the male gametes (sperms) are usually smaller and motile and the female gametes (eggs) are usually larger and nonmotile. The fusion of such gametes is termed **oogamy,** and the resulting diploid cell is termed a **zygote.**

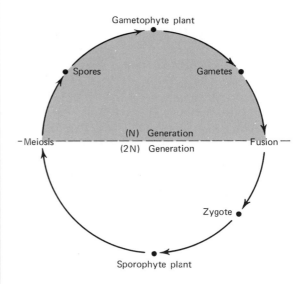

Figure 13-2. Generalized plant life cycle showing the salient features of alternation of generations. The gametophyte generation is shown in green.

Reproduction of plants, with relatively few exceptions, is basically an **alternation of generations,** that is, an asexual stage alternating with a sexual stage. In considering the complete life cycle of a plant, the zygote develops into a multicellular, diploid, asexual body, the **sporophyte,** which is responsible for the production of haploid spores. This is accomplished through meiosis. These spores, in turn, develop into a multicellular, haploid, sexual phase, the **gametophyte.** The essential function of the gametophyte is the production of haploid gametes, which, upon fusion, form a diploid zygote. This generalized cycle of alternation of generations is shown in Figure 13-2.

As representatives of various plant groups are discussed in subsequent chapters, emphasis will be placed on the forms of reproductive structures and their functions, because an analysis of the life cycles of plants is fundamental to an understanding of plant evolution.

CHAPTER IN RETROSPECT

The 357,000 known species of plants are classified according to one of two systems: (1) artificial (based upon certain structural characteristics), and (2) natural (based upon genetic relationships and evolutionary patterns). Although both systems have their inherent weaknesses, botanical taxonomists have tended to devise classification schemes which are more natural than artificial. In such a system a plant has two scientific names, usually derived from Greek or Latin, which consists of a genus and species designation. The species is the fundamental natural unit of classification and represents a group of closely related plants. Genus, family, order, and class represent progressively higher levels of plants until a division is reached, which designates the highest category normally used in this scheme of plant classification. Until

recently plants were grouped into four divisions: Thallophyta, Bryophyta, Pteridophyta, and Spermatophyta. This grouping has proved to be unacceptable in that it actually misrepresents some of the natural relationships of plants. A classification system more widely acceptable by modern botanists divides the plant kingdom into two major subkingdoms (Thallophyta and Embryophyta) and 12 divisions.

The structural and functional aspects of plant reproduction are useful taxonomic indices. Two principal types of reproduction occur: (1) asexual, involving spores and sporangium, and (2) sexual, involving gametes (eggs and sperms) and gametangia. However, in most cases, a diploid asexual body (sporophyte) alternates with a haploid sexual body (gametophyte) in a process called alternation of generations.

Questions and Problems

1. What are the inherent weaknesses in the artificial and natural systems of plant classification? Why is the natural system more widely accepted by modern botanists?
2. Briefly describe the advantages of using scientific names as opposed to common names. Why are scientific names generally derived from Greek or Latin?
3. Compare the traditional and modern systems of plant classification in terms of regrouping such plants as algae, fungi, mosses, and ferns. Check other botany books to determine the variety in plant classification schemes.
4. What are the essential components of the asexual and sexual stages of plant reproduction?
5. What do the following terms signify?

    genus   spore
    species   zoospore
    variety   gamete
    family   archegonium
    class   oogonium

division   antheridium
gametangium   alternation of generations
sporophyte   gametophyte

Suggested Supplementary Readings

Bailey, L. H., *How Plants Get Their Names.* New York: The Macmillan Company, 1933.

———, *Manual of Cultivated Plants.* New York: The Macmillan Company, 1949.

Benson, Lyman, *Plant Taxonomy.* New York: The Ronald Press Company, 1962.

Bold, H. C., *Morphology of Plants.* New York: Harper & Row, Publishers, 1957.

———, *The Plant Kingdom.* Englewood Cliffs, N.J.: Prentice-Hall, Inc., 1964.

Cronquist, Arthur, "The Divisions and Classes of Plants," *Botanical Review,* vol. 26, pp. 425–482, 1960.

Ditmer, H. J., *Phylogeny and Form in the Plant Kingdom.* Princeton, N.J.: D. Van Nostrand Company, Inc., 1964.

Doyle, W. T., *Nonvascular Plants: Form and Function.* Belmont, Calif.: Wadsworth Publishing Co., Inc., 1964.

Lawrence, G. H. M., *An Introduction to Plant Taxonomy.* New York: The Macmillan Company, 1955.

Norman, A. G., "The Uniqueness of Plants," *American Scientist,* vol. 50, pp. 436–449, 1962.

Russell, N. H., *An Introduction to the Plant Kingdom.* St. Louis, Mo.: The C. V. Mosby Company, 1958.

Salisbury, F. B., and R. V. Parke. *Vascular Plants: Form and Function.* Belmont, Calif.: Wadsworth Publishing Co., Inc., 1964.

Whittaker, R. H., "On the Broad Classification of Organisms," *Quarterly Review of Biology,* vol. 34, pp. 210–256, 1959.

14

14
The Origin and Evolution of Plants

INTRODUCTION

Today, biologists clearly recognize the concept that all living forms consist entirely of chemicals in various organizational patterns which function for the maintenance and propagation of the organism. Inasmuch as it is reasonably certain that before life originated there were only chemicals and that living things originated out of chemicals, the starting point to determine possible origins of life should be an analysis of the nature of prebiotic chemicals. Such an analysis must consider the origin of the earth.

THE ORIGIN OF LIFE

Prebiological Conditions
(Chemical Evolution)

There is no totally satisfactory explanation to account for the origin of the solar system, despite the numerous hypotheses and theories that have been proposed. Nevertheless, the age of the earth has been estimated by various methods to be between 4.5 and 5 billion years. The hypothesis most widely accepted today is that the sun and its planets were formed within a massive cloud of cosmic dust and gas in an empty region of our galaxy. It has been estimated that initially the cosmic cloud was 1 light-year in diameter and had a temperature far below 0°C. Most of this material began to condense rapidly into a more compact mass as a result of gravitational forces. The great heat and enormous pressure resulting from condensation initiated thermonuclear reactions

that converted the main condensed mass into a luminous body, the sun. Within the remainder of the dust and gas cloud, lesser centers of condensation were produced, and these became the planets.

Once formed, the earth also began to condense. As a result of condensation, the temperatures of the solid mass rose to 1000 to 3000°C and a selective melting and stratification of its chemicals took place. The heavier molten metals, such as iron and nickel, sank into the core, while the lighter substances migrated nearer the surface. Included in the lighter materials were basalt, granite, and various dehydrated minerals. As the earth cooled, its surface eventually solidified.

The ancient atmosphere probably consisted of an envelope of methane (CH_4), water vapor (H_2O), ammonia (NH_3), and hydrogen (H_2). By comparison, the present atmosphere contains about 78 per cent molecular nitrogen (N_2), 20 per cent molecular oxygen (O_2), and less than 1 per cent (0.03 per cent) carbon dioxide (CO_2), the remainder being distributed as traces of rarer gases such as helium and neon. It is believed that free nitrogen, oxygen, and carbon dioxide were not present in the primitive atmosphere because of the large amounts of hydrogen and the exceedingly high temperatures. Free nitrogen would have combined with hydrogen to form ammonia; carbon would have combined with hydrogen to form methane; and oxygen would have combined with iron, silicon, and aluminum to form many minerals of the earth's crust or with hydrogen to form water vapor.

Most certainly, the simple inorganic gases comprising the ancient atmosphere were not the chemicals from which living forms were directly produced. Any mixture of these compounds is fairly stable, and such gases do not react with each other to form other compounds. It is reasonable to assume that the primary organic molecules from which life arose would be amino acids and nitrogenous bases. Therefore, certain conditions must have existed to transform the simple compounds of the primitive atmosphere into more complex organic substances, the precursors of life.

As the earth cooled, temperatures become low enough that water vapor liquified and began to collect in the basins on the crust. Basins and shallows filled up as torrential rains fell, and the low places on the crust slowly filled with water, forming the first oceans. Dissolved in the seas were atmospheric ammonia and methane deposited by the rains and salts and minerals deposited by rivers flowing down mountain slopes. In addition, volcanic eruptions supplemented the mineral content of the seas.

Logically, if the primitive seas contained the basic materials (simple inorganic compounds) which eventually gave rise to life, some external source of energy must have acted on the mixture to rearrange the atoms to form more complex molecules. One possible source may have been heat energy from the interior of the earth, which could have speeded up the motion of the molecules of the gases, making them collide with sufficient energy to form new combinations of atoms. A second possibility may have been the energy from solar radiation in the form of visible light, ultraviolet light, and X rays bombarding the existing molecules and producing new combinations. A third form of energy which may have rearranged the molecules was electrical energy from lightning.

Experimental data which demonstrated that organic molecules could be produced from a mixture of ammonia, methane, water, and hydrogen were published in 1953. In these experiments, those conditions presumed to exist on the primitive earth were simulated. In the airtight apparatus shown in Figure 14-1, inert methane, ammonia, and hydrogen were circulated past electrical discharges from tungsten electrodes. Boiling water and condensation kept the substances circulating through the closed system. After 1 week,

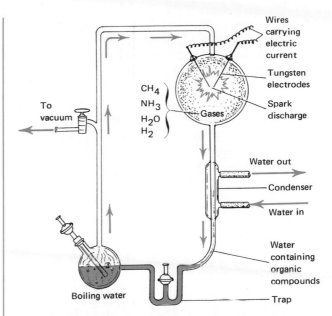

Figure 14-1. A diagram of the Miller spark-discharge apparatus designed for the circulation of methane, hydrogen, and water vapor. (From S. L. Miller, *Journal of the American Chemical Society,* vol. 77, p. 2351, 1955. Copyright 1955 by American Chemical Society.)

the contents of the system, which appeared as a residue collected from a trap, was analyzed. It was found that a variety of organic compounds had been synthesized. Chief among these were several amino acids and various organic acids. Since this classic experiment, other investigators, using a variety of energy sources and regulating the types and amounts of reactants, have synthesized sugars, polysaccharides, fatty acids, glycerol, purines, pyrimidines, and even ATP.

Once chemical evolution progressed to the point where amino acids, sugars, fatty acids, and other simple organic compounds were present, it is probable that they began to join together to produce larger, more complex sugars and polysaccharides such as starch. In other reactions, glycerol combined with various fatty acids to form fats. Amino acids joined end to end to form proteins and enzymes. Another group of complex molecules, the nucleotides, developed from the purines and pyrimidines. These nucleotides, in turn, combined with one another to produce the nucleic acids.

The conditions under which amino acids might be bonded together to form proteins were suggested in a number of experiments in which mixtures of different amino acids were heated for varying periods of time. It was found that dipeptides and some long-chain peptides (proteins) were produced, some of which assumed a tertiary structure, indicating a certain level of order and organization. Elaboration of these experiments by other investigators has shown that other macromolecules may also be formed under the influence of thermal energy. Even though the earth was cooling slowly, its surface would still have retained heat. Moreover, the torrential rains must have become less frequent. Water could have evap-

orated between periods of rain, leaving dry amino acids on the hot surface. Under these conditions amino acids could have bonded together and later rains could have washed the proteins into the oceans.

Assuming that the waters of the primitive earth now contained all the compounds necessary for life, there still remains the problem of organizing the various components into a structural pattern characteristic of even the simplest organism. It has been postulated that one of the first steps in the organization of the first living form was the aggregation of complex molecules, presumably proteins, into complex structural units, and it has been determined that colloidal protein molecules tend to clump together to form more complex units called *coacervates*. These units are clusters of proteins or protein-like substances held together in small droplets within a surrounding liquid (Figure 14-2). Each coacervate is surrounded by

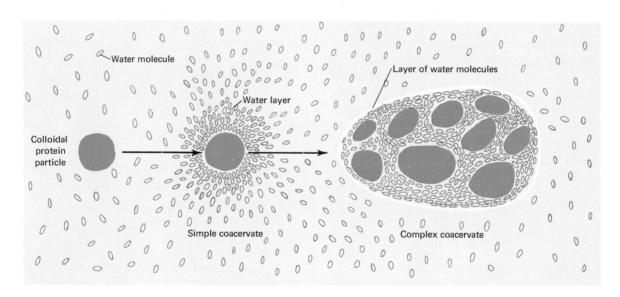

Figure 14-2. Diagrammatic representation of the process of coacervate formation.

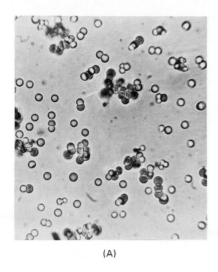

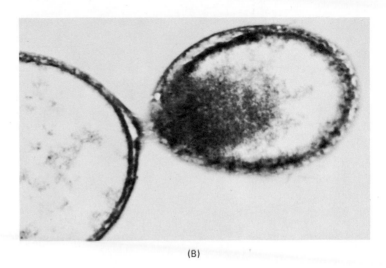

(A)   (B)

Figure 14-3. Microspheres. **(A)** In this photomicrograph note the regular appearance of the spheres and their similarity to spherical bacteria. **(B)** This electron micrograph shows a microsphere containing a double-walled membrane and in the process of "fission." (Courtesy of S. W. Fox, Institute of Molecular Evolution, University of Miami.)

a shell of water, so there is a definite separation between the coacervate droplet and the medium in which it floats. In essence, the water molecules form a membrane around the droplet. Experimental data have shown that coacervates are able to selectively absorb and incorporate various substances from the external medium. Thus the coacervates may "grow." Although such coacervates are not living, they do exhibit certain structural and functional properties ordinarily associated with living organisms.

In 1959 it was demonstrated that when proteins which had been synthesized were treated with hot water, *microspheres* developed when the solution was cooled (Figure 14-3A). These microspheres not only resemble spherical bacteria but also show certain structural properties of cells. For example, they have diameters about the same size as bacterial cells, they have double membranes that are selectively permeable, and they can be plasmo-

lyzed by various salt solutions. Some also have vacuoles and granules. In addition, two microspheres can combine with each other, or one can split apart by a process similar to binary fission (splitting in two) upon reaching a certain critical size (Figure 14-3B).

The experimental evidence derived from subsequent research strongly suggests that coacervates or microspheres could have been the precursors of the first living forms. These data propose mechanisms by which molecular aggregates of different chemical compositions could combine to produce new and better forms. Further, the formation of a surface layer, and later a membrane, concentrated certain chemical substances and increased the probability of chemical reactions taking place; each reaction would influence other reactions in the closed system. As some of the complexes assumed catalytic activities, there was coordination and synchronization of the reactions. Despite all these structural and functional similarities to cells, it must be stressed that neither coacervates nor microspheres are living. They lack a complex structural organization, intricate metabolic pathways, and the genetic code normally associated with living systems. They do, however, reinforce the plausibility of a nonbiological origin for the first living forms.

Thus far in tracing the pattern of chemical evolution, it should be noted that these molecular aggregates lacked both well-developed structural organization and the capacity to reproduce. As the process of chemical synthesis continued, nucleic acids came into existence. It is impossible to say at what stage this occurred, but it was probably quite early. Sooner or later during their development, nucleic acids exhibited the property of duplication. With the appearance of the first nucleic acids, templates or models became available by which more nucleic acids could be produced. For nucleic acids to replicate themselves, various building materials had to be taken in from the surrounding medium. Once a system can reproduce itself from raw materials of the environment, it is living. The principal steps involved in the synthesis of complex aggregates containing nucleic acids are shown in Figure 14-4.

Biological Evolution

Once continuity in the form of genetic control was introduced, nucleic acid systems began to compete for available supplies in the ancient oceans. Such competition was necessary because the surrounding organic compounds were not only used as food but also as building material for nucleic acids. The primitive atmosphere contained no free oxygen, so the respiration of the first cells was anaerobic and these first forms were probably heterotrophic. Those organisms that could use energy most efficiently could reproduce their own kind most rapidly. At this point, electron transfer and energy storage (ATP) probably appeared.

Primitive heterotrophs probably evolved more sophisticated biochemical mechanisms that enabled them to utilize a multiplicity of organic compounds in the environment as energy-providing processes. In addition, some heterotrophs probably evolved saprophytic and parasitic modes of nutrition. In essence, as the first cells multiplied and obtained nutrients, diminishing the available food supply, they also altered the environment in a way that decreased the rate of abiotic synthesis of organic compounds. For example, during anaerobic respiration, great quantities of CO_2 were added to the water and the atmosphere. Abiotic synthesis of organic compounds is more efficient using methane as a raw material rather than CO_2. Furthermore, high concentrations of atmospheric CO_2 would have absorbed some of the solar radiation, depriving the primitive seas of a source of energy for abiotic synthesis.

In the evolutionary history of the first heterotrophs, competition assumed a prominent role. The origin of new forms by mutation certainly enhanced the chances for survival, because life did not become extinct as the supply of organic compounds decreased. Any mutations that would enable an organism to manufacture its own food, and not depend on the dwindling quantities in the oceans, would be considered highly favorable. Among the favorable mutations that did occur were those that made possible the synthesis of light-absorbing pigments and the essential enzymes for photosynthesis. At this stage, a new form of metabolism, photosynthesis, and a new mode of nutrition, autotrophism, were introduced.

In all probability the first photosynthetic mechanism was that of cyclic photophosphorylation. At some point later, the much more complex reactions of noncyclic photophosphorylation would have evolved. Photosynthesis not only pro-

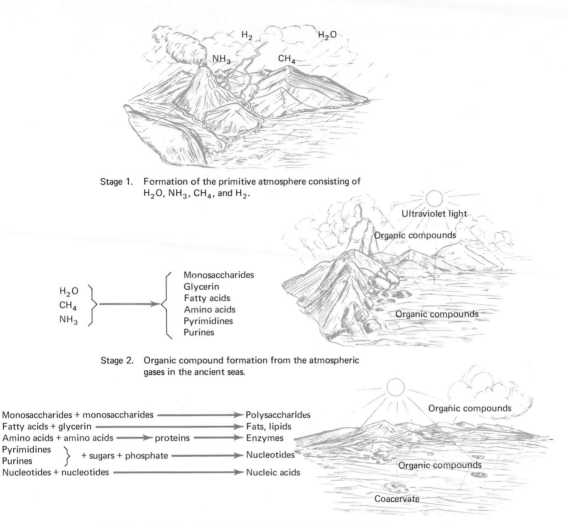

Stage 1. Formation of the primitive atmosphere consisting of H_2O, NH_3, CH_4, and H_2.

$$\left.\begin{array}{l} H_2O \\ CH_4 \\ NH_3 \end{array}\right\} \longrightarrow \left\{\begin{array}{l} \text{Monosaccharides} \\ \text{Glycerin} \\ \text{Fatty acids} \\ \text{Amino acids} \\ \text{Pyrimidines} \\ \text{Purines} \end{array}\right.$$

Stage 2. Organic compound formation from the atmospheric gases in the ancient seas.

Monosaccharides + monosaccharides ⟶ Polysaccharides
Fatty acids + glycerin ⟶ Fats, lipids
Amino acids + amino acids ⟶ proteins ⟶ Enzymes
Pyrimidines ⎫
Purines ⎭ + sugars + phosphate ⟶ Nucleotides
Nucleotides + nucleotides ⟶ Nucleic acids

Stage 3. Formation of complex organic compounds and coacervates.

Figure 14-4. Summary of prebiotic chemosynthesis.

vided a rapid and efficient mechanism for synthesizing organic compounds, but it also brought about significant alterations of the physical environment. One of the by-products of photosynthetic reactions is free molecular oxygen (O_2), a highly reactive gas which combines readily with other substances. In the presence of large quantities of O_2 in the primitive atmosphere, most of the hydrogen, ammonia, and methane were oxidized to other substances. Oxygen probably reacted with methane to produce CO_2 and H_2O, which, with ammonia, converted it to N_2 and H_2O. Thus the new atmosphere consisted of water vapor, carbon dioxide, molecular nitrogen, and large

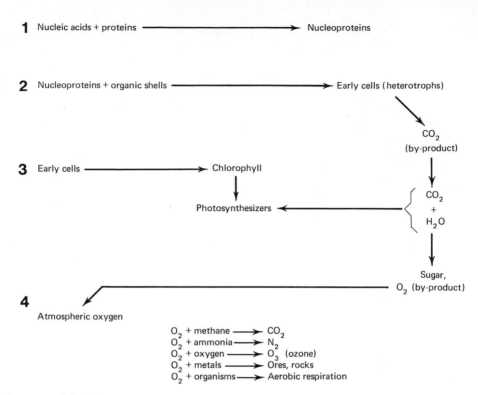

1 Nucleic acids + proteins ⟶ Nucleoproteins

2 Nucleoproteins + organic shells ⟶ Early cells (heterotrophs)

CO_2
(by-product)

CO_2
+
H_2O

3 Early cells ⟶ Chlorophyll

Photosynthesizers ⟵

Sugar,
O_2 (by-product)

4

Atmospheric oxygen

O_2 + methane ⟶ CO_2
O_2 + ammonia ⟶ N_2
O_2 + oxygen ⟶ O_3 (ozone)
O_2 + metals ⟶ Ores, rocks
O_2 + organisms ⟶ Aerobic respiration

Figure 14-5. Summary of the biological evolution of the first living forms.

quantities of free molecular oxygen itself. In addition, at higher altitudes, oxygen molecules combined with each other under the influence of high-energy radiation to form an ozone (O_3) layer. Such a layer absorbs much solar radiation and protects present-day plants and animals from high-energy radiation, which, even in small doses, is lethal. Once living forms arose, they altered the physical environment to such an extent that they destroyed the conditions that had made possible their origin.

The free molecular oxygen supplied through photosynthesis also brought about mutations which made possible a new, much more efficient form of respiration. These mutations provided the biochemical pathways of aerobic respiration, by which far more energy can be obtained from organic molecules than by aerobic respiration alone.

It should be noted that the most advanced organism so far postulated is considerably less complex than the simplest known living organism. There was a period of millions of years of evolution between the complex coacervates and the simplest virus, protozoan, or algal cells. In other words, life did not suddenly appear at a given time under a given set of environmental conditions; rather, life appeared slowly over millions of years (Figure 14-5). Furthermore, the account of the origin and early evolution of life just presented is hypothetical and at best tentative. Any theory regarding the origin of life is highly speculative and subject to continual and often drastic revision. At present, there is no crucial test to ascertain which theory is correct.

Viruses and the Origin of Life

When life began, it probably flickered uncertainly. A living form might have exhibited living characteristics, then reverted to a nonliving state, and, under the influence of certain stimuli, might have come to life again. For example, the earliest cells probably contained aggregates of nucleic acids suspended freely within the cell substance. If the delimiting membrane of some of these cells was ruptured, the nucleic acids could have escaped from the interior of the cell into the primitive seas. In this free state, a nucleic acid aggregate would have been a lifeless, inert chemical substance, for

it lacked its cellular machinery. If, by chance, such inert aggregates contacted living cells and entered them, they would become active again, because the host cell would provide the means for nucleic acid replication. Such nucleic acid aggregates, which escaped from one cell and existed in the inert state, and then entered another cell to become reactivated, may have been the ancestors of modern *viruses*, because viruses lead a similar existence.

Viruses are composed of two principal portions: a protein shell and an inner core of nucleic acid. Viruses, therefore, are not cells. The protein envelope provides protection for the nucleic acid core. The nucleic acid portion of a virus may be either DNA or RNA. Structurally, viruses may be brick-shaped, spherical, or needle-shaped. The needle-shaped virus shown in Figure 14-6 contains RNA and is called *tobacco mosaic virus* (*TMV*). Such

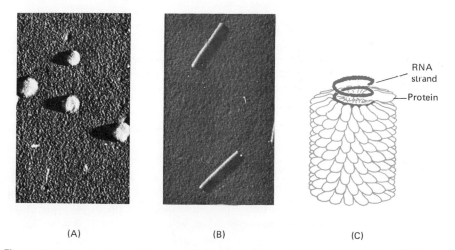

(A) (B) (C)

Figure 14-6. The structure of viruses. (**A**) Electron micrograph of a spherical-shaped influenza virus. ×53,000. (**B**) Electron micrograph of a needle-shaped tobacco mosaic virus. ×46,700. (**C**) Diagrammatic representation of the structure of TMV. (*A* and *B* courtesy of Dr. Robley C. Williams, Virus Laboratory, University of California, Berkeley. *C* from W. M. Stanley and E. G. Valens, *Viruses and the Nature of Life*. New York: E. P. Dutton & Co., Inc., 1961. By permission of the author and the publisher.)

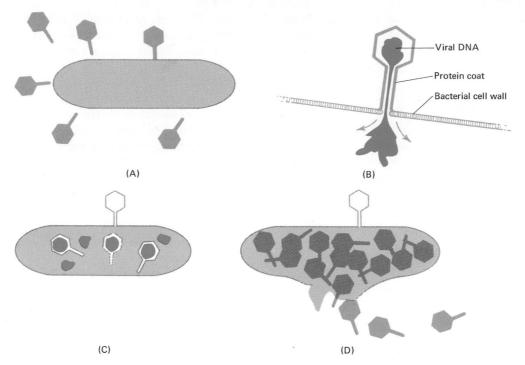

(A)

(B)

Viral DNA

Protein coat

Bacterial cell wall

(C)

(D)

Figure 14-7. Viral replication in a host cell. **(A)** Virus particles surround the host cell and one becomes attached to it. **(B)** The tail-like portion of the virus becomes attached to the cell wall of the host, the viral DNA is injected into the cell, and the protein coat remains outside. **(C)** Inside the host cell new virus DNA molecules are made and new viral protein coats are synthesized. **(D)** After the complete viruses are assembled, the host cell bursts (a phenomenon called lysis), and the viral particles are liberated. Inasmuch as all of the activities proceed at the expense of the host cell, the host dies. (From M. S. Gardiner and S. C. Flemister, *The Principles of General Biology,* 2nd edition. New York: The Macmillan Company, © 1967.)

a virus normally infects the leaves of tobacco plants. In general, viruses which contain a core of RNA attack plants, whereas those that contain DNA infect bacteria and animals, although exceptions do exist.

Viruses are quite inert in the free state because they have no metabolic machinery (enzymes) of their own, and, because they lack raw materials for synthesis, they cannot reproduce. They become reactivated only if they inject their own nucleic acid into a host cell. Inasmuch as this is a process of infection, viruses may be viewed as parasites. Once inside a host cell, the viral nucleic acid assumes genetic control over the chemical machinery of the host. The viral nucleic acid begins to synthesize viral enzymes and proteins, and within a period of a few minutes, the host cell bursts open and liberates a large number of newly formed virus particles. By infecting the host cell, the viral nucleic acid core has reproduced and formed new viral particles, each of which contains a protein coat (Figure 14-7). In time, this infection usually destroys the host cell.

PLANT EVOLUTION

The Nature of Evolution

Evolution, in its simplest and broadest context, means that since their origin, living forms have changed in an orderly succession. It also means that entire species have been altered, some have become extinct, and still others have arisen. The entire concept of evolution is one of biological change according to an orderly series of changing events. Evolution means continuous change, so certain implications necessarily follow. Inasmuch as life had to start somewhere, it is generally believed that life originated only once and, from this point on, species developed through change and the subsequent inheritance of change until the present time. A second implication to be drawn from modern evolutionary theory is that all life comes from life through inheritance. Finally, it can be implied that all life is related in an unbroken chain from the present to the time of its origin.

Most biologists agree that the first living organisms were very simple forms, and, from these, other organisms of greater complexity and diversification have evolved. Although in some groups of organisms processes leading to simplification have occurred, in general, the trend has been to ever-increasing structural complexity and specialization. Evolution implies, therefore, that modern living forms were not the first organisms to inhabit the earth and many that once flourished are no longer alive.

Data That Support the Concept of Evolution

Since the formulation of various evolutionary theories, scientists have sought to gather data to support or refute these theories. The lines of evidence in support of these theories are derived from various sources. A few of these will now be presented.

Comparative Morphological Data. The basic principle of classification is that of similarities and differences of structure. The science of structure, or morphology, provides some important evidence for the facts of evolution. It will be recalled from Chapter 13 that plants are classified into distinct groups based upon structural resemblances. The greater the structural similarities, the more closely related are the organisms. In general, structural similarity indicates common ancestry, and common ancestry implies close relationship. Conversely, pronounced structural differences usually suggest more distant relationships with a common ancestor ages ago. Consider the class Angiospermae. All of these plants are structurally similar in that they produce flowers. Subdivisions of this class exist with regard to the particular type of flower, number of cotyledons, arrangement of fibrovascular tissues, and other structural dissimilarities. Yet angiosperms are more closely related to each other than they are to plants which bear cones rather than flowers. Other structural similarities, such as type of leaf venation, wood anatomy, and seed structure, are also used as morphological evidences to show similarities and close ancestry.

Common ancestry provides a valid explanation for structural similarities. All plants bear some type of relationship to each other, either close or distant. They have *diverged,* or branched out, from ancestral groups of plants. Inasmuch as plants can be arranged into distinct groups from simple to more complex forms, such a classification at least suggests that evolution must have taken place.

Comparative Physiological and Biochemical Data. Biochemical and functional similarities among plants offer evidence for support of evolution, as does comparative morphology. Various chemical criteria have been used as certain taxonomic levels to show these relationships. For example, pigmentation and food reserves are among the principal characteristics used in determining divisional affinities among the algal groups. The presence of resins, such as turpentine, of the pines; the essential oil of the mints; and the latex found in spurges indicate close relationships among plants within these groups. In addition, specific proteins are characteristic of plant groups. The more nearly alike the proteins, the closer is the relationship among the plants. Some investigators have proposed an entire plant classification scheme based upon the similarities and differences of protein characteristics of various plant groups.

In addition to these specific biochemical relationships, general relationships among plants are evidenced by the universality of DNA, ATP, and cytochromes. Furthermore, many plants are related through photosynthesis and all exhibit some pattern of respiration. Thus both biochemical and physiological similarities indicate relationship.

Developmental Data. In many cases, the changes associated with the early development of organisms is an important source of evolutionary evidence. The developing organism often passes through a series of changes, some of which seem to repeat, at least partially, the adult traits of lower and presumably more ancient forms, even though the mature individuals are quite different. These changes are interpreted as a repetition, in very condensed form, of the stages through which the particular species has passed in its evolution from ancestral species. For example, the early stages of a moss gametophyte, the protonema, indicate the algal ancestry of the mosses. Also, the early fern gametophyte, the prothallus, is a filamentous structure resembling a filamentous green alga, suggesting an algal ancestry.

Data from Geographical Distribution. If any single kind of evidence for evolution is most con-

vincing, it is probably the lack of random distribution of plants. Many types of plants are widely distributed over the surface of the earth, but some species exhibit a particular geographic distribution. Moreover, different species of plants are found growing in areas which are geographically distant but which have similar environments. Therefore, climatic differences alone cannot explain plant distributions. Localized distribution can be explained on the assumption that certain plants evolved in the regions which they now inhabit and have been confined by various barriers such as mountain ranges and bodies of water. Plants in these isolated areas develop along divergent lines and become quite different from one another. The longer the period of isolation, the greater the differences among plants of the isolated regions as a result of differences in evolutionary patterns in the separated regions. Thus there are great differences among plants growing under similar environmental conditions but in widely separated and geographically distant portions of the earth. A prime example of plants found in distant regions grown under similar conditions but which have followed divergent evolutionary patterns are those of the desert areas of the southwestern United States and of western Africa. Although both regions are climatically alike (although geographically distant), numerous species of cacti are found in the American deserts and lacking in the African, and spurges are numerous in the African deserts but rare in those of the United States.

Data from Genetics and Domestication. The most convincing evidence that evolution is occurring would be observation of the origin of new forms and an analysis of the mechanisms which brought these new forms into existence. Such evidence has been provided by demonstrating that plants have a potential for change. Simple experiments in hybridization, both natural and artificial, frequently result in the production of new species.

Genetic evidence of evolution is also found in mutation, in which inherited variations give rise to distinct new species. Some biologists are of the opinion that mutation is the principal mechanism of evolution.

Mutations, caused by various agents, have been demonstrated in many groups of plants. Polyploids, such as many cultivated snapdragons, lilies, and petunias, are mutations. The nectarine is a mutant from a peach, and seedless navel oranges are mutant types of seed-producing oranges. It seems likely that spontaneous changes occur in all plant groups. In the domestication of plants, man has controlled the evolutionary process through hybridization and selection and has utilized conspicuous mutations in the development of new varieties. As a result, he has developed plants which exhibit increased yields, better quality, disease resistance, shorter growing seasons, and the ability to grow in diverse environments. Thus, from genetic studies, and through cultivation and domestication, there is ample evidence of the potentiality of living organisms for change, a major theme of evolution.

Fossil and Geological Data. A *fossil* may be defined as any recognizable organic structure or an impression of such a structure. At death, the bodies of plants undergo decomposition by heterotrophic organisms such as bacteria and fungi. In the process, complex organic compounds are broken down into inorganic substances which are incorporated into the various cycles of nature. Sometimes, however, under very special circumstances, some parts may be preserved with varying amounts of modification to form fossils. For this to occur, certain conditions must retard the activities of decay organisms.

For preservation to occur, rapid burial after death in a substance that retards decay and weathering is necessary. Accordingly, fossils are formed in a variety of ways. One such formation is *petrifaction*. Petrifaction occurs when living forests are covered by soil and rock, the weight of the cover pressing the plants. Then by a slow process, mineral-containing water seeps through the covering so that, molecule by molecule, the carbon compounds of the wood are replaced by mineral deposits and the wood is turned to stone. In some cases the intracellular material is replaced and the cell walls of the wood tissue are embedded in stone. As a result of the slow infiltration of the tissues by minerals such as silica, magnesium, or calcium, the cell structure of the original plants is often retained (Figure 14-8A).

Most fossils are *compressions,* which are plant parts flattened by vertical pressure of the overlying rocks. The usual process is for silt to cover the plant or plant parts that have either fallen into or live in water. Steady deposits of silt cover the dead plant and gradually build up a rock cover hundreds of feet thick. Although the plant is compressed, under certain conditions its shape and structure is preserved in such a way that the original plant may be studied.

Finally, *impressions* of various kinds are found which result from the burial of plant parts in soils which eventually harden into rock. Impressions, or prints, especially of leaves, are made in clay and are left when the actual structures are gone. Impressions thus preserve the shape, or external structure, of the plant part, but none of the original plant material remains (Figure 14-8B).

The study of plant fossils, called *paleobotany,* provides one of the most impressive evidences for evolution. To be sure, the fossil record of plants is incomplete. The fact that only certain kinds of organisms are suitable for fossil preservation and the fact that there is no orderly process of fossil formation eliminate any hope that the history of the evolution of *all* organisms can be obtained. Moreover, there are many gaps in the fossil record and therefore never an unbroken chain of evolutionary history. In spite of these inadequacies, paleontological investigations do substantiate the

(A)

(B)

Figure 14-8. Two types of plant fossils. **(A)** Petrifaction. **(B)** Impression. (Courtesy of Field Museum of Natural History, Chicago, Illinois.)

concept of evolution. The study of plant fossils shows that different kinds of fossils occur in various layers of rock, and from this observation it can be concluded that different types of plants have inhabited the earth at different periods during its history. In other words, the plants that have inhabited the earth have been changing constantly. Furthermore, the kinds of fossils found in each rock layer are generally morphologically similar to those found in adjacent strata. Such similarities indicate close relationships among plants and that the plants of various periods have developed from plants of earlier times. In addition, the fossil record indicates that the structure of plants has become more complex and specialized through the history of the earth and that many forms of plants have arisen, lived for a time, and then have become extinct.

Most logically, if fossil specimens are to be of any value, their age must be determined. Geologists have deduced that different geological strata are characterized by unique assemblages of fossils, and that particular strata could be identified by their fossil contents. The various sedimentary rock strata have been deposited in the order of the time of their formation. The oldest rock strata are those found at the bottom of any given sequence of layers and the more recently formed ones occur above these in a definite sequence. Geological dating procedures make it possible to establish the age of various fossils and strata within fairly accurate limits.

One such dating procedure involves the use of radioactive decomposition measurements of isotopes of elements found in rocks. Radioactive elements undergo transformations into more stable forms by the emission of subatomic particles such as neutrons or electrons. This transformation, or decay, continues until all the atoms of a radioactive element have been converted to a stable form. Atoms of ^{238}U (uranium), for example, decay to the stable form ^{206}Pb (lead). The essential feature of such decay is that it proceeds at a constant, measurable rate, and the rate differs for isotopes of different elements. Thus the *half-life,* or period of time required for one half of the atoms of the element to decay for ^{238}U, is 4.5 billion years and the half life of carbon (^{14}C to ^{12}C) is about 5760 years. By comparing the ratio of ^{238}U to ^{206}Pb or the ratio of ^{14}C to ^{12}C in a given rock, its age can be determined.

Inasmuch as uranium is relatively rare as a component of rocks and fossils, its value as a measuring device is somewhat limited. The uranium–lead method, however, has proved to be highly valuable, particularly in determining the age of the earth. For example, the oldest rocks known are estimated by this method to be 3.6 billion years old. The age of the earth is estimated to be between 4 and 5 billion years, probably 4.5 billion. A more satisfactory method of determining the age of fossils, especially recent ones, is *radiocarbon dating.* This procedure depends on the well-established theory that a small but constant proportion of the carbon utilized by living organisms is ^{14}C. This means that a material such as wood, which was at one time part of a living plant, can be analyzed for the amount of radioactive carbon present, thus determining its age. The accuracy of the radiocarbon method is limited to materials that are no more than about 50,000 years old.

By a variety of dating methods, geologists have devised a time scale which categorizes the geologic history of the earth. The scale, called a *geologic time scale,* is shown in Table 14-1. It can be seen that the geologic history of the earth is divided into five successive main divisions of time called *eras.* The three most recent eras are further sub-divided into a number of successive *periods.* The beginning and terminal dates of the eras and periods are designated in terms of major geologic events known to have occurred at those times. The transitions between eras, for example, were characterized by great upheavals consisting of mountain-building processes and severely fluctuating climates. Periods were terminated by minor geologic revolutions.

Relation of Plants to Geologic History

The various eras and periods of the history of the earth are determined not only by certain physical differences among rock formations but also by the types of fossils found in each layer. Accordingly, geologists postulate that rock-formation processes occur at about the same rate in the past as they do now and that the thicker the layer of rock, the more time required for its formation. Moreover, it is assumed that deposition of rocks has occurred in an orderly process, the older layer in a series being the bottom, and that the fossils found in a given stratum represent life of the period during which deposition of the rock occurred. Based on these data, it is implied that the oldest rocks contain fossils of only the most primitive organisms, while the younger formations contain fossils of more highly complex and later-evolved organisms.

Archeozoic ("Ancient Life") Era. Careful analysis of *Archeozoic* rocks indicates that they contain no fossils of either plants or animals. For the most part, Early Archeozoic rocks, which are principally igneous and metamorphic, show no evidence of fossilization, nor do they give any indication of life. In the later years of the Archeozoic, called the Upper Archeozoic, a few sedimentary rocks appeared. In these sediments are accumulations of graphite, and from their form and position some biologists believe that they represent carbon consolidations from the dead bodies of unicellular

TABLE 14-1. Geologic Time Scale

Eras and Duration (millions of years)	Periods (millions of years from present)		Advances of Plant Life as Shown by Fossil Record
Cenozoic (65)		Quaternary (0–2.5)	Increasing dominance of herbs; extinction of many trees.
		Tertiary (2.5–65)	Dwindling of forests; increasing restriction of plant distribution; rise of herbs.
			Worldwide distribution of modern forests; modernization of flowering plants.
Mesozoic (160)	Late	Cretaceous (65–136)	Angiosperms dominant; gymnosperms dwindling.
	Middle	Jurassic (136–190)	First known angiosperms, cycads and conifers dominant; primitive gymnosperms disappear; seed ferns disappear.
	Early	Triassic (190–225)	Increase of cycads, ginkgo, and conifers.
Paleozoic (345)	Late	Permian (225–280)	Dwindling of club mosses and horsetails; first cycads and conifers.
		Carboniferous (280–345)	Extensive coal formation in swamp forests; primitive gymnosperms.
	Middle	Devonian (345–395)	Early land plants; primitive lycopods, horsetails, ferns, and seed plants; primitive gymnosperms.
		Silurian (395–430)	First known land plants; marine algae dominant.
	Early	Ordovician (430–500)	Rise of land plants (?); marine algae dominant.
		Cambrian (500–570)	Some modern algal groups established.
Proterozoic (930)		Precambrian (570–1500)	Bacteria and simple algae (blue-green and green).
Archeozoic (3000)		(1500–4500)	No fossils found; all organisms probably unicellular or very simple.

organisms. Although such data do not constitute a true fossil record, they may be interpreted as an indication of life. The ancient Laurentian mountains of Canada and the innermost gorge of the Grand Canyon in Arizona exhibit exposed Archeozoic strata. Although the Archeozoic era furnishes no definitive evidence of the existence of life, it does supply presumptive evidence of life.

Proterozoic ("Very Early Life") Era. During the *Proterozoic* era, Archeozoic rocks were elevated, eroded, and practically obliterated. Eventually they were lowered below sea level and, on their remains, new strata were laid down. In these rocks, in addition to more graphite of an apparent organic origin and iron ore deposits believed to have been precipitated by bacterial action, there are great layers of limestone. These limestone reefs were formed during the Precambrian. Some of them show a characteristic pattern as though they had been deposited in concentric layers to form large limestone heads called *stromatolites*. Such limestone deposits have been compared with similar formations produced by present-day blue-green algal forms. The similarities are so striking that many biologists believe that Precambrian stromatolites were formed by the activities of blue-green algae that precipitated calcium carbonate from the surrounding waters.

Fossilized remains of organisms in Precambrian strata were discovered in 1954. These cellular remains of simple thallophytes, called *microfossils*, have been dated at about 2 billion years. The preserved specimens consisted of fungal hyphae, spores, and filamentous threads considered to be autotrophic blue-green algae not unlike modern forms (Figure 14-9). About 10 years later, Precambrian fossils bearing the structural remains of bacteria were found. Radioactive determinations dated the rocks containing the bacterial remains at about 3 billion years. This is the oldest record of life on earth. Projecting from this figure, it may be concluded that the origin of life probably dates back much farther than 3000 million years ago.

Fossil remains of the Precambrian, although scanty, do suggest certain conclusions. The most significant is that during Precambrian history both autotrophic and heterotrophic forms of life had evolution. Moreover, the existing organisms were very primitive in structure. Finally, it can be concluded that these organisms were exclusively aquatic forms confined for the most part to shallow seas.

Paleozoic Era. The *Paleozoic* era was the longest in the development of plants, and at its beginning there were no land plants. The climate was

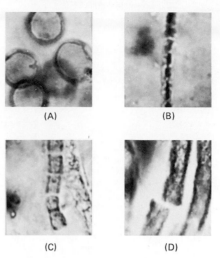

Figure 14-9. Microfossils from the late Precambrian of Central Australia dated between 700 and 900 million years. These fossil microorganisms bear close morphological resemblances to certain thallophytes. ×25,400. **(A)** Loosely clumped spheroidal bodies resembling the zygotes of *Chlamydomonas*. **(B)** Septate filament similar to blue-green algal forms. **(C)** *Ulothrix-like* multicellular filament. **(D)** Nonseptate filament resembling certain yellow-green algae. (Courtesy of E. S. Barghoorn, The Biological Laboratories, Harvard University.)

very hot and there was almost constant rain. The Early Paleozoic (Cambrian and Ordovician periods), which comprises about half of the entire Paleozoic era, furnishes enough evidence to make it clear that algae were already well established. These aquatic plants probably were descended from simpler Precambrian forms. Many of them resembled not only living blue-green and green algae of relatively simple structure but also more highly differentiated types similar to modern red and brown algae. The evolution of the major groups of algae is generally believed to have taken place in the warm seas of the Early Paleozoic, a suitable medium for the extensive development of primitive aquatic plants. Most of these groups probably arose in the seas before the emergence of land plants. During the Ordovician period the

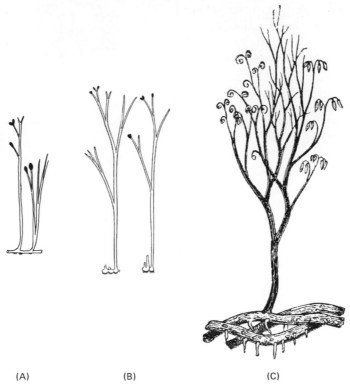

(A)   (B)   (C)

Figure 14-10. Fossil plants of the Middle Paleozoic. **(A)** *Rhynia.* **(B)** *Horneophyton.* **(C)** *Psilophyton.* (*A* from R. Kidston and W. H. Lang, "On Old Red Sandstone Plants Showing Structure From the Rhynie Chert Bed, Aberdeenshire," Pts. 1–5, *Transactions of the Royal Society of Edinburgh,* 1917–1921. *B* and *C* courtesy of Field Museum of Natural History, Chicago, Illinois.)

land was elevated and rains diminished so that various kinds of algae could emerge from the seas. Although no fossils document the transition from an aquatic to a terrestrial habitat, it is generally believed that such an evolution occurred among green algae. It is assumed that green algae developed features that prevented water loss and death by desiccation and later became established on land by the acquisition of structures which afforded anchorage and the maintenance of an erect position.

It is conceivable that during the Ordovician period, more than 400 million years ago, the first land plants evolved. The first fossil record of land plants, however, is not documented until the Silurian period. By the Middle Paleozoic (Silurian and Devonian periods), fossils of many land plants were formed. These plants represent the first true land plants in the fossil record. The first of these plants to evolve were small leafless and rootless forms which in their simple structure were algalike. Unlike the algae, however, their creeping stems and slender erect branches contained primitive vascular tissue. One such form, *Rhynia* (Figure 14-10A), possessed a rhizome with rhizoids, dichotomously upright branches, and terminal sporangia. The leafless stems contained a photosynthetic cortex and an epidermis containing stomata. Quite similar to this plant were *Horneophyton* (Figure 14-10B), which was somewhat smaller and featured tuberous rhizomes, and *Psilophyton* (Figure 14-10C), which contained emergences on its upright branches resembling leaves.

The early land plants apparently did not remain on earth very long, for they disappear from the fossil record before the end of the Devonian. Various types of these extinct plants, however, clearly show that they could have been ancestral to all other evolutionary lines of vascular plants. For example, *Asteroxylon* (Figure 14-11A) possessed rootlike branches arising from the rhizome, vascular tissue in the form of a star, numerous leaflike structures around a straight main stem with irregular dichotomies in the branch stems, and terminal sporangia. All these characteristics indicate trends in the evolution of club mosses. The fossil plant *Hyenia* (Figure 14-11B) possessed forked leaflike structures arranged in whorls and paired spo-

Figure 14-11. Early land plants that may have been ancestral to all other evolutionary lines of vascular plants. **(A)** *Asteroxylon.* **(B)** *Hyenia.* **(C)** *Protopteridium.* (A courtesy of Field Museum of Natural History, Chicago, Illinois. B from S. Leclerc and H. N. Andrews, Jr., "*Calamophytom bicephalum,* a New Species from the Middle Devonian of Belgium," *Annals of the Missouri Botanical Garden,* vol. 47, p. 1, 1960. C After Krausel and Weyland.)

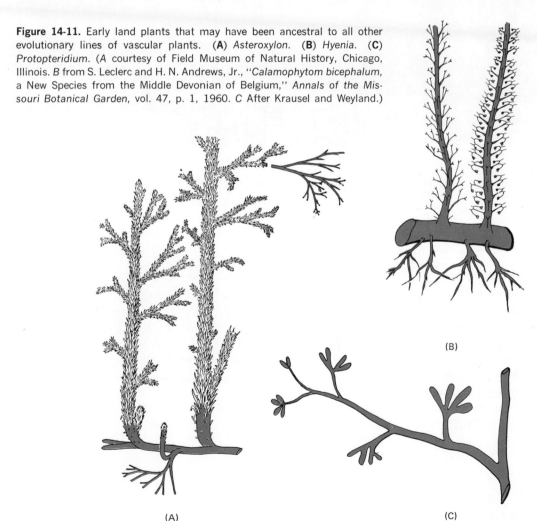

(A)

(B)

(C)

Figure 14-12. Early land plants of the Devonian Period. (Courtesy of Field Museum of Natural History, Chicago, Illinois.)

rangia borne on special stalks, features believed to be characteristic of ancestral horsetails. The pteropsid (fern) direction of evolution is indicated by *Protopteridium* (Figure 14-11C). This fossil plant contains terminal, flattened leaves with a webbing which may have foreshadowed the development of fern fronds. By the end of the Devonian, lycopsid (club mosses), sphenopsid (horsetails), and pteropsid (fern) lines were established and formed a conspicuous portion of the flora (Figure 14-12).

The Carboniferous period was characterized by a mild, moist climate. During this period most of the tracheophytes evolved during the Devonian reached their peak of abundance. Lycopsids and sphenopsids produced huge forests, and ferns grew as tall as trees. These plants, as well as primitive gymnosperms, formed the most extensive swamp flora ever assembled during the history of the earth (Figure 14-13). The death and transformation of these Carboniferous plants led to the formation of extensive coal deposits, which today constitute an important aspect of the economy.

The Permian period was characterized by mountain-building processes, a cooling of the climate, a draining of large swamp areas, and the extinction of many plants. Thus the first great changes in land flora occurred. By the end of the Permian, many species which had flourished during the

sperms, such as cycads, conifers, and *ginkgo,* arose and underwent rapid development. During the Middle Mesozoic (Jurassic period), cycad-like plants, the true cycads, and the ginkgos became the dominant plants. In addition, new groups of coniferous gymnosperms developed, many of which are still living. Some examples are cypresses, yews, pines, and redwoods. The first angiosperms also appeared in the Middle Mesozoic. Although the ancestry of the first angiosperms is unknown, it is generally believed that they evolved from a group of seed ferns. The dispersal and evolution of the angiosperms were so rapid that by the later part of the era (Cretaceous period), they had surpassed the declining conifers as the dominant plants of the earth. Common angiosperms which formed extensive Late Mesozoic forests include oak, fig, willow, magnolia, elm, palm, sassafras, maple, and holly.

Cenozoic Era. If the Mesozoic is characterized as the "age of gymnosperms," the *Cenozoic* might be called the "age of angiosperms," because the angiosperms established themselves as the dominant plants of the earth, while older groups were headed toward extinction. In the early part of the Cenozoic era (Early Tertiary period), which was characterized by a warm, humid climate, the dominant angiosperms were principally trees and shrubs (woody plants). During the Late Tertiary, the flora became restricted in distribution and segregated into climatic types as a general cooling occurred and specific climatic zones were established. As the Cenozoic era continued, the climate became cooler and drier, especially in the polar regions. With the coming of the Great Ice Age during the Quaternary period, many woody plants became extinct, some were restricted to specific geographic ranges, and there was a progressive increase in the numbers of herbaceous plants. In view of this rapid evolution of modern herbaceous

Carboniferous became extinct and were replaced by species that have survived to the present. The closing years of the Paleozoic era also show evidence of extensive gymnosperm forests, which, although now extinct, seem to have been the ancestors of more recent groups of gymnosperms.

Mesozoic Era. The *Mesozoic* era, often called the "age of gymnosperms," was characterized by a climate that became gradually more dry and progressively more severe. During the early portion (Triassic period) occurred a rapid development and diversification of gymnosperms and the disappearance of most of the seed ferns, giant horsetails, and lycopsids of the Paleozoic. The remaining club mosses, horsetails, and ferns resembled those of the present. The higher gymno-

Figure 14-13. A carboniferous swamp forest. (Courtesy of Field Museum of Natural History, Chicago, Illinois.)

angiosperms, it may be said that the most recently evolved angiosperm flora is herbaceous, a product of plant evolution to the present.

Patterns of Evolution

One of the principal aspects of plant evolution has been an increase in structural complexity and specialization, a trend referred to as *progressive evolution*. Although progressive evolution is the most conspicuous trend, all evolutionary advances have not been exclusively progressive. Many examples of *retrogressive evolution,* or the tendency toward decreasing complexity and simplification, may be cited. Among these are the development of structurally simplified types of flowers (grasses) from more complex types (lilylike ancestors); the possible derivation of higher fungus plants from algal ancestors, as a result of loss of chlorophyll;

and the processes of morphological degeneracy accompanying the transition from autotrophic to parasitic modes of nutrition (mistletoe and dodder). Although retrogressive evolution is common in certain plant groups, such changes represent only sidetracks along the main course of progressive evolution and have not appreciably altered the direction of evolutionary advances found among plants.

Another evolutionary trend exhibited by plants is that of *parallel evolution*. According to this pattern, different groups of plants follow common evolutionary patterns, although they are genetically related only very distantly. For example, six distantly related groups of plants independently acquired vessels in their vascular tissues. Vessels, although quite common among angiosperms, have also appeared in a small group of gymnosperms, ferns, horsetails, and club mosses. When parallel evolution proceeds in unrelated plant groups which have adapted to the same type of environment, to the extent that such groups are morphologically very similar, the phenomenon is called *convergent evolution*. This is illustrated by the development of spines and water-storing stems in both cacti and desert euphorbias, two plant groups not closely related to each other.

Evolutionary Mechanisms: The Causes

Biologists generally agree upon the fact of evolution. The basic difficulty arises from differing opinions concerning the mechanisms or causes of the process. Since the time of the Greek philosophers, attempts have been made to identify the causes, some more plausible than others. In this section, a brief historical development of evolutionary thought will be presented.

Lamarck's Theory. The first modern theory which purported to explain evolution was presented in 1809 by Lamarck, who was aware of the effects the environment produced and observed many instances of its operation. For example, he was impressed by the fact that if one plant is grown in a rich soil and another in a poor soil, the former grows larger and more vigorously. He believed that organisms possess a natural ability to adapt advantageously to the environment, so that their structures and functions, modified by the environment, could be passed on to succeeding generations. As part of this adaptation process, he was of the opinion that a particular structure or organ developed or regressed in response to the environment, depending on whether it was used or disused. The use of a structure increases its development to produce a variation which is transmitted to future generations. Conversely, lack of use results in diminution of a part until it disappears completely. The degree to which an organ is used or not used is determined largely by the environment. Lamarck believed that the effects of the environment are cumulative and that through many generations these cumulative effects gave rise to new species.

Although Lamarck's explanation was attractive and plausible in certain respects, experimental evidence was quickly amassed to refute his ideas. For example, it has been shown that acquired characteristics are not inherited if they do not effect genes. Although the environment does influence the development of an organism, this development is not inherited. Plants with the same genotype grown in different environments (light, water, soil, and mineral availability) show different growth patterns. Yet if seeds from both plants are germinated under similar environmental conditions, they will grow into plants which are essentially alike in terms of growth patterns, vigor, and metabolic activities. Thus it is the genotype alone that determines hereditary potentialities, although the environment may influence growth and developmental patterns. As a whole, the theory of Lamarck receives little support from the academic community.

Darwin's Theory. As a result of many years of careful study and observation, Darwin, in 1859, presented a theory of evolution which laid the groundwork for many modern theories. His theory is predicated on three principal observations. First, a given population of organisms is capable of producing far more offspring than can possibly survive under conditions of limited space and available metabolites necessary to support them. A single orchid flower may produce over 1 million seeds, and a single mushroom may disperse over 2 billion spores. Second, considering the problem of overpopulation, Darwin observed that, to survive, organisms compete with each other for sufficient space and available metabolites. This competition exists not only among organisms of different species but also among individuals of the same species. As a result, only a small proportion of the total offspring survive the competition because they have developed favorable variations. The unfit organisms, which comprise the great majority of individuals, perish because they have not adapted favorable variations for survival. Finally, Darwin noted that individuals of the same species are not exactly alike but differ from each other in varying degrees. Under a given set of environmental circumstances, successful competitors possess highly favorable variations which help them to adjust advantageously, whereas unsuccessful ones have less favorable variations and are thus at a disadvantage. The favorable mutations are transmitted to succeeding generations, further variations are produced, followed by competition and selection of the best adapted, and so on. In this way, according to Darwin, the continued operation of natural selection over long periods of time produces forms sufficiently different to be called new species.

Most biologists accept the Darwinian concepts of overproduction, competition, and variation which result in natural selection. Criticism of the theory stems from those points concerning the inheritance of variation and the degree to which natural selection can lead to significant changes in organisms. A population of a species often shows many variant forms. Some of these are the result of *environmental variations,* which are induced by environmental factors acting on the vegetative tissues of plants. Such variations are not heritable, because they are not under the influence of the genes in reproductive cells. Some variations, however, called *genetic variations,* are transmitted to offspring irrespective of environmental influences. Whereas environmental variations are not significant in evolution, genetic variations are an important aspect in the evolution of new species. In formulating his theory, Darwin failed to distinguish between heritable and nonheritable variations, and he assumed that nearly all variations are heritable. A given plant species contains a number of genetic variations by which its members are distinguishable from one another on the basis of features that have little or no adaptive significance. Such features are called *nonadaptive characters.* Darwin's theory of natural selection accounts for the evolution of *adaptive characters* but ignores the nonadaptive traits.

Another criticism of Darwin's theory is that it failed to distinguish between variations of small magnitude, commonly within the range of normal variability, and variations which are so large as to categorize organisms as new species. It was Darwin's view that evolutionary changes were brought about as a result of an infinite number of small, more or less continuous and cumulative variations. Critics of this view maintain that small variations could not appreciably effect evolutionary change, and they emphasize the importance of larger, discontinuous variations as the principal evolutionary mechanism.

De Vries's Theory. Darwin's ideas concerning evolution do not explain the mechanism of variations. He assumed that they occurred, and that, once present, natural selection became operative. Another attempt to explain the cause of evolutionary change was proposed by de Vries in 1901. His mutation theory held that new species of organisms are produced by sudden unexplained changes of considerable magnitude and that, once formed, the new species are distinct from the very first moment of their origin. Changes of such magnitude de Vries called *mutations,* which he considered the basic units of evolution on which natural selection may act. Whereas Darwin's concept of variations was that of continuity in an unbroken, gradually merging series, de Vries maintained that variations were discontinuous and discrete, unconnected by series of intermediate forms. It should be emphasized that de Vries offered no explanation for the causes of these mutations. He did, however, suggest the role of mutations in the evolutionary process.

Current Evolutionary Thought. To be sure, all the facts of evolution have not yet been explained. During the past 50 years, many data have been gathered concerning the physical and chemical basis of heredity, variation, and its inheritance. As more knowledge concerning the nature of the hereditary material is acquired, more facts of evolution will be explained. At present it appears that the mechanisms of evolution are the result of at least four operative phenomena. These are mutation, genetic recombination, natural selection, and reproductive isolation.

The medium of evolution is the population, and the raw materials of the evolutionary process are the inheritable variations which appear among individuals in a population. A population is a geographically localized group of organisms of the same species in which the individuals interbreed. All populations form a *gene pool,* a collective designation for all the genes of an interbreeding population. In the course of successive sexual generations, the gene pool of a population may become thoroughly shuffled and reshuffled. Evolution operates through the gene pools of populations. As a result of mutations and gene recombinations, some individuals featuring new hereditary variations may appear in a population. Some mutations, which arise as a result of internal activities of the organism and independent of the environment, are the result of alteration of the chromosome number or chromosome aberration during mitosis. Other mutations arise as a result of changes in genes. Still other variations may develop as a result of gene recombinations through independent assortment, crossing over, and random fertilization. Hybridization between species also enhances variation, provided the hybrids are fertile.

Such mutations and recombinations, after being introduced into a gene pool, are capable of spreading out to many or all members of the population. Whether or not this occurs depends on natural selection. Many of the variations, which are produced by mutations and recombined during sexual reproduction, will be less fitted to the existing environment and the variants will perish. Others, however, which may be better adapted, will survive and may completely replace the parental type. Thus there occurs a survival of the fittest mutations, which may be slowly accumulated and transformed into new species. Evolution operates, therefore, through the appearance of inheritable variations by mutation and sexual recombination, and by the spreading out of these variations through a population by natural selection.

In this context, environmental forces are con-

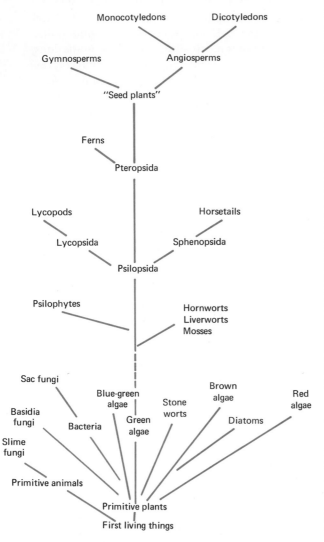

Monocotyledons   Dicotyledons

Gymnosperms   Angiosperms

"Seed plants"

Ferns

Pteropsida

Lycopods   Horsetails

Lycopsida   Sphenopsida

Psilopsida

Psilophytes   Hornworts
Liverworts
Mosses

Sac fungi   Brown
algae   Red
algae
Blue-green
algae   Stone
worts
Basidia
fungi   Green
algae   Diatoms
Bacteria
Slime
fungi

Primitive animals

Primitive plants
First living things

Figure 14-14. One possible arrangement for the evolutionary tree of the principal plant groups.

sidered to be directive rather than causative factors of evolutionary change. Variations are brought about by mutations, and the nature of the environmental circumstances determines which mutations will adapt and therefore survive. Thus the environment directs the course of evolutionary development; it does not cause evolution.

The evolution of a species begins with the differentiation of an existing population into two or more new ones along directions controlled by natural selection. If these new populations become isolated, so that interbreeding no longer exists, they become new species. Isolating mechanisms may be *spatial* (oceans, mountains), *temporal* (variations in flowering or reproductive cycles), or *genetic* (gamete incompatibility or sterility). When isolation occurs, variations within a population cannot spread out to other populations. Moreover, similar mutants are not likely to occur in different isolated populations, because their environments may be somewhat different. Prolonged periods of isolation may cause the original species to develop into two or more new ones because of the accumulation of new variations and the failure of the new genetic forms to interbreed. These new forms, through the agency of natural selection, may evolve into distinct species.

Course of Plant Evolution

The plants of today are but a small fraction of those which have existed in the past. Most have flourished and become extinct. Unfortunately, a complete picture of plant evolution is not possible, because there are many unfilled gaps in their evolutionary history. Botanists, however, have reached certain tentative conclusions concerning the probable path of evolution in the plant kingdom. Accordingly, an evolutionary tree has been designed to summarize these conclusions (Figure 14-14). It should be emphasized that the representation of plant evolution presented here is but one possible arrangement based upon the facts of plant relationships as they are now known. No doubt, as more evolutionary data become available, a more complete picture showing other relationships of plants will be developed.

The phylogenetic tree, derived principally from the morphology of fossils and living plants, represents the probable origins and relations of the main plant groups. From the first living things, the products of chemical evolution, both primitive plants and primitive animals probably evolved. The first primitive plants to evolve may have resembled modern-day bacteria. These bacteria-like organisms may have developed the various groups of algae and higher fungi. The lower fungi (Myxomycophyta) may have developed from a living form intermediate between primitive plants and primitive animals. It appears that the main line of evolution from the thallophytes to higher land plants is from the green algae (Chlorophyta) to primitive Psilopsida forms. From this line, the green algae are believed to have also given rise to the Bryophyta. Hornworts, liverworts, and mosses are believed to be terminal evolutionary groups, that is, ones which did not give rise to more complex forms. Much evidence suggests that the Psilopsida groups were the first true land plants and these, in turn, through certain morphological adaptations, led to the evolution of the Lycopsida, Sphenopsida, and Pteropsida. Most botanists feel that modern seed plants developed from a common ancient, extinct fern stock. The dicots probably developed earlier than the monocots, which are believed to have evolved from a dicotyledonous ancestor.

The phylogenetic tree of evolutionary development, although far from complete, emphasizes the fact that evolution is not a straight-line process but rather a complex branching pattern. Many of the branches represent terminal groups, some of which have become extinct, others of which

have remained unchanged to the present time and have not given rise to other groups (Bryophyta). Moreover, other branches indicate that some groups of plants which seem to have appeared in succession have not survived to the present (Psilopsida). Others have lived to the present unchanged (Cyanophyta). Nowhere on the earth can living representatives of a continuous evolutionary pattern be found.

Evolutionary Trends among Plants

In the course of plant evolution over millions of years, striking evolutionary changes, through adaptation and variation, have taken place. There is no way of determining what the first plants were really like, although hypotheses have been suggested to account for the origin and general structure of the first living forms. At a very early period these primitive forms probably developed into an extremely simple plant type, consisting of a single mass of protoplasm and displaying some resemblance to the most primitive plants now in existence. Given such a starting point, it is possible to identify the most important trends which have led to the production of the highly complex and diversified plants of today. Among the more obvious and more significant of these evolutionary trends are:

1. The gradual evolution of colonial and multicellular plants of increasing size and structural complexity from unicellular forms.

2. The progressive differentiation of protoplasm from a relatively uniform mass into a highly specialized complex leading to a division of labor within multicellular plant bodies.

3. The progressive change in reproduction from an asexual type to a sexual mode, which made possible variation in successive generations.

4. The origin and evolution of alternation of generations with the gradual dominance of the sporophyte generation.

5. The migration of plants from an aquatic to a terrestrial habitat.

6. The evolution of vascular, storage, strengthening, protective, and other tissues, which afforded a successful transition to land.

7. The differentiation of the multicellular plant body into vegetative and reproductive organs.

8. The origin of the seed habit as a result of retention of gametophytes and sporophytes produced from them within sporangial walls.

9. The decreasing dependence upon water for fertilization through the development of pollen tubes in seed-producing plants.

In the chapters that follow, these trends, as well as others, will be traced within the major plant groups. In this context, emphasis will be placed on form and function, a product of evolutionary tendencies.

CHAPTER IN RETROSPECT

The earth is probably between 4.5 and 5 billion years old and arose initially from a cosmic cloud along with other components of the solar system. Recent experiments relating to the possible origin of life have indicated that organic molecules, identical with those found in living systems, could be produced when conditions presumed to have existed on the primitive earth were simulated. The problem of organizing the various organic molecules into a structural pattern characteristic of even the simplest organism was apparently solved by the formation of more complex units called coercervates. Later steps in the pattern of chemical evolution included the development of nucleic acids, which possess the living property of self-replication. These complex aggregates containing nucleic acid may have been ancestral to modern viruses or other simple organisms which possess similar chemical characteristics. Once living organisms were established on the primitive earth they began to change in orderly succession by a process called evolution. Comparative morphology, genetics, fossil findings, biochemical and functional similarities, and other data offer evidence to support the concept of evolution. Evolution may proceed along four paths: (1) progressive (the most common), (2) retrogressive, (3) parallel, and (4) convergent. According to current evolutionary thought, the mechanism of evolution consists of mutation, genetic recombination, natural selection, and reproductive isolation.

Various geological eras (Paleozoic, Mesozoic, and Cenozoic) were characterized by plants which exhibited increasing degrees of complexity. However, evolutionary development is not a straight-line process but has a complex branching pattern. No present-day plant group can be traced backward to show a continuous evolutionary pattern. Some of the more obvious evolutionary trends include progressive differentiation of protoplasm, development of alternation of generations, migration of plants from an aquatic to a terrestrial habitat, and the origin of the seed habit.

Questions and Problems

1. What was the probable effect of photosynthesis on the gases of the primitive atmosphere? (Indicate your answer by the use of chemical equations.)
2. Discuss the composition of the atmosphere after a long period of photosynthesis by primitive autotrophs.
3. Outline the hypothetical course of chemical evolution in the formation of the first living organisms.

What evidence exists to either support or refute this hypothesis?

4. Briefly describe the concept of biological evolution in terms of supportive evidence. What is meant by (a) progressive evolution, (b) retrogressive evolution, (c) parallel evolution, and (d) convergent evolution?

5. List the dominant plants which characterize the various geological eras.

6. Contrast the evolutionary theories of Darwin, Lamarck, and de Vries. Which of these theories is most compatible with current evolutionary thought? Explain your answer.

7. Which evolutionary trends can you identify in the plants that characterized the various geological eras?

8. Indicate the various mechanisms by which evolution occurs. Which means do you believe accounts for the greatest amount of biological change? Give reasons for your opinion.

9. What is meant by radiocarbon dating? Can you name other means by which a geologic time scale may be established?

Suggested Supplementary Readings

Barghoorn, E. S., and S. A. Tyler, "Micro-Organisms from the Gunflint Chert," *Science*, vol. 147, pp. 563–577, 1965.

Crow, J. F., "Ionizing Radiation and Evolution," *Scientific American*, September 1959.

Delevoryas, T., *Morphology and Evolution of Fossil Plants.* New York: Holt, Rinehart and Winston, Inc., 1962.

Dobzhansky, T., "The Genetic Basis of Evolution," *Scientific American*, January 1950.

——, "The Present Evolution of Man," *Scientific American*, September 1960.

Dodson, E. O., *Evolution, Process and Product.* New York: Reinhold Publishing Corporation, 1960.

Eiseley, L. C., "Charles Darwin," *Scientific American*, February 1956.

Heiser, C. B., and D. M. Smith, "The Origin of *Helianthus multiflorus*," *American Journal of Botany*, vol. 47, pp. 860–865, 1960.

Kulp, J. L., "Geologic Time Scale," *Science*, vol. 133, pp. 1105–1114, 1961.

Laurence, W. L., "Seaweed—Man's Common Ancestor?", *Science Digest*, vol. 43, pp. 6–8, 1958.

Lessing, L. P., "Coal," *Scientific American*, July 1955.

Libby, W. F., "Radiocarbon Dating," *Endeavour*, vol. 13, pp. 5–16, 1954.

Ryan, F. J., "Evolution Observed," *Scientific American*, October 1953.

Sporne, K. R., "On the Phylogenetic Classification of Plants," *American Journal of Botany*, vol. 46, pp. 385–394, 1959.

Stebbins, G. L., *Variation and Evolution in Plants.* New York: Columbia University Press, 1950.

Tyler, S. A., and E. S. Barghoorn, "Occurrence of Structurally Preserved Plants in Pre-Cambrian Rocks of the Canadian Shield," *Science*, vol. 119, pp. 606–608, 1954.

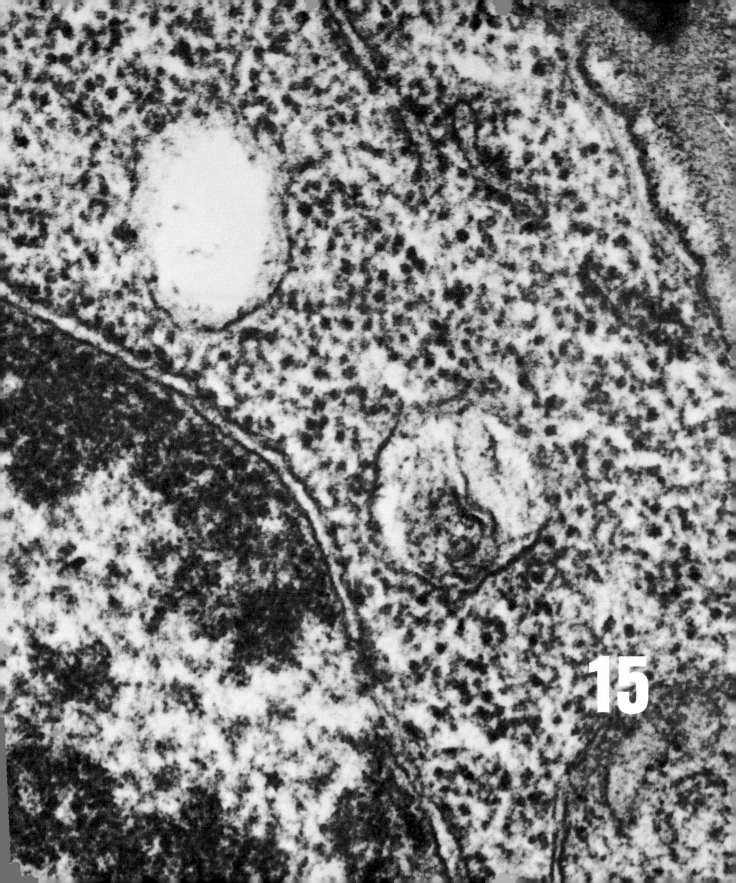

15
Monera: Fission Plants

INTRODUCTION

Within the world of living things, organisms exhibit enormous diversity in form and function. For many years, it has been customary to divide all organisms into two large groups, the animal kingdom and the plant kingdom, and most of the larger, complex forms are clearly recognizable as belonging to one of the two kingdoms. For example, there is little difficulty in deciding whether advanced organisms such as oak trees and dogs are plants or animals. Problems arise, however, when an attempt is made to classify primitive organisms, that is, those closely related to ancestral types which gave rise to more advanced forms. Such primitive forms have neither distinctly plant features nor distinctly animal features but features from both kingdoms simultaneously. Most likely, the first living forms possessed neither plantlike nor animal-like characteristics. The distinction of being a plant or an animal appears to have been acquired in separate lines of descent after a relatively long period of evolutionary history.

THE INTERRELATIONS OF ALL LIVING THINGS

The scheme of classification presented in Chapter 13 is based upon a natural system in which plants are treated as a kingdom separate from animals. Such a classification shows relationships among plant groups but not their interrelations with animals or ancestral types. In this context,

then, such a scheme is somewhat generalized, although convenient. This has led a number of taxonomists to establish alternative classifications. One of these recognizes four principal kingdoms of living organisms: *Monera, Protista, Metaphyta,* and *Metazoa* (Figure 15-1). The first two are composed of primitive organisms and the last two of more advanced types.

The Monera and Protista, which go back farthest in evolutionary history, are believed to have been descended, independently, from the first living forms. Monera include all organisms in which the cells do not contain nuclear membranes and therefore no true nuclei. In addition, all are basically unicellular. Modern representatives of this group include members of the *Division Schizophyta* (fission plants)—the bacteria (*Schizomycetes*) and the blue-green algae (*Schizophyceae*).

The Protista kingdom includes both plantlike and animal-like organisms, all of which possess true nuclei. Some living Protista are unicellular, others are multicellular. In addition to acquiring nuclei in their evolutionary development, Protista also evolved true chromosomes and many evolved chloroplasts, autotrophic and heterotrophic modes of nutrition, and ameboid and flagellary locomotion. As a result of these evolutionary adaptations, some modern Protista are both plantlike and animal-like, others show distinctive animal traits, and still others exhibit plant characters. The four major modern Protista groups are the fungi, the algae (except the blue-green), the slime molds, and the protozoa.

It is generally believed that the Metaphyta and Metazoa were derived from ancient Protista and the Metaphyta were descended from an ancient line of green algae. The Metaphyta includes higher multicellular plants which have complicated life cycles and produce embryos. Most are photosynthetic. Included in this kingdom are members of the *Division Bryophyta* and *Division*

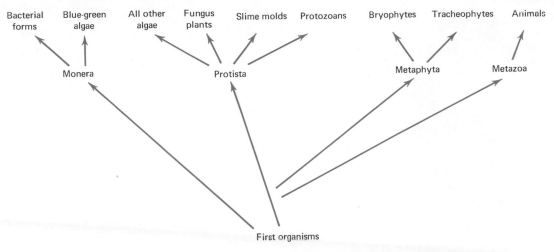

Figure 15-1. The possible interrelations of all living forms. According to this tentative arrange-ment, four principal kingdoms are recognized. Protozoa and all Metazoa are regarded as animals and all other groups are considered to be plantlike. It should be emphasized, how-ever, that certain of the plant groups also possess distinctively animal-like characteristics. Typical of these are the slime molds and euglenoids.

Tracheophyta. Metazoa are multicellular, non-photosynthetic organisms that are obviously ani-mals and are usually motile.

MONERA (DIVISION SCHIZOPHYTA)— THE FISSION PLANTS: Bacteria and Blue-Green Algae

Living Monera are descendants of possibly the most ancient and most primitive types of living organisms. Morphologically they are distinguished by the absence of many features found in the cells of other organisms (Figure 15-2). For example, a typical cell of the Monera lacks an organized nuclear body with a well-defined nuclear mem-brane. The DNA molecules, which contain the genetic materials, are not found as part of chromo-somes as in other organisms but assume the form of one or more irregular clumps. The cytoplasm

is characterized by the absence of vacuoles and an inability to undergo cyclosis. Photosynthetic monerans (blue-green algae) do not contain orga-nized photosynthetic units (chloroplasts) but struc-tures resembling a single layer of a single granum called *chromatophores,* which are dispersed through-out the cytoplasm.

Living representatives of the Monera comprise two classes: the *Schizomycetes* (bacteria) and the *Schizophyceae* (blue-green algae), both of which belong to the *Division Schizophyta* (fission plants). Although the exact evolutionary relationship of the bacteria to blue-green algae is somewhat obscure, certain data suggest that the division as a whole is a naturally related category.

Class Schizomycetes (Bacteria)

Form and Function. Bacteria are probably the simplest and smallest living organisms possessing cellular organization. Individual bacteria are visi-

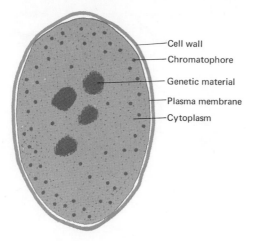

Cell wall
Chromatophore
Genetic material
Plasma membrane
Cytoplasm

Figure 15-2. Diagrammatic representation of a typical moneran cell. In addition to the absence of distinct nuclear bodies delimited by membranes and cellular organelles, these cells show no evidence of cyclosis or distinct vacuoles.

ble only with the aid of a microscope and are usually less than 5 μ in length and may be as small as 0.5 μ in width. Inasmuch as bacteria exhibit few conspicuous morphological features, especially the unicellular forms, the most distinguishable characteristics visible with light microscopy are cell shape and size. The commonly observed forms are cocci (spherical), bacilli (rod-shaped), and spirilla (helically shaped) (Figure 15-3A). The first two types may also be joined to form colonies or filaments, the larger colonies being visible to the unaided eye. The various morphological forms which bacteria assume are often useful in identifying a particular species. Other data, however, are generally necessary for identification. These include bacterial reactions to various stains, the type of nutrient material most favorable to growth, metabolic wastes, and the color and form of colonies.

Structures common to many bacterial species are shown in the composite diagram in Figure 15-4

of a bacillus type. In most bacterial cells, a rigid cell wall surrounds the protoplast. These walls differ from those of higher plants because their rigidity is derived from the presence of substances called *mucopeptides* (amino acid and amino sugar molecules) and not cellulose or related compounds. The effectiveness of penicillin as an antibiotic is in part related to its ability to prevent the formation of mucopeptides and hence to interfere with bacterial multiplication. Very often, a slime layer of varying thickness may accumulate on the outer surface of the cell wall. This gelatinous covering, which consists of polysaccharides and organic acids, is called a *capsule,* and its presence seems to be related to the ability of a bacterium to cause disease and resist the defenses of the host organism.

The protoplast of bacterial cells is surrounded by a plasma membrane which, as in other cells, is selectively permeable. Like other Monera cells, there is no organized nucleus or delimiting nuclear membrane. The genetic material occurs as dispersed clumps, and no vacuoles are present, although cytoplasmic granules and droplets consisting of various polysaccharides, proteins, and lipids are present.

All known spirilla and many bacilli possess flagella and are motile. Many bacilli also produce *endospores,* which are small sporelike structures formed within the parent cell wall (Figure 15-5). They may be spherical, ellipsoid, or oval in shape and may occupy a terminal or central position inside the parent cell. An endospore forms when a portion of the protoplast containing some of the cytoplasm and genetic material is enclosed by a membrane or wall inside the cell. Eventually, the part of the cell external to the endospore degenerates. The endospore is extremely resistant to heat, chemicals, desiccation, and other environmental adversities. Under favorable conditions, the endospore germinates (wall ruptures) and it gives rise

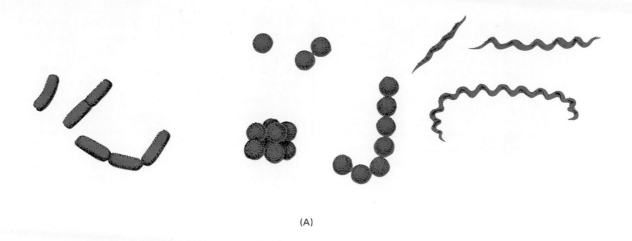

(A)

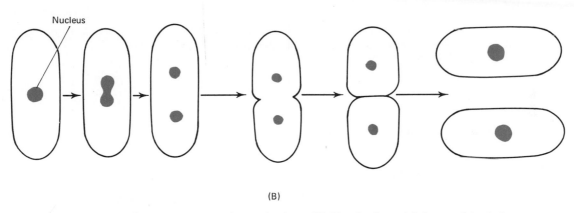

Nucleus

(B)

Figure 15-3. Bacterial forms and reproduction. **(A)** The fundamental forms of bacteria: bacillus (*left*), coccus (*center*), and spirillum (*right*). **(B)** Asexual reproduction of a single bacterium by simple fission.

to a single vegetative cell. There is no increase or decrease in the number of individuals during endospore formation. Very few disease-causing bacteria produce endospores.

Reproduction in most bacteria is by simple **binary fission** (Figure 15-3B). In the process the nuclear material divides, a cross wall develops from the periphery inward, and the newly formed cells separate. Under optimum growth conditions, fission can occur in some species as frequently as once every 20 minutes, and under these conditions enormous masses weighing tons could be produced. In nature, however, such rapid multiplication is restricted by competition, availability of

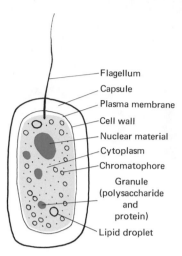

Figure 15-4. The structure of a typical bacterium.

nutrients, accumulation of toxic metabolic by-products, and by variations in moisture and temperature.

Some bacteria engage in processes that give end results similar to the sexual reproductive activities of other plant groups. These processes involve a unidirectional transfer of genetic material resulting in genetic recombination (Figure 15-6). One such process is *conjugation*, in which a donor cell transfers a portion of its genetic material to a recipient cell (Figure 15-6A). The descendents of the recipient cell often possess genetic material that differs from the original preconjugation parental type. A second type of recombination is called *transformation*, in which a living recipient cell picks up DNA fragments that have been released into the medium by a dead donor cell (Figure 15-6B). In the process donor cells pass certain traits to recipient cells which appear in subsequent generations. A third example of recombination, termed *transduction*, involves bacterial viruses called *bacteriophages* (Figure 15-6C). This process consists of the passage of DNA fragments from one bacterial cell (donor) to another (recipient) by virus particles. As a result, the newly acquired characteristics appear in the offspring.

Most bacteria are heterotrophic, living as either

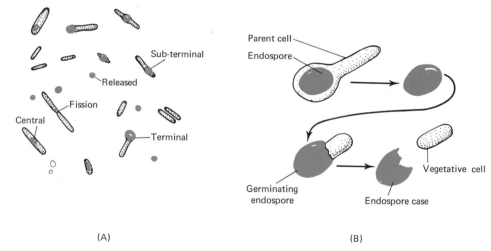

(A)   (B)

Figure 15-5. Endospore formation and germination. (A) Various sizes and shapes of bacterial endospores showing their relative positions within parent cells. (B) Endospore germination following release from the parent cell.

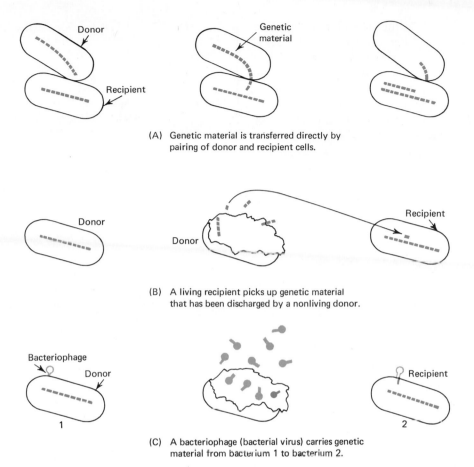

(A) Genetic material is transferred directly by pairing of donor and recipient cells.

(B) A living recipient picks up genetic material that has been discharged by a nonliving donor.

(C) A bacteriophage (bacterial virus) carries genetic material from bacterium 1 to bacterium 2.

Figure 15-6. Types of genetic recombination demonstrated in certain bacterial forms. **(A)** Conjugation. **(B)** Transformation. **(C)** Transduction.

saprophytes or parasites, and the majority of these are *aerobic;* that is, they require molecular oxygen for respiration. However, some obtain their energy through **fermentation** only and are killed in the presence of molecular oxygen. These bacteria are called *obligate anaerobes.* Some bacterial species are autotrophic and thus capable of synthesizing their own organic compounds. Autotrophic bacteria may be separated into two groups: *chemosynthetic*

autotrophs, which oxidize inorganic compounds (ammonia, nitrate, sulfur, hydrogen, or ferrous iron) and trap the released energy for synthesis, and *photosynthetic autotrophs,* which utilize light energy for their synthetic reactions.

Among the photosynthetic bacteria are the *purple sulfur bacteria* and the *green sulfur bacteria.* During photosynthesis both groups utilize hydrogen sulfide (H_2S) instead of water as a hydrogen source and liberate sulfur instead of oxygen. In these respects they differ from higher photosynthetic autotrophs. They also differ from higher plants in terms of the type of photosynthetic pigment present. In higher plants, the principal pigment is chlorophyll *a*. The purple sulfur bacteria contain a special type of chlorophyll called *bacteriochlorophyll*, whereas the green sulfur bacteria have other special types of chlorophyll (*chlorobium chlorophyll*) which differ somewhat from bacteriochlorophyll. The general equation for photosynthesis of both these types of bacteria may be represented as follows:

$$CO_2 + H_2S + pigment \xrightarrow{\text{light}} photosynthate + S$$

Classification and Importance of Bacteria. Among all living forms, the taxonomic subclassification of the bacteria is at best extremely vague and tentative. This stems from the fact that nonmorphological characteristics are used to identify bacterial species and that there is a great deal of uncertainty concerning the value of these characters in determining relationships. Nevertheless, a number of orders are recognized by bacteriologists (Table 15-1).

Order Pseudomonadales (*Pseudomonads*). Bacteria of the order *Pseudomonadales* are probably the most primitive organisms on the earth, inhabiting freshwater or marine habitats. Morphologically, this order contains members representing the three basic bacterial forms and most are locomoted by flagella. Some of the more important pseudomonads are the photosynthetic and chemosynthetic bacteria and many heterotrophic species. Included among the chemosynthetic Pseudomonadales are the nitrifying bacteria, which oxidize ammonia to nitrites and nitrites to nitrates (*Nitrosomonas* and *Nitrobacter*). It will be recalled

TABLE 15-1. Classification of the Bacteria

Kingdom Monera
  Division Schizophyta (fission plants)
    Class Schizomycetes (bacteria). About 1600 species.
      Order Pseudomonadales (pseudomonads). *Nitrobacter, Nitrosomonas, Pseudomonas.*
      Order Chlamydobacteriales (sheathed bacteria). *Sphaerotilus, Crenothrix.*
      Order Hyphomicrobiales (budding bacteria). *Hyphomicrobium, Rhodomicrobium.*
      Order Eubacteriales (true bacteria). *Azotobacter, Rhizobium, Clostridium, Salmonella, Escherichia, Bacillus, Streptococcus, Staphlococcus.*
      Order Actinomycetales (branching bacteria). *Mycobacterium, Streptomyces.*
      Order Caryophanales (filamentous bacteria). *Caryophanon.*
      Order Beggiatoales (gliding bacteria). *Beggiatoa, Thiothrix.*
      Order Myxobacteriales (slime bacteria). *Myxococcus, Chondromyces.*
      Order Spirochaetales (spiral bacteria). *Treponema, Borrelia.*
      Order Mycoplasmatales (pleuropneumonia organisms).
      Order Rickettsiales (rickettsiae). *Rickettsiae.*
      Order Virales (viruses). Although placed here for convenience, the viruses cannot be considered as closely related to bacteria.

that both of these genera are important components of the nitrogen cycle (Chapter 7).

Order Eubacteriales (*True Bacteria*). The *Eubacteriales* constitute the largest and most comprehensively studied group and the major portion of the introductory discussion concerning form and reproduction is based on these studies. Many eubacteria are abundant in soil, water, sewage, decaying plant and animal materials, foods, and the atmosphere. Others are found living parasitically in the bodies of animals and plants. As a group, these bacteria are of considerable economic importance because of their beneficial or harmful activities.

Among the Eubacteriales are several important genera capable of nitrogen fixation, an important activity of the nitrogen cycle. These include *Azotobacter, Rhizobium,* and *Clostridium.* The species of *Azotobacter* and *Clostridium* are free-living organisms common in the soil. Species of *Rhizobium,* by contrast, are symbiotic nitrogen-fixing bacteria found in the root nodules of leguminous plants.

Decay is also an important activity of several eubacteria. By decomposition, these bacteria break down complex organic molecules of dead plants and animals into simpler inorganic substances. As a result the elements are released to the environment and are incorporated into the various natural cycles, where they may once again be returned to living organisms as metabolites for growth and development.

A large bacterial flora, including many eubacteria, also inhabit the intestines of animals. In this regard some are potential pathogens, whereas others are necessary for the well-being of the host. *Escherichia coli* is a common nonpathogenic species that is essential for the proper functioning of the digestive tract of man. *Salmonella typhosa,* which often spreads through fecal contamination of water supplies, causes typhoid fever. *Streptococcus pyogenes* is the causative agent in scarlet fever, sore throat, and tonsilitis.

Some Eubacteriales are important in food processing and food spoilage. Butter, cheese, yogurt, and sauerkraut are examples of foods in which processing involves bacterial activity. *Lactobacillus, Streptococcus,* and *Leuconostoc* genera of bacteria are important in such processing. These and other

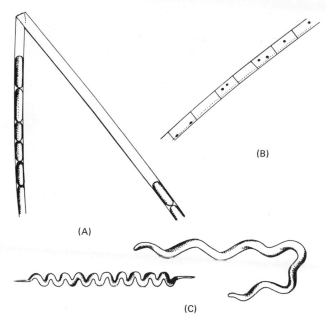

Figure 15-7. Diversity of bacterial forms. **(A)** Order Chlamydobacteriales: *Sphaerotilus natans* showing cells inside sheath. **(B)** Order Beggeatales: *Beggiatoa alba* showing details of a single filament. **(C)** Order Spirochaetales: *Treponema (left), Borrelia (right).*

common bacteria also cause the souring and spoilage of milk. Species of *Clostridium* are often responsible for tetanus (lockjaw) and botulism.

Order Chlamydobacteriales (Sheathed Bacteria). The *Chlamydobacteriales* are similar in structure to the true bacteria. Many cells, however, are bounded by a cylindrical sheath, forming a long filament (Figure 15-7). These bacteria are aquatic organisms, some of which are responsible for the deposition of bog iron ore. One genus, *Sphaerotilus* (Figure 15-7A), commonly found in organic matter such as sewage, is often used as an indicator of pollution.

Order Beggiatoales (Gliding Bacteria). The *Beggiatoales* are flexible, gliding cells which form filamentous colonies. Two common genera, *Beggiatoa* (Figure 15-7B) and *Thiothrix,* are often desig-

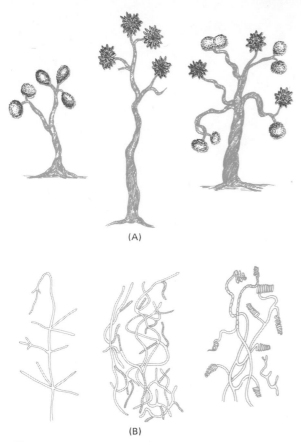

(A)

(B)

Figure 15-8. Diversity of bacterial forms. **(A)** Order Myxobacteriales. Various types of fruiting bodies. **(B)** Order Actinomycetales. A few representative forms showing branched coenocytic filaments.

nated as colorless blue-green algae because of their resemblance to the filamentous blue-green alga *Oscillatoria.* Both are chemosynthetic autotrophs that oxidize sulfur compounds in their metabolism, as do the purple sulfur bacteria. Filaments of *Beggiatoa* are free-floating and capable of gliding and flexing movements identical with the movements of *Oscillatoria. Beggiatoa* and *Oscillatoria* are also similar in the structure of the colonies and the manner in which they are formed. The principal difference between these genera is that the

former is nonphotosynthetic, whereas the latter is a photosynthetic autotroph. So many similarities exist between *Beggiatoa* and *Oscillatoria* that genera such as *Beggiatoa* are often regarded as colorless blue-green algae that have evolved from pigmented algae by the loss of pigments.

Order Spirochaetales (Spiral Bacteria). Members of the *Spirochaetales* are unicellular forms in which the cells are long, slender, helically coiled, and motile. Some reach lengths up to 0.5 mm. The cell walls, which are very delicate and flexible, are surrounded by a bundle of spirally wound filaments. Locomotion is probably caused by the flexing of these filaments. One of the best-known spirochetes is *Treponema pallidum* (Figure 15-7C, left), which causes syphilis. Another form, *Borrelia vincenti* (Figure 15-7D, right), is associated with trench mouth.

Order Myxobacteriales (Slime Bacteria). Organisms of the *Myxobacteriales* occur as single rod-shaped forms with thin walls. No flagella are present, but the organisms are capable of either individual or mass gliding movements on slimy materials which they secrete. The most remarkable feature of the slime bacteria is their production of communal fruiting bodies (Figure 15-8A). In the formation of these fruiting bodies many cells migrate into a mass, stop moving, become embedded in secreted slime, and the entire aggregate forms a fruiting body. This behavior closely resembles that of certain slime molds. Myxobacteria are abundant in certain types of soils, rotting vegetation, and marine habitats. Some are capable of killing other bacteria and fungi; others decompose cellulose.

Order Actinomycetales (Branching Bacteria). The *Actinomycetales* typically form branched *coenocytic* (many nuclear bodies and no cross walls) filaments of hyphae superficially resembling those found in many true fungi (Figure 15-8B). Reproduction is through fragmentation, budding, fission, conidia, or by motile sporangiophores. Some genera are quite familiar through the diseases they cause. *Mycobacterium tuberculosis,* for example, is the causative agent in tuberculosis, and *Mycobacterium leprae* causes leprosy. Other genera produce extremely important antibiotics, such as streptomycin, aureomycin, and actinomycin.

Class Schizophyceae (Blue-Green Algae)

Form and Function. The blue-green algae are unicellular, colonial, or filamentous organisms occupying a variety of habitats. They are distributed in fresh and marine water and in moist subaerial environments throughout the world. Extensive growths of blue-green algae frequently occur in areas which are subjected to intermittent wetting, such as tree bark, flowerpots, tidal banks, stream banks, and the periphery of pools, ditches, and bodies of stagnant water. Some blue-greens grow on snow in the cold of arctic pools; others are found in hot springs where the temperature may exceed 85°C. Such species are responsible for the characteristic color around Old Faithful in Yellowstone National Park. Despite the designation blue-green algae for this group, some in fact consist of a variety of shades, such as black, purple, red, yellow, green, blue, and intermediates of these. The Red Sea, for example, is said to have been named because of the abundance of *Trichodesmium,* a red blue-green algal form.

The protoplast of a typical blue-green algal cell is relatively simple and shows very little differentiation (Figure 15-9). It consists of an inner colorless portion and an outer pigmented region. Conspicuously absent are vacuoles, mitochondria, endoplasmic reticula, Golgi bodies, and organized nuclei. The term **procaryotic** is used to distinguish cells which lack organized nuclei and nuclear membranes. The nuclear material, which occurs as a loose network, is generally located in the central region of the cell. The outer cytoplasm contains chromatophores which are similar to

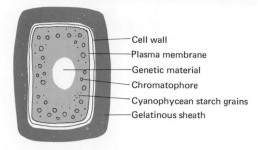

Figure 15-9. Diagrammatic representation of a typical blue-green algal cell.

those of the photosynthetic bacteria. Thus there are no chloroplasts. The photosynthetic pigments consist of chlorophyll *a,* the same pigment found in higher plants, carotene, various xanthophylls, *phycocyanin* (blue pigment), and *phycoerythrin* (red pigment). The last two are unique cyanophyte pigments. Numerous inclusions also occupy the outer cytoplasm. Among these are food reserves stored as carbohydrates (*cyanophyte starch*) and proteins (*cyanophycin granules*). Cyanophyte starch, different from the starch found in higher plants, is also unique among the blue-green algae.

The cell walls differ from those of most bacteria in that they contain cellulose and frequently pectin. Sometimes a prominent outer gelatinous *sheath* is also secreted by the cell. There are no flagellated cells, either vegetative or reproductive. A few filamentous forms lacking an encapsulating sheath, however, seem to undergo some movement. The movement may be simple gliding or combined with rotation around a longitudinal axis somewhat similar to that of the Beggiatoales. Cell division is by fission, as in bacteria. Reproduction also frequently occurs by fragmentation of filaments. There is now some evidence, similar to that discovered in certain bacteria, that genetic recombination apparently occurs, as shown in changes in the resistance of some blue-green algae

TABLE 15-2. Classification of the Blue-Green Algae

Kingdom Monera.
Division Schizophyta (fission plants).
Class Schizophyceae (blue-green algae). About 1500 species.
Order Chroococcales. *Gloeocapsa, Chroococcus.*
Order Chamaesiphonales. *Chamaesiphon.*
Order Oscillatoriales. *Nostoc, Oscillatoria.*

(*Anacystis nidulans*) to the antibiotics penicillin and streptomycin.

Classification and Diversity. Similar to bacteria, blue-green algae are basically unicellular and the cells frequently do not separate after division. As a result, loose associations of colonies or tightly joined filaments may be formed. Inasmuch as these algae lack cellular differentiation and are simple in form, the arrangement of cells in an aggregate and the manner in which newly produced cells are added to the aggregate are used to distinguish higher levels of classification (Table 15-2).

The simplest members of the division may be considered as unicellular free-floating forms which, upon division, produce two new individuals. Such unicellular forms often produce loose colonies held together by the gelatinous slime sheaths which they secrete. After cell division, the sheath may not rupture and daughter cells produced by successive divisions become associated and are retained within a common sheath. From such isolated individuals or indefinite associations, colony formation and filament formation may be viewed as developing along two different lines. One line has led to the evolution of colonial forms and the other to the filamentous types, both having arisen from the same primitive unicellular stock. Based upon these relationships, three orders of blue-green algae are recognized: *Chroococcales, Chamaesiphonales,* and *Oscillatoriales* (*Hormogonales*). The first two are composed of solitary or colonial

forms, and the third consists of distinctly filamentous types.

Order Chroococcales. All the Chroococcales are unicellular, but the cells often form colonies held together by a common gelatinous sheath. Two genera, *Gloeocapsa* and *Chroococcus,* form simple types of colonies in which the cells aggregate in a rather definite, spherical mass (Figure 15-10A). In both, the plane of division is random. In *Merismopeida,* the cells divide regularly in only two planes and the colony appears as a flat, single-layered sheet of cells (Figure 15-10B). If regular division occurs in three planes as in *Eucapsis,* a cubical colony is produced (Figure 15-10C).

Order Chamaesiphonales. Members of the Chamaesiphonales, like the Chroococcales, are either unicellular or colonial, but they differ by producing true spores. In this regard, any vegetative cell may function as a unicellular sporangium. All such spores produced are called **aplanospores,** indicating that they are nonmotile. Lack of flagella is characteristic of all blue-green algae.

Order Oscillatoriales (*Hormogonales*). All members of this order are filamentous forms, both unbranched and branched (pseudobranched), in which cell divisions are always restricted to a single plane (Figure 15-11). The filaments may be free or aggregated into various types of macroscopic, gelatinous masses in the form of cushions or balls. In some of the filamentous forms, as in the genera *Nostoc, Cylindrospermum,* and *Anabaena,* certain cells, called **heterocysts,** somewhat larger

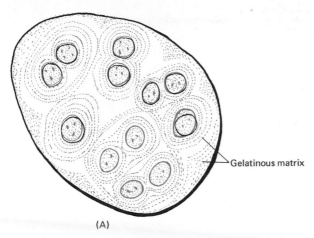

Gelatinous matrix

(A)

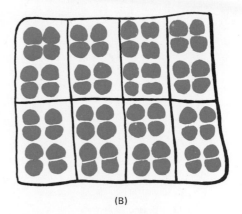

(B)

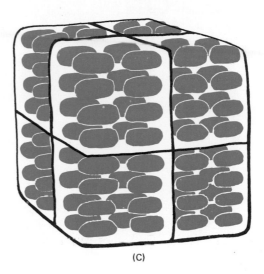

(C)

Figure 15-10. Representative colonial blue-green algae. Order Chroococales. **(A)** *Gloeocapsa,* **(B)** *Merismopedia,* and **(C)** *Eucapsis.* Note that the plane of division ultimately determines the shapes of the colonies.

than the vegetative cells, separate the filament into sections called *hormogonia.* Some investigators believe that heterocysts serve as centers of stored food material. Others hold that they are remnants of reproductive organs which have degenerated in the course of evolution. Still others are of the opinion that they represent weak points in the filament which make fragmentation possible. Some cells of a filament of a few species also form nonmotile, thick-walled spores called **akinetes** (Figure 15-11C). Unlike heterocysts, which are associated with a mechanical breaking of the filament, akinetes represent the enlargement and development of a thick wall around a cell in response to unfavorable conditions, the wall ruptures, and the cell assumes its normal metabolic activities. These spores, unlike those of higher plants, then, are not truly asexual reproductive cells, because there is no increase in numbers. They do, however, represent a simple type of specialization.

Oscillatoria is another fairly common representative of this order. In some species the free surface of the apical cell is broadly convex and the filament tapers near the end. No heterocysts are present, but the hormogonia are delimited by dead cells known as **concave cells,** or separation discs, which may function in fragmentation.

Economic Importance. The direct utilization of blue-green algae by man is at present very

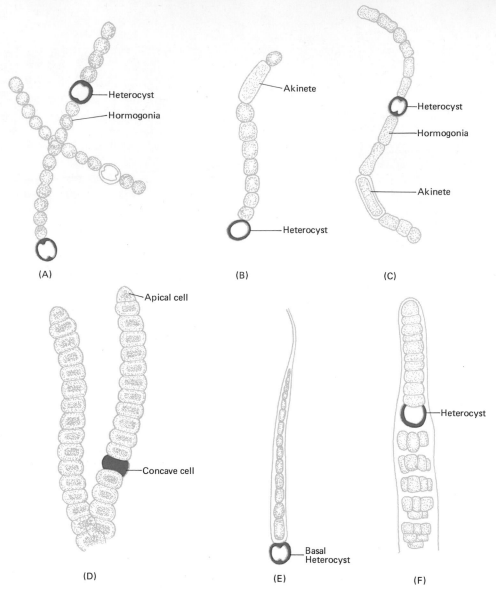

Figure 15-11. Representative filamentous blue-green algae. **(A)** *Nostoc,* **(B)** *Cylindrospermum,* **(C)** *Anabaena,* **(D)** *Oscillatoria,* **(E)** *Rivularia,* and **(F)** *Stigonema.*

limited. Inasmuch as they are photosynthetic autotrophs, they are most important as primary food producers in all the various environments in which they occur. Moreover, certain filamentous forms assume special significance because they are capable of nitrogen fixation and therefore help to maintain soil fertility. Their presence in water-clogged fields also prevents deterioration in aeration and, by increasing the available oxygen to roots, minimize disease susceptibilty. Extensive malformations of blue-green algae also help to bind the soil, reducing soil erosion.

If allowed to grow unchecked, algae often become so abundant in reservoirs and other bodies of water that they form large populations called "blooms." These blooms not only seriously decrease visibility through the water, but they also kill fish by starving them of oxygen as a result of their respiration. Blooms also give the water an objectionable odor and clog the filtering systems. Controlled treatment with copper sulfate often kills the algae in sufficient quantities to clear the water.

Relationships of the Schizophyta

Relative simplicity of cell structure (procaryotic), small size, simple form, and tremendous nutritional diversity seem to suggest the primitive character of bacteria. Moreover, the antiquity of the bacteria is indicated by certain types of iron deposits contained in the Precambian strata. In cellular structure and size the blue-green algae appear to merge into the bacteria. Both groups lack organized nuclei, well-defined organelles, and any degree of cellular specialization. In addition, binary fission is the common means of reproduction, although genetic recombination has been described. Considering these data, it has been postulated that if the first autotrophs were similar to modern green photosynthetic bacteria, two groups may have diverged from them: (1) the heterotrophic bacteria by degeneration, and (2) blue-green algae (Figure 15-12). According to this hypothesis, the blue-green algae are considered to be advanced bacteria.

Bacteria and blue-green algae differ in a num-

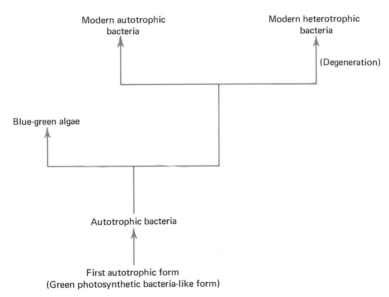

Figure 15-12. Possible relationship between bacteria and blue-green algae. According to this tentative postulation, it is assumed that the blue-green algae are advanced bacteria, being derived from a green photosynthetic bacterial stock.

ber of respects. For example, there is considerable variation in the types of photosynthetic pigments, mechanisms of photosynthesis, presence of flagella, and means of locomotion. These differences, which suggest a distinct ancestral relationship, lead to the conclusion that blue-green algae and bacteria are not closely related but are at the same evolutionary level of organization. In this regard they may have evolved from different ancestors during the same geologic time period. It seems clear that additional studies of the chemistry, reproduction, and fine structure of these two groups are needed before there can be any certainty of their true relationships.

CHAPTER IN RETROSPECT

The fission plants (Monera) include bacteria (Schizomycetes) and blue-green algae (Schizophyceae). Bacteria are probably the simplest and smallest organisms possessing any definite cellular organization and structural form. The most commonly observed forms are spherical (cocci), rod-shaped (bacilli), and helical (spirilla). Like other fission plants, bacteria do not possess an observable nuclear membrane; the genetic material occurs as dispersed clumps. Under optimum conditions, nuclear material divides by simple binary fission (reproduction) at the rate of once every 20 minutes. Most bacteria are either saprophytes or parasites, although a few autotrophic representatives can obtain their nutritional needs by oxidizing inorganic compounds (chemosynthesis) or by converting solar energy to chemical energy (photosynthesis). Bacteriologists recognize several orders of bacteria, the largest of which is called Eubacteriales (true bacteria). Eubacteria are frequently found in soil, water, sewage, decaying plant and animal materials, foods, and the atmosphere. A few representatives of this group are parasitic and can cause such diseases as scarlet fever (*Streptococcus pyogenes*), tetanus and botulism (*Clostridium*), and typhoid fever (*Salmonella typhosa*). Bacteria from other taxonomic groups are the causative agents of syphilis (*Treponema pallidum*), trench mouth (*Borrelia vincenti*), tuberculosis (*Mycobacterium tuberculosis*), and leprosy (*Mycobacterium leprae*). However, the vast number of bacteria are beneficial to the living world. For example, some genera produce extremely important antibiotics; other groups are involved in nitrogen fixation, decomposition, various symbiotic relationships, and food processing.

The blue-green algae are morphologically and physiologically more similar to bacteria than to other algae. They are widely distributed in marine and freshwater environments but are also frequently found on tree barks and the periphery of bodies of water. All blue-green algae contain photosynthetic pigments which permit them to photosynthesize. Flagellated motion is absent because flagella are never present, although some gliding or rotating motion can be observed. Three orders of blue-green algae are recognized: Chroococcales, Chamaesiphonales, and Oscillatoriales.

Unlike bacteria, the economic importance of blue-green algae is relatively limited. Blue-greens are photosynthetic organisms, so they are important as primary food producers, although many other algal groups are much more significant in this respect. Other activities of some biological importance include nitrogen fixation, soil binding, water "blooms," and oxygen production.

Questions and Problems
1. Briefly describe the methods by which bacteria may engage in genetic recombinations. What are the advantages of genetic recombinations as compared to other means of bacterial propagation?
2. Compare the nutritional patterns of bacteria and blue-green algae. In which ways are they alike? How do they differ?

3. What are the structural characteristics of various orders of bacteria? Briefly describe those activities which are unique to bacteria representing the major taxonomic groups.

4. Why are bacteria regarded as one of the most important groups of plants? List examples of particular bacteria to illustrate their importance.

5. What are the essential structural differences existing between bacteria and blue-green algae? Why are they grouped together?

6. Contrast the reproductive and locomotive habits of blue-green algae and bacteria. How are such activities related to the survival of these simple organisms?

7. Why are blue-green algae considered to be of less economic importance than bacteria? How do blue-green algae relate ecologically to other organisms?

8. Compare the pigments found in bacteria and blue-green algae. How are these pigments especially adapted for autotrophic activities?

9. What are the relative merits of placing plants and animals into four major kingdoms: Monera, Protista, Metaphyta, and Metazoa. Why are bacteria and blue-green algae placed in the first kingdom?

Suggested Supplementary Readings

Burdon, K. L., and R. P. Williams, *Microbiology*. New York: The Macmillan Company, 1968.

Clayton, R. K., and M. Delbruck, "Purple Bacteria," *Scientific American,* November 1951.

Clifton, C. E., *Introduction to the Bacteria*. New York: McGraw-Hill, Inc., 1958.

Hotchkiss, R. D., and E. Weiss, "Transformed Bacteria," *Scientific American,* November 1956.

Oginsky, E. L., and W. W. Umbreit, *An Introduction to Bacterial Physiology*. San Francisco: W. H. Freeman and Company, 1959.

Pelczar, M. J., and R. D. Reid, *Microbiology*. New York: McGraw-Hill, Inc., 1958.

Sistrom, W. R., *Microbial Life*. New York: Holt, Rinehart and Winston, Inc., 1962.

Stanier, R. Y., M. Doudoroff, and E. A. Adelberg, *The Microbial World*. Englewood Cliffs, N.J.: Prentice-Hall, Inc., 1957.

Wollman, E. L., and F. Jacob, "Sexuality in Bacteria," *Scientific American,* July 1956.

Zinder, N. D., "Transduction in Bacteria," *Scientific American,* November 1958.

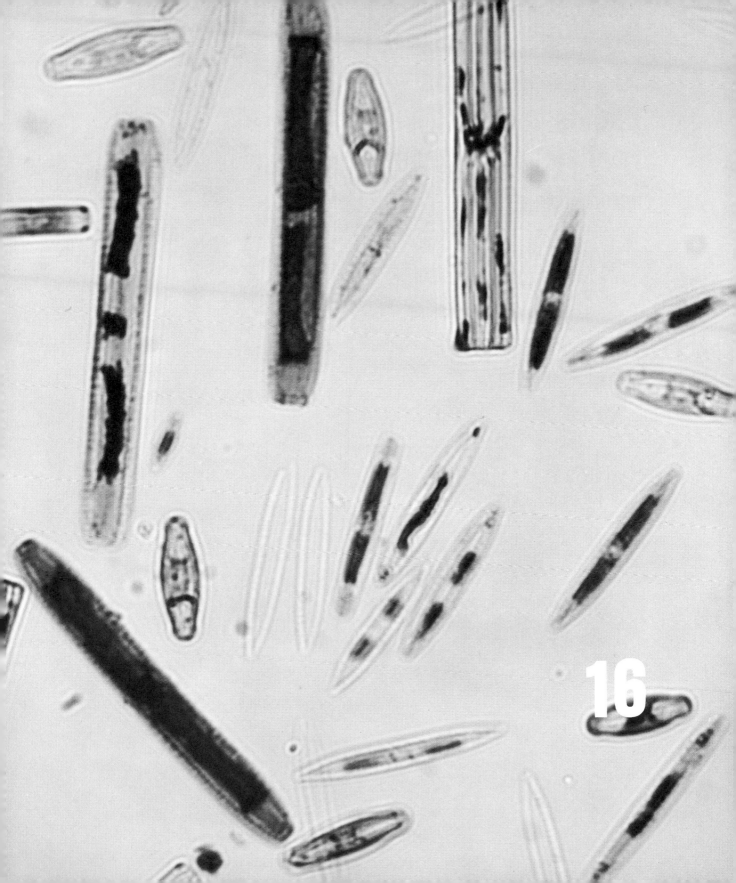

16
Photosynthetic Protista: Algal Groups

INTRODUCTION

The term *algae* is a collective taxon used to designate a vast array of organisms which exhibit extreme diversity with respect to habitat, size, form, organization, function, and reproduction. As such, this term represents an artificial grouping of plants. Recent biochemical and physiological studies, however, suggest that algae may be classified into distinct divisions, each of which has sufficient characteristics to distinguish it from other algal groups as well as from other plant divisions.

This chapter will be concerned with an analysis of various algal groups with regard to increasing complexity of form and function, the products of evolutionary advances. These modifications will be considered not only as evolutionary advances over fission plants, but also as modifications which, at some later time, gave rise to more complex land plants. It is far beyond the scope of this book to identify all these adaptations of form and function, so only selected organisms will be studied as representatives of the major evolutionary trends. In this regard, emphasis will be placed on the green algae (Chlorophyta), which are generally considered to be progenitors of land plants and which constitute the main line of evolution leading to bryophytes and vascular plants.

Following the evolution of the first cellular form, it appears that two principal lines of divergence developed. One of these, the *moneran* line, gave rise to modern bacteria and blue-green algae. The other line, the *Protista,* gave rise to a basic cell type, which, after considerable exploitation and modification, produced all other living forms. These include living Protista (algae, fungi, slime molds, and protozoa) and Metaphyta and Metazoa. Unfortunately, there is no fossil evidence to document the existence of such as ancestral form. Moreover, no living representatives are known. Despite this lack of direct evidence, there is substantial indirect evidence to suggest the probable form of the ancestral protistan cell. Most of this evidence comes from an analysis of the basic structure and physiological activities which Protista, Metaphyta, and Metazoa have in common. In other words, by analyzing the common characteristics of all Protista now in existence, the basic form of the hypothetical ancestral protistan cell can be inferred.

PROTISTAN FORM AND FUNCTION

In general, unicellular protistans are larger than unicellular monerans, averaging about 10 μ in diameter, and correspondingly they have evolved a higher degree of cellular complexity (Figure 16-1). Not only have Protista advanced to the cellular level of organization, many also exhibit a distinctive tissue level of development. In all cells of the Protista the genetic material exists as distinct structures, the **chromosomes,** that are in turn delimited within a true **nucleus** bounded by a well-defined **nuclear membrane.** *Nucleoli* are also universally present. Such organized nuclei are termed **eucaryotic,** as contrasted with the **procaryotic,** or disorganized, "nuclei" of moneran forms.

The presence of chromosomes made possible two new types of cell division: mitosis and meiosis. Both of these processes, which originated with the Protista, are more complex cellular processes than the characteristic amitotic binary fission of the Monera. The development of meiosis also provided the potential for the production of genetically dissimilar offspring, a phenomenon not possible through fission.

The cytoplasm of Protista, capable of **cyclosis,** contains rather prominent **vacuoles** that function in a storage, excretory, or water-balancing capacity. None of these features is found in the Monera. Photosynthetic protists (algae) also contain true plastids, particularly **chloroplasts,** which are far more complex in structure than chromatophores of the Monera. In addition to the photosynthetic pigment, chlorophyll *a,* that is found in the Monera, autotrophic protistans contain one or more varieties of chlorophyll (*b, c, d,* and *e*) as well.

In response to adverse environmental conditions, most protista are capable of undergoing transformation from the vegetative state to dormant stages. Under conditions of drought, temporary lack of water, fluctuations of temperature, or other adversities, a *hygroscopic* (water-holding), pectin-containing, gelatinous material is secreted (*palmelloid stage*) or a heavy membrane or wall is produced (*encapsulated stage*).

Reproduction in the Protista is by various methods, sexual and asexual. Some also exhibit a definite alternation of generations as part of their life cycles, in which haploid and diploid cells are produced during various stages. In general, single cells function as reproductive cells (gametes or spores) in Protista, and these cells have few, if any, protective structures. In higher plants (Metaphyta), reproductive structures are always multicellular and are protected within specialized reproductive organs.

Protista exist in various vegetative states. Some are unicellular or colonial *flagellates.* The number, relative length, insertion, and arrangement of flagella are often used as a basis for classifying

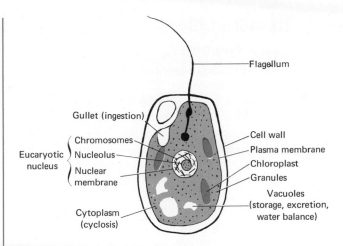

Figure 16-1. The principal structural features of a typical protistan cell.

motile flagellated protists. Other Protista, by loss of flagella, locomote by characteristic amoeba-like movements, either as unicellular or colonial forms. The *amoeboid* vegetative state is found in certain protozoa and slime molds. Still other protists exist in *sessile* (nonmotile) vegetative states as unicellular, colonial, or filamentous organisms.

Protista have evolved many processes for nutrient procurement. Some are photosynthetic, others are heterotrophic, subsisting as parasites, saprophytes, or bulk feeders by ingesting food. Many unicellular forms are capable of obtaining nutrients by a combination of these methods.

In considering the Protista as a whole, all possess a higher degree of cellular complexity as contrasted with Monera. Protista differ, however, in modes of nutrition, reproductive patterns, and their existence in a number of vegetative states. Based upon these similarities and differences, four groups of Protista are recognized: algae, slime molds, fungi, and protoza. This chapter will be concerned with an analysis of algal forms, the photosynthetic protista.

THE ALGAL GROUPS

General Characteristics

The term algae is applied to a vast assemblage of plants, referring to about 1800 genera with 21,000 species, which differ enormously in form and function. Most algal forms are aquatic, inhabiting either fresh or marine water. Still others are terrestrial and are found growing in or on moist soil, tree trunks and branches, rocks, or on other plants and animals. Although considered to be "primitive" plants, algae have adapted to almost every conceivable type of environment. In this regard, the term "primitive" refers to a close relationship with ancestral forms and a lack of structural complexity rather than to any lack of adaptation to various habitats.

Unicellular algae are found in all algal divisions except one (Phaeophyta) and are considered to be the basic cell types from which other body types have arisen. This is especially true of the flagellated unicell, which is typical of the algal cell in general

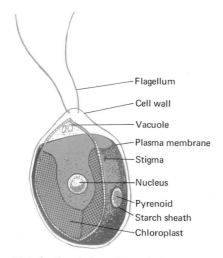

Figure 16-2. Sectional view of a typical algal cell (*Chlamydomonas.*)

— Flagellum
— Cell wall
— Vacuole
— Plasma membrane
— Stigma
— Nucleus
— Pyrenoid
— Starch sheath
— Chloroplast

(Figure 16-2). The cell walls of most algae consist primarily of carbohydrates, in which cellulose is almost always a constituent. The walls of many algae also contain gelatinous materials, often pectinaceous, which protect against desiccation and help to bind individual cells together into colonial aggregates. The protoplast, delimited by a plasma membrane, contains contractile vacuoles, which apparently function in regulating osmotic balance; nuclei; an extensive endoplasmic reticulum; mitochondria; and Golgi bodies.

One of the most prominent organelles is the chloroplast, which consists of photosynthetic lamellae confined within membranes as in higher plants. Chloroplasts of many algae contain one or more specialized regions called **pyrenoids.** These proteinaceous areas in green algae are associated with starch synthesis.

Motile cells of algae may be flagellate and/or amoeboid. In flagellated forms motility is due to the activity of one or more (usually two) **flagella** (Figure 16-3). All algal flagella are composed of 11 fibrils, enclosed within a flagellar sheath continuous with the plasma membrane. Some flagella are smooth and lack external hairs (*whiplash*); others contain fine hairs along their length and are called *tinsel flagella.* In terms of length, flagella may be equal (*isokontan*) or unequal (*heterokontan*). With regard to position, flagella may be *apically* inserted or *laterally* inserted. Flagellated cells of the green algae have two whiplash flagella; those of the euglenoids have one (or more) tinsel flagella; those of the brown algae have one whiplash and one tinsel.

Flagellated cells of many algae are capable of reacting to light stimuli, a phenomenon called *phototaxis.* Such a reaction is associated with the presence of a light-absorbing structure, a *stigma,* or eye spot. The stigma is a modified chloroplast which contains a red or orange carotenoid pigment.

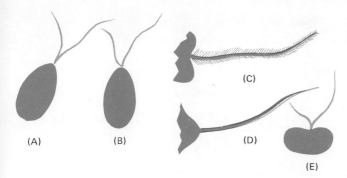

Figure 16-3. Types of flagella found among algae. **(A)** Isokontan, apical insertion. **(B)** Heterokontan, apical insertion. **(C)** A tinsel flagellum. **(D)** A whiplash flagellum. **(E)** Heterokontan, lateral insertion.

The plant body, or **thallus,** of algae is highly varied, ranging in size from microscopic, unicellular plants to multicellular organisms over 150 feet in length. In addition to considerable variation in size, a wide range of patterns of organization is evidenced among the algal divisions. Some are either motile or nonmotile *unicellular* forms, generally uninucleate, although a few are multinucleate. In many species, cells are grouped into aggregations called *colonies.* Such colonies may be *filamentous* (threadlike) or they may assume a variety of *nonfilamentous* forms. Nonfilamentous algal colonies are usually spherical or flat aggregations of cells in which cell divisions may be unlimited or determinate. In either case, the cells may be independent of one another or they may be continuous by strands of cytoplasm. Frequently, such colonies, either motile or nonmotile, are surrounded by a common gelatinous covering. filamentous colonies, usually multicellular, consist of an end to end placement of cells which form branched or unbranched networks. The cells of filaments may also be either uninucleate or multinucleate. In general, there is little differentiation of cells into tissues and organs in colonial algae. In this regard, none of the algal divisions possess a true vascular system, and therefore no true roots, stems, or leaves are present. It should be noted that in one of the more complex algal groups (Phaeophyta), some multicellular genera have rootlike, stemlike, and leaflike parts as well as striking similarities to phloem and meristematic tissue of vascular plants.

Reproduction among various groups of algae is varied and extremely diverse. The reproductive structures of algae are unicellular with no layers of sterile cells, and zygotes rarely develop into multicellular zygotes while still within the female plant. Algal reproduction is of two kinds: asexual and sexual. The principal asexual means consists of cell division, fragmentation, and the formation of various types of spores. Many unicellular algae reproduce by simple *cell division,* in which the resultant daughter cells may separate immediately or may remain together as part of colonies. Reproduction of *fragmentation* involves the segmentation of mature, nonfilamentous colonies into two or more portions. Asexual reproduction by spore formation is quite common, and a number of types of spores are distinguished. **Zoospores** are flagellate motile cells that are produced by nonmotile organisms. They vary with respect to flagella number, although usually one or two are present. Nonflagellated spores are referred to as **aplanospores.**

Sexual reproduction involves the production and union of gametes and is either isogamous, heterogamous, or oogamous. Algae which reproduce sexually exhibit three fundamental types of *life cycles,* that is, the cyclical series of events which take place starting with one individual until similar individuals are again produced. The salient features of these life cycles are gametic union, meiosis, and the nature of the plant between these events (Figure 16-4). The first two of these are called **haplobiontic,** indicating that only a single form of the plant (either haploid or diploid) occurs

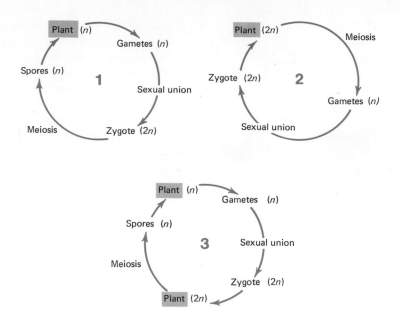

Figure 16-4. Three basic types of algal life cycles. Cycles 1 and 2 are haplobiontic, that is, only one form of the plant appears in the life cycle. In cycle 1, the haploid plant is dominant. The diploid plant characterizes cycle 2. Cycle 3 is diplobiontic, in which both haploid and diploid plants are produced and both phases alternate with each other. (Adapted from C. J. Alexopoulos and H. C. Bold, *Algae and Fungi*. New York: The Macmillan Company, 1967.)

in the life cycle. In several groups of algae, especially the green, yellow-green, and golden-brown, the zygote is usually the only cell of the diploid generation. The third cycle, called **diplobiontic,** is one in which both haploid and diploid plants are formed and these phases alternate with each other. The Chlorophyta is the only algal division in which all three types of life cycle occur.

Classification of Algal Divisions

The primary criteria used in distinguishing the various algal divisions are biochemical and involve characters such as pigmentation, storage products, and the chemical composition of the cell walls. Another important criterion is the presence of absence of flagella, their type, insertion, and number. Additional distinguishing features, especially for grouping classes within the divisions, are whether cells are uninucleate or multinucleate, the pattern of sexual reproduction, and the type of alternation of generations. Based upon these characteristics, particularly biochemical ones, the various algal forms are divided into seven principal divisions (Table 16-1). A more detailed classification of the algal divisions is shown in Table 16-2.

A Brief Survey of the Algal Divisions

The principal evolutionary trends within the algae will be analyzed with emphasis on Chlo-

TABLE 16-1. Comparative Aspects of Algal Divisions*

Division	Types of Photosynthetic Pigments		Storage Products				Cell Wall	Flagellation	Body Form
	Chlorophylls	Other Pigments	Starch	Other Polysaccharide	Fat	Alcohol			
Euglenophyta	*a* and *b*	Carotenes, xanthophylls	Paramylum	–	+	–	Usually none none	One or more tinsel	Unicellular
Chlorophyta	*a* and *b*	Carotenes, xanthophylls	+	–	–	–	Cellulose, pectins	Usually two or more whiplash	Unicellular, colonial, multicellular
Xanthophyta	*a* and *e?*	Carotenes, xanthophylls	–	Chrysolaminarin	–	Sitosterol	Cellulose, pectins, or lacking	Heterokontan	Principally unicellular
Chrysophyta	*a* and *c*	Carotenes, xanthophylls (lutein, fucoxanthin)	–	Chrysolaminarin	+	–	Cellulose, pectin, silica, or none	Absent or one tinsel or one tinsel and one whiplash	Principally unicellular
Pyrrophyta	*a* and *c*	Carotenes, xanthophylls	+	+	+	–	Cellulose	Heterokontan	Mostly unicellular, some colonial
Phaeophyta	*a* and *c*	Carotenes, xanthophylls (fucoxanthin)	–	Laminarin	–	Mannitol	Cellulose pectin, and algin	One tinsel, one whiplash	All multicellular
Rhodophyta	*a* and *d*	Carotenes, xanthophylls (lutein, phycoerythrin, phycocyanin)	Floridean starch	–	–	–	Cellulose	None	Mostly multicellular

*+, present; –, absent.

TABLE 16-2. Classification of the Algae

Kingdom Protista.
  Division Euglenophyta (euglenoids). About
      450 species.
    Class Euglenophyceae.
      Order Euglenales. *Euglena, Peranema,
        Astasia.*
      Order Colaciales. *Colacium.*
  Division Chlorophyta (green algae and
      stoneworts). About 7000 species.
    Class Chlorophyceae (green algae).
      Order Volvocales. *Chlamydanonas,
        Pandorina, Volvox.*
      Order Tetrasporales. *Tetraspora.*
      Order Ulotrichales. *Ulothrix, Protococcus,
        Caleochaeta, Stigeoclonium.*
      Order Ulvales. *Ulva, Schizomeris,
        Enteromorpha.*
      Order Schizogoniales. *Schizogonium,
        Prasiola.*
      Order Cladophorales. *Cladophora,
        Sphaeroplea.*
      Order Oedogoneales. *Oedogonium.*
      Order Zygnematales. *Zygnema, Spirogyra.*
      Order Chlorococcales. *Chlorella,
        Hydrodictyon, Pediastrum.*
      Order Siphonales. *Codium, Bryopsis,
        Acetabulum.*
      Order Siphonocladales. *Valonia.*
      Order Dasycladales. *Acetabularia.*
    Class Charophyceae (stoneworts).
      Order Charales. *Nitella, Chara.*
  Division Xanthophyta (xanthophytes). About
      400 species.
    Class Xanthophyceae (yellow-green algae).
      Vaucheria, Botrydium, Tribonema.
    Class Chloromonadophyceae (chloromonads).
      Horniella, Rickertia, Gonyostomum.
  Division Chrysophyta (chrysophytes). About
      5800 species.
    Class Chrosophyceae (golden algae).
      Chromulina, Chrysamoeba, Hydrurus.
    Class Bacillariophyceae (diatoms).
      Pinnularia, Navicula.
  Division Pyrrophyta (pyrrophytes). About
      1000 species.
    Class Dinophyceae (dinoflagellates).
      Peridinium.
    Class Cryptophyceae (cryptomonads).
      Cryptomonas.
  Division Phaeophyta (brown algae). About
      1500 species.
    Class Isogeneratae. *Ectocarpus.*
    Class Heterogeneratae. *Laminaria.*
    Class Cyclosporae. *Fucus.*
  Division Rhodophyta (red algae). About 4000
      species.
    Class Rhodophyceae (red algae).
      Subclass Bangiophycidae. *Porphyridium,
        Goniotrichum.*
      Subclass Florideophycidae. *Nemalion,
        Cumagloia.*

rophyta. It should be noted, however, that similar trends occurred within other algal divisions as well. Although these trends are not as apparent, and in some cases not as well developed, in other algal groups, we feel that the beginning student should gain some understanding of the diversity of form in these divisions. The following section provides a cursory survey of the seven algal divisions in terms of morphological diversity.

Division Euglenophyta (Euglenoids). *Euglenophyta* are considered by some to be animals and are typically green, although some are colorless (Figure 16-5). All are unicellular forms with definite nuclei and most have plastids. Euglenoid cells lack a cell wall but are enclosed by a rigid or flexible membrane called a *periplast.* One of the principal characteristics of this division is extreme motility, a function facilitated by flagella, which are whiplike appendages. The number of flagella varies among different species but most contain only one. Reproduction is asexual by longitudinal fission. Some genera, such as *Euglena* (Figure 16-5B), also contain a *gullet,* an organelle that accomplishes the ingestion of food.

Division Chlorophyta (Green Algae and Stoneworts). The *Chlorophyta* comprises a wide variety of forms with plastids containing pigments similar

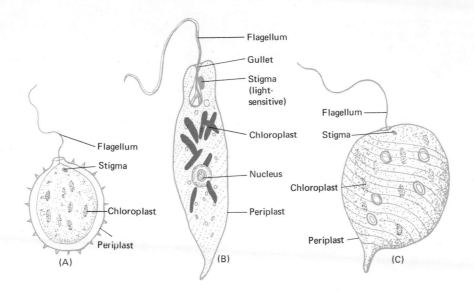

Figure 16-5. Euglenoids. These organisms exhibit a number of plantlike and animal-like characteristics. They resemble plants in that many are chlorophyllous cells capable of photosynthesis. They are animal-like in that a cell wall is absent, they are highly motile, and those that lack chlorophyll are heterotrophic. **(A)** *Trachelomonas,* **(B)** *Euglena,* and **(C)** *Phacus.*

to those found in higher plants. Most are freshwater inhabitants; some are marine. In general, marine forms are quite conspicuous. With regard to form, the plant body of green algae may be unicellular, colonial, or multicellular. Vegetatively, they are relatively simple, but they exhibit tremendous diversity in life history and in reproduction. Chlorophyta may be either motile or nonmotile. The two classes which comprise the division are *Chlorophyceae* (green algae) and *Charophyceae* (stoneworts). Representatives from each of the classes are shown in Figure 16-6.

Reproduction may be asexual or sexual. Asexual reproduction in the green algae may be by fission, fragmentation, or zoospores, motile spores locomoted by hairlike appendages. Sexual reproduction in the green algae is accomplished by isogamy, heterogamy, or oogamy.

Division Xanthophyta (Xanthophytes). Most *Xanthophyta* occur in freshwater habitats, including soil; a few are marine. The distinctive color of xanthophytes is due to the presence of yellow-green plastids, many of which are found in each vegetative cell. Although the majority of genera are unicellular, nonmotile forms, some are filamentous. Motile cells, both vegetative and reproductive, are characterized by two anterior flagella of unequal length. The two classes included in this division are *Xanthophyceae* (yellow-green algae) and *Chloromonadophyceae* (chloromonads) (see Figure 16-7).

The principal method of reproduction is by fission, but various types of spores, such as zoospores and aplanospores, may be formed. Sexual reproduction is not widespread, but when it does occur it is usually isogamous.

Figure 16-6. Representative Chlorophyta. **(A–F)** Chlorophyceae. **(A)** *Pleurococcus*. This nonmotile, typically unicellular form, may appear as a thin green coating on tree trunks, stone walls, and flower pots usually in shaded areas. **(B)** *Chlamydomonas* is a unicellular flagellate found in standing fresh water and damp soil. **(C)** *Volvox* is a colonial algal form consisting of 500 to 60,000 vegetative cells and occurs in temporary and permanent pools of fresh water. At maturity some cells enlarge and specialize as reproductive cells. **(D)** *Ulothrix* is an unbranched, filamentous form that occurs in fresh water, often in flowing streams. The basal cell is slightly modified into a holdfast for anchorage to the substrate. **(E)** *Oedogonium*, also an unbranched filament, occurs in small permanent pools and slow streams. **(F)** *Ulva*, or sea lettuce. **(G, H)** Charophyceae. **(G)** *Nitella* and **(H)** *Chara*. Charophyceae are distinguished by apical growth and are differentiated into nodal and internodal regions. Note the whorls of branches at the nodes.

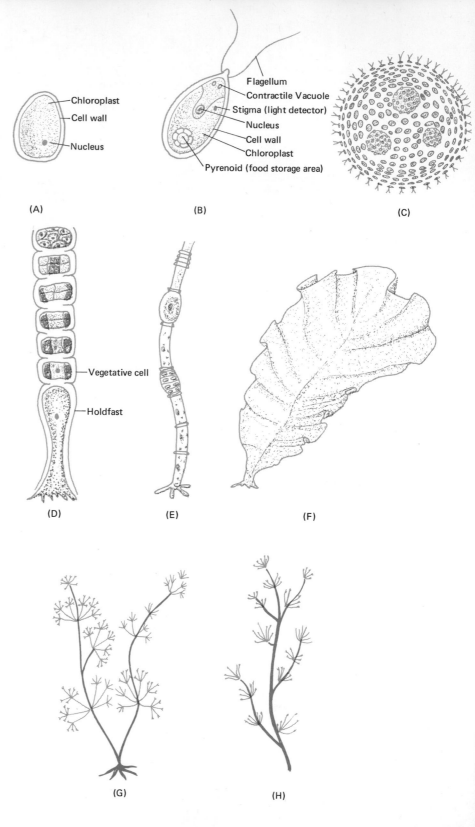

Figure 16-7. Representative Xanthophytes. **(A–C)** Yellow-green algae. **(A)** *Tribonema* is a multicellular, unbranched filament in which the individual cells are barrel-shaped. Note the spore forming in the terminal cell. **(B)** *Vaucheria* appears as green felt on the surface of damp soil and consists of a long, tubular, somewhat branched thallus. **(C)** *Botrydium* is a terrestrial alga that appears as a dense, green coating on damp soil. The entire plant consists of one cell containing many nuclei and plastids. **(D, E)** Chloromonads. **(D)** *Gonyostomum.* **(E)** *Rickertia.*

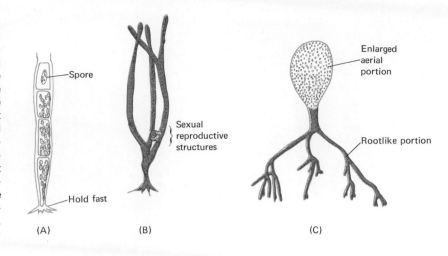

(A) (B) (C)

Spore

Hold fast

Sexual reproductive structures

Enlarged aerial portion

Rootlike portion

(D)

(E)

Division Chrysophyta (Chrysophytes). Two classes, *Chrysophyceae* (golden-algae) and *Bacillariophyceae* (diatoms), comprise the *Chrysophyta.* Chrysophytes inhabit damp soil and fresh and marine waters. The vast majority of Chrysophyta are unicellular or colonial, although some have simple, multicellular plant bodies (Figure 16-8). Some forms are motile, others are not. If flagella are present, they are of unequal length. Cell walls of diatoms and some golden algae are composed of silica.

Reproduction is usually asexual by cell division or by the formation of motile and nonmotile spores. Sexual reproduction, when it occurs, is generally oogamous.

Division Pyrrophyta (Pyrrophytes). Most *Pyrrophyta* are dinoflagellates, a variety of unicellular, motile species usually possessing cell walls of cellulose. The division has both colorless and pigmented forms, the photosynthetic species having a yellowish-green to brown color due to the presence of abundant caratenoids. The chief

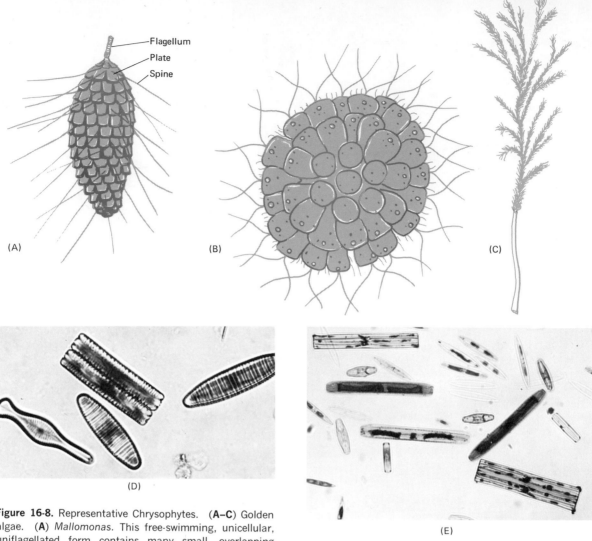

Flagellum
Plate
Spine

(A)

(B)

(C)

(D)

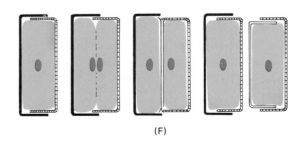

(E)

(F)

Figure 16-8. Representative Chrysophytes. **(A–C)** Golden algae. **(A)** *Mallomonas.* This free-swimming, unicellular, uniflagellated form contains many small, overlapping siliceous plates that bear spines. **(B)** *Synura,* a colonial golden-brown alga is found in fresh and brackish water. **(C)** *Hydrurus* is a colonial genus of fast mountain streams. The plant body of large colonies may be as much as three feet long and contain hundreds of thousands of cells. **(D, E)** Diatoms. These important members of the Chrysophyta are the most abundant of the minute plants of plankton of fresh and salt waters. Their cell walls consist of two halves called valves, one of which fits over the other like a box cover. **(F)** Reproduction in diatoms is by fission, each half carrying with it one of the valves of the parent cell and producing one new one. (Photos courtesy of Carolina Biological Supply Company.)

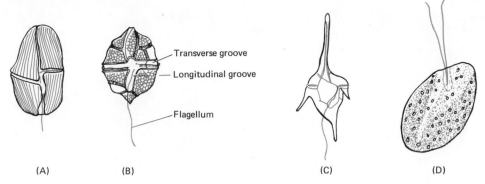

Figure 16-9. Representative Pyrrophytes. (**A–C**) Dinoflagellates. (**A**) *Gymnodinium*. (**B**) *Peridinium*. (**C**) *Ceratium*. Note the conspicuous transverse and longitudinal grooves. (**D**) *Cryptomonas*, a representative cryptomonad.

dinoflagellate characteristic is the presence of two external grooves (furrows) located on a cell wall composed of interlocking plates. Each groove contains a single flagellum. One groove encircles the cell transversely and the other runs longitudinally along one side. Dinoflagellates belong to the class *Dinophyceae* (Figure 16-9). The remainder of the division consists of members of the class *Cryptophyceae,* the cryptomonads. Most of these are unicells containing two flagella (Figure 16-9).

The usual method of reproduction is cell division. Sexual reproduction is rare. Pyrrophytes are chiefly marine, although some species occur in freshwater.

Division Phaeophyta (Brown Algae). *Phaeophyta* are almost exclusively marine forms. Some so-called seaweeds are members of this division. Chlorophylls *a* and *c* are usually masked by a brown pigment, **fucoxanthin,** which gives these plants their characteristic brown color. All brown algae are multicellular and are composed of branching filaments (Figure 16-10). Many of the larger forms have *bladders,* which are gas-filled structures that lift the forked branches toward the surface. The principal food reserves are sugars and fats.

Asexual reproduction is occasionally by fragmentation but more often by flagellated zoospores. When sexual reproduction occurs, it is either isogamous or oogamous.

Division Rhodophyta (Red Algae). The color of the plants of the *Rhodophyta* is due chiefly to the presence of the red pigment **phycoerythrin.** Chlorophylls *a* and *d,* as well as phycocyanin, are usually masked by the red pigment. Most red algae are marine forms, with a multicellular plant body consisting of branching filamentous structures and are attached to a substratum (Figure 16-11).

Asexual reproduction is by nonmotile spores. Sexual reproduction is oogamous by nonmotile gametes. Flagellated cells do not occur in the division.

Principal Evolutionary Trends in Algal Groups

In general, at least five distinctive evolutionary trends are apparent among algae. These evolutionary advances are considered to be elaborations of form and function of the fission plants and are developed even further in land plants. Many of these trends are found in all algal divisions. Some

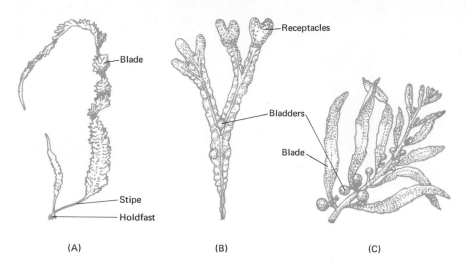

Figure 16-10. Representative genera of brown algae. **(A)** *Laminaria* (devil's apron). These plants, also called kelp, may attain great lengths, although most commonly they are one foot long. The plant body consists of a holdfast, a stipe (stemlike structure), and a blade, or leaf-like structure. **(B)** *Fucus* (rack weed). The body of this alga is a flat, ribbonlike and repeatedly forking form containing air bladders and terminal receptacles, which bear the sexual organs. The bladders are filled with gas and serve as floats. **(C)** *Sargassum* (gulf weed) is a tropical genus, several feet long, which is found in the Sargasso Seas, a large area in the Atlantic Ocean between the West Indies and North Africa.

are peculiar to only a few. Inasmuch as the green algae are believed to be the forerunners of higher plants, the Chlorophyta will be preferentially analyzed as representatives of these trends. Based upon a multiplicity of evolutionary data, particularly fossil evidence, botanists have identified a more or less hypothetical picture of the evolution of the algae. The following are considered to be some of the more important evolutionary developments in this plant group: (1) the evolution of cellular complexity, (2) the development of multicellularity, (3) differentiation of the thallus, (4) the evolution of definite patterns of sexual reproduction, and (5) the evolution of alternation of generations.

Cellular Complexity. One of the most obvious evolutionary innovations of the typical algal cell, as contrasted with that of the moneran type, is the degree of structural complexity. The eucaryotic nature of the algal cell implies that photosynthetic protists have advanced to the *cellular* level of organization, whereas procaryotic moneran cells, although somewhat organized, may be viewed as more primitive. Perhaps one of the most important algal adaptations is the possession of chromosomes, which provided the basis for mitosis and meiosis. Other adaptations, cited earlier in this chapter, afford algal cells with an extraordinary potential for diversification in function as well as in form. Once cellular organization and complexity evolved, algal cells exploited many environments and progressed along various lines of evolution. One such line led to the development of multicellularity.

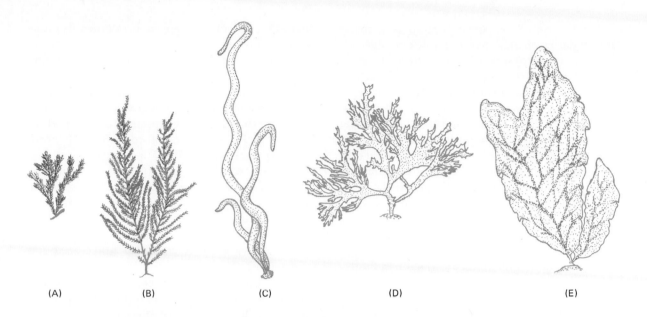

Figure 16-11. Representative red algae. **(A)** *Antithamnion.* **(B)** *Bonnemaisonia.* **(C)** *Nemalion.* **(D)** *Callophyllis.* **(E)** *Polyneura.*

Multicellularity. Multicellularity, or the aggregation of unicellular forms into discrete units, is quite conspicuous among the various algal divisions. It is generally believed that multicellularity had its origin in the inability of unicells to separate after division. Initially, primitive unicellular organisms evolved colonies (loose aggregations of similar cells), which in turn gave rise to multicellular plant bodies, in which there is some degree of differentiation among the component cells. The individual cells in a colony may be grouped in spherical masses, in filaments, or in sheets. Through a still closer union between aggregated cells, these colonies gradually developed into definite multicellular plants, probably much like some of the simpler seaweeds of today. Thus the potential was established for the production of large and complex plant bodies which form a conspicuous portion of modern flora.

Chlorophyta is representative of many divergent evolutionary trends beginning with walled and flagellated unicells and culminating with multicellular filaments and even three-dimensional leaflike plant bodies. To be sure, the development of such a condition of multicellularity is speculative, but there are a few trends found among green algae that suggest certain of the stages that may have been involved. There are three distinct main lines of evolution within the Chlorophyta, which may have led to various types of multicellularity. The first of these, the *volvocine* line, or series, led to the production of a motile colonial form. The second, called the *tetrasporine* line, produced uninucleate filamentous plant forms, and the third, the *siphonous* line, gave rise to, among other lines of development, a filamentous plant body with few cellular partitions and multinucleate cells.

The Volvocine Series. Chlamydomonas is a living unicellular green alga that probably resembles the ancestral form from which the entire plant kingdom originated (Figure 16-2). It is a biflagellate motile cell similar in form and function to an organism in the nonmotile state, becoming embedded in a gelatinous matrix and dividing mitotically to produce many daughter cells. In this state, called the *palmelloid* state, the individual cells are loosely aggregated and there is no organization or connection between them (Figure 16-12A). In the volvocine line, a series of colonial forms has evolved from the *Chlamydomonas*-like type, and, although a variety of forms exist, possibilities are limited. In this regard, the volvocine series has apparently been a blind alley and has had limitations in evolutionary potential.

In the colonial forms, each cell resembles a motile vegetative *Chlamydomonas* cell. As the cell divides, the daughter cells are oriented according to a definite pattern and remain aggregated by a common mucilaginous matrix. In some cases, the cells are interconnected by cytoplasmic strands. In the simpler colonies, the cells are vegetatively similar and possess the same reproductive potentiality. More advanced colonies contain certain cells which remain vegetative and others which are specialized for reproduction. In general, the volvocine series is characterized by a gradual increase in the number of cells with a tendency toward a simplified division of labor.

In *Gonium* (Figure 16-12B), an example of the simplest colonial stage, a flat platelike colony of 4, 8, 16, or 32 cells is produced, and all cells are reproductive. Every cell swims in unison, so the entire plant moves as a unit. *Pandorina*, a genus of slightly more complex colonial forms (Figure 16-12C), contains 8 to 32 cells arranged in a single layer about the periphery. Its advances over *Gonium* include a certain degree of regional differentiation and an interdependence of the vegetative cells on one another.

A still more advanced genus is *Pleodorina* (Figure 16-12D), in which the cells (32 to 128) of the spherical colonies are conspicuously differentiated into vegetative and reproductive types. The ultimate in this line of evolution is represented by the genus *Volvox*, where 500 to 50,000 cells form a hollow spherical colony with only a few scattered reproductive cells (Figure 16-12E). It should be noted that in sexual reproduction there is also an apparent evolutionary series. *Gonium* is isogamous, which is considered to be a primitive sexual reproductive pattern. *Pandorina* exhibits a simple heterogamy, whereas *Pleodorina* and *Volvox* are oogamous.

In considering the volvocine series as a whole, a number of trends toward multicellularity are apparent. First, there is a distinct change from a unicellular to a colonial existence, with a progressive increase in the number of cells in the colonies. Second, there is a gradual dependence of cells upon one another. Finally, there is an increasing division of labor in which certain cells remain vegetative while others specialize for reproduction. Such a series is of considerable interest in that it demonstrates a line of development followed by green algae through which multicellular individuals may have evolved from motile unicells. In tracing this series it should not be thought that each genus gave rise to a subsequent one. Consideration should be given to the fact, however, that each genus evolved from an ancestor that resembled the modern genus placed just before it.

The Tetrasporine Line. Many Chlorophyta consist of a multicellular plant body in which the thallus may be either nonbranching or branching. Starting with a palmelloid prototype of nonmotile unicells such as *Chlamydomonas,* there is a series of

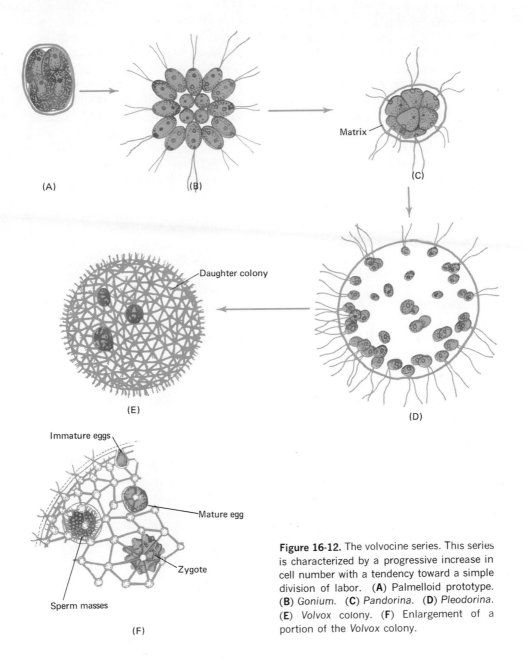

(A)

(B)

Matrix

(C)

Daughter colony

(E)

(D)

Immature eggs

Mature egg

Zygote

Sperm masses

(F)

Figure 16-12. The volvocine series. This series is characterized by a progressive increase in cell number with a tendency toward a simple division of labor. **(A)** Palmelloid prototype. **(B)** *Gonium.* **(C)** *Pandorina.* **(D)** *Pleodorina.* **(E)** *Volvox* colony. **(F)** Enlargement of a portion of the *Volvox* colony.

forms in the tetrasporine line ranging from the unicellular, nonmotile condition to more complex multicellular thalli. In contrast to the volvocine line, the tetrasporine line of evolution exhibits limitless possibilities and the main feature is the uninucleate condition of the cells.

If the nonmotile vegetative cell of *Chlorococcum* (Figure 16-13A) is used as a starting point, it can be seen that repeated division of cells in the vegetative condition can form large macroscopic colonies of undifferentiated cells, such as those in *Tetraspora* (Figure 16-13B). A further development involves a more closely knit *filament* in which the cells divide in the same plane. Such divisions produce a simple unbranched filament consisting of a single row of cells **(uniseriate).** This condition is exemplified by *Ulothrix* (Figure 16-13C). Essentially, *Ulothrix* is an unbranched filament of haploid cells in which the thallus is attached to the substratum by a specialized *holdfast* cell. Horizontal division of cells causes an increase in length. In the filament, adjacent cells are closely aligned in that they share common end walls and are connected by cytoplasmic strands. Thus *Ulothrix* represents a true filament, a basic step in the evolution of multicellularity in the algae.

From the basic uniserate, unbranched, filamentous forms such as *Ulothrix,* platelike plants might be formed by cell divisions in two planes. One

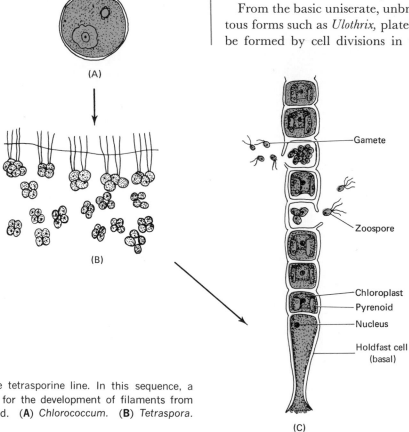

Figure 16-13. The tetrasporine line. In this sequence, a proposed pattern for the development of filaments from unicells is outlined. (A) *Chlorococcum*. (B) *Tetraspora*. (C) *Ulothrix*.

such genus is *Ulva* (Figure 16-14A) or sea lettuce, which consists of an expanded leaflike thallus two cells thick. A further elaboration of the filamentous habit in algae is the development of branches, which are considered to be precursors to certain characteristics of land plants. The most extensive elaboration of the filamentous body form is seen in *Fritschiella,* in which erect branches and rhizoids are present (Figure 16-14B).

The Siphonous Line. Along a third line of evolution the multinucleate nature of the cell is a prominent feature. This line of development,

paralleled in other plant groups, results in the formation of filamentous thalli with relatively few body partitions. Among simple unicellular forms, such as *Centrosphaera* (Figure 16-15A), the multinucleate condition may have arisen by failure of septa to form after nuclear division. Advanced siphonous forms such as *Pediastrum* (Figure 16-15B), a colonial thallus, *Urospora* (Figure 16-15C), a simple filamentous form, and *Codium* (Figure

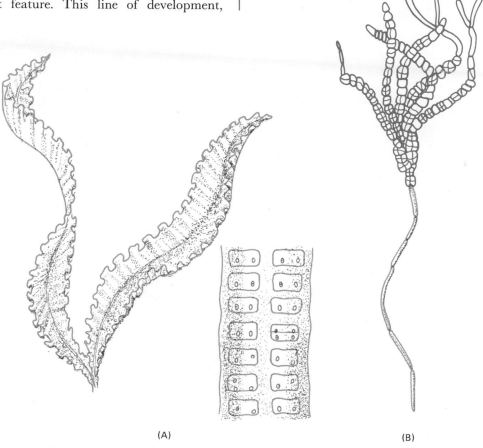

(A)   (B)

Figure 16-14. The tetrasporine line. (**A**) *Ulva* with transverse sectional view of thallus. (**B**) *Fritschiella* showing erect branching thallus.

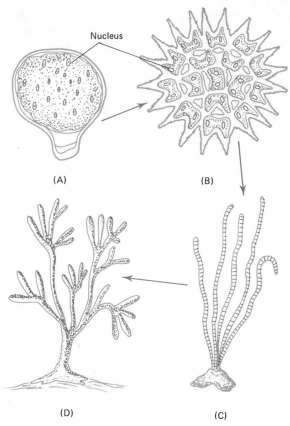

Nucleus

(A)

(B)

(D)

(C)

Figure 16-15. The siphonous line. From a simple unicellular, multinucleate form, this line of development results in the formation of filamentous thalli with relatively few body partitions. (**A**) *Centrosphaera.* (**B**) *Pediastrum.* (**C**) *Urospora.* (**D**) *Codium.*

16-15D), a branched filamentous form, exemplify the evolutionary development of this series. Although rather elaborate algal thalli are found within the siphonous series, this habit is carried to further extremes in certain fungal groups.

Thallus Differentiation. The evolution of multicellular algae was soon followed by another equally important advance, the beginning of differentiation. Unicellular algae perform all

physiological activities within the confines of the cell wall. Soon after multicellular forms arose, a division of labor among cells became a dominant theme, so that instead of a single cell carrying on all functions, certain cells became specialized. In this regard, some assumed one function and others assumed different functions along with structural specialization. It seems likely that the first specialized activity was reproduction. Instead of every cell dividing and giving rise to new individuals, only certain ones specialized as spores or gametes. Two examples of this are seen in *Oedogonium* and *Vaucheria* (Figure 16-16). In both of these genera, sperms are formed in antheridia and eggs in oogonia. Specialization of cells for reproduction in *Oedogonium* is a simpler type than that in *Vaucheria.* In *Oedogonium,* cells of the ordinary filament have become transformed into antheridia and archegonia, whereas in *Vaucheria* the reproductive structures are produced on specialized branches.

The basal cell of a *Ulothrix* filament (Figure 16-13C) is an example of vegetative specialization in which a *holdfast* has become differentiated for anchorage. Another vegetative specialization is the *bladder* of certain brown algae. In fact, the internal organization of the brown algae is the most elaborate and highly specialized of all the algal groups. In habit, certain brown algae superficially resemble higher vascular plants in that they consist of a rootlike holdfast, a stemlike *stripe,* and a leaflike *blade.* One such example is *Laminaria* (Figure 16-17). The stripe of some kelps has an outer surface tissue, the epidermis; a middle tissue, the cortex; and a central core tissue, the medulla. In addition, some species have a meristematic layer similar to the cambium of higher vascular plants as well as elongated cells closely paralleling phloem cells. From the preceding description of brown algae, it might appear that they are the precursors to land plants. They do not, however, contain chlorophyll *b,* a large percentage of

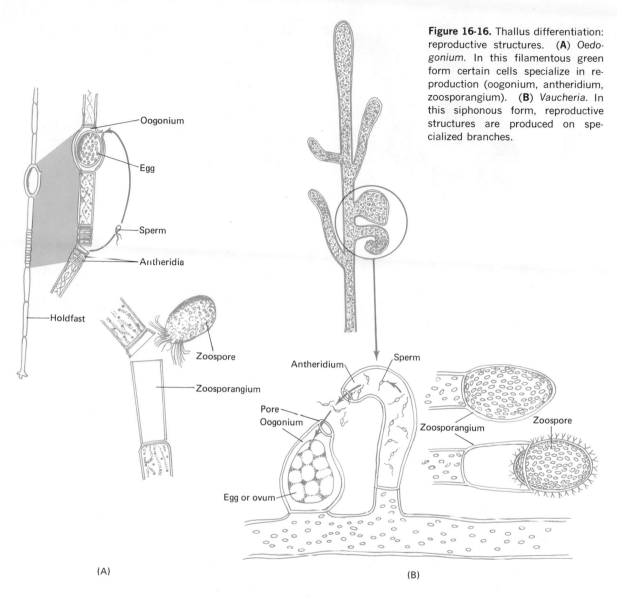

Figure 16-16. Thallus differentiation: reproductive structures. **(A)** *Oedogonium*. In this filamentous green form certain cells specialize in reproduction (oogonium, antheridium, zoosporangium). **(B)** *Vaucheria*. In this siphonous form, reproductive structures are produced on specialized branches.

(A)

(B)

xanthophylls, protective layers of sterile jacket cells around reproductive organs, multicellular embryos within archegonia, cuticles, and xylem tissue. They do, however, represent an example of parallel evolution in more than one group of unrelated organisms. It will be seen later that the process of differentiation has progressed steadily during the evolution of the plant kingdom and has resulted in the development of complex individuals found in higher groups. In these plants,

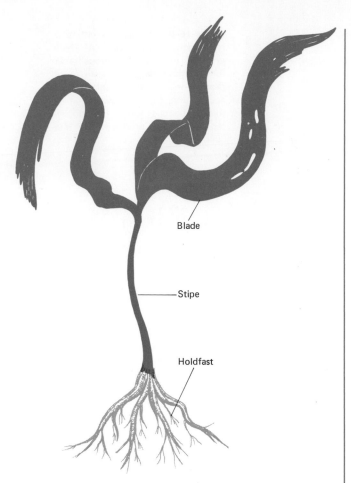

Blade

Stipe

Holdfast

Figure 16-17. Thallus differentiation: vegetative structures. Certain brown algae, such as *Laminaria,* possess highly differentiated thalli that superficially resemble those of higher plants. Note the rootlike holdfast, the stemlike stipe, and the leaflike blade. Internally, this plant is also differentiated into tissues, some of which are similar to those of vascular plants.

cells are grouped into tissues, tissues into organs, and organs into systems. Essentially, differentiation has made possible the existence of higher plants.

Sexual Reproduction. Certain stages in the evolution of reproduction can be inferred from a study of the algae. The evolutionary sequence leads from algae without sexual reproduction to algae reproducing sexually by means of similar gametes to those reproducing sexually with dissimilar gametes. In one algal division (Euglenophyta) reproduction is entirely asexual—longitudinal division of the cell into two. Sexual reproduction, if it does occur, is rare in euglenoids and is not well authenticated. Such a process of cell division is a little more advanced than the fission of moneran cells. Cell division is common in multicellular algae, in which there is an increase in the size of the plant. In these types, such as *Ulothrix,* any cells may divide usually with the exception of the holdfast cell.

A more specialized and advanced type of asexual reproduction is by means of spores, a phenomenon which occurs in all Chlorophyta. In practically all cases, any vegetative cell may function directly as a unicellular sporangium. In some cases, the entire content of the cell is liberated as a spore, but, more commonly, a cell subdivides through successive divisions and gives rise to many spores. Such spores may be aplanospores or zoospores.

Sexual reproduction, characterized by the union of gametes, represents a more advanced type of reproduction. All green algae undergo sexual reproduction. In terms of an evolutionary sequence, there has been a general progression from isogamy in primitive types, to heterogamy in more advanced types, to oogamy in very advanced types. In most green algae, any vegetative cell may function directly as a unicellular gametangium. In some the sexes are separate; in others they are both present in a single thallus.

The manner in which sexuality arose is well documented in the Volvocine, Tetrasporine, and Siphonous lines of evolution, and for the most part similar patterns exist in all three lines. For purposes of simplicity. The evolution of sexual repro-

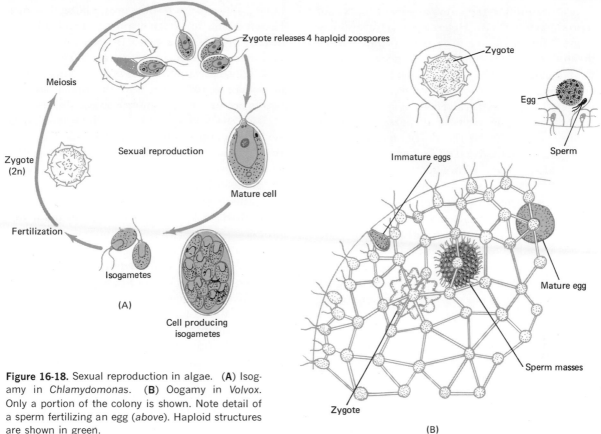

Figure 16-18. Sexual reproduction in algae. (A) Isogamy in *Chlamydomonas*. (B) Oogamy in *Volvox*. Only a portion of the colony is shown. Note detail of a sperm fertilizing an egg (*above*). Haploid structures are shown in green.

Within the figure the following labels appear:

Zygote releases 4 haploid zoospores

Meiosis

Zygote (2n)

Sexual reproduction

Fertilization

Mature cell

Isogametes

(A)

Cell producing isogametes

Zygote

Egg

Sperm

Immature eggs

Mature egg

Sperm masses

Zygote

(B)

duction will be traced in the Volvocine series only. In most species of *Chlamydomonas,* sexual reproduction is at a relatively simple level. There are no separate male and female individuals and the gametes formed are all alike; there is no distinction between male gametes (sperms) and female gametes (eggs). This condition, in which morphologically similar gametes are produced, is termed **isogamy** and probably represents a primitive level of sexual reproduction. In sexual reproduction the protoplast produces 16 to 32 isogametes, which are indistinguishable from vegetative cells. They may be regarded as small vegetative cells which fuse and act as gametes under certain conditions. They fuse in pairs, forming zygotes, which develop thick walls and resist adverse environmental conditions. The diploid zygote undergoes meiosis, forming four haploid zoospores which become mature plants when liberated (Figure 16-18A). In light of numerous studies, it appears that gametes probably evolved from vegetative cells.

Pandorina, consisting of a hollow sphere of 8 to

32 cells, exhibits **heterogamy** as part of the sexual reproductive pattern. Heterogamous, or anisogamous, reproduction involves two morphologically dissimilar gametes, small male gametes and large female gametes. Although an advance over isogamy, heterogamy is still somewhat primitive in that both gametes involved possess flagella and are free-swimming. Heterogamy is regarded as an intermediate stage in the differentiation of sex.

Oogamy is considered to be an advanced type of sexual reproduction in which a smaller motile male gamete, the **sperm,** unites with a larger nonmotile gamete, the **egg,** producing a diploid zygote. Sexual reproduction in *Volvox* is always oogamous (Figure 16-18B). Specialized cells of the parent colony enlarge without dividing and form nonmotile eggs. Other cells divide many times and produce numerous motile sperms. After fertilization, the zygote formed within the egg-producing cell enlarges and develops a thick wall. Upon release from the parent colony, the zygote, after a period of dormancy, germinates. Following meiosis, haploid nuclei are formed which give rise to new daughter colonies.

Oogamous reproduction foreshadows the pattern of sexual reproduction of green land plants in several significant aspects. First, the egg and sperm have been evolved. Second, the egg is formed within the protective covering of the female reproductive cell. Finally, there is a tendency, although only for a short time, for the zygote to be retained in the female reproductive structure. In the life history of any organism, the gametes are always the most delicate structures and any protection afforded them would be very advantageous. As will be seen later, the basic pattern of oogamy developed within various algal groups is prerequisite to survival of plants on land.

The causes which led to the development of sexual reproduction are not known, but its virtual universality at least implies its special significance. Sexual fusion not only results in increased vigor but increased variability, the raw material of evolutionary advances. Many activities of plants are directed toward the successful accomplishment of sexual reproduction and the better adapted plants are usually the more successful ones.

Alternation of Generations. The term **alternation of generations** means that there are two distinct phases in the life history of the plant, each of which gives rise to the other. One phase, called the **gametophyte,** is a haploid multicellular plant that produces gametes. The other phase, the **sporophyte,** is a multicellular diploid plant that produces spores. The gametophyte generation forms gemetes which fuse and give rise to the sporophyte generation. The sporophyte in turn produces large numbers of spores, each of which develops into a new gametophyte. Thus there is an alternation of generations.

In a **haplobiontic** life cycle only a single form of the plant occurs in nature. In other words, the conspicuous plant, or generation, is either a haploid plant or a diploid plant. If the haplobiontic life cycle is characterized by a haploid adult, the cycle is called **haplotonic.** If, by contrast, the haplobiontic cycle consists of a diploid adult, it is termed **diplotonic.** In other life cycles, called **diplobiontic,** both haploid and diploid individuals represent the species and there is a definite pattern of alternation of generations.

Chlamydomonas is haploid as an adult and exhibits a haplotonic life cycle (Figure 16-19). In the cycle, the zygote is the only diploid stage, and the haploid adult is formed by *zygotic meiosis.* Any spores produced are formed by mitotic divisions and, like the adults which produce them, they are haploid. In many algae the haploid individual may produce both gametes and spores, and often the production of either is determined by environmental factors. Such a cycle occurs in all primitive and many advanced groups among green, yellow-brown, and several other algae divisions, and in many fungi. This type of life cycle was probably characteristic of the first sexually reproducing unicellular organisms, and it may thus be the prototype from which all other types originated.

Other algal forms, such as *Fucus,* consist of diploid adults in which meiosis occurs during the maturation of gametes. Such a cycle is diplotonic and the adults are formed from the fusion of haploid gametes. In this regard, meiosis may be termed *gametogenic meiosis.* The tips of the diploid thallus develop swollen reproductive structures called *receptacles* whose surface is packed by numerous openings that lead into cavities (*conceptacles*) where the sex organs are located (Figure 16-20). In the male gametangium, the microspores (male spores) in the antheridium undergo meiosis and form numerous biflagellated sperm. In the female gametangium, the megaspores (female spores) in the oogonium undergo meiosis and give rise to haploid eggs. Note that gametes are formed directly from spores and that the gametes undergo meiosis during their maturation. The gametes are released into the water, fertilization occurs, and diploid zygotes result. As the zygote divides and develops into a mature plant, the cells remain diploid and the adult *Fucus* is diploid, including gametangia and gamete-producing cells (spores). In this case the gametophyte generation is reduced

Figure 16-19. Haplotonic life cycle of *Chlamydomonas.* The salient features of this reproductive pattern are the formation of haploid adults and zygotic meiosis. (Haploid cells are shown in green.)

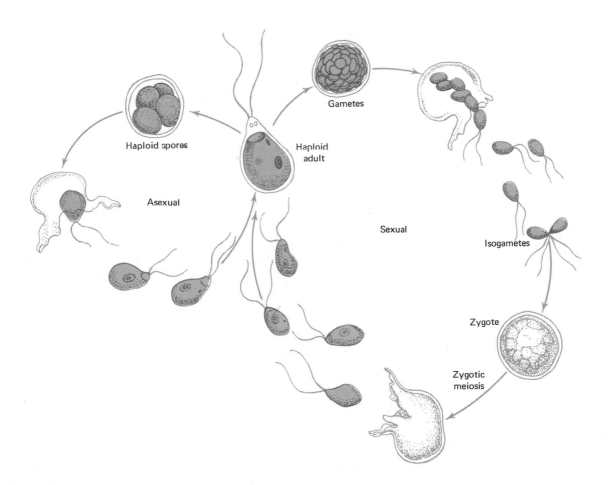

Haploid spores

Gametes

Asexual

Haploid adult

Sexual

Isogametes

Zygote

Zygotic meiosis

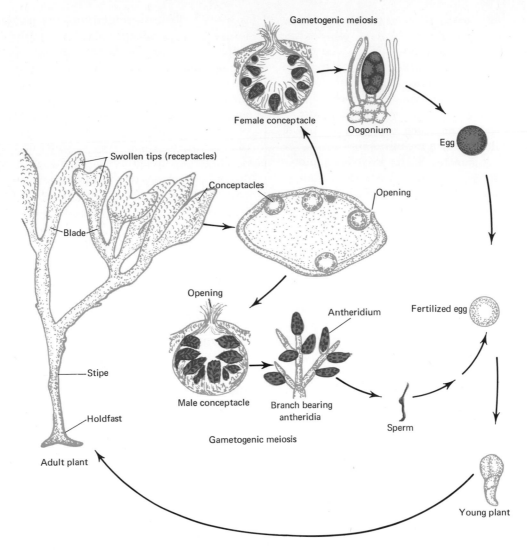

Figure 16-20. Diplotonic life cycle of *Fucus*. Note that the adult plant is diploid and meiosis occurs during gamete maturation. Haploid structures are shown in green.

to merely the gametes. If a plant with a diplotonic life cycle produces spores, they originate from diploid cells through mitosis, and the spores consequently are diploid as well.

It is generally believed that the diplotonic life cycle is derived from the haplotonic pattern. In the haplotonic life cycle, meiosis takes place in the zygote at the start of the life cycle. By contrast, in diplotonic types meiosis is postponed to the time of gamete formation at the end of the cycle. The adaptive advantage to postponing meiosis is that diploid adults contain paired chromosomes and therefore paired genes. If, through mutation, one gene of a pair is altered, the other still retains the genetic integrity of a specific trait.

Among the most complex life cycles found in algae are those characterized by *sporogenic meiosis* and alternation of generations. Such cycles, called **diplobiontic,** postpone meiosis until the spore-producing stage and the cycles are divided into two generations, each represented by a separate adult. The zygote develops into a diploid adult, the sporophyte generation, which produces haploid spores as a result of sporogenic meiosis. These haploid spores give rise to a haploid adult, the gametophyte generation, which produces gametes. At fertilization, these gametes form zygotes which give rise to diploid adults, the sporophyte generation. Thus this type of cycle is characterized by a definite alternation of generations. Diplobiontic cycles are quite widespread, occurring in many Chlorophyta and Chrysophyta; in all Phaeophyta, except *Fucus;* in some Rhodophyta; in all slime molds; in many fungal groups; and in all Metaphyta.

Two algal genera will be considered as representatives of the diplobiontic cycle. In one of these (*Ulva*), the sporophyte and gametophyte generations are morphologically indistinguishable. Such an alternation is termed *isomorphic.* In the second genus (*Derbesia*), the two sequential generations differ in form and the alternation is called *heteromorphic.*

The gametophyte generation of *Ulva* consists of either male or female plants (Figure 16-21). Gametes produced by the gametophyte plants are haploid and at fertilization they fuse to form diploid zygotes. After many cell divisions, the zygotes produce diploid plants which represent the sporophyte generation. The sporangia of these diploid plants form cells (spore-mother cells)

which undergo meiosis and produce **meiospores.** There is a reduction division in their formation, so the meiospores are haploid. Upon germination, the meiospores, through successive mitotic division, plants, each of which is also haploid. Thus the alternating cycle is completed. Note that meiosis occurs within spores of the sporophyte generation.

In the diplobiontic life cycle of the genus *Derbesia,* the alternating generations are quite distinct (Figure 16-22). The gametophyte generation consists of spherical thalli, separately sexed, in which there is an attachment by coenocytic rhizoid-like branches. The haploid gametes fuse in the water and form a coenocytic, multinucleate, branched thallus which constitutes the sporophyte generation. At maturity, the diploid plant forms well-defined sporangia in which meiosis occurs. The spores within the sporangia (spore mother cells) undergo meiosis and meiospores are produced. When the meiospores are released, they give rise to male and female gametophyte plants, completing the cycle.

All sexually reproducing algae have an alternation of generations. In many, the gametophyte generation is larger, with the sporophyte represented by a single diploid cell, the zygote. In others, it is almost all sporophyte, essentially as in animals, so that the relative development of these two generations varies greatly. In still other algae, both generations may be nearly the same size, either similar or different in appearance. Bryophytes exhibit conspicuous and dominant gametophyte generations, whereas in ferns and their allies it is the sporophyte generation which dominates. In seed-bearing plants, the sporophyte is also dominant, but the gametophyte, instead of remaining independent and free-living as in ferns, is greatly reduced and is embedded in the tissues of the sporophyte. In general, once alternation of generations became established in primitive plants, there has been a progressive tendency

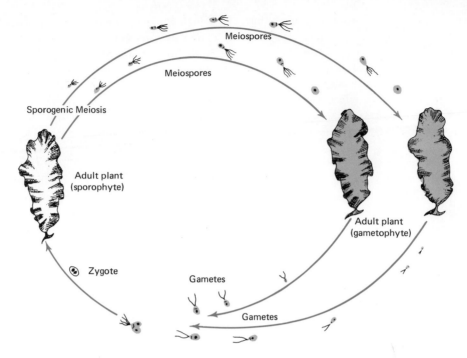

Figure 16-21. Diplobiontic life cycle: isomorphic. In *Ulva,* both sporophyte and gametophyte adults are morphologically similar. Note the point at which sporogenic meiosis occurs in the cycle. (Haploid plants and cells are shown in green.)

for the sporophyte to become dominant and for the gametophyte to become greatly reduced and dependent upon the sporophyte.

Summary of Algal Evolutionary Trends

In considering the algae as a group, it may be seen that evolution has progressed to a relatively advanced stage and that some algae have almost reached the level of the vascular plants in their development. Among the algal advances which were passed on to vascular plants are cellular specialization, mitotic division, progressive specialization in body type (vegetative and reproductive), diverse modes of sexual reproduction, and the development of haploid and diploid stages. Not all these trends are found in any one alga, but the potentialities are present in the algae as a whole. Furthermore, the fact that in a given alga not all characters are in the same stage of evolution is direct evidence that evolution progresses at different rates in different plant groups.

Most of the algae are not capable of further evolutionary advance, especially in adapting to terrestrial habitats. A few, however, made a successful transition, and it is generally accepted that certain green algae were the forerunners of land plants. Certain modern genera of Chlorophyta illustrate features to support the hypothesis that all green plants (vascular and nonvascular) evolved from a green alga or green-alga-like

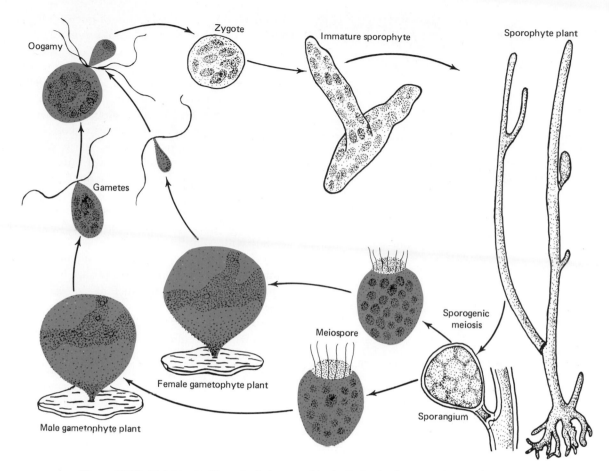

Figure 16-22. Diplobiontic life cycle: heteromorphic. In *Derbesia*, the sporophyte and gameto-phyte plants are morphologically distinguishable. (Haploid plants and cells are shown in green.)

percursor. Figure 16-23 illustrates a tentative scheme of the possible interrelationships of the various algal divisions.

Economic Aspects of Algae

Inasmuch as all algae are photosynthetic, they are the primary food source in the diet of fish and other aquatic organisms. Essentially, aquatic algae are to aquatic organisms what green land plants are to herbivorous organisms, although such a relationship is not as obvious or as direct. Nevertheless, it is of equal importance in the economy of nature.

Red algae are used directly by man and are important for certain industrial and domestic uses and as food. *Chondrus crispus,* the "Irish moss," is

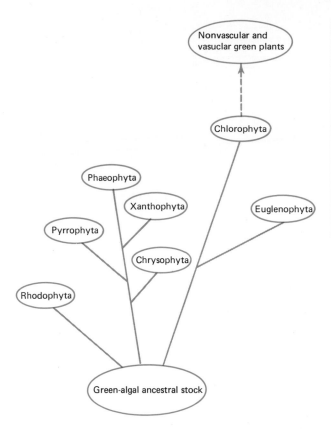

Figure 16-23. Possible interrelationships of the various algal divisions.

made into a dessert, "sea moss farina," as well as other puddings. Dulse, or sea kale (*Rhodoenia*), is eaten like candy, used as a relish with potatoes, or cooked in soups such as "St. Patrick's soup." *Porphyra*, known as purple laver, is also used widely as a food. It is often used by Europeans in making soup, and in Japan it is used in the preparation of macaroni, soups, and sauces. The Japanese have artificially cultivated beds of *Porphyra* for many years as part of one of their most valuable maritime industries. *Gelidium*, along with other red algae, is the commercial source of agar, which is

widely used in the preparation of microbiological media for culture work, as a solidifying agent in desserts, as a mild laxative, as a capsule for antibiotics, and as a sizing material for textiles. Other red algae are sources of gelatin, or jellies, which are used as bases for shoe polish, water-base paint, hair dressing, ice cream, shaving cream, and cosmetics. Although coral animals are often cited as the primary builders of reefs, the *coralline*, or calcareous, red algae actually contribute more to the building up of these formations than do the animals.

The brown algae are also directly important to man. In the past, they have sometimes been used as a fertilizer and as a source of iodine and potassium. These uses have been largely discontinued because cheaper sources of supply are now available. Kombu, a product of *Laminaria* and other kelps, has been used as food in the Orient since 1730, as a vegetable, steeped as a drink, or coated with sugar and eaten as a confection. Algin, a derivative of cell walls, has led to the development of valuable algin industries in Britain and California. Algin is used as a stabilizer or as a moisture retainer in ice cream, paint, pharmaceuticals, ink, soap, and shampoo.

Euglenophyta are especially important in the phytoplankton freshwater areas. Although they are of little direct use to man, certain euglenoids are used as bioassay organisms (especially vitamin B_{12}) and thus play a useful role in precise physiological and nutritional studies. Several genera probably participate in the general breakdown of complex organic molecules in sewage disposal tanks.

Among the Chrysophyta, the yellow-green and golden-brown algae are an initial link in the plant to fish food chain. Otherwise, they have little economic significance. The diatoms, however, make up a large proportion of the vegetable plankton of both fresh and salt water and are considered to be the primary food supply of the sea. Under natural conditions, when diatoms die, their siliceous walls drift to the bottom of the body of water in which they occur and accumulate as massive deposits of *diatomaceous earth.* One of the largest of these deposits is in Lompac, California, where the accumulation is 12 miles square and may reach a depth of 3000 feet. It is believed that such deposits were raised above sea level by geologic activity. Diatomaceous earth is a soft, light, crumbly material used in filters or as an abrasive in toothpaste and polishes, as insulation shielding, in dynamite as an absorbent, and in the building industry in concrete and brick.

Dinoflagellates of the division Pyrrophyta are one of the most important ultimate sources of food in the warmer parts of the ocean. On the negative side, they cause sporadic red tides that kill large numbers of fish and other marine animals. These red tides are caused by a red pigment secreted by certain dinoflagellates (*Glenodinium, Gymnodinium,* and *Gonyaulax*) which are poisonous.

Green algae, in addition to being primary food sources, especially in freshwater, may also be used directly by man. Species of marine genera such as *Ulva* and *Monostroma* have been used as food (laver) in some countries (for example, Japan and Wales). In India, dried *Spirogyra* and *Oedogonium* are similarly used. Some Chlorophyta may become contaminants on the surface of reservoirs, lakes, and ponds and give a disagreeable flavor to the water. Green algae are also important inhabitants of oxidation ponds in sewage-treatment plants, providing oxygen for rapid decomposition of raw sewage by bacteria.

CHAPTER IN RETROSPECT

Algae represent a diverse group of chlorophyllous plants which exhibit significant evolutionary advances over fission plants. Most algal organisms are aquatic, inhabiting either fresh or marine water, although quite a few species can be found growing on moist soil, tree trunks and branches, rocks, and even animals. The plant body of algae may range in size from microscopic unicellular forms to huge multicellular organisms over 150 feet in length. However, even these larger algae do not possess a true vascular system, leaves, stems, or true roots. The principal asexual means of algal reproduction consists of cell division, fragmentation, and spore formation. Sexual reproduction involves the production and union of gametes and may be isogamous, heterogamous, or oogamous.

Based principally upon biochemical, locomotor, and reproductive characteristics, algae are classified into seven plant divisions, as indicated in Table 16-2.

It is generally believed that primitive green algae were the forerunners of higher land plants. A study of the history of green algae indicates the following evolutionary advancements: (1) cellular complexity, (2) multicellularity, (3) definite patterns of sexual reproduction, (4) cellular differention and specialization, and (5) alternation of generations. Once alternation of generations had become established in higher vascular plants, the sporophyte phase dominates over the gametophyte phase.

It should be noted that the importance of algae to fish and other aquatic organisms is comparable to the importance of grass and other chlorophyllous vegetation to grazing animals. In addition, many algae, especially red and brown algae, are consumed directly by humans in certain parts of the world, and some are found in various food preparations such as ice cream, jellies, and soup. Algae are also employed in the manufacture of numerous pharmaceutical products, cosmetics, shoe polish, water-base paint, and other industrial products.

Questions and Problems

1. What do the following terms signify?

   hygroscopic   fucoxanthin
   sessile   volvocine line
   diplobiontic   meiospores
   flagella   palmelloid state
   fragmentation   diatomaceous earth
   zoospores   holdfast
   alternation of   thallus
      generations   heterogamy
   isogamy   oogamy

2. Briefly describe the habitat of most algae. What are the problems that aquatic algae would face if they were transported to a terrestrial environment? In your response, consider reproduction, nutrition, translocation of essential materials within the organism, mechanical support, variations in temperature, and other environmental factors.

3. What is the nature of asexual reproduction in algae? What are the various types of asexual reproduction? Describe each.

4. Contrast the biological significance of asexual and sexual reproduction. How does sexual reproduction help algae to adapt to a changing environment?

5. What criteria are used to classify algae into various divisions? Describe the relative merits of each criterion as a taxonomic tool.

6. Why are euglenoids sometimes regarded as animals rather than plants? What are the most reliable means of distinguishing plants from animals? Consider nutrition, locomotion (vegetative and reproductive phase), life cycles, and cellular organization in your response.

7. Describe alternation of generations in *Fucus*. In your description, use a word diagram and consider the following terms: conceptacles, receptacles, male and female gametagium, microspores, megaspores, haploid, diploid, meiosis, fertilization, gametophyte, and sporophyte generations.

8. Compare the life cycle of the genus *Derbesia* with that of *Fucus*. What are the similarities and differences?

9. How have algae become more advanced over fis-

sion plants? What evolutionary contributions have algae made to vascular land plants?

10. What is the value of algae to man and other animals?

Suggested Supplementary Readings

Blum, J. L., "The Ecology of River Algae," *Botanical Review*, vol. 22, pp. 291–341, 1956.

Chase, F. M., *Useful Algae*. Smithsonian Institution Annual Report, Publication 3667, pp. 401–452, 1941.

Dawson, E. Y., *How to Know the Seaweeds*. Dubuque, Iowa: William C. Brown Company, Publishers, 1956.

Fritsch, F. E., *Structure and Reproduction of the Algae*, vols. 1 and 2. New York: Cambridge University Press, 1935 and 1945.

Jackson, D. F., *Algae and Man*. New York: Plenum Publishing Corporation, 1964.

Krauss, R. W., "Mass Culture of Algae for Food and Other Organic Compounds," *American Journal of Botany*, vol. 49, pp. 425–435, 1962.

Palmer, C. M., *Algae in Water Supplies*. Washington, D.C.: U.S. Public Health Service Publication 657, 1959.

Pringsheim, E. D., *Pure Cultures of Algae*. New York: Hafner Publishing Company, 1964.

Round, F. C., *The Biology of the Algae*. New York: St. martin's Press, Inc., 1965.

Scagel, R. F., et al., *An Evolutionary Survey of the Plant Kingdom*. Belmont, Calif.: Wadsworth Publishing Co., Inc., 1965.

Schwimmer, M., *The Role of Algae and Plankton in Medicine*. New York: Grune & Stratton, Inc., 1955.

Smith, G. M., *Cryptogamic Botany*, vol. 1. New York: McGraw-Hill, Inc., 1955.

———, *The Freshwater Algae of the United States*. New York: McGraw-Hill, Inc., 1950.

Tiffany, L. H., *Algae, The Grass of Many Waters*. Springfield, Ill.: Charles C Thomas, Publisher, 1958.

Tilden, J. E., *The Algae and Their Life Relations*. Minneapolis, Minn.: University of Minnesota Press, 1935.

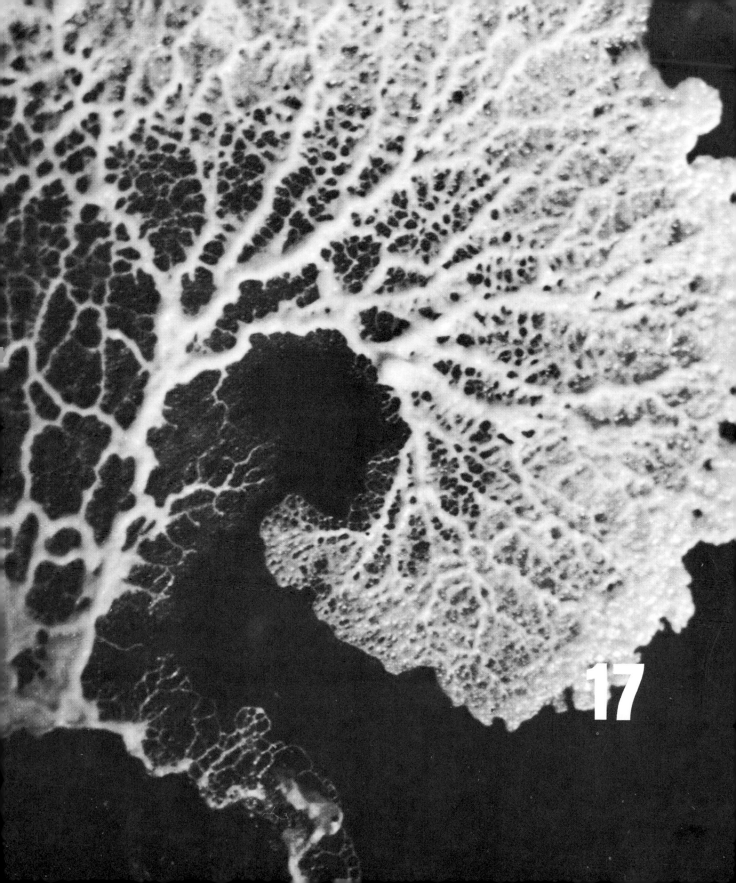

17
Nonphotosynthetic Protista: Slime Molds and Fungi

INTRODUCTION

Until quite recently, the slime molds, true fungi, and bacteria were placed together in a single subdivision referred to as the fungi. Simplicity of structure, the absence of chlorophyll, and storage of food reserves other than starch were the principal criteria for such a grouping. At present, however, these three groups are categorized into separate divisions—the *Myxomycophyta* (slime molds), the *Eumycophyta* (true fungi), and the *Schizophyta* (bacteria and blue-green algae). The term fungi, like algae, no longer has taxonomic status as a group name but is still retained as a convenient designation.

The Myxomycophyta and Eumycophyta differ from bacteria in several important aspects. Their cells, which contain eucaryotic nuclei, are considerably larger than those of bacteria, the thalli of most species are multicellular, most reproduce according to some type of sexual process, and the production and structure of their spores are usually different from bacterial spores. The slime molds differ from the bacteria and from most fungi in their naked (lack of cell walls), amoeboid assimilative stages, and **holozoic** (food-ingesting) mode of nutrition. They resemble the Eumycophyta, however, in producing walled spores, usually borne in characteristic fruiting bodies. There is still some uncertainty as to the relationship between the true fungi and slime molds. It is customary to study the slime molds along with the true fungi because of possible interrelationships, but they also show strong affinities with protozoa.

DIVISION MYXOMYCOPHYTA: Slime Molds

The slime molds, or slime fungi, are a unique group of plants having practically no economic significance. They are of considerable importance to biologists, however, because their vegetative phase of growth exhibits animal-like characteristics, while their reproductive phase is distinctly plantlike. They also provide relatively large masses of naked protoplasm for intensive experimental investigation. Often they are found growing on rotting logs, damp soil, or other decaying organic mater as glistening viscous masses of slime.

The division Myxomycophyta includes organisms of two very distinct types: the class *Myxomycetes,* or *true slime molds;* and the *cellular slime molds.* Referred to in the latter group are two classes, the *Acrasiomycetes* and *Labyrinthulomycetes.* Neither is related closely to the other or to the true slime molds. Class distinction is based primarily on the differences in structure of the fruiting bodies and details of the life cycles. The classification of slime molds is shown in Table 17-1.

Class Myxomycetes (True Slime Molds)

The adults of the myxomycetes consist of a naked mass of amoeboid protoplasm, the *plasmodium,* which contains hundreds of diploid nuclei that are not delimited by cell boundaries (coenocytic). This vegetative phase of the life cycle moves in amoeboid fashion and ingests solid food particles in the same manner as an amoeba. Migrating plasmodia typically are fan-shaped and

TABLE 17-1. Classification of the Slime Molds

Kingdom Protista.
  Division Myxomycophyta: About 500 species.
    Class Myxomycetes (true slime molds).
      Order Ceratiomyxales. *Ceratiomyxa.*
      Order Liceales. *Dictydium, Lycogala.*
      Order Trichiales. *Trichia.*
      Order Echinostelium. *Echinostelium.*
      Order Stemonitales. *Stemonitis, Diachea.*
      Order Physarales. *Physarum, Didymium.*
    Class Acrasiomycetes (cellular slime molds).
      Order Acrasiales. *Dictyostelium.*
    Class Labyrinthulomycetes (net slime molds).
      Order Labyrinthulales. *Labyrinthula,*
      Labyrinthoriza.

are composed of a network of veinlike strands (Figure 17-1). The plasmodium of many slime molds is white; other plasmodia are yellow, orange, or red.

For the most part, the plasmodium phase is animal-like in behavior. When a myxomycete plasmodium reproduces, it flows together into one or more heaped mounds and grows upright stalks and bulbous tips. Reproductive structures are referred to as fruiting bodies, the most common type being the sporangium (Figure 17-2) supported on upright stalks. Under certain conditions, the plasmodium becomes stationary and develops these fruiting bodies. In this state, the reproductive phase, the appearance and behavior of the organisms are plantlike. Within the fruiting bodies, cleavage of the multinucleate protoplasm results in the formation of uninucleate haploid spores, each containing a rigid cell wall. The spores escape, and, if environmental conditions are suitable, they germinate and divide mitotically to produce naked motile gametes know as *swarmers.* These fuse in pairs to form zygotes, which soon lose their flagella and become amoeboid. As this amoeboid form moves along the substratum, its diploid nucleus undergoes successive divisions (mitotic) without accompanying cytokinesis. Essentially, the, the zygote develops into a coeoncytic plasmodium. The life cycle, which proceeds from a diploid amoeboid plasmodium to a stationary spore-producing organism, to haploid spores, to flagellated gametes, to zygote, and back to amoeboid plasmodium is shown in Figure 17-3.

Class Acrasiomycetes (Cellular Slime Molds)

Although members of the acrasiomycetes resemble the true slime molds in many respects, they have several distinct characteristics. Spores produced by fruiting bodies do not develop into flagellated gametes as in Myxomycetes, but instead give rise to free-living amoeboid cells, each with a single haploid nucleus (Figure 17-4). These amoebae feed on bacteria and other organic matter, grow, and divide both by mitosis and cytokinesis until a rather extensive population results. When the food supply is exhausted, some of these amoebae secrete a hormonal substance, *acrasin,* which attracts other amoebae toward a center of aggregation. This adult vegetative body,

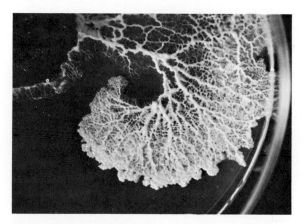

Figure 17-1. Plasmodia of *Physarum gyrosum* showing fan-shaped form and internal network of veinlike strands. Approx. ×1.3. (Courtesy of Dr. C. J. Alexopoulos.)

(A)

(B)

Figure 17-2. Fruiting bodies of the true slime molds. (**A**) *Didymium minus* showing upright stalks and bulbous tips, the sporangia. (Courtesy of Dr. C. J. Alexopoulos.) (**B**) *Stemonitis.* (Courtesy of Carolina Biological Supply Company.)

Figure 17-3. Life history of a true slime mold. See text for amplification. (Haploid generation is shown in green.)

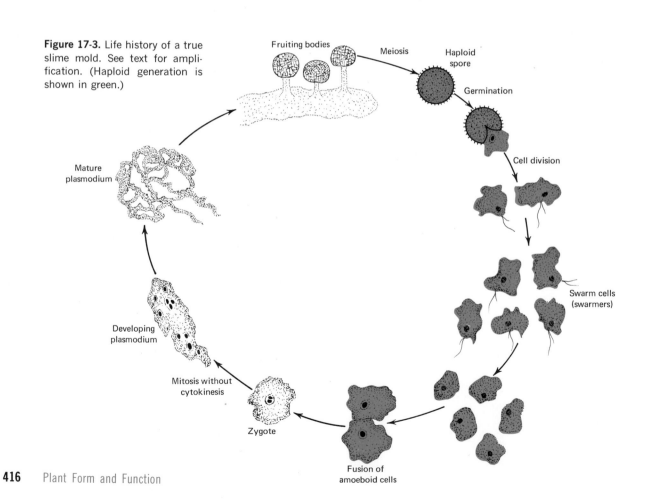

Fruiting bodies

Meiosis

Haploid spore

Germination

Cell division

Swarm cells (swarmers)

Mature plasmodium

Developing plasmodium

Mitosis without cytokinesis

Zygote

Fusion of amoeboid cells

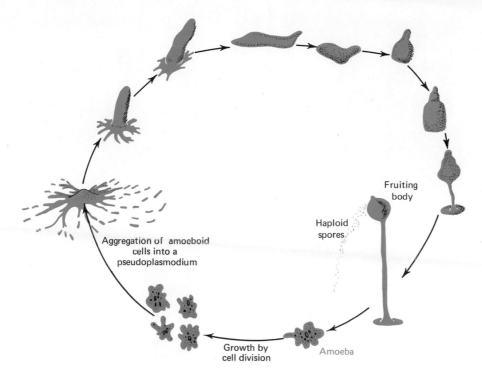

Figure 17-4. Life history of a cellular slime mold. (From C. J. Alexopoulos, *Introductory Mycology,* 2nd edition. New York: John Wiley & Sons, Inc., 1962.)

Labels within figure:

Fruiting body

Haploid spores

Aggregation of amoeboid cells into a pseudoplasmodium

Growth by cell division

Amoeba

called a *pseudoplasmodium,* is a cellular colony composed of hundreds and even thousands of uninucleate amoeboid cells, each of which retains its individuality. A more advanced vegetative stage, the *slug,* reaches a certain point of organization and becomes physiologically differentiated for sporulation. As part of this process, the slug becomes sedentary and forms a nipple-like structure at one end, which in time develops into a stalk and a bulbous tip. Thus the fruiting body is formed. Spores formed within the sporangia are haploid and, when released, give rise to free-living amoebae, completing the cycle.

Organisms of the class *Labyrinthulomycetes,* which are exclusively aquatic, have not been sufficiently studied to provide a complete life history. The class is characterized by multicellular pseudoplasmodia and motile unicellular swarmers. Fruiting bodies have not been observed.

DIVISION EUMYCOPHYTA (TRUE FUNGI)

True fungi differ from slime molds in that most have a filamentous organization with rigid cell walls and the vegetative thallus can only absorb

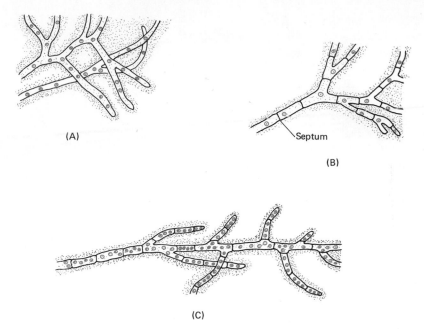

Figure 17-5. Types of hyphae found in true fungi. **(A)** Coenocytic (nonseptate). **(B)** Septate with uninucleate cells. **(C)** Septate with multinucleate cells. (From C. J. Alexopoulos and H. C. Bold, *Algae and Fungi,* New York: The Macmillan Company, 1967.)

dissolved nutrients (by producing exoenzymes), whereas a plasmodium is capable of ingestion. All Eumycophyta are heterotrophic; most are saprophytes, although some species are parasites of man, other animals, or plants. Symbiotic relationships with plants (lichens) or animals are not uncommon. Members of this division are highly diversified and occur on plant and animal remains of all types in most environments, and many are of major economic or medical significance to man. Primitive fungal groups are aquatic and produce flagellated reproductive cells. The advanced fungal groups are terrestrial and form nonmotile reproductive cells.

Thallus Differentiation

The body of a true fungus is either unicellular or, more commonly, filamentous. The individual tubular filaments, called **hyphae,** branch and rebranch to form an extensive system or mass, the **mycelium.** In some cases, mycelia are loosely arranged in an irregular network, whereas in more complex forms they are compacted into orderly patterns, such as mushrooms. Hyphae are distinguished in three morphological types: coenocytic, septate with uninuclear cells, and multinucleate septate mycelia (Figure 17-5). Cell walls are composed principally of cellulose, chitin, or a combination. The fungal cell itself, like other protistan cells, contains a variety of organelles, including eucaryotic nuclei and, in general, is structurally similar to those of higher plants and animals.

Reproduction is both asexual and sexual. Among the asexual methods are cell division, budding (as in yeasts), fragmentation, and spore formation.

Most fungi form two or more types of spores in the life cycle. Asexual spores may be borne in sporangia or at the ends of specialized hyphae **(conidiophores)** either single or in chains. Spores borne in sporangia are called **sporangiospores** and may be motile; those developed directly on hyphae are termed **conidia** and are never motile. Sexual reproduction, which involves a fusion of nuclei, is varied, as in the algal groups. Sexual spores, produced only after sexual reproduction, are diploid. Many fungi produce a third type of spore formed as a product of meiosis. These spores are called **meiospores.** Meiospores are characteristic of the so called higher fungi.

Classification of Fungi

About 80,000 species of fungi are known. As in the case of all other protists, their evolutionary origin is obscure. Traditional hypotheses purport that either red algae (Rhodophyta) or protozoa represent the ancestral stock from which fungi arose. Other hypotheses suggest that other algal groups or slime molds are possible ancestors. It would appear that additional data are needed before the problem of fungal ancestry can be resolved with any degree of certainty. In any case, the Eumycophyta are divisible into four classes based upon differences in the sexual stage of the life cycle (Table 17-2). Three of these, the *Phycomycetes, Ascomycetes,* and *Basidiomycetes,* are characterized by pronounced morphological differences and variations in their reproductive cycles. The

TABLE 17-2. Classification of the Fungi

Kingdom Protista.
   Division Eumycophyta (true fungi). About 80,000 species.
     Class Phycomycetes (algal fungi).
      Subclass Chytridromycetidae.
       Order Chytridiales. *Physoderma.*
       Order Blastocladiales. *Allomyces.*
       Order Monoblepharidales. *Monoblepharus.*
      Subclass Domycetidae.
       Order Lagenidiales. *Olpidiopsis.*
       Order Saprolegniales. *Saprolegnia, Achlya.*
       Order Leptomitales. *Sapromyces.*
       Order Peronosporales. *Albugo, Pythium, Phytophthora.*
      Subclass Zygomycetidae.
       Order Mucorales. *Rhizopus, Mucor.*
       Order Entomophthorales. *Entomophthora.*
       Order Zoopagales. *Endocochlus.*
     Class Ascomycetes (sac fungi).
      Subclass Hemiascomycetidae.
       Order Endomycetales. *Saccharomyces.*
       Order Ascoideales. *Ascoidea.*
       Order Taphrinales. *Taphrina.*
      Subclass Euascomycetidae.
       Order Eurotiales. *Talaromyces, Penicillium, Aspergillus.*
       Order Erysiphales. *Erysiphe, Phyllactinia.*
       Order Xylariales. *Neurospora, Xylaria.*
       Order Hypocreales. *Claviceps.*
       Order Pleosporales. *Pleospora.*
       Order Helotiales. *Pseudopeziza, Monilinia.*
       Order Pezizales. *Peziza, Morchella.*
       Order Tuberales. *Tuber.*
     Class Basidiomycetes (club fungi).
      Subclass Heterobasidiomycetidae.
       Order Tremellales. *Tremella.*
       Order Uredinales. *Puccinia.*
       Order Ustilaginales. *Ustilago.*
      Subclass Homobasidiomycetidae.
       Order Polyporales. *Clavulina.*
       Order Agaricales. *Agaricus, Armanita, Caprinus.*
       Order Hymenogastrales. *Gasterella.*
       Order Lycoperdales. *Lycoperdon, Arachnion.*
       Order Phallales. *Phallus, Mutinus.*
       Order Nidulariales. *Nidularia, Crucibulum.*
     Class Deuteromycetes (imperfect fungi).
      Order Sphaeropsidales. *Phoma.*
      Order Melanconiales. *Pestalotia.*
      Order Moniliales. *Sporobolomyces.*
      Order Mycelia sterilia. *Sclerotium.*
     Class Lichenes (lichens).
      Subclass Ascolichenes (ascomycete symbiont).
      Subclass Basidiolichenes (basidiomycete symbiont).

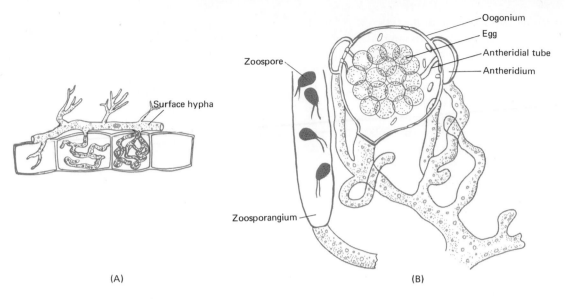

(A)

(B)

Figure 17-6. Representative Phycomycetes. (A) A downy mildew penetrating the epidermis of a leaf. (B) *Saprolegnia* showing details of asexual and sexual reproductive structures. (C) *Rhizopus,* the common bread mold. Note the conspicuous coenocytic hyphae, a characteristic of all Phycomycetes. (Photo courtesy of Carolina Biological Supply Company.)

fourth class, *Deuteromycetes,* includes fungi for which no sexual reproductive cycle has been observed. Fungi in this artificial class are known only by their conidial (asexual) stages. When one of the Deuteromycetes is proved to be an asexual stage of a known species of one of the other three classes, it passes automatically to that group. In making such transfers, the number of Ascomycetes has increased substantially, that of Basidiomycetes somewhat less, and that of the Phycomycetes not at all. The lichens (*Class Lichenes*) may, for convenience, be placed adjacent to the true fungi because the greater proportion of the plant body consists of the fungal component. Their exact

(C)

Figure 17-7. Representative Ascomycetes. (**A**) *Saccharomyces*, yeast. (**B**) *Peziza*, a cup fungus. (**C**) *Morchella*, a morel. (**D**) *Aspergillus*, a blue-green mold. (**E**) Enlarged aspect of asci and ascospores and their relationship to the vegetative hyphae.

position in the plant kingdom, however, is not clear.

As with the algal divisions, before analyzing the evolutionary trends within the Eumycophyta, a brief survey of the fungal classes will be presented so that an understanding of the diversity of form can be gained.

Class Phycomycetes (Algal Fungi). This class is so named because, in terms of structure and reproductive patterns, its members are quite similar to certain green algae. Most algae fungi contain branching, nonseptate hyphae with haploid nuclei. A few simple forms are unicellular. Cross walls, or septations, appear only during the formation of reproductive structures (Figure 17-6).

Class Ascomycetes (Sac Fungi). These fungi constitute one of the largest classes, ranging from unicellular yeasts to rather complex cup fungi (Figure 17-7). The vegetative hyphae of these plants, unlike those of the Phycomycetes, are *septate* (separated by cross walls), with each com-

partment usually containing a single nucleus. The plant body of the larger forms consists of a much-branched mycelium, extending throughout the substratum, and a definite fruiting body (*ascocarp*) developed at the surface as a result of sexual reproduction. The body of the ascocarp consists of a mass of sterile hyphae, partially or completely surrounded by a protective envelope of compact mycelia. Despite enormous diversity in their fruiting bodies, all ascomycetes are similar in that they form a reproductive structure called an **ascus** during the sexual cycle (Figure 17-7E).

Class Basidiomycetes (Club Fungi). A number of fairly large and conspicuous fungi such as mushrooms, puffballs, toadstools, and bracket fungi belong to this class (Figure 17-8). The fruiting bodies of these plants, although apparently a solid mass of tissue, is nevertheless composed of hyphae. The principal distinguishing characteristic of this group is a club-shaped reproductive structure, the **basidium,** the swollen terminal cell

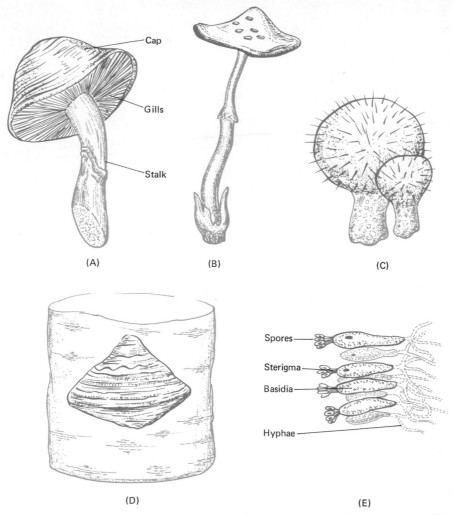

Figure 17-8. Representative Basidiomycetes. (**A**) *Agaricus*, common edible mushroom. (**B**) *Amanita*, poisonous mushroom, or toadstool. (**C**) *Lycoperdon*, puffballs. (**D**) *Fomes*, bracket fungus. (**E**) Enlarged aspect of basidia and associated structures.

of a hypha, which typically bears four basidio-spores (Figure 17-8E). Each basidium is supported on a delicate stalk called the *sterigma*.

Class Deuteromycetes (Imperfect Fungi). This is a large, miscellaneous, artificial group of fungi in which sexual reproduction has not been identified. Most are saprophytes, but a large number

are parasites, particularly on the leaves and fruits of plants. Inasmuch as many species of this group are the asexual stages of ascomycetes, they are classified as such.

Class Lichenes (Lichens). For convenience, these plants are placed adjacent to the true fungi. They are widely distributed on barks of trees, rotting wood, rocks, and on the soil. Lichens represent a symbiotic relationship in which algae (Chlorophyta or Schizophyceae) are entangled in mycelia of fungus plants (usually ascomycetes). According to the form of the plant body, lichens are of three general types: *crustose* (crustlike), *foliose* (leaflike), and *fruticose* (shrublike). These are shown in Figure 17-9A.

Asexual reproduction is by means of **soredia,** small masses of hyphae enclosing a few algal cells

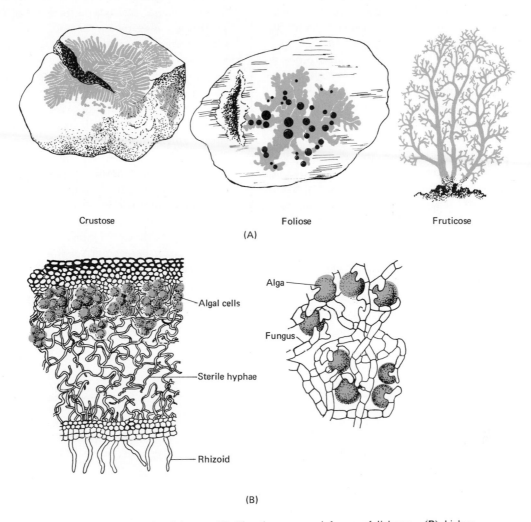

Crustose

Foliose

Fruticose

(A)

Algal cells

Sterile hyphae

Rhizoid

Alga

Fungus

(B)

Figure 17-9. Representative lichens. **(A)** The three general forms of lichens. **(B)** Lichen thallus and reproduction. *Left:* Section through the thallus showing the mass of mycelial filaments and algal cells embedded in the mycelium. *Right:* Enlarged aspect of a soredium.

(Figure 17-9B). Soredia appear as minute powdery masses on the surface of a lichen. Each soredium is capable of producing a new plant.

The association between the fungus and the alga in a lichen has prompted much speculation as to the relationship which might exist. They probably aid each other in nutrition, the fungus providing water and essential elements and the alga producing food by photosynthesis.

Lichens are of primary natural importance as pioneers in plant succession on bare rock (Chapter 21). They gradually degrade the surface of the rock, thus initiating soil formation. Eventually, when adequate amounts of soil are accumulated, other plants may become established.

In the following sections, the Phycomycetes, Ascomycetes, and Basidiomycetes will be treated with respect to distinguishing characteristics, evolutionary trends, possible relationships, and economic importance.

Form and Function of the Algal Fungi. The members of the class *Phycomycetes* are generally regarded as the most primitive class in the Eumycophyta. As their name implies, algal fungi bear certain resemblances to the green algae, particularly in methods of reproduction, although they are much simpler in these respects than the higher fungi. The phycomycetes occur in both aquatic and terrestrial habitats and range from microscopic unicellular organisms to those composed of hyphae. The species containing hyphae are coenocytic, and cross walls form only when reproductive structures are produced. Phycomycetes obtain nutrients as either parasites or saprophytes. Parasitic species grow on the bodies of other fungi, algae, seed plants, and animals, particularly insects and fish. Saprophytic organisms grow in damp soils, dung, and upon dead plant or animal remains.

Methods of sexual reproduction are quite similar to those of the algae in that isogamy, heterogamy, and oogamy are of common occurrence. All species reproduce asexually by spores (sporangiospores). In aquatic species, zoospores are produced, whereas terrestrial organisms give rise to nonflagellated spores (aplanospores).

Within the Phycomycetes a number of evolutionary trends have developed. Among these were the development of complex thalli which differentiated for vegetative and reproductive processes, and the transition from aquatic to terrestrial habitats. As will be noted later, the type and position of flagella on zoospores may be used to separate algal fungi into various taxa and are

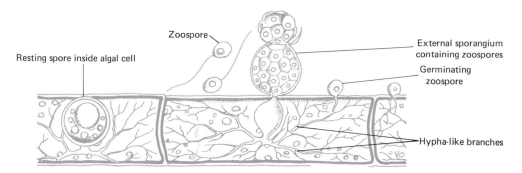

Resting spore inside algal cell

Zoospore

External sporangium containing zoospores

Germinating zoospore

Hypha-like branches

Figure 17-10. *Chytridium lagenaria* parasitizing an algal filament shown in various stages of the life cycle.

important in determining the relationships of the various fungal groups.

Members of the order *Chytridiales,* the chytrids, are among the smallest and simplest fungi known (Figure 17-10). The thallus consists of a single cell, sometimes bearing hyphalike branches. Most of these organisms are parasites of algae, aquatic fungi, microscopic animals, insects, and seed plants. In fact, some cause serious desease of higher plants such as potato-wart disease, in which dark wartlike outgrowths develop on the tubers; a leaf and fruit disease of cranberries; and a brown-spot disease of corn.

Chytrids are all either aquatic or have a dependence on water for their dispersal. Sexual reproduction seems to be rather infrequent in this group and is extremely varied. When it does occur, however, there is a fusion of motile isogametes (entire thalli) which give rise to zygotes. The zygotes develop into thick-walled resting spores, or sporangia, because these organisms are unicellular. Within the sporangia meiosis occurs, and upon germination the zoospores are released. The chytrid zoospores bear a single posterior whiplash flagellum. Essentially, the zoospores and motile gametes are similar in appearance, and in many cases they are interchangeable in function. The primitive character of Chytridiales is apparent in both thallus structure and life history.

In some of the more complex Phycomycetes, the thalli show a more advanced form both in vegetative and reproductive characters. The order *Saprolegniales* (water molds) contains some representative genera that illustrate these advances. These fungi are entirely aquatic and exhibit a closer relationship to algal forms than any other order of Phycomycetes, especially in certain morphological features. Its members live chiefly as saprophytes on dead leaves and twigs, dead insects, and other animal bodies. A few are parasitic, attacking fish and amphibians.

A representative member of this order is *Saprolegnia* (Figure 17-11). This fungus possesses an extensive coenocytic mycelium, the walls of which contain cellulose. Some of the hyphae are slender tapering threads which attach to the substratum; others, more blunt in form, spread out freely into the water. At the terminal portions of these free-floating filaments cross walls partition elongated sporangia, the *zoosporangia.* Within the zoosporangia, many biflagellate zoospores are produced. Upon their release from a pore at the top, they locomote for a time, settle down, lose their cilia, and develop a surrounding wall. Eventually, these spores germinate to form new hyphae.

Sexual reproduction is heterogamous. Oogonia appear as enlarged cells containing a number of spherical, uninucleate eggs. The antheridia, containing sperms, grow toward the oogonia, penetrate them by means of delicate fertilization tubes, and discharge their contents. The male nuclei fertilize the eggs and form zygotes, which germinate to produce new hyphal filaments. Inasmuch as a single thallus bears both male and female sex structures, plants such as *Saprolegnia* are termed **homothallic.** Other genera of this order, such as *Achlya,* are **heterothallic;** that is, antheridia and oogonia are produced on separate individuals.

The more primitive members of Phycomycetes produce biflagellate zoospores; the more advanced forms are terrestrial and develop nonmotile spores. Many algal fungi have become quite successful in terrestrial habitats, and modifications which make these organisms adaptable to the land, have given rise to thalli which diverge considerably from aquatic forms. The transition from aquatic to amphibious to terrestrial habitats is well illustrated by a series of genera such as *Saprolegnia* (Saprolegniales), *Pythium* (Peronosporales), Phytophthora (Peronosporales), and Rhizopus (Mucorales).

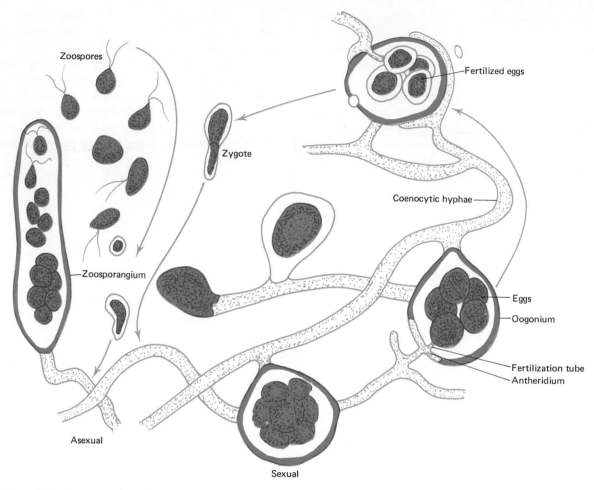

Figure 17-11. *Saprolegnia.* Asexual and sexual reproduction.

Labels in figure: Zoospores, Zygote, Fertilized eggs, Coenocytic hyphae, Zoosporangium, Eggs, Oogonium, Fertilization tube, Antheridium, Asexual, Sexual

Saprolegnia, as stated earlier, is a water mold, and although it typically inhabits calm freshwater, it may also be found in well-irrigated soils. In this regard, it may be viewed as an amphibious fungus in which there is a strong tendency toward the aquatic environment but there is also a terrestrial tendency. Representative genera of the order *Peronosporales* exhibit various amphibious characters which later gave rise to strictly terrestrial adaptations. Members of this order, also called blights and downy mildews (Figure 17-6A), are chiefly parasitic on higher plants and have thalli composed of whitish, variably branching hyphae. Their spores germinate on the surface of a stem or leaf and the hyphae enter the tissues. Once inside the tissues, the hyphae absorb food by slender projections **(haustoria)** and, quite often, the host dies.

Pythium is a genus more or less intermediate between the Saprolegniales and the *Peronosporales.*

It is found in soil more often than in water and lives both parasitically and saprophytically. It attacks various species of higher plants, causing rotting of roots and a seedling disease called "damping off." *Phytophthora infestans,* a fungus that causes a serious potato disease, produces sporangia containing biflagellated zoospores when conditions are favorable, but, in the absence of environmental water, the sporangia may become detached and grow directly into a new mycelium. The conidia of *Phytophthora infestans* are borne on specialized conidiophores which emerge from the potato leaf through the stomates. Depending on environmental conditions, the conidium may function as a sporangium and produce zoospores, or it may germinate directly, thus functioning as a true conidium. Such a phenomenon illustrates one advance in the transition from an aquatic to a terrestrial habitat, because there is less of a dependence upon water for spore dissemination.

The culmination of the evolutionary trend from aquatic to terrestrial habitats is well illustrated by members of the *Mucorales,* which produce nonmotile, encapsulated spores well adapted to disperal on dry land. In such terrestrial forms, motile spores are absent, and there is a great dependence upon small, light, airborne spores. Fungi of this order are common on damp organic matter, such as bread, rotting fruit, and other foodstuffs. The common bread mold *Rhizopus* is the most familiar example (Figure 17-12).

The mycelium of *Rhizopus* consists of nonseptate multinucleate hyphae which are designated according to position and form. The absorptive hyphae which penetrate the bread are called **rhizoids;** those directly over the surface of the bread and running more or less parallel to it are termed **stolons;** those arising from the horizontal stolons are called **sporangiophores.** Each sporangiophore terminates as a swollen globulose **sporangium** which begins as a swelling into which many nuclei migrate. The central portion becomes separated from the peripheral zone by the deposition of a wall and is designated the **columella.** The protoplasm surrounding the columella is partitioned into small portions which develop into **sporangiospores.** The sporangium breaks open rather irregularly, releasing the spores, each of which is capable of forming new hyphae. This asexual cycle may be repeated indefinitely.

Sexual reproduction occurs only when mycelia of two morphologically indistinguishable but physiologically distinct strains (plus and minus) come into contact. At contact, a hormonal mechanism becomes operative and the tips of adjacent hyphae (progametes) are delimited by walls into single, multinucleate cells, the **gametangia.** These contiguous gametangia fuse to form a thick-walled **zygospore** which enters a period of dormancy from one to three or more months. At germination, meiosis occurs and the zygospore breaks open and produces a single sporangiophore bearing a sporangium.

Form and Function of the Sac Fungi. The *Ascomycetes* constitute a myriad and varied group of fungi, many of which are economically important in agriculture, medicine, and industry. Most are typically land-inhabiting plants and include both parasitic and saprophytic species. Their thalli vary from unicellular yeasts to powdery mildews and cottony molds to complex cup fungi. The vegetative hyphae are septate, unlike those of the Phycomycetes, and are therefore multicellular filaments. None of the species produce motile spores or gametes, even though a few are aquatic.

Most Ascomycetes form conidia, and after sexual reproduction all sac fungi produce a distinguishing saclike structure, the **ascus,** within which usually eight **ascospores** are produced. A group of asci is generally embedded in a mass of sterile hyphae and partially or completely surrounded by a protective covering consisting of compact mycelium. Such a fruiting body is termed an **ascocarp.**

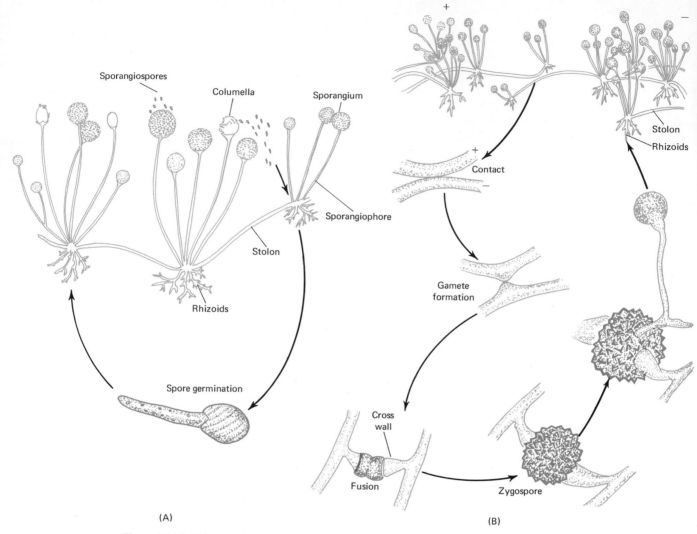

Figure 17-12. *Rhizopus nigricans* (black bread mold). **(A)** Vegetative mycelia showing various types of hyphae and method of spores production. **(B)** Sexual reproduction.

An ascus usually develops as a result of a sexual process (Figure 17-13), and the events leading to its formation are fairly complicated. Hyphae from heterothallic, haploid mycelia develop multinucleate sex organs. The female sex organ, the

ascogonium, consists of a globose, multinucleate basal cell and an elongated terminal portion, the *trichogyne.* The male sex organ, the **antheridium,** is somewhat elongated and slender and contains haploid nuclei. The trichogyne grows to the

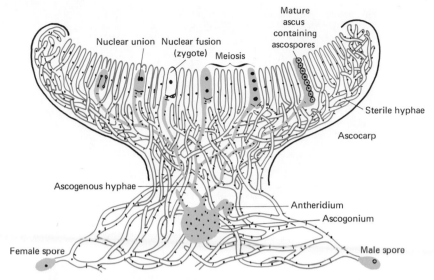

Figure 17-13. Life cycle of a typical Ascomycete as seen in a magnified section through a cup fungus. See text for amplification. All haploid structures are shown in green. (From *Fundamentals of Cytology*, by L. W. Sharp, copyright 1943, McGraw-Hill, Inc. Used by permission of McGraw-Hill Book Company.)

ascogonium, the walls between them dissolve, and the antheridial nuclei pair with those of the ascogonium. Inasmuch as the nuclei do not fuse, this occurrence is not considered to be a true sexual process. After one or more nuclei are paired in the ascogonium several filaments, called *ascogenous hyphae,* are produced. Eventually the paired nuclei of terminal cells of each of these hyphae fuse to form zygotes. Coincident with this fusion, there is septation of the hyphae. The diploid nucleus within the zygote then undergoes meiosis, producing four haploid nuclei, each of which divides mitotically. The eight nuclei thus formed are the ascospores, and the zygote is now designated as an ascus. At maturity, the asci open and release the haploid spores. Each ascospore, at germination, may form hyphae and ultimately a new fungus plant.

The simpler Ascomycetes, including yeasts and related forms, are placed in the subclass *Hemiascomycetidae.* These fungi form a single ascus and the thallus is a loose network of hyphae. In addition, the ascus is born free at the tip of a hypha, without any protecting or supporting mycelium. A second, larger subclass, the *Euascomycetidae,* consists of fungi with filamentous thalli and all produce ascocarps.

Subclass Hemiascomycetidae. Some of these fungi are parasitic, but the vast majority are saprophytic as inhabitants of water and soil, on fruits, and in exudates from injured plant tissues. Two important genera are *Saccharomyces* and *Taphrina.*

The *Saccharomyces,* or yeasts, are the simplest Ascomycetes. The plant body is unicellular (Figure 17-14A) and a few do produce hyphae, although the formation of hyphae is apparently reduced. Yeasts are economically important in that they ferment sugars to alcohol and CO_2. The produc-

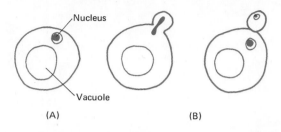

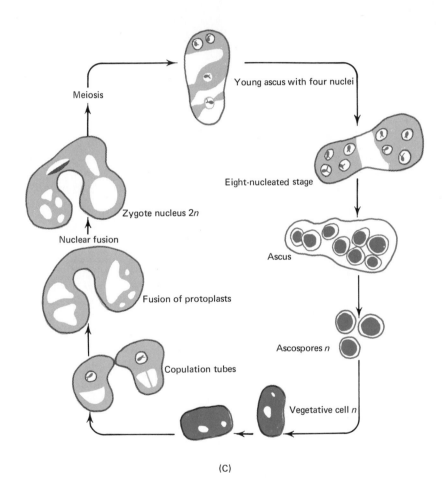

(C)

Figure 17-14. *Saccharomyces.* (**A**) Vegetative cell. (**B**) Asexual reproduction by budding. (**C**) Sexual reproduction.

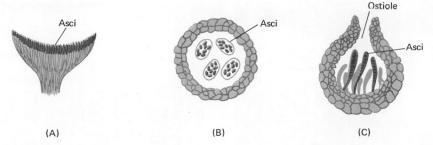

Figure 17-15. Types of ascocarps (fruiting bodies) found among Ascomycetes. **(A)** Apothecium. **(B)** Cleistothecium. **(C)** Perithecium.

tion of wines and beers and the baking of leavened bread are intimately associated with the products of alcoholic fermentation. Reproduction in yeasts is most commonly asexual by budding (Figure 17-14B). In this process, a small protuberance forms, expands, and separates from the mother cell. During bud formation, the nucleus divides and one daughter necleus migrates into the developing bud. Sexual reproduction may occur when two adjacent cells form short processes and their contents fuse, producing a diploid zygote nucleus. The cell with the fused contents functions as an ascus in which meiosis and mitosis occur. The meiotic division gives rise to four haploid cells, and a further mitotic division produces eight cells, the ascospores. Each spore is capable of forming a new yeast cell (Figure 17-14C).

Taphrina deformans, leaf-curl fungus, is a parasite of vascular plants, particularly on the leaves of peach trees. The hyphae of the fungus penetrate between cells of the host, which typically induces abnormal growth of host tissues. The host cell increases both in size and number, and, as a result, there is a curling of the leaves. The fungal hyphae bear asci and an infected leaf develops surface wounds where asci erupt through the epidermis.

Subclass Euascomycetidae. All members of this subclass form prominent fruiting bodies (ascocarps) which may be categorized into three general types (Figure 17-15). An **apothecium** is an open cup-shaped ascocarp bearing an exposed layer of asci. A **cleistothecium** is a spherical fruiting body without external openings, containing asci arranged at varying levels. A **perithecium** is a flask-shaped ascocarp with a pore at the top, the *astiole.*

Apothecial fruiting bodies are characteristic of the *cup fungi* (Figure 17-16). These fleshy fungi of varied sizes and forms are found commonly in soils and upon dead trunks, branches, and stumps of trees. The ascocarps are often brightly colored, especially in shades of red and orange. In *Peziza* (Figure 17-16A), the apothecium is the conspicuous portion of the plant, but absorptive hyphae penetrate the substratum, forming extensively branched mycelial masses, from which the visible reproductive bodies develop. Asci are produced along the inner surface of the cup-shaped fruiting body and ascospores are discharged when a lid (*operculum*) on each ascus opens.

Related to the cup fungi are the morels (*Morchella*), also called the sponge fungi. These develop large spongelike ascocarps in which each depression is comparable to the cup-shaped ascocarp of *Peziza* (Figure 17-16B). Both morels and

(A)   (B)

Figure 17-16. Apothecial ascocarps of (**A**) *Peziza* (a cup fungus) and (**B**) *Morchella* (a sponge fungus). (Courtesy of Carolina Biological Supply Company.)

truffles, which produce subterranean, tuber-like fruiting bodies, are edible fungi and are prized as delicacies.

Among the cleistothecial-type ascocarps are genera such as *Aspergillus, Penicillium,* and *Erysiphe.* The first two genera belong to an order known as the blue and green molds (*Eurotiales*). The last is a member of the *Erysiphales,* or the powdery mildews. *Aspergillus* and *Penicillium* are common molds which appear on almost any damp organic substrate. The abundant conidial masses usually form radiating chains and are typically bluish or greenish in color. In *Aspergillus,* the conidia are borne on globose hyphal tips (Figure 17-17A). These fungi, which appear as green and yellow growths, cause the decay of many types of stored fruits and vegetables and the deterioration of leather goods and fabrics. Several species are responsible for ear and lung infections in domesticated mammals and man. *Aspergillus* species are

also used in the production of alcohol from rice starch and in the manufacture of organic acids, including citric and oxalic acid.

In *Penicillium,* the conidiophores are branched and each branch bears numerous chainlike conidia (Figure 17-17B). Growths of this fungus appear as blue molds that are responsible for the spoilage of bread, fruits, paper, leather, and lumber. Some of the beneficial attributes of *Penicillium* include the ripening of various types of cheese (Roquefort and Camembert) and the source of the antibiotic penicillin.

Powdery mildews are chiefly parasitic on the leaves and, less commonly, the stems of higher plants such as grapes, grasses, roses, apples, and various cereal plants (Figure 17-17C). The fruiting body is usually black and bears hyphal appendages. Some of the hyphae are branched and extend into space; others (haustoria) penetrate the host tissue and absorb food. The powdery mildews

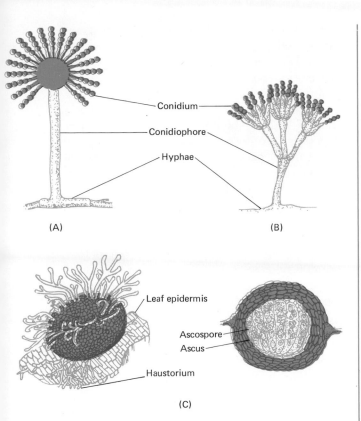

(A)

(B)

Conidium

Conidiophore

Hyphae

Leaf epidermis

Ascospore

Ascus

Haustorium

(C)

Figure 17-17. Cleistothecial-type ascocarps. **(A)** *Aspergillus.* **(B)** *Penicillium.* **(C)** Powdery mildew. Hyphae and cleistothecium on the surface of a host leaf. *Right:* Section through the cleistothecium showing asci and ascospores.

appear as white, dustlike patches on the surface of the host and are generally only mildly parasitic. A few species, however, do cause rather extensive crop losses.

Perithecial fruiting bodies are formed by such genera as *Claviceps, Neurospora, Ceratocystis, Venturia,* and *Endothia.* The ergot disease of rye is caused by *Claviceps purpurea* (Figure 17-18), in which the grains are enlarged into purplish bodies filled with hyphae. These bodies contain a toxic substance, *ergotinine,* which poisons and kills organisms con-

suming the grains. Paradoxically, ergot is an important drug used to check hemorrhaging. *Neurospora,* the pink bread mold, produces numerous pink or orange conidia within the oval or pear-shaped perithecia. Much knowledge concerning the biochemical aspects of genetics has been derived from studies of *Neurospora crassa.* Of considerable economic importance are perithecial types such as *Ceratocystis ulmi,* the fungus responsible for the Dutch elm disease; *Venturia inaequalis,* the causative agent of apple scab; and *Endothia*

Figure 17-18. Perithecial ascocarp of *Claviceps.* Logitudinal section through fruiting body showing perithecia along periphery. (Courtesy of Carolina Biological Supply Company.)

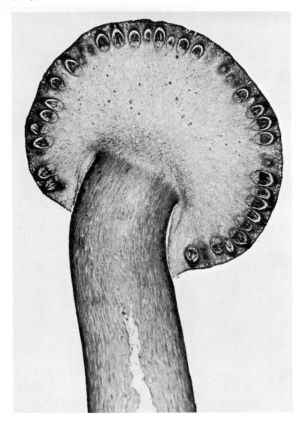

parasitica, the organism responsible for chestnut blight.

Form and Function of the Club Fungi. All members of the *Basidiomycetes* are characterized by a special reproductive structure, the **basidium.** A basidium is typically an enlarged terminal, club-shaped cell of a hypha, bearing four **basidiospores** on small stalks called the *sterigmata.* Whereas ascospores develop inside an ascus, basidiospores develop on the outside of a basidium. Some botanists consider a basidium to be a modified ascus, and there is some evidence to suggest that club fungi have evolved from ascomycete ancestors. In most members of this class, septate hyphae produce conspicuous fruiting bodies **(basidiocarps)** after sexual reproduction has occurred.

Sexual reproduction results from the fusion of uninucleate, septate hyphae (Figure 17-19). As in the Ascomycetes, nuclear fusion is delayed, so that the binucleate cell produced from two different hyphae carries one nucleus from each parent, and the nuclear pair in each cell is a dicaryon ($n + n$). The end cells of various hyphae become swollen and club-shaped and the two haploid nuclei fuse to form a diploid ($2n$) zygote, which then undergoes meiosis to form four haploid (n) nuclei (Figure 17-19). The club-shaped basidium produces four protuberances into each of which one of the nuclei passes. When a nucleus reaches the apex of a protuberance, the tip of the protuberance becomes walled off. Enlargement of the structures in which the nuclei are now located completes the formation of basidiospores. At maturity, the basidiospores are released and, upon germination, form hyphae of uninucleate cells. The cells of these primary hyphae, after a short period of growth, become binucleate either by fusion of cells of different mycelia (in heterothallic species) or by fusion of cells of the same hypha in homothallic organisms. At the time of basidiospore formation,

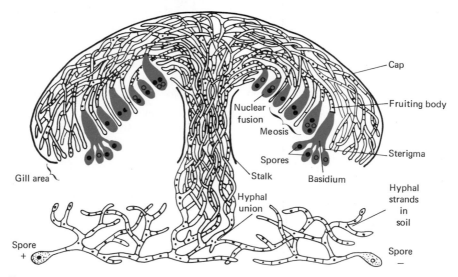

Figure 17-19. Life cycle of a typical Basidiomycete as seen in a magnified section through a mushroom. (From *Fundamentals of Cytology,* by L. W. Sharp, copyright 1943, McGraw-Hill, Inc. Used by permission of McGraw-Hill Book Company.)

the two nuclei fuse and the events proceed as described above.

The class Basidiomycetes is divided into two subclasses, the *Heterobasidiomycetidae* and the *Homobasidiomycetidae.* These are separated from one another on the basis of basidial structure, basidiospore germination, and other features. In the Heterobasidiomycetidae the basidia are partitioned transversely or longitudinally into two, three, or four cells. Each cell produces one basidiospore. The basidia frequently arise from thick-walled resting spores. Basidia of the Homobasidiomycetidae are relatively uniform, nonseptate, and more or less club-shaped. In addition, the basidium is unicellular and usually produces four basidiospores. Upon germination, basidiospores of most Homobasidiomycetidae form a germ tube or hypha; those of many Heterobasidiomycetidae produce conidia of one or more types.

Subclass Heterobasidiomycetidae. Among the Heterobasidiomycetidae are the jelly fungi (*Tremellales*), smuts (*Ustilaginales*), and rusts (*Uredinales*). The term "jelly fungi" is given to species of the order Tremmellales, a name derived from the gelatinous basidiocarp texture found in many species. These fungi contain hyphae within the jelly from which basidia are produced and are found as saprophytes on dead tree branches and decaying logs.

From an economic standpoint, the rusts and smuts are the most important Heterobasidiomycetidae. Both are parasites of vascular plants. Smuts are a large group of fungi so named because of their dark-colored spores. They attack floral organs and associated structures, chiefly among members of the grass family. All known species require a single host (*autoecious*) for completion of their life cycle. Smuts parasitize a number of economically important plants, principally cereals. Chief among these are *Ustilago maydis,* the common corn smut, and *Ustilago tritici,* the loose smut of wheat.

Rusts have the most complicated histories of any of the fungi. None produce basidiocarps, although as many as five different kinds of spores may be formed by a single species. Because of the rusty leaves and stems which they cause on the host plant, they are called rusts. The rust fungi are virulent parasites of various ferns and seed plants. Many are capable of parasitizing two unrelated species of hosts, alternating between one host and another (*heteroecious*). Probably the most important rust to man is *Puccinia graminis* (Figure 17-20), the causative agent of the black stem rust of wheat. This fungus, found everywhere that wheat is grown, destroys chlorophyll and uses the sugars synthesized by the wheat, resulting in decreased grain size and reduced per acre yield.

Subclass Homobasidiomycetidae. Within the subclass Homobasidiomycetidae are included a number of fungi that are fairly well known (Figure 17-8). Some familiar representatives are mushrooms (toadstools), bracket fungi, and puffballs. *Agaricus* (Figure 17-8A) is the common edible mushroom, which is a saprophytic plant. The fleshy basidiocarp consists of a *stipe* (stalk), an umbrella-shaped cap, or *pileus,* the underside of which contains *gills.* Unicellular club-shaped basidia are exposed along the surfaces of the gills. After the basidiospores are shed, they germinate to form new hyphae. Nuclear fusions occur among the hyphae and, after further vegetative enlargement of the mycelia, new basidiocarps are produced.

The bracket fungi are similar in many respects to *Agaricus,* differing chiefly in the structure of the fruiting body and in the greater tendency to grow on wood. The genus *Fomes* (Figure 17-8D) includes tree-inhabiting species, the hyphae of which produce exoenzymes that aid in the digestion of bark and wood. After a period of vegetative activity, the hyphae produce basidiocarps which, in many species, are shelflike growths on the sides of trees or logs.

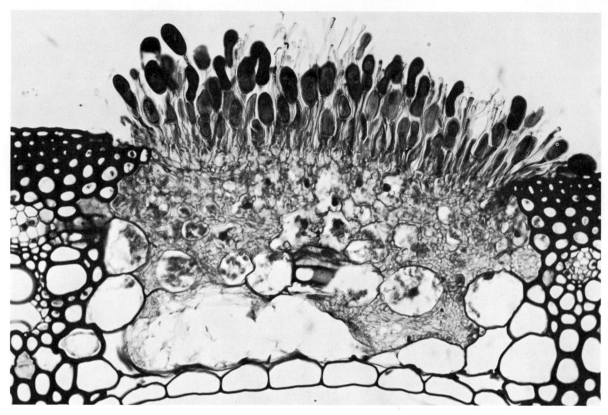

Figure 17-20. Heterobasidiomycetidae. *Puccina graminis* (wheat rust) in the spore-producing stage on a wheat stem. (Courtesy of Carolina Biological Supply Company.)

The basidiocarp in puffballs is a globular structure which in some species becomes very large and is usually edible when young (Figure 17-8C). The fruiting bodies of puffballs vary greatly in size (some exceed 4 feet), color, form, and texture. Spore release occurs when the fruiting body decomposes or is mechanically disturbed. A single puffball may produce as many as 7500 billion spores.

Phylogenetic Relationships among Fungi

As noted earlier in this chapter, the origin of the fungi is still quite obscure. In many ways the fungi parallel the algae, and it would appear that at least some of the fungal groups had algal ancestors or evolved from the same kinds of ancestors as did some algae. Yet there is good evidence to suggest that not all fungi had the same ancestry and that at least some had origins among protozoans.

With regard to class relationships, the structure and reproduction of the Phycomycetes are very similar to those of certain algae, especially siphonous Chlorophyta. Such parallels suggest an algal ancestry for the algal fungi. The presence of amoeboid movement in zoospores of some chytrids may indicate a relationship with Myxomycophyta.

Two general theories have been advanced to explain the origin of Ascomycetes. First, some have suggested that Ascomycetes evolved from certain red algae by the loss of photosynthetic pigments. In these two plant groups, the form of the reproductive structures, as well as certain details of the sexual cycle, appear to be similar. The more commonly accepted viewpoint, however, is that the Ascomycetes evolved from the algal fungi. Once again, support of the second theory is derived from similarities of reproductive structures.

Basidiomycetes and Ascomycetes show a considerable amount of similarity during the development of the basidium and basidiospores and the ascus and ascospores. The dicaryon condition of certain hyphae and the formation of conidia in both groups suggest a close relationship. Moreover, the general development of the basidium is like that of the ascus. Finally, it might be noted that fusion of male and female nuclei, followed by meiosis, occurs immediately before the formation of basidiospores and ascospores. It is generally believed that a basidium is an evolutionary development from an ascus, the number of spores being reduced from eight to four, and the matured spores having migrated to an external position on the stcrigmata of a basidium.

CHAPTER IN RETROSPECT

The slime molds (Myxomycophyta) and true fungi (Eumycophyta) represent two distinct divisions of nonchlorophyllous, spore-bearing plants which are traditionally referred to as fungi. Blue-green algae and bacteria, also grouped with fungi, differ from slime molds and true fungi in that they are considerably smaller (microscopic) and exhibit different patterns of reproduction and nutrition. Virtually all fungi are parasites or saprophytes,

because they lack a photosynthetic mechanism. Slime molds possess a holozoic mode of nutrition which bears a remarkable resemblance to the food-getting process of amoebae. However, the vegetative thallus of true fungi can only absorb dissolved nutrients. The pattern of sexual reproduction is relatively complex and differs somewhat in each of the fungal groups. All fungi form spores. In the case of the common puffball, one large fruiting body may produce billions of spores.

As a group, slime molds have very little economic significance, although they are interesting subjects for various experimental investigations. Slime molds are grouped into three classes: (1) true slime molds (Myxomycetes), (2) cellular slime molds (Acrasiomycetes), and (3) net slime molds (Labyrinthulomycetes). Class distinction is based primarily on differences in the structure of fruiting bodies and details of the life cycles. By contrast, certain representatives of Eumycophyta are both biologically and economically significant. For instance, morels, truffles, and mushrooms are tasty foods; wheat rusts, smuts, and mildews cause serious economic losses; and lichens may serve as pasturage for numerous grazing animals as well as pioneers on bare-rock plant successions. There are approximately 80,000 species of Eumycophyta grouped into five classes, as shown in Table 17-2. Some common representatives include mushrooms, puffballs, toadstools, smuts, rusts, and bracket fungi (Basidiomycetes); yeasts, mildews, morels, truffles, *Penicillium*, pink bread mold, and leaf-curl fungus (Ascomycetes); and common bread mold and water mold (Phycomycetes).

Questions and Problems

1. Compare the structure, nutrition, and sexual behavior of the three classes of slime molds (Myxomycetes, Acrasiomycetes, and Labyrinthulomycetes). Why are slime molds sometimes classified as ani-

mals? Give reasons why you think they are more plantlike than animal-like or vice versa.

2. Consider the economic significance of slime molds and true fungi and compare them with other plant groups. Which of the fungal representatives is most important to the economy of man and to the economy of nature?

3. List the essential differences that may exist in the reproductive patterns of algal fungi and sac fungi. Compare the life cycles of Myxomycophyta and Eumycophyta.

4. Would you describe the relationship between the fungus and the alga in a lichen as symbiotic? Explain.

5. Describe the evolutionary origin of nonphotosynthetic protista. Are all members of this group phylogenetically related? Give evidence to support your belief.

6. Compare the nutrition of slime molds with that of true fungi. How does this compare with the algae (exclude blue-green algae)?

7. What is the meaning of the following terms?

plasmodium	haustoria	apothecium
mycelium	stolons	basidiospore
meiospores	zygospore	pileus
ascocarp	perithecium	stipe

Suggested Supplementary Readings

Ahmadjian, V., "Lichens," *Annual Review of Microbiology,* vol. 19, pp. 1–20, 1965.

Ainsworth, G. C., and A. S. Sussman (editors), *The Fungi,* vols. I and II. New York: Academic Press, Inc., 1965 and 1966.

Alexopoulos, C. J., *Introductory Mycology.* New York: John Wiley & Sons, Inc., 1962.

———, "The Myxomycetes II," *Botanical Review,* vol. 29, pp. 1–78, 1963.

———, and H. C. Bold, *Algae and Fungi.* New York: The Macmillan Company, 1967.

———, and J. L. Koevenig. *Slime Molds and Research* (BSCS Pamphlet 13, American Institute of Biological Sciences). Boston, Mass.: D. C. Heath and Company, 1963.

Bonner, J. T., *The Cellular Slime Molds.* Princeton: Princeton University Press, 1959.

Emerson, R., "Molds and Man," *Scientific American,* January 1952.

Lamb, I. M., "Lichens," *Scientific American,* October 1959.

Robinson, R. F., "Food Production by Fungi," *Scientific Monthly,* vol. 75, pp. 149–154, 1952.

18

18
Metaphyta: Nonvascular Land Plants

INTRODUCTION

In Chapter 15 it was noted that in terms of a natural system of classification, living organisms are divisible into four principal kingdoms. Two of these, Monera and Protista, have been the subject of previous chapters. In general, organisms in these two kingdoms are at a relatively low level of organization, the Protista being somewhat more advanced. The third kingdom, the Metazoa, contains organisms that are distinctly animal-like and will not be treated in this text. The final kingdom, the Metaphyta, consists of plantlike organisms which for the most part are terrestrial species that exhibit complex patterns of form and function. Within the kingdom Metaphyta are two major divisions, Bryophyta and Tracheophyta. The bryophytes will be analyzed in this chapter and the tracheophytes (vascular plants) will be treated in Chapters 19 and 20.

Based upon considerable supportive data, it is generally believed that Metaphyta as a whole are descended from algalike plants, probably certain Chlorophyta. In the course of this transition from aquatic to terrestrial habitats, it appears that two different lines of evolution emerged. One of these was characterized by the development of conducting and supporting tissue—that is, a vascular system. The products of this course of evolution are the *Tracheophyta* (vascular plants). In the sec-ond line of evolution no specialized conducting or supporting tissue developed. Modern descendents of this latter line of evolution are the liverworts, hornworts, and mosses—the *Bryophyta*.

DIVISION BRYOPHYTA
General Characteristics

Bryophytes are the most primitive of the green land plants. Although they undoubtedly evolved from green algae, there is no evidence to support the conclusion that they are intermediate evolutionary forms between aquatic plants and primitive vascular plants. The Bryophyta form an isolated group of plants, and, as far as can be determined, they are not ancestors of complex modern vascular plants. In this regard, bryophytes constitute an evolutionary dead end.

As a group, bryophytes have a characteristic life cycle, in which the sporophyte has no direct connection to the substratum but is attached to and dependent upon the gametophyte for its nutrition. The gametophyte generation is always dominant; that is, the gamete-producing plant lasts longer, is larger in form and more conspicuous than the sporophyte, and is nutritionally independent. The common moss plant represents the dominant gametophyte generation.

Mosses and their allies are relatively small plants, either thallose or leafy, ranging in size from $\frac{1}{16}$ inch to about 2 feet, most being less than 6 inches high. Because of their small size, there is no need for an elaborate conducting and supporting system. None of the bryophytes has true roots, stems, or leaves. The rootlike organs (*rhizoids*) of the gametophytes penetrate the substratum and function principally in anchorage, although they may serve in an absorptive capacity. The main shoot (*caulid*) is stemlike in appearance and the "leaves" (*phyllids*) are flattened and generally a single cell thick.

Gametophyte plants may bear both antheridia and archegonia (**monoecious**), or the sex organs may be located on different plants (**dioecious**). In either case, the sex organs are extremely small (generally microscopic) and are multicellular structures which have an outer layer of protective sterile cells. By contrast, sex organs in lower plant groups usually consist of only a single cell.

The male sex organ (**antheridium**) is generally attached to the gametophyte by a stalk and usually appears as an elongated or spherical sac. Within the sac, surrounded by an outer jacket of sterile cells, is a central mass of cells called *androcytes.* Each androcyte develops into a slender, elongate, biflagellate sperm. When the antheridium is mature and wet, the jacket bursts under turgor pressure, releasing hundreds of sperm, one of which fertilizes the egg.

The female sex organ (**archegonium**), also borne on the gametophyte plant, is generally flask-shaped and is differentiated into two distinct portions. The first of these, the *venter,* comprises the swollen base. The second portion, the *neck,* is a continuation of the venter and appears as the slender upper portion of the archegonium. In mosses and liverworts, the archegonia are typically surface structures, but in the hornworts they are almost completely embedded in the gametophyte, with only the tip of the neck exposed. The neck consists of an outer layer of jacket cells around an elongate cavity, or canal. At the base of the canal is a large *egg,* a smaller *ventral canal cell* above this, and then a row of *neck canal cells.* When the archegonium is mature, the neck canal cells and the ventral canal cell disintegrate to form a slimy mass which fills the cavity. This gelatinous mass absorbs water, swells, and ruptures the jacket at the tip of the neck. Once external, the slime attracts sperm and the egg is fertilized. The diploid, fertilized egg (*zygote*) marks the beginning of the sporophyte generation. The sporophyte plant, produced by the zygote, is typically an unbranched structure, attached to the gametophyte by a basal region called the *foot.* Spores are borne in a terminal sporangium termed the **capsule,** which is usually elevated above the gametophyte plant by means of a slender stalk called the *seta.* The spores, which are haploid cells, represent the first stage of the gametophyte generation. After being liberated from the capsules, they germinate, and eventually give rise to green, photosynthetic gametophyte plants.

The life cycle of bryophytes, in which there is an alternation of gamete-producing and spore-producing generations, exhibits the same pattern that is found in the algae. In most algae, however, the fertilized egg, the first cell of the sporophyte, becomes separated from the gametophyte. In all bryophytes and tracheophytes the egg is retained in the tissues of the gametophyte after fertilization. Whereas the gametophyte is the dominant plant in bryophytes, the sporophyte is the dominant plant of the vascular groups.

Classification

The Bryophyta are generally divided into three classes: *Hepaticae* (liverworts), *Anthocerotae* (hornworts), and *Musci* (mosses), (see Table 18-1). The Hepaticae are represented by about 9000 species and 240 genera, the Anthocerotae by 5 genera and approximately 100 species, and the Musci by about 670 genera and nearly 14,500 species.

Class Hepaticae (Liverworts). The *Hepaticae* are the simplest of the Bryophyta. Most are plants of moist environments but only in rare cases are they entirely aquatic. All grow prostrate, or practically so, on various substrates including soil, rotted wood, or bark of trees. The plant body is either *thallose,* which is flat and dichotomously branched, or *leafy,* in which the thallus is composed of an axis (*caulid*) bearing leaflike expansions, the *phyllids* (Figure 18-1). In both forms the sym-

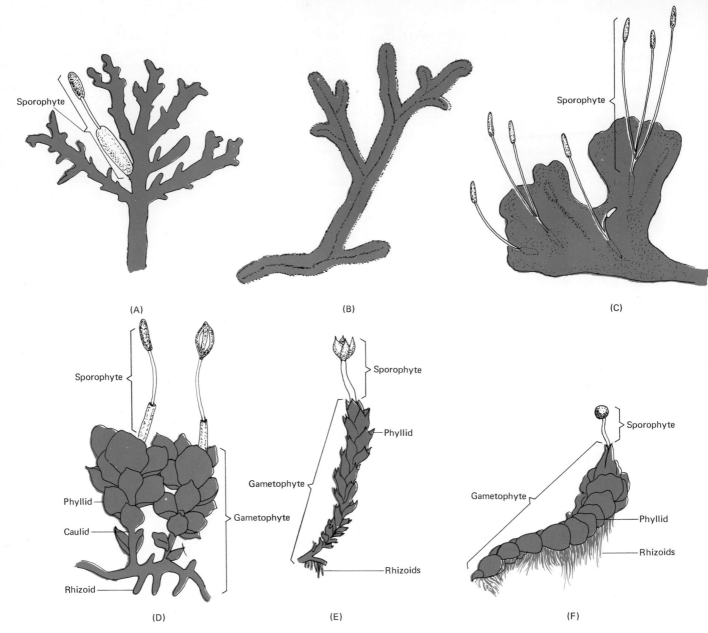

Figure 18-1. Representative liverworts showing various thallus forms. (**A–C**) Thallose types. (**A**) *Riccardia.* (**B**) *Metzgeria.* (**C**) *Monoclea.* (**D–F**) Leafy types. (**D**) *Calobryum.* (**E**) *Marsupella.* (**F**) *Plectocolea.* Note the position of the sporophytes relative to the gametophyte plants.

TABLE 18-1. Classification of the Bryophyta

Kingdom Metaphyta.
  Division Bryophyta (mosses and allies). About
      23,600 species.
    Class Hepaticae (liverworts).
      Order Calobryales. *Calobryum, Takakia.*
      Order Jungermanniales. *Porella, Marsupella.*
      Order Metzgeriales. *Pellia, Riccardia.*
      Order Sphaerocarpales. *Sphaerocarpos,
          Gethallus.*
      Order Marchantiales. *Marchantia, Riccia.*
    Class Anthocerotae (horned liverworts).
      Order Anthocerotales. *Anthoceros.*
    Class Musci (mosses).
      Subclass Sphagnidae.
        Order Sphagnales. *Sphagnum.*
      Subclass Andreacidae.
        Order Andreaeales. *Andreaea.*
      Subclass Tetraphidae. *Tetraphis,
          Tetrodonitium.*
      Subclass Polytrichidae.
        Order Polytrichidales. *Polytrichum,
          Dawsonia.*
      Subclass Buxbaumiidae. *Buxbaumia.*
      Subclass Bryidae.
        Order Fissidentales. *Fissidens.*
        Order Dicranales. *Dicranella.*
        Order Pottiales. *Hymenostrylium.*
        Order Grimmiales. *Gremmia.*
        Order Funariales. *Funaria.*
        Order Schistostegales. *Schistostega.*
        Order Tetraphidales. *Tetraphis.*
        Order Eubryales. *Mnium.*
        Order Isobryales. *Lontinalis.*
        Order Hookeriales. *Hookeria.*
        Order Hypnobryales. *Thiudium.*

metry of the plant is *dorsiventral;* that is, there is a distinct dorsal (upper) and ventral (lower) surface. Rhizoids are developed on the lower surface of the liverworts, whereas the sex organs are borne on the upper surface. Inasmuch as the thallose liverworts exhibit certain features that evolved in early nonvascular land plants, they will be treated as transitional organisms of the bryophytes.

A representative thallose liverwort, *Marchantia* (Figure 18-2A and B), demonstrates the transitional nature of Bryophyta. In many features (vegetative) it is adapted to a terrestrial habitat, but in others (especially reproductive) it is basically an aquatic plant. The gametophyte is a flat, ribbon-shaped dichotomously branching thallus commonly 2 to 3 inches long. It increases in size by growth of one or more apical cells that border the pointed area on each side of notches at the apex of each branch of the thallus. Rhizoids, located on the ventral surface, attach the plant to its substratum.

Transition to a terrestrial habitat implies the evolution of certain adaptive characters. Chief among these land modifications are the evolution of compact, multicellular plant bodies able to prevent desiccation; modification of photosynthetic tissues for carbon dioxide diffusion; and specialized structures for water intake. A section through the thallus of *Marchantia* demonstrates the nature of these terrestrial adaptations (Figure 18-2C). Internally, the thallus body reveals its multicellular construction. The adaptive advantage of such an arrangement is that, of the hundreds of thousands of cells present, only a small percentage is exposed to the environment. A relatively large amount of cell surface of the outer cells is impregnated with cutin, a waxy, waterproofing material found covering the epidermis of leaves of higher plants.

The upper epidermis, usually chlorophyllous, is provided with pores or openings, each of which is surrounded by a group of specialized epidermal cells and which opens into air chambers. The pores are analogous to stomates of higher plants and the specialized nonchlorophyllous epidermal cells superficially resemble guard cells. The pores permit the inward diffusion of CO_2, which is absorbed by the moist surfaces of the photosynthetic cells arising from the bottoms of the air chambers. Each chamber contains numerous branched and

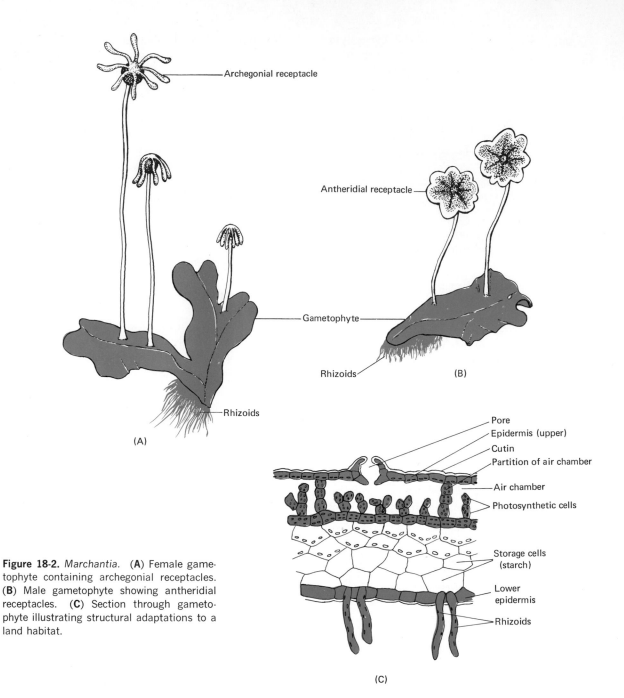

- Archegonial receptacle

(A)

- Antheridial receptacle

- Gametophyte

- Rhizoids

(B)

- Pore
- Epidermis (upper)
- Cutin
- Partition of air chamber

- Air chamber

- Photosynthetic cells

- Storage cells (starch)

- Lower epidermis

- Rhizoids

(C)

Figure 18-2. *Marchantia.* **(A)** Female gametophyte containing archegonial receptacles. **(B)** Male gametophyte showing antheridial receptacles. **(C)** Section through gametophyte illustrating structural adaptations to a land habitat.

- Rhizoids

chainlike chlorophyllous cells in which photosynthesis takes place. The arrangement of these cells and the presence of pores facilitate gaseous exchange for respiration as well as photosynthesis.

Although the epidermal pores aid in the diffusion of atmospheric gases, they also cause *Marchantia* to lose substantial quantities of water through evaporation. Unlike the stomata of more complex land plants, the mechanism for opening and closing the pores is much less efficient. Excessive water loss in *Marchantia* is prevented by the presence of rhizoids, which are single-celled, hairlike structures that originate from the outer cells of the lower epidermal layer. These structures greatly increase the surface for absorption thus increasing water uptake. Most certainly, any interaction of factors which permits free gaseous exchange while compensating for excessive water loss must have survival value.

Sexual reproduction in *Marchantia* occurs only under wet environmental conditions, a phenomenon that illustrates the transitional nature of these plants. The dorsal surface of the gametophyte thallus develops stalks upon which discs are borne. These discs, or *receptacles,* differ in appearance on male and female gametophytes (Figure 18-3A and B). The *archegonial receptacles* are somewhat expanded and umbrella-like, typically containing nine fingerlike lobes around the margin. Microscopic examination reveals that the **archegonia** are developed in rows between these lobes with their necks projecting downward (Figure 18-3C). The *antheridial receptacles,* by contrast, are disclike, with scalloped edges. **Antheridia** are attached to the base of the receptacles by slender stalks (Figure 18–3D). Sperms produced by the antheridia are attracted by chemicals given off by the archegonium. If water is not present, fertilization cannot occur. Frequently, several sperm penetrate the neck of the archegonium, but normally only one fertilizes the egg. Each gamete is haploid, so that the union of sperm and egg produces a diploid zygote, the first cell of the sporophyte generation.

After fertilization, the stalks of the receptacles increase in length. The development of the young sporophyte, the **embryo,** proceeds from the growth of the single-celled zygote (Figure 18-3E). Initially, the embryo is spherical, but as development continues it differentiates into three distinct portions (Figure 18-3F): the *foot,* which grows into the gametophyte tissues and functions as an absorbing organ; the *stalk* (seta); and a *capsule* (sporangium) that is enclosed by a sterile jacket of cells. The capsule contents differentiate into spores and *elaters,* which are elongate spindle-shaped structures with internal spiral thickenings (Figure 18-3G). These elaters are hygroscopic and twist and bend in response to changes in moisture content. The various movements of the elaters help to force the spores out of the capsule. Following meiosis, the spores are discharged and, upon germination, give rise to gametophyte plants. Thus the alternation of generations is completed.

Asexual reproduction occurs in *Marchantia* by the formation of specialized structures called *gemmae* (Figure 18-4). These gemmae are minute spherical bodies borne on stalks and contained within *cupules* (cuplike formations). The cupules, about $\frac{1}{4}$ inch in diameter, are formed on the dorsal surface of the gametophyte thallus. Under conditions of favorable moisture, the cupules fill with water and the gemmae are washed free of the thallus. Upon germination the gemmae develop rhizoids and eventually form new gametophytes. This type of asexual reproduction may be considered vegetative reproduction because a portion of the parent plant gives rise to a new generation.

Class Anthocerotae (Hornworts or Horned Liverworts). The class *Anthocerotae* represents a very small group of bryophytes. Chief among the

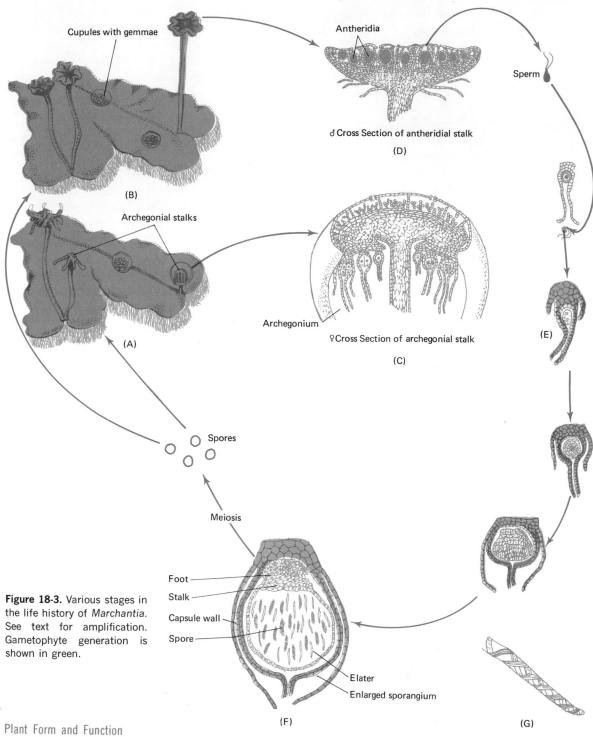

Cupules with gemmae

Antheridia

Sperm

♂ Cross Section of antheridial stalk

(D)

(B)

Archegonial stalks

(E)

Archegonium

(A)

♀ Cross Section of archegonial stalk

(C)

Spores

Meiosis

Foot

Stalk

Capsule wall

Spore

Elater

Enlarged sporangium

Figure 18-3. Various stages in the life history of *Marchantia*. See text for amplification. Gametophyte generation is shown in green.

(F)

(G)

(A)

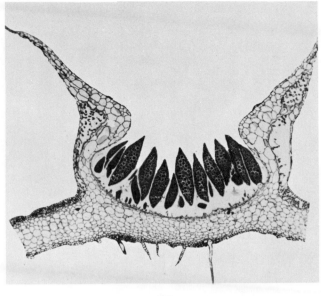

(B)

Figure 18-4. Sexual reproductive structures of *Marchantia*. **(A)** Gametophytes containing cupules on dorsal surface. **(B)** Section through a cupule showing gemmae. (Courtesy of Carolina Biological Supply Company.)

prominent structural characters of this class are the very simple form and structure of the thallus, deeply embedded antheridia and archegonia, and the structure and development of the sporophyte. The gametophyte thallus of *Anthoceros* is small with scalloped margins and is typically circular in outline (Figure 18-5A). The thallus surface is void of midribs or furrows of any type. On the ventral surface are located numerous smooth-walled rhizoids. Internally there is little or no differentiation of tissue, although there are mucilage-filled cavities occupied by colonies of *Nostoc*. The cells of the thallus are parenchymatous, and each possesses a single large chloroplast and a conspicuous pyrenoid. *Pyrenoids* are characteristically present in the chloroplasts of the cells of algae but are unknown elsewhere except in this case. This type of cellular anatomy contrasts

sharply with that of other bryophytes, which possess numerous chloroplasts and do not contain pyrenoids. The hornwort thallus is further characterized by the absence of air chambers and air pores.

Both archegonia and antheridia are developed within the tissues of the thallus rather than superficially, and they are typically borne on the same thallus rather than on separate ones. It should be noted that except for a tier of neck cells there is no sterile jacket, and the adjacent cells of the gametophyte protect the egg and the canal cells (Figure 18-5B). In this regard, the archegonium is like that of certain pteridophytes and differs from other bryophytes.

The structure of the sporophyte is the feature of greatest interest in this class. In direct contrast to the simple gametophyte, the sporophyte of

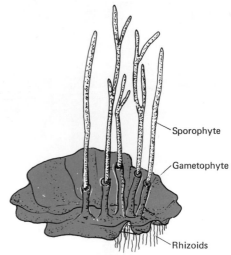

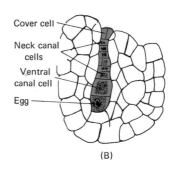

Figure 18-5. *Anthoceros.* (**A**) Gametophyte thallus containing sporophytes. *Left:* A cluster of plants. *Right:* Diagrammatic representation. (**B**) Archegonia of *Anthoceros* contained within the tissue of the gametophyte. (Photo courtesy of Carolina Biological Supply Company.)

Anthoceros is complex (Figure 18-5A). It develops from the fertilized egg and appears as a cylindrical, spikelike capsule, slightly tapering toward the apex. The base of the capsule is attached to the gametophyte by means of a *foot* and is enveloped by a collar-like sheath formed from the thallus tissue (Figure 18-6A). Growth of the capsule is accomplished by elongation of cells produced by a meristematic region at the base. This generative region remains active during the life of the gametophyte.

The structure of the capsule, similar to that of moss plants, is an example of convergent evolution. Sectional views (Figure 18-6A and B) of a mature capsule reveal a central tissue composed of sterile cells called the *columella.* Surrounding the columella is a cylinder of *sporogenous* tissue containing spores and elaters. Outside of this tissue is a zone of cortical cells, which in turn is covered by a well-developed epidermis with stomata and a cutin layer. Both the cortical layer and the epidermis contain chloroplasts, so the sporophyte is capable of photosynthesis, although dependent upon the gametophyte for metabolites. When the sporophyte is mature it splits longitudinally into two hornlike strips and releases the haploid spores.

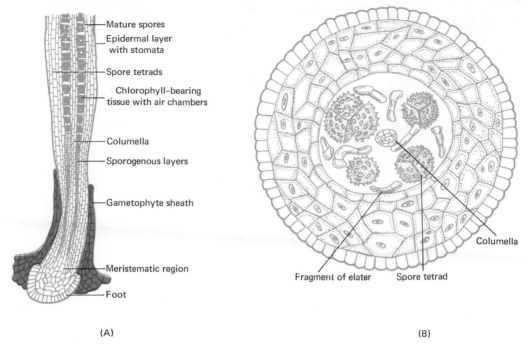

Mature spores
Epidermal layer
  with stomata
Spore tetrads
Chlorophyll–bearing
  tissue with air chambers
Columella
Sporogenous layers
Gametophyte sheath
Meristematic region
Foot

Columella

Fragment of elater   Spore tetrad

(A)   (B)

Figure 18-6. Sporophyte of *Anthoceros*. (**A**) Longitudinal section. (**B**) Transverse section.

The meristematic zone at the base continues to produce cells which give rise to new spores and sporophyte tissue. Thus the growth of the capsule is continued over an extended period of time.

Class Musci (Mosses). The *Musci*, or mosses, are a large group of plants that constitute the higher bryophytes. "True" mosses are frequently confused with other plants such as "sea moss" (red algae), "reindeer moss" (lichens), "spanish moss" (flowering plants), and "club mosses" (primitive vascular plants). None of these, however, are regarded as strict botanical designations, and the term moss should be applied only to plants of the class Musci. Mosses are common inhabitants of wooded areas and moist environments. They usually grow in dense stands forming extensive carpets or mats and are seldom more than 6 to 8 inches high.

An individual moss plant, isolated from others in a stand, consists of a more or less erect caulid ("stem") which bears expanded phyllids ("leaves"). The major axis is anchored to the substratum by means of rhizoids. In most moss plants, antheridia and archegonia are borne at the apex or on the side of the leafy axis. Some mosses are dioecious, others are monoecious. After fertilization, the sporophyte develops at the apex or on the side of the gametophyte, depending upon the position of the archegonium. The outstanding features which distinguish mosses from liverworts and hornworts are the algalike protonema, which are a product of spore germination, the radial symmetry of the thallus, and the elaborate capsule of the mature sporophyte.

The conspicuous adult moss plant, which represents the gametophyte generation, develops from

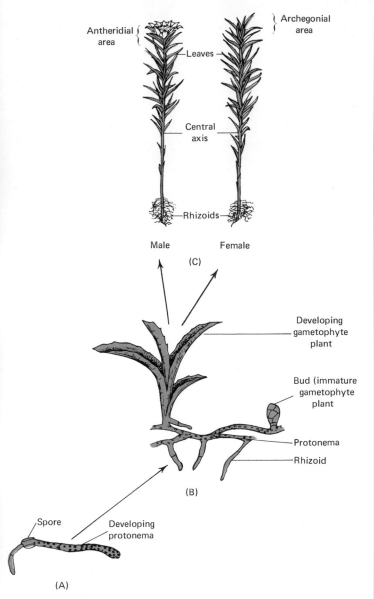

Antheridial area

Leaves

Central axis

Archegonial area

Rhizoids

Male

Female

(C)

Developing gametophyte plant

Bud (immature gametophyte plant)

Protonema

Rhizoid

(B)

Spore

Developing protonema

(A)

Figure 18-7. Gametophyte generation of a moss plant. **(A)** Germinating spore forming a protonema. **(B)** Mature stage of protonema development showing rhizoids, buds, and erect "leafy" gametophyte plant. **(C)** Mature gametophyte plants (male and female).

a spore (Figure 18-7). Assuming that environmental conditions are favorable, a moss spore germinates to form a **protonema,** a branched filament of elongated cells, somewhat resembling a green, branched filamentous alga. The portion of the protonema growing over the surface of the ground consists of chlorophyllous cells containing cross walls. Some filaments of the protonema also penetrate the soil, lose their chlorophyll, and function as rhizoids. A few of the terminal cells of the green portion of the protonema become swollen and form budlike structures, which after division of a single apical cell, eventually produce the "leafy" gametophyte plants.

Sex organs are borne at the apices of the gametophyte plants. In *Polytrichum,* several antheridia are located at the tip of the leafy shoot and are separated by sterile hairs called *paraphyses* (Figure 18-8A). Each antheridium consists of a short *stalk,* a layer of sterile *jacket cells,* and numerous cuboidal cells, which eventually form mature *sperm.* Each sperm is helical in form and is biflagellate, bearing a superficial resemblance to flagellated organisms. Archegonia, formed on other plants, also contain *paraphyses* and consist of a *stalk,* a *venter* surrounding the egg, and a long *neck* within which a *canal* is located (Figure 18-8B).

Immediately after fertilization, the zygote initiates the development of the sporophyte (Figure 18-9). The embryonic sporophyte grows downward into the stem of the gametophyte, where it forms a foot. This basal portion of the embryo facilitates the passage of water and nutrients from the gametophyte. Continued growth of the embryo gives rise to an elongated stalk, or *seta,* which bears a terminal spore case, the *sporangium.* In *Polytrichum,* a cap, or *calyptra,* covers the capsule until it is mature. The calyptra is a portion of the neck and venter of the archegonium which becomes detached and is carried upward by the

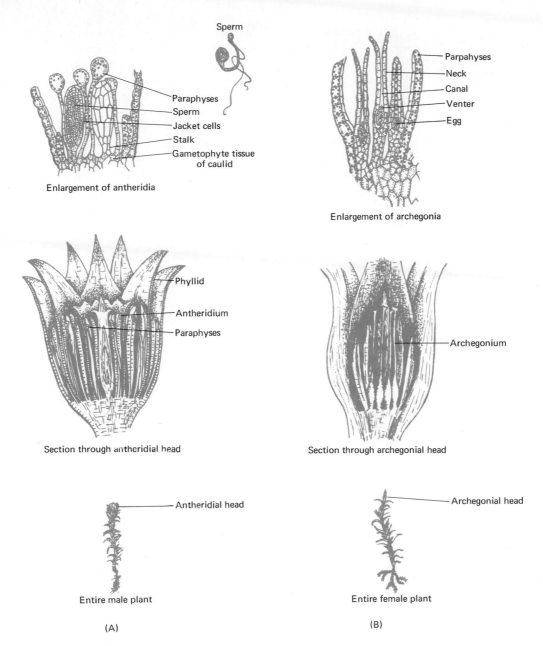

Sperm

Paraphyses
Sperm
Jacket cells
Stalk
Gametophyte tissue
of caulid

Enlargement of antheridia

Parpahyses
Neck
Canal
Venter
Egg

Enlargement of archegonia

Phyllid
Antheridium
Paraphyses

Section through antheridial head

Archegonium

Section through archegonial head

Antheridial head

Entire male plant

(A)

Archegonial head

Entire female plant

(B)

Figure 18-8. Sex organs of *Polytrichum.* (**A**) Various aspects of antheridia. (**B**) Various aspects of archegonia. All of these structures are haploid.

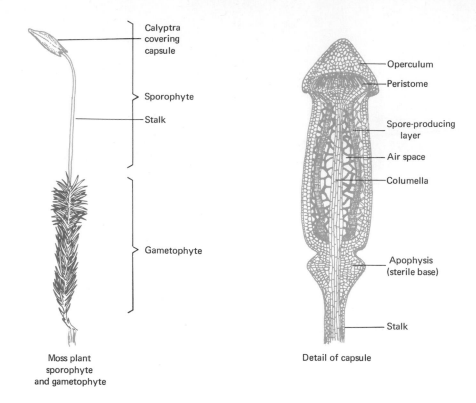

Labels for left figure (Moss plant sporophyte and gametophyte):
- Calyptra covering capsule
- Sporophyte
- Stalk
- Gametophyte

Moss plant
sporophyte
and gametophyte

Labels for right figure (Detail of capsule):
- Operculum
- Peristome
- Spore-producing layer
- Air space
- Columella
- Apophysis (sterile base)
- Stalk

Detail of capsule

Figure 18-9. *Polytrichum.* Note the details of capsule and upper portion of the stalk of the sporophyte generation. The gametophyte plant is shown in green.

apical growth of the embryo. The capsule itself sheds the calyptra at maturity and may be either erect or hanging. Structurally the capsule consists of an upper lid, or *operculum,* and a swollen base of sterile tissue, the *apophysis.* Beneath the operculum there is a ring of teeth collectively designated the *peristome.* There is considerable differentiation of tissues inside the capsule. The covering consists of a well-defined epidermis with stomata and guard cells. Just beneath the epidermis is a region of two or more layers of cells that form the capsule wall. Within the capsule are two portions, an inner column of sterile tissue, the *columella,* and a cylindrical spore-producing layer.

The *sporogenous layer* gives rise to haploid spores through meiosis.

As maturity approaches, the tissues of the interior of the capsule break down, the operculum is shed, and the spores are discharged. The release of spores is facilitated by the peristome, the teeth of which are hygroscopic. When the moisture content of the air is high, they bend into the capsule, whereas dry atmospheric conditions cause them to bend outward. Thus these movements cause and control the release of spores. Upon germination the spores produce protonema.

The life cycle of a moss plant is shown in Figure 18-10. It will be noted that the life cycle is divided

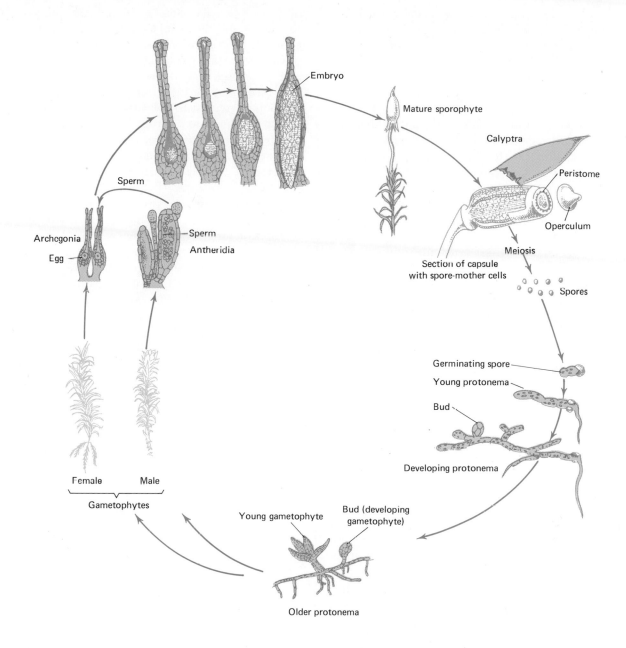

Figure 18-10. Life history of a moss plant. All haploid structures are shown in green.

Embryo

Mature sporophyte

Calyptra

Peristome

Operculum

Meiosis

Section of capsule
with spore-mother cells

Spores

Sperm

Sperm

Antheridia

Archegonia

Egg

Female   Male

Gametophytes

Germinating spore

Young protonema

Bud

Developing protonema

Young gametophyte

Bud (developing
gametophyte)

Older protonema

into two distinct phases: the gametophyte generation and the sporophyte generation. The gametophyte (haploid) begins with the formation of spores, which upon germination produce the protonema. Subsequent growth, development, and differentiation of the protonema give rise to the leafy, radially symmetrical gametophyte moss plants. Gamete formation by the antheridia and archegonia signals the end of the gametophyte generation, and the fertilization of the egg (zygote) initiates the sporophyte (diploid) generation. Development of the zygote ultimately produces the sporophyte, which consists of the foot, seta, and capsule containing diploid spores. The spores represent the last cells of the sporophyte generation. Essentially there is a definite alternation of generations in which the gametophyte produces gametes, the gametes fuse to form zygotes, and the zygotes form sporophyte plants. Haploid spores from the sporophyte germinate to produce gametophytes.

Moss plants exhibit certain characteristics that distinguish them as transition plants between aquatic and terrestrial environments. In some respects they retain certain algal features; in others they develop more advanced provascular traits. Consider the fact that water is required for fertilization. Vegetatively, moss plants are fairly well adapted to a terrestrial mode of life, yet they depend on water for sexual reproduction. In this regard, mosses, and all bryophytes as well, may be viewed as amphibious plants. Considering that a dependency upon water for fertilization is a primitive characteristic of land plants, more advanced plants eventually evolved a modification of this phase of the life cycle. Directly related to the necessity of water for reproduction is the relative size and independence of the sporophyte and gametophyte generations. In the moss (and bryophyte) life cycle the gametophyte is the larger and independent generation, whereas the sporophyte is somewhat smaller and almost entirely dependent on the gametophyte to which it is attached. In higher land plants the situation is reversed; the sporophyte shows an increase in size and independence, whereas the gametophyte shows a decrease in these respects. It appears that the sporophyte, which evolved the vascular system, is better adapted to a terrestrial environment, in that reproduction by spores is well suited to that habitat. By contrast, the flagellated gametes of the gametophyte generation are better suited to an aquatic mode of life.

In one very important respect mosses seem to retain a close algal relationship. This is evidenced by the structure of the protonema, which is very similar to that of certain green algae. Some botanists have suggested that the protonema represents a portion of the thallus of an ancient algal ancestral type in which the erect portion gave rise to the "leafy" gametophyte and the prostrate portion has been retained as the protonema. This resemblance has led some to believe that mosses have evolved from the green algae.

In the course of their evolution mosses established at least two critical features which later were more fully developed by land vascular plants. The first of these features is the retention and protection of the embryo within parental tissues. The zygote is formed inside the female reproductive organ and an embryo develops from the dividing zygote still inside the protective coverings of the archegonium. The chances of survival are appreciably increased by such an evolutionary modification because the embryo is less likely to sustain mechanical injury or desiccation. A second important structural character evidenced in the mosses is the presence of a strand of tissue occupying the center of the stemlike axis. A transverse section through the axis shows a relatively simple structure and a slight specialization of cells. Externally there is an epidermal layer under which is a chlorophyllous cortical region. In the center of the axis, the cells are smaller in diameter but vertically elongate, with somewhat thickened walls. This central strand gives the axis a certain degree of mechanical support, and some botanists believe that it also facilitates conduction of materials. If these assumptions are true, such a tissue may represent a type of provascular tissue which later evolved into a true vascular tissue containing tracheids and sieve cells.

Relationships of the Bryophyta

As stated earlier in the chapter, most botanists agree that bryophytes evolved from a green algal form. The modern view is that bryophytes are derived from the green algae through prepsilophyte ancestors in which both the gametophyte and the sporophyte were green and physiologically independent. In the earliest transitional forms of all three bryophyte classes, both generations were probably independent and green, with erect radially symmetrical shoots. Eventually the sporophyte became attached to, and partially parasitic upon, the gametophyte, and thus was reduced in form. In this regard, the sporophyte of *Anthoceros,* which is large and somewhat independent, is considered to be the most primitive sporophyte among bryophytes. Smaller reduced sporophytes are considered to be more advanced. A similar reduction occurred among the gametophyte plants. The more primitive forms (mosses) have "leafy" radially symmetric gametophytes, whereas gametophytes of liverworts and *Anthoceros* are dorsiventral and are interpreted as derived from ancestrally erect, radial forms. Thus in both generations there has been a reduction to simpler structures, and all living bryophytes are believed to represent an evolutionary dead end from which no higher plants were evolved. The earliest vascular plants (psilophytes) are found in Devonian and Silurian strata, whereas bryophytes probably did not appear until later.

Economic Importance

In general, bryophytes are economically less important than most other plant groups. In the process of succession (Chapter 21) mosses act as soil formers following lichens or other lower forms of plant life on bare rock. Through various chemical and mechanical activities, they convert the top surface of rock, or similar substrate, into soil. Once the soil is established, higher forms of plants, such as ferns, flowering plants, shrubs, and trees, inhabit these areas. Inasmuch as mosses are so prolific that they form masses or carpets covering the soil, they also help to retard erosion by holding soil firmly in place and preventing it from washing away. *Sphagnum,* or peat moss, has an exceptionally high water-holding capacity and because of this property has been used as a surgical dressing and as a packing material by florists and horticulturists. The most important economic product of mosses is peat, a fuel material, derived from *Sphagnum* and other mosses.

CHAPTER IN RETROSPECT

Metaphyta are descended from algalike plants, probably certain chlorophyta, and consist of two major divisions: (1) Bryophyta (nonvascular plants), and (2) Tracheophyta (vascular plants). Liverworts, hornworts, and mosses represent the three classes of bryophytes. These plants are characterized by a relatively long-lived gametophyte generation which is nutritionally independent, and a less conspicuous sporophyte, which is attached to and dependent upon the gametophyte.

The liverworts (Hepaticae) are the simplest bryophytes, consisting of either a thallose or leafy plant body. A representative thallose liverwort, *Marchantia,* demonstrates a typical flat, ribbon-shaped, dichotomously-branched thallus commonly 2 to 3 inches long. This species is well adapted to a terrestrial habitat and possesses rhizoids which attach it to a substratum. A relatively small percentage of cells are exposed to the environment, and these cells are usually impregnated with a waxy, waterproofing material (cutin). Hornworts (Anthocerotae) are numerically a very small group of bryophytes whose gametophyte generation usually contains only a single chloroplast with a pyrenoid body, a unique feature of this class. The largest group of bryophytes are mosses (Musci). An individual moss plant is comprised of an erect "stem" which bears "leaves." They can be distinguished from liverworts and hornworts by the presence of a protonema, an algalike structure produced by spores. The life cycle of a moss plant begins with the formation of spores, which upon germination give rise to the leafy gametophyte. At maturity the gametophyte produces gametes within the antheridia and archegonia. Subsequent fertilization initiates the sporophyte (diploid) generation, which ultimately produces the sporophyte plant. This structure consists of the foot, seta, and capsule containing diploid spores. Essentially, the alternating of spore formation and gamete production represents a typical pattern of alternation of generations.

Bryophytes are economically less important than vascular plants and fission plants, although they do contribute substantially, as do lichens, to the process of bare-rock plant successions. *Sphagnum,* or peat moss, is used as a packing material and as an inexpensive fuel.

Questions and Problems

1. Define the following terms:

monoecious	protonema
dioecious	operculum
vascular plants	calyptra
antheridium	columella
archegonium	thallose

2. Describe the general life cycle of bryophytes. How does this pattern differ from that of Eumycophyta?
3. Compare the sporophyte and gametophyte of hornworts, liverworts, and mosses. Which is the most complex? Name each structure and use a diagram to explain your answer.
4. By way of a word diagram, describe the pattern of alternation of generations in the moss plant *Polytrichum.* The following terms should be incorporated in your response: gametophyte, sporophyte, antheridium, archegonium, sterile jacket cells, sperm, egg, sporangium, diploid, haploid, calyptra, operculum, columella, spores, and zygote.
5. Why are the bryophytes regarded as the most primitive land plants? How does their level of complexity compare with true fungi and algae?
6. Compare the economic significance of bryophytes with fission plants and true fungi.
7. What special means of adaptation have made it possible for bryophytes to live in a land environment? Consider life cycle, structural characteristics, and mode of nutrition in your answer. How do these adaptations compare with green algae?

Suggested Supplementary Readings

Allen, C. E., "The Genetics of Bryophytes II," *Botanical Review,* vol. 11, pp. 260–287, 1945.

Bodenberg, E. T., *Mosses.* Minneapolis, Minn.: Burgess Publishing Company, 1954.

Conrad, H. S., *How to Know the Mosses and Liverworts.* Dubuque, Iowa: William C. Brown Company, Publishers, 1956.

Lewis, K. R., "The Genetics of Bryophytes," *Transactions British Bryol. Society,* vol. 4, pp. 111–130, 1961.

Thieret, J. W., "Bryophytes as Economic Plants," *Economic Botany,* vol. 10, pp. 75–91, 1955.

Watson, E. V., *The Structure and Life of Bryophytes.* London: Hutchinson University Library, 1964.

19
Metaphyta: Lower Vascular Plants

INTRODUCTION

Tracheophytes, or vascular plants, are the largest group of plants on earth. All the plant groups considered in earlier chapters (15 through 18) have no distinct vascular tissue, although, as indicated earlier, mosses do possess a central core of tissue that may assume supporting and conducting functions. In contrast, the tracheophytes are characterized by the presence of tracheary elements and sieve-tube elements organized into xylem and phloem in the adult sporophytes. All lines of evidence suggest that tracheophytes developed directly from one of the algal groups, probably the Chlorophyta. Unfortunately, an extremely large gap exists between the algae and vascular plants, and there are no living plants which bridge the gap. Bryophytes are somewhat intermediate but are closer to algae than to even the simplest tracheophytes, and it appears that the intermediate plants have become extinct.

TRACHEOPHYTES

General Characteristics

In addition to the presence of vascular tissue, tracheophytes are identified by independent and dominant sporophytes and greatly reduced gametophytes. This condition is in direct contrast to that found in bryophytes, in which the gametophyte is dominant. Another unique feature of this division is that the sporophytes of most are differentiated into roots, stems, and leaves. It

should also be noted that vascular plants are typically land plants, although a few (water ferns, water lilies, and so on) exhibit certain secondary, aquatic tendencies. In this regard, tracheophytes, like bryophytes, have evolved terrestrial adaptations, many of which will be noted in the following discussion.

Classification

The Tracheophyta includes both fossil and living vascular plants of the subdivisions *Psilopsida, Lycopsida, Sphenopsida,* and *Pteropsida* (Table 19-1). For convenience of study, the Tracheophyta will be subdivided into lower tracheophytes and higher tracheophytes. The lower tracheophytes (psilopsids, lycopsids, sphenopsids, and ferns) will be the subject of this chapter. Chapter 20 will be concerned with an analysis of the higher tracheophytes (gymnosperms and angiosperms).

Evolution in the Lower Tracheophytes

It seems reasonable to assume that the transition from an aquatic to a terrestrial habitat was a gradual process occurring over millions of years. Those plants that evolved adaptations permitting growth in a dry environment established themselves as the first land plants. The magnitude of this evolutionary process from sea to land is best understood if the kinds of modifications necessary for survival on land are considered. The basic problem confronting early land plants was that of water availability. An aquatic green algal form consisting of a relatively simple thallus is completely surrounded by water which contains nutrient salts. Thus for the submerged alga there is no major problem with respect to metabolite procurement and distribution. By contrast, a tracheophyte has no such limitless supply of water and minerals. A successful land plant must possess an efficient mechanism for absorbing materials from the soil and transporting them throughout

TABLE 19-1. Classification of the Lower Tracheophytes

Kingdom Metaphyta.
  Division Tracheophyta (vascular plants). About 300,000 species.
    Subdivision Psilopsida (psilopsids).
      Class Psilopsidae.
        Order Psilophytales. *Rhynia, Psilophyton, Hyenia, Horneophyton, Asteroxylon.*
        Order Psilotales. *Psilotum, Tmesipteris.*
    Subdivision Lycopsida (lycopsids).
      Class Lycopodinae.
        Order Lycopodiales. *Lycopodium.*
        Order Selaginellales. *Selaginella.*
        Order Lepidodendrales. *Lepidodendron.*
        Order Isoetales. *Isoetes.*
        Order Pleuromeiales. *Pleuromeia.*
    Subdivision Sphenopsida (sphenopsids).
      Class Equisetinae.
        Order Equisetales. *Equisetum.*
        Order Hyeniales. *Calamophyton.*
        Order Sphenophyllales. *Sphenophyllum.*
        Order Calamitales. *Calamites.*
    Subdivision Pteropsida (ferns and seed plants).
      Class Filicineae (ferns).
        Order Cladoxyles. *Pseudosporochnus.*
        Order Colnopteridales. *Stauropteris.*
        Order Ophioglossales. *Ophioglossum.*
        Order Marattiales. *Marattia.*
        Order Filicales. *Polypodium, Osmunda.*
        Order Marsileales. *Marsilea, Pilularia.*
        Order Salviniales. *Salvinia, Azolla.*

the plant. Once obtained from the environment, the water must be conserved.

One of the most interesting aspects of tracheophyte evolution is the origin of plants capable of inhabiting the land. To be sure, the solution to the problem of the origin of land plants is extremely difficult and at best only speculative. In 1903 the French paleobotanist O. Lignier proposed a theory of the origin of land plants. He suggested that the green algal type of plant body that preceded land plants consisted of a three-

dimensional dichotomously branched thallus, probably similar to that of *Fucus* (Figure 19-1A). One of the lateral branch systems bent over, penetrated the soil, and became transformed into a root system (Figure 19-1B). Other branches straightened out, thus forming the main axis (stem with branches). Some of these branches flattened out and became leaves. Further modifications of the plant body gave rise to cuticle over the aerial portions and an internal vascular system, which facilitated both conduction and support. In subsequent pages of this chapter the probable evolution of the lower tracheophytes based on these

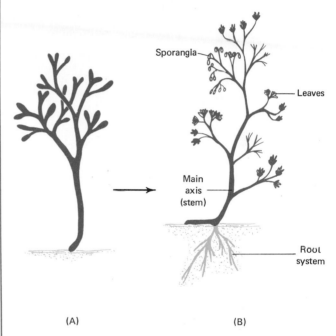

(A)   (B)

Figure 19-1. Diagrammatic representation of Lignier's hypothesis of the origin of land plants: **(A)** Dichotomizing green algal ancestor. **(B)** Hypothetical simple vascular plant in which one portion has penetrated the soil to become the root system and the terminal portions have straightened out to become the main axis with branches. Some of the branches have flattened out to become leaves, others have specialized sporangia at the tips.

evolutionary trends will be traced. It will be seen that there are many data in support of Lignier's hypothesis.

Subdivision Psilopsida (Psilopsids). The Psilopsida is the oldest and most primitive group of vascular plants that ever existed on earth. Included are the extinct order *Psilophytales,* found as Silurian and Devonian fossils, and the order *Psilotales,* which contains two living genera. Psilophytes were the simplest kinds of vascular plants, having evolved from a primitive stock that probably also gave rise to the other major tracheophyte groups.

The sporophyte body in the members of the Psilopsida shows little differentiation, consisting of a dichotomously branched axis (stem) without true roots or leaves. Expanded appendages of the stem function as photosynthetic organs. The horizontal, underground stem may bear rhizoids but has no true roots. Internally, the vascular tissue appears as a **protostele,** that is, an arrangement in which there is a solid core of xylem (no pith) surrounded by a cylinder of phloem. This is generally considered to be the most primitive kind of vascular cylinder (**stele**) by plant anatomists. In the primitive forms xylem is extremely simple and homogeneous, consisting only of annular tracheids. The more advanced members contain further elaborations of annular tracheids in the form of spiral and scalariform types. Inasmuch as none of the Psilopsida possesses a vascular cambium, the entire sporophyte consists of primary tissues, and there is no secondary growth. Sporangia are borne terminally on the main branches or lateral branches, and, because

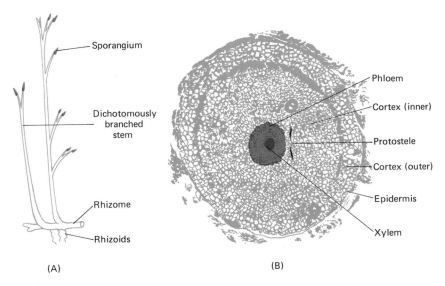

(A)

(B)

Figure 19-2. *Rhynia.* **(A)** Sporophyte showing dichotomous branching, leafless axis, and rootless subterranean portion. **(B)** Section through the stem showing internal arrangement of tissue and centrally located, cylindrical protostele. (From R. Kidston and W. H. Lang, "On Old Red Sandstone Plants Showing Structure from the Rhynie Chert Bed, Aberdeenshire," Pts. 1–5, *Transactions of the Royal Society of Edinburgh,* 1917–1921. By permission.)

only one kind of spore is produced, the plants are *homosporous.*

Order Psilophytales. Members of this order are known only through fossil remains of the sporophytes in Silurian and Devonian strata throughout the world. No gametophytes are known. Perhaps the simplest of all vascular plants is *Rhynia* (Figure 19-2A), named after the village Rhynie in northern Scotland, where the first good fossils were found. It is a small rushlike plant consisting of an upright dichotomously branched stem anchored to the substratum by a rhizome bearing rhizoids. Some of the aerial branches are entirely vegetative; other branches terminate in a single sporangium containing numerous spores with cutinized walls. A transverse section through any part of the axis reveals the prominent protostele (Figure 19-2B), which is composed entirely of annular tracheids. The phloem is surrounded by a thick cortex, which is enclosed, in turn, by a cutinized epidermis containing stomata. The structure of *Rhynia* is the simplest known of any vascular plant because there are no roots or leaves and it does not possess an elaborate spore-producing apparatus. It is not difficult to imagine how a plant such as *Rhynia* might have evolved from an algal ancestor as proposed by Lignier.

Within the Psilophytales a number of evolutionary lines of development are exhibited by certain genera. *Horneophyton* (Figure 19-3A), another common fossil in the Rhynie beds, is basically similar to *Rhynia* but has thick, tuberous rhizomes and a well-developed columella in the sporangium. *Psilophyton* (Figure 19-3B) is rootless and dichotomously branched, but the stem bears numerous, small, spinelike appendages which are regarded by some paleobotanists as incipient leaves. Another genus, *Asteroxylon* (Figure 19-3C), shows still more structural complexity both externally and internally. The horizontal, underground rhizome of *Asteroxylon* is naked and some of the branches bear superficial resemblance to roots of higher plants. From the rhizome arises aerial branches which are densely covered with scalelike emergences, or "leaves." These appendages are not vascularized, so they are not designated true leaves. Internally, the stele, although still simple, is larger than that of *Rhynia,* and the xylem is lobed so that the protostele is star-shaped. Originating from the arms of the stele are vascular bundles, which terminate in the cortex but are not continuous with the "leaves." This situation may be interpreted as an early stage in the evolution of true leaves in which the vascular bundle has not yet formed the vein of the leaf.

The discovery and reconstruction of these early plants suggested that *Rhynia* is indeed the simplest type of vascular plant. Such features as protostelic conducting tissue, rhizomes, dichotomous branching, absence of true leaves, and terminal sporangia have generally been regarded as characteristics of simple vascular plants that evolved from aquatic ancestors. As a result, *Rhynia,* and other psilophytes, are considered as the stock from which all other vascular plants evolved.

Order Psilotales. This order has two living genera, *Psilotum* (Figure 19-4A) and *Tmesipteris* (Figure 19-4B), and is of special interest because its members more closely resemble the Psilophytales than any other plant group. Because *Psilotum* is the more common genus, it may be taken as a representative example of Psilotales to indicate the most important features of the group.

The mature sporophyte consists of a dichotomously branched, green aerial shoot emerging from a horizontal rhizome. The erect branches are devoid of true leaves but bear irregularly scattered and paired scales which are considered to be emergences and not vascularized leaves. In the axil of each pair is a three-lobed, thick-walled sporangium which in its development seems originally to have terminated a short branch (Figure 19-4A).

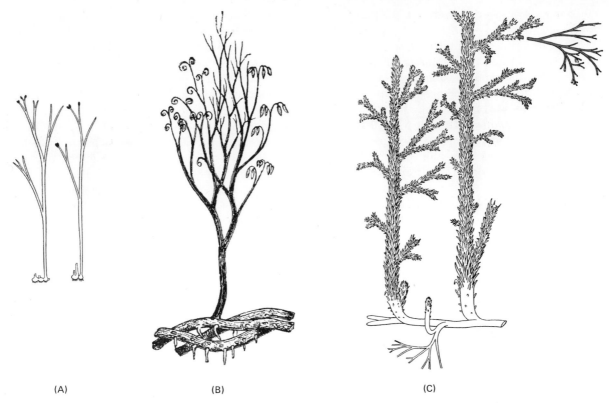

(A) (B) (C)

Figure 19-3. Sporophytes of various Psilophytales showing increasing complexity of form. (**A**) *Horneophyton.* (**B**) *Psilophyton.* (**C**) *Asteroxylon.* (Courtesy of Field Museum of Natural History, Chicago, Illinois.)

The spores are all alike, so *Psilotum* is homosporous. Internally, the rhizome has a protostele, but the aerial branches contain a pith, so the vascular arrangement in the stem is called a **siphonostele** (Figure 19-5A). Thick-walled sclerenchyma functions as pith, and the xylem is composed of annular and scalariform tracheids. The stele of the aerial branches is surrounded by an endodermis with cells whose radial walls have prominent thickenings called **Casparian strips.** These strips regulate the passage of water and solutes between the cortex and the stele. External to the endodermis is a parenchymatous cortex and a single-layered epidermis with a cuticle and stomata.

The gametophyte (Figure 19-5B), formed by germination of a spore, is nonphotosynthetic and subterranean. It resembles a rhizoid, is brownish, elongated, and cylindrical, and may branch dichotomously. There are numerous scattered rhizoids, each of which is a projection of a single

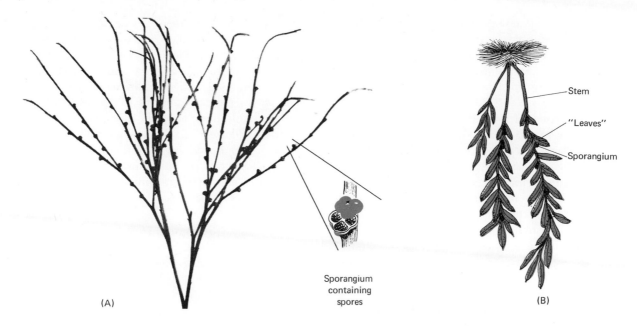

Figure 19-4. Sporophytes of Psilotales. **(A)** *Psilotum*, showing entire plant and enlargement of a sporangium. **(B)** *Tmesipteris*, a hanging Psilotales. (Photo courtesy of Carolina Biological Supply Company.)

surface cell. Some of the internal cells are densely filled with mycorrhizal fungous filaments which probably aid in nutrition of the gametophyte. Antheridia and archegonia are found in large numbers all over the gametophyte surface and are similar in structure to those found in the bryophytes. After fertilization an embryo develops that is composed of an absorbing foot and a mass of

Figure 19-5. *Psilotum.* **(A)** Cross section of the stem showing siphonostele and internal tissues. **(B)** Details of a gametophyte. (From A. A. Lawson, "The Gametophyte Generation of the Psilotaceae," *Transactions of the Royal Society of Edinburgh*, 1917. By permission.)

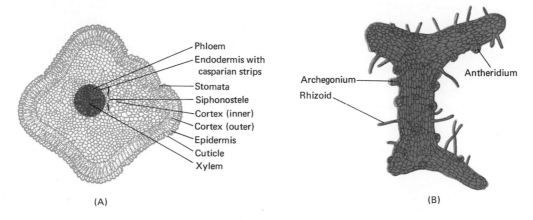

463

embryonic cells that gives rise to the shoot. One interesting feature to be noted is that some larger *Psilotum* gametophytes contain a slender strand of vascular tissue with annular tracheids. Inasmuch as tracheids are invariably found only in sporophytes, *Psilotum* is the only known plant in which vascular tissue normally develops in the gametophyte. This may indicate that the sporophyte and gametophyte generations are basically similar.

Relationships and Importance of Psilopsida. The evolutionary significance of the Psilophytales is that they are the most primitive known group of vascular plants and are considered to be a transitional group linking the aquatic algae with more advanced tracheophytes (club mosses, horsetails, and ferns). In some respects, the Psilophytales substantiate the hypothesis for the evolution of the root as proposed by Lignier. In other respects they are of particular interest because they suggest how club mosses, horsetails, and ferns may have originated by various leaf modifications and developments. Certain psilophytes, such as *Asteroxylon* (Figure 14-11A), with small "leaves" developed as emergences may have given rise to the Lycopsida line. Others, such as *Hyenia* (Figure 14-11B), with whorled branches may have been the forerunners of the Sphenopsida line. The fern line may well have evolved from such forms as *Pseudosporochnus* and other types in which the branch tips were flattened, possibly leading to the evolution of large leaves.

Subdivision Lycopsida (Lycopsids). The *Lycopsida* include extinct forms, known only from their fossil remains, and living forms. The fossil orders include *Lepidodendrales* and *Pleuromeiales* and will not be discussed. Living orders include *Lycopodiales, Selaginellales,* and *Isoetales.* All the living forms are relatively small, whereas some of the extinct relatives were rather large conspicuous trees of the Carboniferous period (see Figure 14-13).

All sporophytes of members of the Lycopsida, unlike those of Psilopsida, are differentiated into true roots, stems, and leaves. The roots and stems branch dichotomously and the leaves are usually spirally arranged and very small (*microphyllous*). Such leaves are extensions of the outer tissue of the stem, containing only one vascular bundle or leaf trace (Figure 19-6). No leaf gaps are present. Sporangia are borne singly on the upper surfaces of leaves (*sporophylls*) and are usually arranged in the form of a cone or **strobilus.**

Order Lycopodiales (Club Mosses). Some of the common species of this order are called club mosses, trailing pine, ground hemlock, and ground pines. However, these forms are similar to mosses, hemlocks, and young pines only in appearance and have no close relationship with them. Of the two living genera in this order, *Lycopodium* will serve as the representative member (Figure 19-7A). The sporophyte always has a much branched stem that in some species is erect and in others is creeping. Both roots and stems exhibit dichotomous branching, and the stem is covered by many small, spirally arranged leaves. Internally, the structure of *Lycopodium* differs from that of any

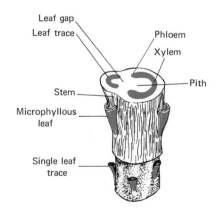

Figure 19-6. Vascular supply to a microphyllous leaf.

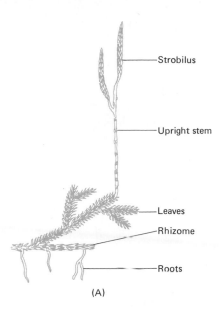

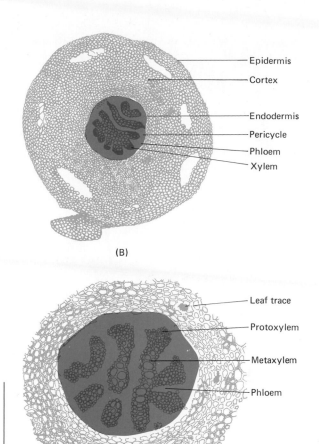

Figure 19-7. *Lycopodium.* (**A**) Entire sporophyte plant. (**B**) Cross section of a stem (*above*) and an enlargement of the protostele (*below*).

other vascular plant (Figure 19-7B). The stele is a protostele which often has a lobed appearance and consists of alternating bands of xylem and phloem. The **protoxylem** (first-formed primary xylem) is located nearest the cortex and the **metaxylem** (later-formed primary xylem) is nearer the center of the axis. Such an arrangement of xylem is termed **exarch,** a characteristic of primitive vascular plants. From each protoxylem group a small vascular bundle is pinched off and departs to a leaf. Other tissues comprising the sporophyte stem are the pericycle, endodermis, cortex, and epidermis. All tissues are primary, and stomata occur on the leaves and stems.

In some species single sporangia are borne in leaf axils, which may be located at any point on the stem (Figure 19-8A). Such sporangia-bearing leaves are termed **sporophylls.** In other species,

the sporophylls are smaller and occur at restricted regions at the tips of branches forming cones or strobili (Figure 19-8B). A single cone (Figure 19-8C) is actually a portion of a branch with an aggregation of sporophylls and their attached sporangia often raised above the leafy shoots by an elongated stalk. The spores are all of one type; that is, the plant is homosporous.

The spores, upon germination, usually give rise to subterranean, colorless gametophytes which take many years to reach maturity. The gameto-

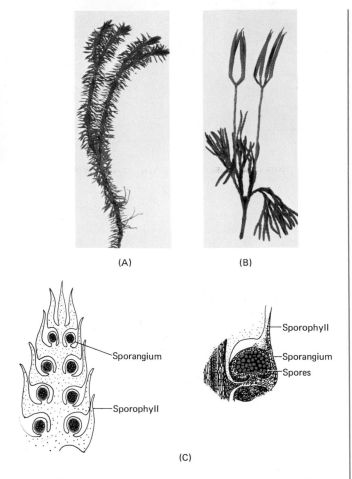

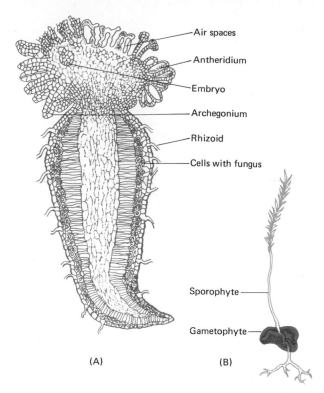

Figure 19-9. *Lycopodium.* (**A**) Diagrammatic representation of a longitudinal section through the gametophyte. (From H. Bruchmann, "Die Keimung der Sporen und die Entwicklung der Prothallien von *Lycopodium clavatum* L., *L. annotinum* L. und *L. Selago* L.," *Flora,* vol. 101, p. 220, 1910. By permission). (**B**) Young sporophyte developing from the gametophyte.

Figure 19-8. *Lycopodium* sporophylls and sporangia. (**A**) *Lycopodium lucidulum,* containing single sporangia borne in leaf axils. No strobilus is present. (**B**) *Lycopodium complanatum,* showing sporophylls grouped in a strobilus. (**C**) Details of a single cone. (Photos courtesy of Carolina Biological Supply Company.)

phyte, or **prothallus,** of some species, however, is green and grows on the soil surface. Generally the common, mature gametophyte bodies are of two types—a carrot-shaped form that matures at the surface of the ground (Figure 19-9A), and a rather lumpy, irregularly convoluted type formed under-

neath the ground. Rhizoids are attached to the gametophyte and some of the internal cells contain fungal mycelia. Antheridia and archegonia are embedded in the surface. The sperms, which are biflagellate, are similar to those produced by mosses. After fertilization, the zygote, through divisions, forms an embryo within the archegonium. The young leafy sporophyte remains attached to the gametophyte for a short time and then the latter dies (Figure 19-9B). The life cycle of *Lycopodium* is shown in Figure 19-10.

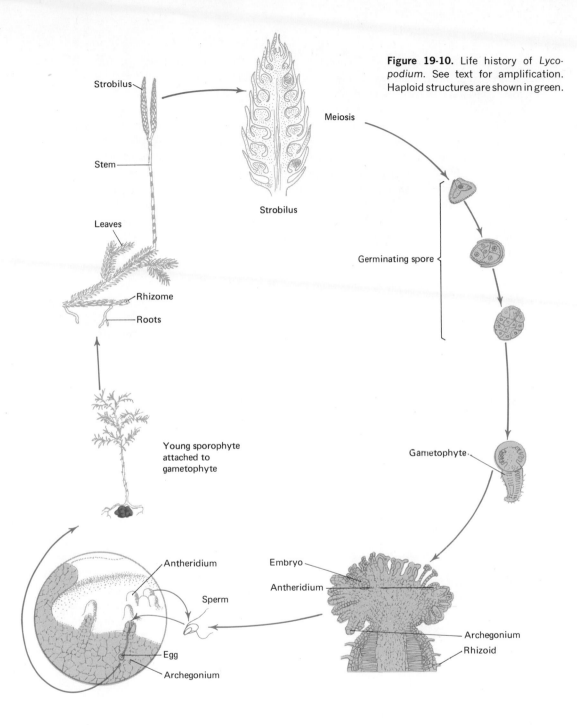

Strobilus

Stem

Leaves

Rhizome

Roots

Strobilus

Figure 19-10. Life history of *Lyco-podium.* See text for amplification. Haploid structures are shown in green.

Meiosis

Germinating spore

Gametophyte

Young sporophyte
attached to
gametophyte

Antheridium

Sperm

Egg

Archegonium

Embryo

Antheridium

Archegonium

Rhizoid

Order Selaginellales (Small Club Mosses). The mature sporophyte of *Selaginella,* although similar to that of *Lycopodium,* is usually more delicate in appearance and the stems in many cases are prostrate (Figure 19-11A). The stems branch dichotomously and the leaves occur in four rows, two of large leaves and two of smaller ones. At the base of each leaf projecting from its upper surface there is a small scalelike outgrowth, the *ligule,* which is probably a vestigial structure. Very young sporophytes develop roots like other lycopsids, but practically all later roots are produced from leafless branches, or *rhizophores,* which grow from the stem into the ground and there bear adventitious roots. The vascular system of the stem is usually a protostele, although in some species it is a siphonostele. There is no vascular cambium and the bundles are closed (Figure 19-11B).

Selaginella is heterosporous. In the strobili produced at the tips of the branches, two kinds of sporangia and spores are borne (Figure 19-12). The larger sporangia (**megasporangia**), borne on the lower sporophylls (**megasporophylls**), contain four large, thick-walled spores (**megaspores**). The smaller sporangia (**microsporangia**), borne on the upper sporophylls (**microsporophylls**), contain a large number of small spores (**microspores**). The developmental history of the spores after germination is quite different. The megaspores give rise to female gametophytes (bearing only archegonia), whereas the microspores produce male gametophytes (bearing only antheridia).

While still retained in the megasporangium, the megaspore germinates to form a gametophyte, the *megagametophyte.* A sectioned megaspore (Figure 19-12) reveals the structure of the megagametophyte, several layers of cells thick, in which archegonia are embedded. As the megaspore develops, the spore wall ruptures in the archegonial area. The microspore also germinates to form a *microgametophyte* while still in the microsporangium. A mature microgametophyte (Figure 19-12), which is completely enclosed by the microspore wall, consists of a single *prothallial* cell, believed to represent the vestige of a once more abundant vegetative tissue; one antheridium with its layer of jacket cells; and the *antherozoids,* which eventually produce biflagellate sperm.

When the microsporangial walls rupture, some of the microspores may land near the megasporangia. Free water is necessary to break open the microspores to release the sperm. The sperms swim to the megaspores while the megaspores are still retained in their megasporangia, enter the archegonia, and fertilize the eggs. The zygote undergoes divisions and forms a suspensor and an embryo. After the megaspore falls to the ground, the embryo forms a root, a stem, two cotyledons, and

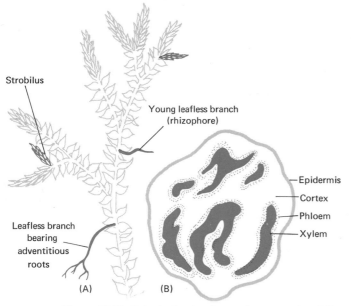

Figure 19-11. *Selaginella* showing various aspects of the sporophyte. (**A**) Entire plant. (**B**) Transverse section of a stem showing internal arrangement of tissue.

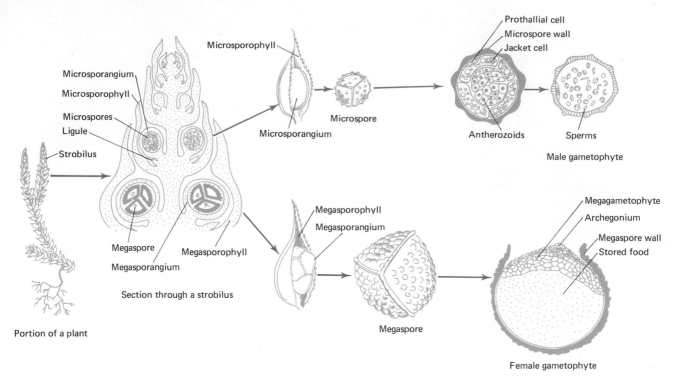

Figure 19-12. *Selaginella* strobilus and details of heterosporous germination. (Reprinted from "A Study of the Sporangia and Gametophytes of *Selaginella apus* and *Selaginella rupestris*," by F. M. Lyon, *Botanical Gazette*, vol. 32, p. 124, 1901. By permission of the University of Chicago Press.)

leaves and develops into an independent sporophyte (Figure 19-13).

Selaginella exhibits numerous advances in reproduction over other plant groups studied thus far, and some of these features in certain species have been thought to illustrate how the seed habit may have arisen from heterosporous ancestors. Among these features are the growth of the megagametophyte while still upon the parent plant and within the spore wall, fertilization of the egg prior to discharge from the parent plant, and a few early stages of development of the sporophyte while still

attached to the strobilus. It should be noted that, although certain species of *Selaginella* suggest early stages of seed evolution, *Selaginella* should not be regarded as an ancestor to seed plants, but as a representative of a group of plants in which certain trends toward seed production have persisted for a long period of time. Plants which have attained the seed habit are characterized by integumentary coverings around the seeds, food supplies within the seeds not derived from the parent plant for the development of the embryo, and a definite resting stage between embryo for-

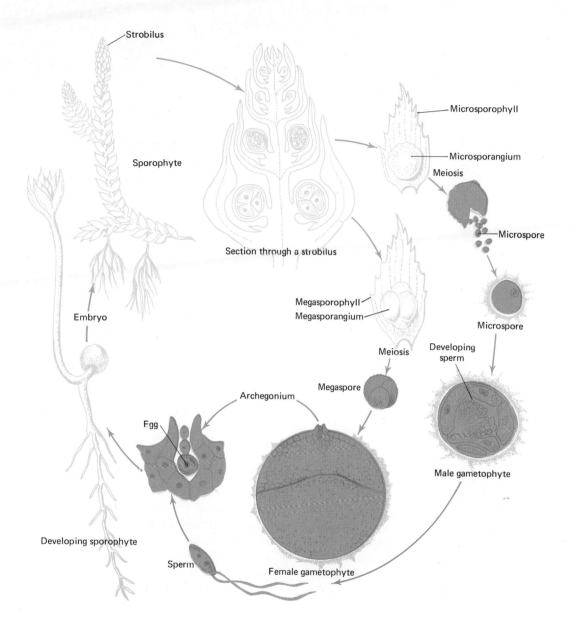

Figure 19-13. Life cycle of *Selaginella*. (Haploid generation is shown in green.)

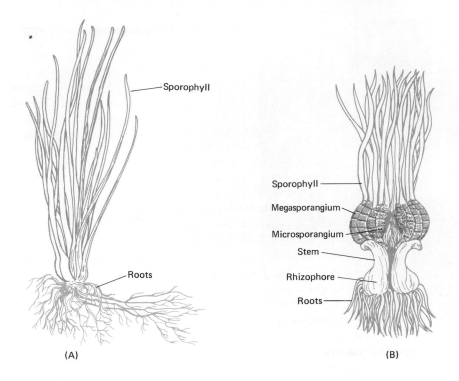

Figure 19-14. *Isoetes.* **(A)** Entire plant. **(B)** Longitudinal section through plant.

mation and its subsequent growth into a sporophyte. Inasmuch as a few species of *Selaginella* do not possess these characteristics, the selaginellas are regarded as a group of plants that have evolved parallel to ancient seed plants, but they never attained the seed habit.

Order Isoetales (Quillworts). This order contains two living genera, *Isoetes* (Figure 19-14) and *Stylites.* Most species of *Isoetes* are aquatic. The plants consist of a tuft of quill-like leaves attached to a lobed, cormlike axis. The upper part of the axis is the stem and bears leaves; the lower root-producing part of the axis is referred to as the *rhizophore.* Each leaf contains a ligule and is potentially a microsporophyll or megasporophyll. The outer leaves bear single megasporangia at

their bases and the inner leaves contain microsporangia. Quillworts of the genus *Isoetes* are the only living lycopsids capable of secondary growth, lateral expansion being due to the activity of a cortical cambium.

Relationships and Importance of Lycopsida. As has been noted earlier, the lycopsida are believed to have evolved from the Psilophytales. Although members of the Lycopsida are of no real economic significance, the group does illustrate some noteworthy evolutionary advances over the Psilopsida. For example, the presence of true roots and leaves, increased development of the vascular tissue, and the organization of sporophylls into cones are all viewed as advances over Psilopsida. For some reason not clearly understood, the Lycopsida

Figure 19-15. *Equisetum* sporophytes shown growing in clusters in various stages of development. (Courtesy of Carolina Biological Supply Company.)

whorls at the nodes of the hollow stems, which are conspicuously ribbed longitudinally. Internally, the stems typically contain siphonosteles, with branch gaps but without leaf gaps. Epidermal cells, in addition to being cutinized, also contain large amounts of silica. The sporangia are borne on stalked umbrella-like structures, the **sporangiophores,** and are commonly aggregated into strobili.

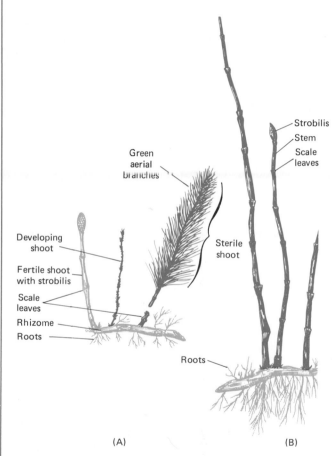

Figure 19-16. *Equisetum* sporophytes. **(A)** *Equisetum arvense* showing separate sterile and fertile aerial shoots. **(B)** *Equisetum prealtum* showing only one type of aerial stem (fertile) bearing an apical strobilus.

apparently have not given rise to more advanced groups of plants. Some have developed secondary growth and exhibit early stages of the seed habit.

Subdivision Sphenopsida (Sphenopsids). Like the lycopsids, the sphenopsids attained their evolutionary peak during the Carboniferous period. All modern sphenopsids belong to a single genus, *Equisetum,* the species of which are generally herbaceous and usually do not exceed 3 feet in height (Figure 19-15). Jointed stems are characteristic of all Sphenopsida. Microphyllous leaves occur in

The sporophyte of some species of *Equisetum* (horsetail or scouring rush) produces aerial stems of two distinct types. One is short-lived, unbranched, without chlorophyll, and produces spores. The other is sterile and green and persists throughout the growing season (Figure 19-16A). In other species there is only one type of aerial stem, a green structure bearing a cone at its apex (Figure 19-16B). The stems may have slender lateral branches or they may be unbranched and erect. All species have a rather extensive perennial development of underground rhizomes from which aerial branches arise. Stems of *Equisetum* are cylindrical, conspicuously jointed, and marked with longitudinal grooves or furrows. The scalelike microphyllous leaves are borne in whorls at the nodes.

Internally (Figure 19-17), the stem of horsetails consists of an epidermis, cortex, and vascular bundles forming a thin cylinder surrounding a prominent central cavity. There is no secondary growth, because a vascular cambium is absent. In addition to the central canal, two other types of longitudinal canals are typically present: *vallecular canals,* which are located in the cortex and are associated with a longitudinal groove of the stem; and *carinal canals,* associated with the vascular bundles. Outside of each vallecular canal in the cortex there is a mass of green photosynthetic tissue (chlorenchyma). Sclerenchyma fibers, the walls of which are impregnated with silica, occur in the outer cortex and provide mechanical support of the stem. The vascular system consists of collateral bundles arranged as a siphonostele. In the internodes the bundles are distinct and widely spaced, whereas in the nodal regions the bundles form a continuous ring of tissue and from this mass a small bundle departs to each leaf. Each bundle consists of primary xylem and phloem, the xylem being sparce and poorly developed. The cells of the epidermis have highly silicified walls

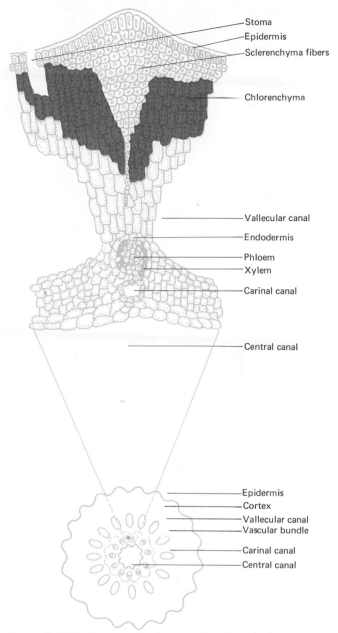

Figure 19-17. *Equisetum* sporophyte, section through stem showing arrangement (*below*) and details (*above*) of internal tissue.

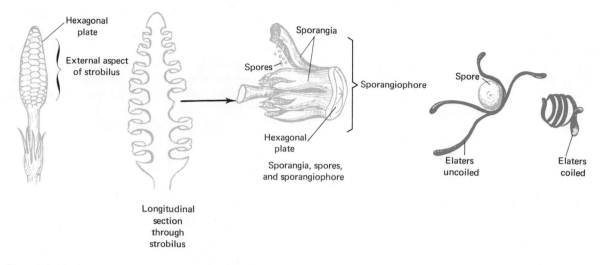

Figure 19-18. *Equisetum.* Strobilus, sporangiophore, sporangia, and spores.

with stomates usually confined to the groove between the ridges.

The spore-bearing organs of *Equisetum* are aggregated in definite strobili at the apex of the stem (Figure 19-18). Each sporangiophore consists of a short stalk and a hexagonal plate which is attached to the stalk at a right angle. Sporangia are located on the inner surfaces of each plate. At the time of spore discharge, the sporangia open by means of a longitudinal slit on the inner side. After spore liberation the fertile shoots die. The spores of the horsetails are all alike (homosporous) and contain numerous chloroplasts. The outer spore wall opens up in the form of four delicate bands with flattened, spoon-shaped tips called **elaters** (Figure 19-18). These elaters are hygroscopic and coil around the spore when the humidity is high but uncoil and spread when it is dry. It is believed that the elaters aid in the discharge of spores from the sporangium, or else they play a role in spore dispersal.

The spores germinate, forming green gametophytes which are circular or irregular in form and give rise to vertical, ribbon-like lobes. Antheridia and archegonia are borne on the same thallus under usual conditions of growth. Archegonial development usually precedes that of the antheridia. Numerous multiciliate, spirally coiled sperm are produced in each antheridium. Each archegonium contains a single egg. The product of fertilization is a new sporophyte generation. The life history of *Equisetum* is shown in Figure 19-19.

Relationships and Importance of Sphenopsida. These plants are of little economic importance, although together with the Lycopsida, they contributed their vegetative parts to the formation of coal during the Carboniferous period. They are considered to be a separate line derived from the Psilophytales which did not give rise to other plant groups.

Subdivision Pteropsida: Class Filicinae (Ferns). The *Pteropsida,* which includes all living and fossil ferns and seed plants, is the largest tracheophyte subdivision and includes the great majority of land plants today. Ferns were probably primitive, and an ancestral group of ferns then gave rise to the seed plants. The pteropsid line is distinguished, as are the other lines, by specific morphological and anatomical features. All pteropsid sporophytes are dominant and possess roots, stems, and large leaves. These large leaves, or *megaphyllous leaves,* are believed to have evolved from the flattening and transformation of a whole branch system. Thus they are not merely extensions of the outer tissues of the stem as represented by the microphyllous leaves of other lower vascular plants. Megaphylls contain multiple vascular bundles which show *leaf gaps* where they branch off from the stele (Figure 19-20). Sporangia are usually borne on the lower surfaces or, in a few cases, on the margins of sporophylls. Included in this subdivision are three classes: *Filicinae* (ferns), *Gymnospermae* (conifers and allies), and *Angiospermae* (flowering plants). The first of these groups, the Filicinae, will be discussed in this chapter. The others, the seed plants, will be treated in Chapter 20.

Filicinae are characterized by large leaves, or *fronds,* that are typically pinnately compound and uncoil as they develop, a condition termed *circinate vernation.* The stem in many cases is relatively weak and inconspicuous and in most species is a subterranean rhizome. Roots usually arise adventitiously from the horizontal rhizomes. Most ferns

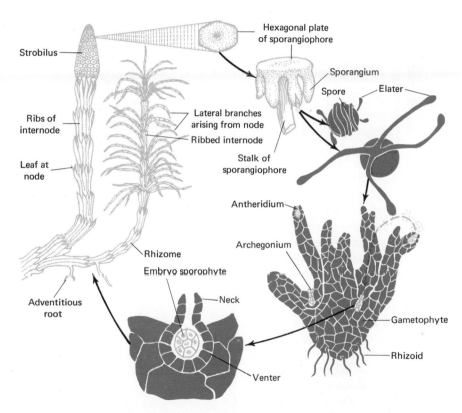

Figure 19-19. The life cycle of *Equisetum*. (Haploid generation is shown in green.)

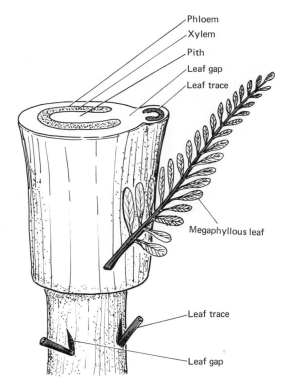

Phloem
Xylem
Pith
Leaf gap
Leaf trace

Megaphyllous leaf

Leaf trace

Leaf gap

Figure 19-20. Vascular supply to a megaphyllous leaf. Note the prominent leaf gaps.

are homosporous and none produce seeds. There is a well-defined alternation of generations in which the sporophyte is dominant. Sperms require water as a medium of transport before fertilization. The ferns are subdivided into four orders, of which only the *Filicales* (true ferns) will be discussed.

The mature sporophyte with roots, stems, and leaves bearing sporangia is the recognized fern plant (Figure 19-21). The stem, generally a small, creeping, underground structure, is perennial and produces new leaves each year. Quite frequently the stem is covered with fibrous remnants of leaf bases and a hairy outgrowth. The root system is generally small and wiry, consisting of many

hairlike adventitious roots arising from the stem. The fronds of common ferns vary in size and form, but all are characterized by circinate (unrolling) growth during early stages of development. The leaf consists of two principal parts, the **stipe** (stalk) and the **blade.** If the leaves are compound, the subdivisions of the blade are referred to as **pinnae.**

In many instances, the internal anatomy of members of the Filicales is similar to that of higher tracheophytes, especially the flowering plants. In a few simple ferns the vascular cylinder is a protostele, but much more commonly it appears as a siphonostele which may be continuous or broken into a network of discrete bundles (Figure 19-22). The siphonostele is an evolutionary modification of the protostele in which the center fails

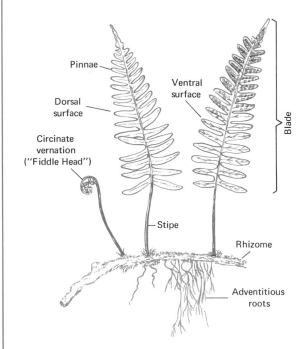

Pinnae

Dorsal surface

Ventral surface

Circinate vernation ("Fiddle Head")

Blade

Stipe

Rhizome

Adventitious roots

Figure 19-21. Sporophyte of a fern showing rhizome, adventitious roots, young leaf uncoiling, and mature leaves with sporangia on ventral surface.

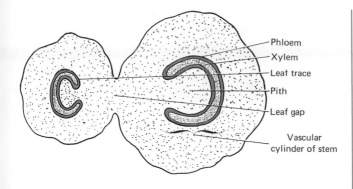

Phloem
Xylem
Leaf trace
Pith
Leaf gap
Vascular
cylinder of stem

Figure 19-22. Siphonostele of a fern stem showing leaf trace and leaf gap in transverse section.

to mature into xylem and remains as pith instead. The vascular tissue is generally bounded by a pericycle and a well-defined endodermis. The xylem consists of pitted or scalariform tracheids and the phloem is largely made up of sieve tubes with slanting end walls and scattered seive areas.

The vascular connection between stem and leaf consists of the passage of a segment of the vascular tissue of the siphonostele into the base of the petiole as the leaf trace (Figure 19-22), leaving a break (leaf gap) in the stele. Leaf gaps are conspicuously absent in all other lower tracheophytes and appear to be related to the large size of the leaves. Once the strand or strands of vascular tissue enters the leaf, it subdivides into finer structures as in the blades of higher plants. The structure of fern leaves is also similar to that of higher plants. The differentiation of the mesophyll into a palisade region and a spongy region is usually evident, but these regions are not as clearly defined as they are in many flowering plants. Considering that ferns live mostly in the shade, the leaf is usually thin and contains large intercellular spaces.

Roots of the mature fern sporphyte are entirely adventitious and, like those of other vascular plants, are protostelic. The three primary meristems—protoderm, ground meristem, and procambium—originate from cells derived from the pyramid-shaped apical cell (apical meristem). The apical cell is covered by a root cap.

In many ferns, two distinct types of leaves may be recognized: sterile vegetative leaves and fertile spore-bearing leaves (sporophylls). This condition in which reproductive and vegetative leaves are borne on the same plant is termed *dimorphism*. The spores of a fern are always produced within multicellular sporangia which usually develop in clusters on the lower surface of a leaf. Each cluster, or group, of sporangia is referred to as a **sorus,** or "fruit dot" (Figure 19-23). The sori occupy definite positions on the leaf surface and are usually found on the veins or at vein endings. Associated with the sori of many ferns are thin membranous outgrowths of the leaf surface which cover the sori. These are called **indusia** (Figure 19-23). The form of the indusium varies with different genera and species and is often used as a character in identification. An individual sporangium commonly consists of a stalked and somewhat flattened case, with a wall, a single cell layer in thickness (Figure 19-23). Flattened cells form the sides of the sporangium. Connecting the sides and arching over about two thirds of the sporangium is a row of cells, the **annulus,** in which the cells are thick-walled next to the sporangium and thin-walled on the outer edge. The lower third of the sporangial wall consists of thinner walled cells extending to the stalk. These cells are termed *lip cells.* The annulus is hygroscopic and at certain times it bends back, suddenly flips forward, and the spores are projected some distance.

Upon germination (Figure 19-24), the spore produces a short filament of cells resembling the protonema of a moss. Later, this filament, with rhizoids, grows into a small ($\frac{1}{8}$ to $\frac{1}{3}$ inch in diameter), flat, heart-shaped gametophyte (prothallus).

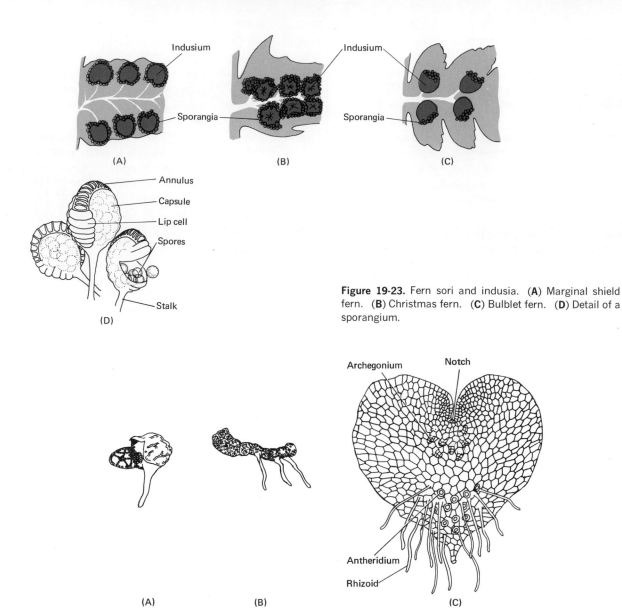

Figure 19-23. Fern sori and indusia. (**A**) Marginal shield fern. (**B**) Christmas fern. (**C**) Bulblet fern. (**D**) Detail of a sporangium.

Figure 19-24. Gametophyte formation. (**A**) Germination of a spore. (**B**) Filamentous stage with rhizoids. (**C**) Mature prothallus showing location of archegonia and antheridia on ventral surface.

Antheridia and archegonia are produced on the ventral surface of the gametophyte. The antheridia are located among the rhizoids, and the archegonia are found just back of the notch where the apical cell is located. Fertilization occurs when water is present between the ventral surface of the prothallus and the soil. Motile sperm are released from the antheridia and travel in the water to the egg cell in the archegonium. The developing embryo soon differentiates into the body of a young sporophyte consisting of a foot, first leaf, stem, and primary root (Figure 19-25). The foot absorbs food from the gametophyte and passes it on to the remainder of the sporophyte. Once the primary root penetrates the soil and the first leaf is developed, the sporophyte is established as an independent plant and the gametophyte dies. In time, the primary leaf is replaced by sporophytic fronds and the primary root gives way to adventitous roots, so an adult plant bearing sporangia is produced. The life cycle of a fern is shown in Figure 19-26.

Relationships and Importance of the Ferns. Ferns and other pteropsids are believed to have evolved from the Psilophytales. Certain fossil ferns resemble the Psilopsida rather closely. The seed plants (Gymnospermae and Angiospermae) are considered to have been derived from certain of the extinct Filicinae.

Modern ferns are of little economic importance. From an aesthetic point of view, they are grown extensively for decorations. The fiddleheads and young foliage of many species are edible and are used as green vegetables in the Orient. A drug is derived from the rhizomes of some ferns and is used to expel worms, especially tapeworms, from the intestinal tract of man. Fossil ferns also contributed to the deposits of the Carboniferous period, which were transformed into coal.

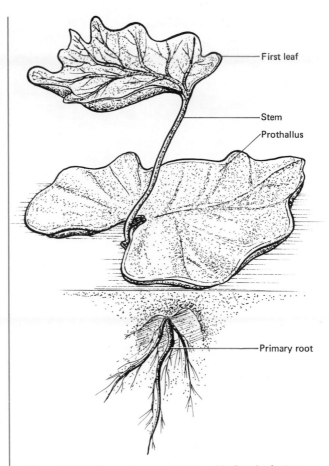

Figure 19-25. Young fern sporophyte with first leaf, stem, and primary root attached to gametophyte.

CHAPTER IN RETROSPECT

Tracheophytes are characterized by the presence of vascular tissue, a terrestrial habit, and the dominance of the sporophyte. Inasmuch as there are over 300,000 species of tracheophytes, they represent the largest group of plants. The more primitive members of this large group, sometimes

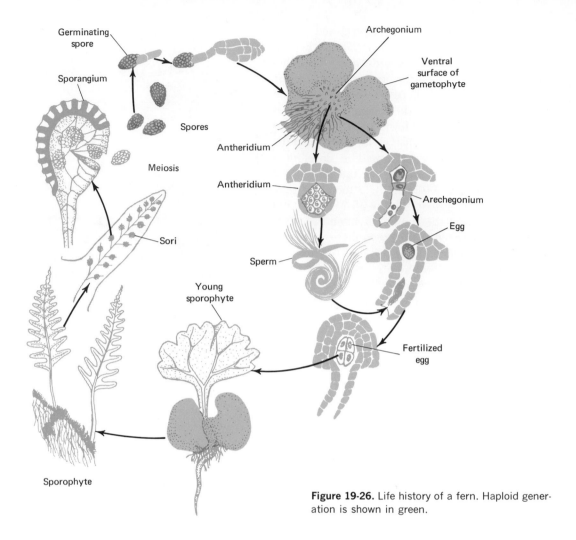

Germinating
spore

Sporangium

Spores

Meiosis

Sori

Sporophyte

Archegonium

Ventral
surface of
gametophyte

Antheridium

Antheridium

Arechegonium

Egg

Sperm

Young
sporophyte

Fertilized
egg

Figure 19-26. Life history of a fern. Haploid generation is shown in green.

referred to as the lower tracheophytes, consist of the psilopsids, lycopsids, sphenopsids, and ferns. Probably the most primitive vascular land plants to have ever inhabited the earth belong to the subdivision Psilopsida. Plants from within this ancient group are transitional and are regarded as the link between aquatic algae and the more advanced vascular plants (clubmosses, horsetails, and ferns). Unlike the psilopsids, all sporophytes of the lycopsids (extinct and living) are differentiated into true roots, stems, and leaves. This group, which includes the club mosses and quillworts, possess very small and spirally arranged leaves (microphyllous) which bear sporangia on their upper surfaces. Representatives of the order *Selaginellaes* (small club mosses) exhibit certain incipient advances toward the attainment of the seed habit.

The sphenopsids were dominant plants during the Carboniferous period and are known principally as fossils, although about 25 living species of *Equisetum* (horsetail) still remain. Pteropsida is the largest tracheophyte subdivision and includes all living and extinct ferns as well as the seed plants (gymnosperms and angiosperms). The ferns (Fillicinae) are characterized by large leaves or fronds that are pinnately compound and display circinate vernation. Both reproductive (spore-bearing) and vegetative leaves are borne on the same plant, a condition known as dimorphism.

Like other lower tracheophytes, the ferns are not economically significant, although some have ornamental value and a few species have been used as foods and medicines.

Questions and Problems

1. What are the salient points of Lignier's theory regarding the origin of land plants? What data exist to support Lignier's hypothesis?
2. What is the evolutionary relationship of *Rhynia* to other vascular plants?
3. Compare *Psilotum, Selaginella,* and *Lycopodium* with regard to structures (internal and external) and methods of reproduction.
4. What are the characteristics of plants which have attained the seed habit of reproduction? Describe how certain species of *Selaginella* displayed early trends toward the seed habit.
5. Contrast the evolutionary advances of Psilopsida, Lycopsida, Sphenopsida, and the ferns. Give specific examples to illustrate each point.

6. Briefly describe the nature of the vascular system in horsetails (*Equisetum*) and ferns. Which is more advanced? Why?
7. Outline the life cycle of a true fern. Use a diagram and include appropriate labels.

Suggested Supplementary Readings

Andrews, H. N., "Evolutionary Trends in Early Vascular Plants," *Cold Spring Harbor Symposia,* vol. 24, pp. 217–234, 1960.

———, *Studies in Paleobotany.* New York: John Wiley & Sons, Inc., 1961.

Bold, H. C., *Morphology of Plants.* New York: Harper & Row, Publishers, 1957.

Delevoryas, T., *Morphology and Evolution of Fossil Plants.* New York: Holt, Rinehart and Winston, Inc., 1962.

———, *Plant Diversification.* New York: Holt, Rinehart and Winston, Inc., 1966.

Foster, A. S., and E. M. Gifford, *Comparative Morphology of Vascular Plants.* San Francisco: W. H. Freeman and Company, 1959.

Salisbury, F. B., and R. V. Parke, *Vascular Plants: Form and Function.* Belmont, Calif.: Wadsworth Publishing Co., Inc., 1964.

Scagel, R. F., et al., *An Evolutionary Survey of the Plant Kingdom.* Belmont, Calif.: Wadsworth Publishing Co., Inc., 1965.

Sporne, K. R., *The Morphology of Pteridophytes.* London: Hutchinson and Co. (Publishers) Ltd., 1962.

Stewart, W. N., "More About the Origin of Vascular Plants," *Plant Science Bulletin,* vol. 6, pp. 1–5, 1960.

Wagner, W. H., and A. J. Sharp, "A Remarkably Reduced Vascular Plant in the United States," *Science,* vol. 142, pp. 1483–1484, 1963.

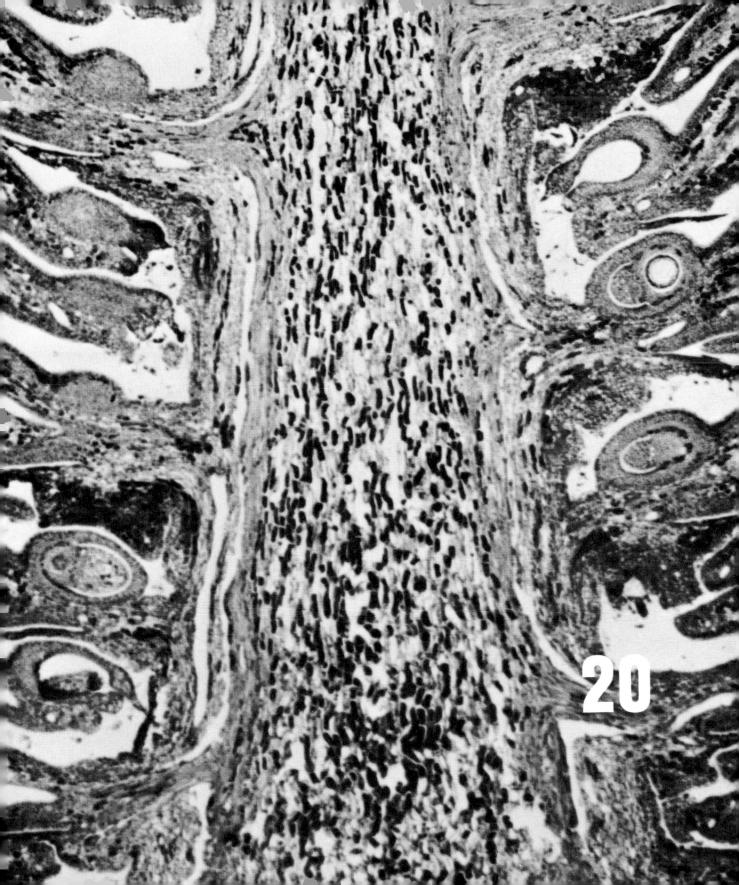

20
Metaphyta: Higher Vascular Plants

INTRODUCTION

In considering the Bryophyta and Tracheophyta up to this point it should be noted that, although these plants have evolved various adaptations to terrestrial life, water is still needed for the *microgamete* (sperm) to swim to the *macrogamete* (egg) for fertilization. In this regard, then, these plants are essentially aquatic as far as sexual reproduction is concerned. The higher Tracheophyta (Gymnospermae and Angiospermae) have developed two major evolutionary advances in their reproductive cycles that have freed them from the ancestral aquatic environment. These advances are the development of a male gametophyte (and male gamete) that does not require water as a medium of fertilization, and the development of seeds. Reproduction by seeds is one of the principal distinctions between higher and lower tracheophytes and is generally viewed as a significant factor in the success of the higher groups. The origin of the seed habit in vascular plants from a non-seed-bearing prototype involved the following evolutionary stages: (1) the development of heterospory, (2) eventual reduction of the number of spores in a megasporangium to one, (3) development of the megagametophyte inside the megaspore, (4) retention of the megaspore inside the sporangium, and (5) enclosure of the sporangium by protective seed coats (integumentary system).

GENERAL CHARACTERISTICS

All seed plants are heterosporous and the greatly reduced gametophytes that develop from the spores are always parasitic (dependent) upon the sporophyte, which is quite conspicuous and dominant. In some cases both microspores and megaspores are produced by the same plant, in which case the sporophyte is referred to as **monoecious.** In others, (**dioecious** plants), microspores and megaspores are produced by different sporophytes. Ultimately, microgametophytes, still enclosed by the microspore wall, are released and carried by wind or animals, especially insects, to megasporangia. At the stage in which they are shed from the microsporangium, the young male gametophytes are called **pollen grains.** Instead of falling to the ground and germinating there, the pollen grains are carried and deposited in contact with megasporangia, a process known as **pollination.** The megasporangium, or nucellus, which contains the female gametophyte, is protected by a covering (**integument**) and the whole structure—the potential seed—is called an **ovule.** Inasmuch as the macrogametophyte is buried inside the ovule, the sperms of most seed plants are delivered to the egg through an outgrowth from the pollen grain called the **pollen tube.** The pollen tube digests its way through the tissue enclosing the female gametophyte and breaks open, delivering the sperm. After fertilization the seed is formed and is ready to leave the parent sporaphyte. Under favorable conditions the seed germinates and the embryo within it develops into a new sporophyte plant.

CLASS GYMNOSPERMAE

The *gymnosperms* include ancient lines of plants, some of which extend back to the Lower Carboniferous period. For many years the gymnosperms were considered to form a natural taxonomic group of closely related plants derived from a common ancestral stock. The evidence now available, however, suggests that gymnosperms include two evolutionary lines that arose at different times and are genetically independent of each other (see Table 20-1). The first of these subclasses, the *Cycadophytae,* consists of smaller plants, usually with pinnate leaves and unbranched stems, possessing a large pith region, little wood, and a thickened cortex. The Cycadophytae in turn is subdivided into three orders—*Cycadofilicales* (*Pteridospermae*), *Bennettitales,* and *Cycadales.* The second evolutionary line, the subclass *Coniferophytae,* contains larger plants which branch considerably and have simple leaves, and stems with a small pith area, abundant wood, and a relatively small cortex. The orders *Cordaitales, Ginkgoales, Coniferales,* and *Gnetales* comprise this subclass.

In addition to the characters already given for the Pteropsida, gymnosperms typically produce naked seeds usually borne on the upper side of open scales that are often produced in cones. All are heterosporous and the sporophytes consist of true roots, stems, and leaves. The sporophyte body is usually large (mostly trees) and woody with an active vascular cambium producing considerable secondary xylem and phloem. Woody tyssues are composed mostly of single-celled elements, the tracheids. All are perennial and most are evergreen. With regard to reproduction, gymnosperms are characterized by various methods of pollination, the production of pollen tubes, and independence from water as a medium for fertilization.

Cycadofilicales—The Seed Ferns

The earliest fossil remains of seed plants are found in Carboniferous strata, largely as compressions of fronds. Among the oldest recognized fossil seed plants are some that greatly resemble ferns. In fact, the resemblance is so great that initially they were thought to be ferns, but a thorough study of the fossils revealed that there were seeds attached to the fernlike foliage. Thus these plants were called the *seed ferns.*

Although seed ferns varied greatly in structure and habit, many were small trees (about 30 feet) with an unbranched stem and a crown of large leaves (Figure 20-1A). Others were smaller, with slender stems, both branched and unbranched. Such plants are probably intermediate between ferns and "typical" gymnosperms. They differed from true ferns not only in the nature of their reproductive structures but also in pattern of venation, which was of the primitive forking, or dichotomous, type, rarely found in living ferns. In addition, seed ferns had a vascular cambium that formed a small amount of secondary wood. The pollen sacs and seeds were borne not in cones (strobili) but directly upon the fronds. In some cases, possibly in the more primitive forms, the seeds were produced on branched stalks as in Psilopsida. In more advanced genera, seeds were produced on the surfaces of the leaves. Fertilization was probably accomplished by swimming sperms. It appears that higher gymnosperms evolved from these ancient fernlike seed plants.

TABLE 20-1. Classification of the Gymnospermae

Kingdom Metaphyta.
 Division Tracheophyta.
  Subdivision Pteropsida.
   Class Gymnospermae (gymnosperms).
    Subclass Cycadophytae.
     Order Cycadofiliciales. *Neuropteris.*
     Order Bennettitales. *Weilandiella.*
     Order Cycadales. *Zamia, Cycas.*
    Subclass Coniferophytae.
     Order Cordaitales. *Cordaites.*
     Order Ginkgoales. *Ginkgo.*
     Order Coniferales. *Pinus, Picea, Cedrus, Sequoia, Larix, Tsuga.*
     Order Gnetales. *Gnetum, Ephedra, Welwitschia.*

(A)

(B)

(C)

Figure 20-1. Representative Cycadophytae. **(A)** A seed fern (*Neuropteris*) of the order Cycadofilicales. **(B)** A cycadeoid (*Weilandiella*) of the order Bennettitales. **(C)** A cycad (*Zamia*) of the order Cycadales. (Courtesy of Field Museum of Natural History, Chicago, Illinois.)

Bennettitales—The Cycadeoids

These plants were common throughout most of the Mesozoic and probably arose from the seed ferns. Some had short stems that bore large fernlike or plamlike leaves; others had tall, slender branching stems with small-bladed leaves (Figure 20-1B). At one time it was believed that the most significant feature of these plants was their production of both microsporophylls and megasporophylls in a spiral-shaped structure bearing a close resemblance to an angiosperm flower. Around the entire structure was a series of *bracts* (reduced or modified leaves) which were thought to be forerunners of the calyx and corolla. Their seeds contained an embryo, two cotyledons, and small amounts of endosperm. Although the vegetative

(A)

(B)

bodies of the cycadeoids and angiosperms are quite dissimilar, botanists at one time suggested that the flowering plants evolved from the Bennettitales. Most of these ideas relating the Bennettitales to flowering plants have been abandoned.

Cycadales—The Cycads

The *Cycadales* are represented both as mesozoic fossils and as living forms comprising many genera. In general habit, the extant forms closely resemble palm trees and tree ferns (Figure 20-1C) and they are the most primitive living seed plants. The stem is typically unbranched, with a crown of large, fernlike pinnate leaves. Some of the 100 species have columnar trunks; others have short, tuberous stems. All cycads are dioecious and,

Figure 20-2. Cordaitales. **(A)** Habit of *Cordaites*. **(B)** Portion of a branch showing simple, parallel-veined, straplike leaves and strobili. (Courtesy of Field Museum of Natural History, Chicago, Illinois.)

except for one genus (*Cycas*), the pollen sacs and ovules are arranged in cones. The motile sperm of the cycads is one of their most primitive features.

Cordaitales

This extinct group of fossil plants known from Paleozoic strata is the first of several orders placed in the subclass Coniferophytae. They were tall trees, reaching a height of 100 feet or more, and contained long, narrow, undivided leaves with parallel veins (Figure 20-2). Internally the stems of more primitive genera resembled cycads and

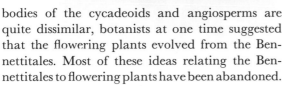

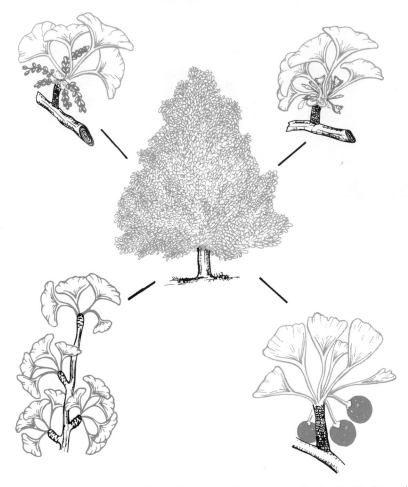

Figure 20-3. Ginkgoales. *Ginkgo biloba* shown in various aspects. *Center:* Habit of tree. *Lower left:* Section of shoot showing spirally arranged leaves. *Upper left:* Shoot with microstrobili. *Upper right:* Shoot with ovules. *Lower right:* Shoot with mature seeds.

consisted of a large pith, scanty wood, and a large cortex. Advanced genera contained stems with a small pith, much wood produced by extensive secondary growth, and a thin cortex. Such forms have stems that closely resemble conifers. In addition, both microsporangia and megasporangia were borne in cones rather than on vegetative leaves, another characteristic suggestive of the conifers.

Ginkgoales—The Ginkgoes

Ginkgo biloba, the maidenhair tree, is the only modern representative of this group, which is well documented in fossil strata as far back as the Permian (Figure 20-3). *Ginkgo* is a fairly large tree, known to reach heights of 100 feet or more. It has been cultivated for centuries in Asiatic countries and has been transported to numerous cities in the United States and Europe, where it is widely planted in parks, on campuses, and along boulevards. Structurally, *Ginkgo* produces a vertical axis with horizontally held branches, and the characteristic leaves are broadly fan-shaped and bilobed with dichotomous venation. Unlike most other gymnosperms, the maidenhair tree is deciduous. The stem and root are anatomically much like those of conifers.

Ginkgo is strictly dioecious. The male sporophyte has a loose, short strobilis containing a catkin-like aggregation of paired microsporangia. By contrast, the female reproductive organ bears little resemblance to a strobilus. The ovules are borne in twos or threes at the top of a long stalk. Following pollination, pollen tubes with motile sperm are produced and fertilization occurs. The products of fertilization are yellowish seeds, which at full maturity have an objectionable odor. For this reason the male trees rather than the megasporangiate trees are preferred in cultivation.

Order Coniferales—The Conifers

The order *Coniferales* is the largest group of modern coniferophytes and its members include many well-known genera such as *Pinus* (pine), *Picea* (spruce), *Cedrus* (cedar), *Sequoia* (redwood), *Cupressus* (cypress), *Larix* (larch), *Taxus* (yew), *Abies* (fir), and *Tsuga* (hemlock). Most conifers are trees; however, some are shrubs. They vary in size from small creeping plants, to shrubby bushlike plants, to the big trees of California. All conifers are characterized by the production of conelike strobili—thus their designation cone-bearing plants. Most of them are monoecious, although a few species are dioecious. In none are both male and female organs borne in the same cone.

Life History of the Pine. Inasmuch as it is impossible to discuss all, or even a few, representative conifers, only the pine will be treated. *Pinus* is perhaps the best known and, in many

regions, the most widely distributed conifer, and its life cycle in its main outlines illustrates the characteristic reproductive processes of all other conifers. In general, the anatomy and developmental processes of the sporophyte body of a pine tree are similar to those of woody dicots, discussed in Chapter 4. A major distinction, however, is that in coniferous wood there are no vessels, only tracheids, and there are scattered resin cells as well.

The leaves of conifers are always simple and usually narrow and needlelike, ranging in size from $\frac{1}{2}$ inch to several inches long and less than $\frac{1}{8}$ inch wide. In the pines the leaves are produced in clusters of two to five leaves in a fascicle on short lateral branches (Figure 20-4A). The number of leaves in a fascicle depends upon the species involved. In cross section (Figure 20-4B) it can be seen that a pine leaf is adapted for xeric conditions. Typically, there is a heavy cuticle surrounding a thick epidermis with sunken stomata. Immediately inside the epidermis occurs a *hypodermis* of one to three layers of thick-walled cells. Internal to this layer is the mesophyll, which consists largely or wholly of parenchyma, often with one or more resin ducts. The vascular bundles, usually one or two, are surrounded by a well-defined endodermis.

Two types of cones are produced in pines—seed (megasporangiate) cones and pollen (microsporangiate) cones—and both are borne on the same tree. The pollen cones or male cones are rather small and inconspicuous and endure for only a few weeks in the spring. They are produced in clusters and each is composed of a central axis to which is attached a number of spirally arranged microsporophylls (Figure 20-5A). The scalelike microsporophylls (cone scales) bear two elongated microsporangia on their lower surface, in which numerous microspore mother cells undergo meiosis to produce four haploid microspores. The

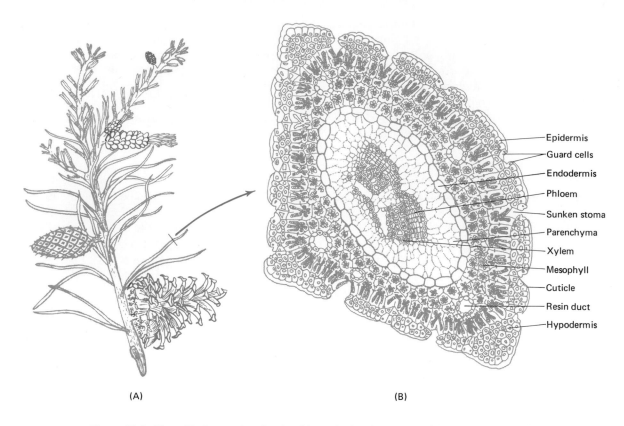

(A)   (B)

Epidermis
Guard cells
Endodermis
Phloem
Sunken stoma
Parenchyma
Xylem
Mesophyll
Cuticle
Resin duct
Hypodermis

Figure 20-4. Pine. (**A**) Composite sketch of branch showing clustered leaves and cones in various stages in development. (**B**) Transverse section through a pine needle (leaf) showing arrangement of internal tissues.

microspores develop an inner and an outer wall over most of their surface, the outer wall expanding to form air-filled, winglike appendages. While still inside the microsporangia each microspore germinates to form mature pollen grains (Figure 20-5B). The pollen grain consists of a greatly reduced male gametophyte, that is, two *prothallial cells* which quickly degenerate, a small *generative cell* around which a wall forms, and a larger *tube cell* in the remaining cytoplasm. At maturity the microsporangia, now called pollen sacs, break open and release the pollen grains. Massive amounts of

pollen are carried by the wind to the female cones.

The seed cones, or female cones, are the more conspicuous of the two types of cones borne on conifers. The megasporangiate cone of *Pinus* consists of a large central axis which bears a number of spirally arranged bracts (Figure 20-6A). In the axil of each bract is a thick, woody scale containing two ovules on the upper surface. Each ovule (Figure 20-6B) consists of a megasporangium surrounded by and united with the integument and containing a single megaspore cell. There is

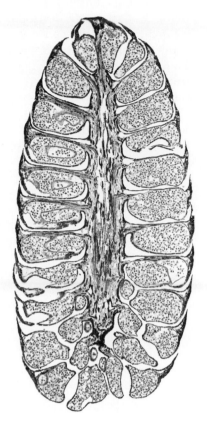

Figure 20-5. Male reproductive structures in pine. **(A)** Micro-sporangiate cone in longitudinal section and a diagrammatic enlargement of a single microsporophyll (viewed from below) containing two sporangia. **(B)** Development of microspores into mature pollen grains. (Photo A courtesy of Carolina Biological Supply Company. *B* From C. J. Chamberlain, *Gymnosperms: Structure and Evolution,* Chicago: University of Chicago Press, 1935. By permission of Mrs. Charles J. Chamberlain.)

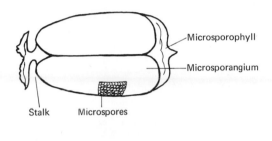

Microsporophyll

Microsporangium

Stalk   Microspores

(A)

Degenerate prothallial cells
Generative cell
Tube cell
Wings

Microspore mother cell

Developing microspore

Mature pollen grain

(B)

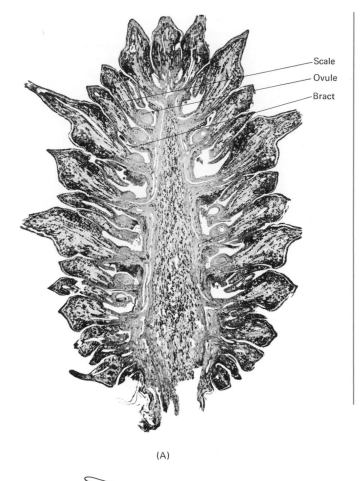

Scale

Ovule

Bract

(A)

a small tubular opening in the integument, the **micropyle.** Early in the development of the megasporangium, the megaspore mother cell undergoes meiosis and gives rise to four megaspores arranged in a row. Three of these disintegrate and the one farthest from the micropyle remains as the functional megaspore (Figure 20-6C).

The functional megaspore germinates within the megasporangium and there is enlargement and free division of the nucleus with the formation of cell walls around each. The resulting multicellular body is the female gametophyte, the megagametophyte, with two to five archegonia located at the micropylar end (Figure 20-7A). Essentially, the gametophyte is comparable in function to the endosperm of angiosperm seeds. The mature archegonium itself is greatly reduced, consisting chiefly of neck cells and a large egg.

In the process of pollination the young female cones stand erect at the ends of branches and the scales are spread apart. The pollen grains sift down among the scales of the female cone, some landing in a sticky resinous secretion near the open micropylar end of an ovule. After the pollen grains are caught in the sticky fluid the secretion with-

Figure 20-6. Female reproductive structures in pine. (**A**) Megasporangiate cone in longitudinal section. (**B**) Enlargement of ovule before meiosis. (**C**) Ovule after meiosis. (Photo courtesy of Carolina Biological Supply Company.)

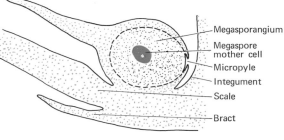

Megasporangium

Megaspore mother cell

Micropyle

Integument

Scale

Bract

(B)

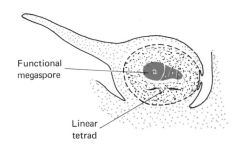

Functional megaspore

Linear tetrad

(C)

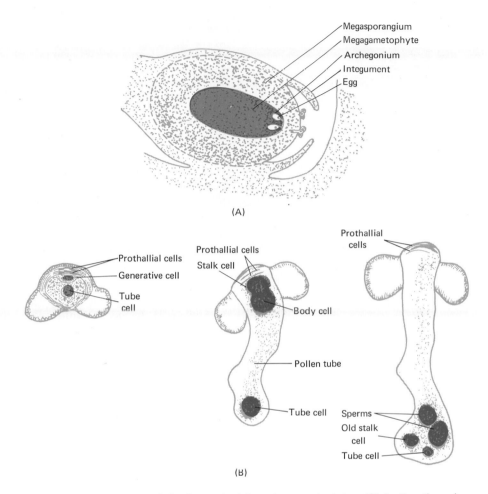

Megasporangium
Megagametophyte
Archegonium
Integument
Egg

(A)

Prothallial cells
Generative cell
Tube cell

Prothallial cells
Stalk cell
Body cell
Pollen tube
Tube cell

Prothallial cells
Sperms
Old stalk cell
Tube cell

(B)

Figure 20-7. Pollination and development of the microgametophyte. **(A)** Section through a pine ovule showing archegonia and the initiation of pollen tube formation. **(D)** Details of the development of the microgametophyte. (*B* from C. J. Chamberlain, *Gymnosperms: Structure and Evolution,* Chicago: University of Chicago Press, 1935. By permission of Mrs. Charles J. Chamberlain.)

draws, carrying the pollen grains to the tissues of the megasporangium. The cells of the integument adjacent to the micropyle become swollen and close the opening.

After a variable period of time, the pollen grain germinates to form a pollen tube which penetrates the tissues of the megasporangium (Figure 20-7A). Early in the development of the pollen tube, the tube nucleus migrates from the pollen grain to the tip of the tube (Figure 20-7B). Later the generative cell divides into two cells, one the *stalk cell* and the other the *body cell*. The stalk cell,

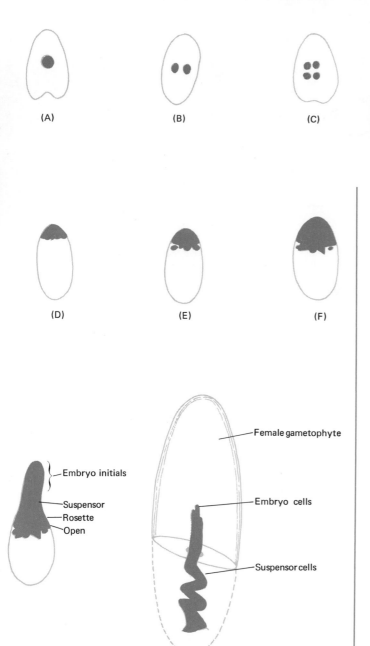

(A)

(B)

(C)

(D)

(E)

(F)

(G)

— Embryo initials
— Suspensor
— Rosette
— Open

(H)

— Female gametophyte

— Embryo cells

— Suspensor cells

Figure 20-8. Stages in the development of a pine embryo. **(A)** The egg nucleus and the sperm nucleus combine during fertilization. **(B)** First division of the zygote. **(C)** Second division of the zygote resulting in a flat plane of four cells. **(D–F)** Formation of the proembryo. **(G)** Development of tiers. **(H)** Elongation of suspensor cells and penetration of the embryo into the gametophyte (endosperm). (Reprinted from "Suspensor and Early Embryo of *Pinus*," by J. T. Bucholz, *Botanical Gazette*, vol. 66, p. 185, 1918. By permission of the University of Chicago Press.)

probably vestigial, is sterile and passes down into the pollen tube as it continues to elongate. The body cell again divides to form two nonmotile male gametes, the sperms. The pollen tube next penetrates between the archegonial neck cells and discharges its contents into the archegonium. Fertilization is accomplished by the fusion of the nucleus of one sperm (the other degenerates) with the nucleus of the egg. All other remaining nuclei within the gametophyte disintegrate.

Following fertilization in *Pinus,* which occurs a year after pollination, the zygote nucleus divides (Figure 20-8). The two resulting nuclei also divide, forming four nuclei, and become arranged in a flat plane of four cells. Subsequent nuclear divisions produce a *proembryo* consisting of 16 cells, in four tiers of four cells each. Of the four tiers of cells, one tier gives rise to four embryos; the cells of the next tier elongate to many times their original size to form the *suspensors;* the third tier, the *rosette tier,* may give rise to about a dozen additional cells; and the last, the *open tier,* eventually disintegrates. Only one embryo produced survives and it is pushed into the female gametophyte tissue by the elongation of the suspensor cells. The cells of the female gametophyte are utilized by the embryo for subsequent growth and development.

The mature embryo of pine consists of the **radicle,** the **hypocotyl,** the **cotyledons** (usually eight), and the **plumule** (Figure 20-9A). The

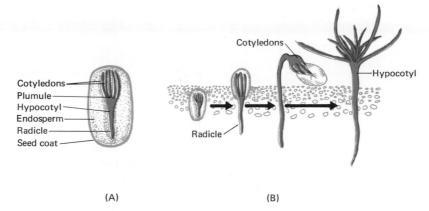

Figure 20-9. Mature pine seed and germination. (**A**) Longitudinal section through a pine seed showing parts of the embryo plant. (**B**) Stages in the germination of a pine seed.

radicle is the embryonic root, the hypocotyl is the transition region between the stem and the root, the cotyledons are embryonic leaves, and the plumule is the embryonic shoot. The mature embryo is embedded in the female gametophyte, which is the principle food storage tissue of the seed. The embryo and the gametophyte are in turn surrounded by a seed coat developed from the integument of the ovule. Essentially, the *seed* represents a ripened ovule containing an embryo. In a number of pines the seed coat extends out into a thin wing at one end of the seed, which aids in wind dispersal. When the female cone opens, the seeds are liberated.

Under suitable environmental conditions, the seed imbibes water and begins its germination (Figure 20-9B). The process is very similar to seed germination in angiosperms (Chapter 5). The radicle elongates, pushes through the seed coat, and penetrates the soil as the primary root. The hypocotyl, at first arched, straightens as it grows above the soil level, carrying the seed coat and cotyledons upward. The cotyledons, typically green and needle-like, absorb food from the seed, the integument drops away, and the plumule

develops into the shoot. After a number of years of vegetative growth, the tree produces cones each year throughout its lifetime. The complete life cycle of pine is summarized in Figure 20-10.

In discussing the evolutionary sequence of the principal gymnosperm groups, pine, as a representative example, may be viewed as an intermediate between a fern and an angiosperm. The life cycle of a pine features certain evolutionary advances over the fern life cycle, which may be summarized as follows:

1. In pine, there are two types of strobili, more complex and well developed (than in lower tracheophytes), which contain microsporangia and megasporangia.

2. These sporangia produce microspores and megaspores, which germinate into microgametophytes and megagametophytes, respectively.

3. The gametophytes are much further ruduced than those of the ferns and are retained within the protective tissues of the cone during a portion of their development or are permanently retained.

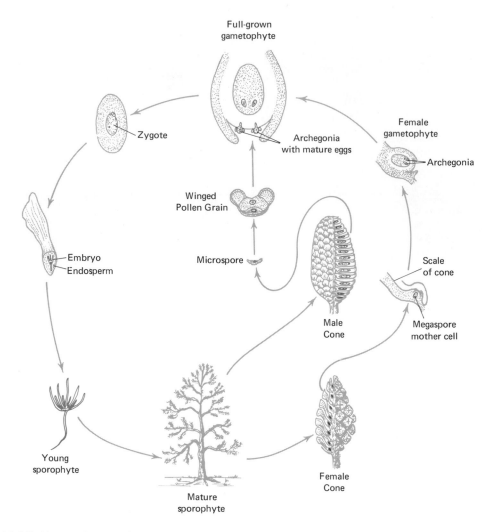

Figure 20-10. Life history of a pine. See text for details.

4. The gametophytes are dependent entirely upon the sporophyte.

5. The gametes are not exposed to adverse environmental conditions. The sperm move to the egg through a pollen tube and the egg remains embedded in the gametophyte tissue.

6. Fertilization is independent of outside conditions of moisture; thus the sperms typically are not flagellated.

7. The integument of the megasporangium affords protection for the egg and developing embryo against injury and dessication and enables

(A)

(B)

Figure 20-11. Representatives of the Gnetales. **(A)** *Welwitschia mirabilis.* Note the large, woody, tuberous stem from which the straplike leaves emerge. Strobili appear on slender, upright branches. **(B)** *Ephedra trifurca,* a shrublike plant that grows in arid parts of the world, including the southwestern United States. (Courtesy of Field Museum of Natural History, Chicago, Illinois.)

the seed to retain its stored food. The seed habit and dormancy are considered to represent major evolutionary advances in the plant world.

Order Gnetales

This final order of the gymnosperms contains no known ancestors and, compared to other gymnospermous plants, is a modern group. The three genera which comprise this group are of uncertain relationship but appear to be gymnospermous in character (Figure 20-11). Collectively, *Gnetales* exhibit certain angiosperm characters, including

vessels in the secondary wood (their ontogeny is different, however, from that of angiosperm vessels), vascular rays, two cotyledons, the absence of archegonia, broad net-veined leaves, ovules with two integuments, and the resemblance of some strobili (male) to angiosperm flowers. Based upon these structural features the Gnetales were at one time thought to be more or less directly ancestral to the angiosperms. However, the absence of a significant fossil record and the morphological isolation of the group has led most botanists to conclude that the Gnetales are a very highly specialized group (or groups) of gymnosperms, and any relationship to angiosperms is remote.

Relationships of the Gymnosperms

The gymnosperms, as they have been considered here, consist of two groups, the Cycadophytae and the Coniferophytae. Both groups apparently originated from some ancestral heterosporous tracheo-phytes such as psilophytes. The lines of development separated early and both may have evolved via the Cycadofilicales, or seed ferns. Although both subclasses are of similar age, the Conifero-phytae have advanced further in their evolutionary series. Such advances include nonmotile sperms, simple leaves, small pith and cortex, a freely branched stem, and emphasis on vascular tissue.

Economic Importance of the Gymnosperms

The gymnosperms, particularly the conifers, are an important source of lumber and lumber products. Among the principal timber-producing trees are pine, Douglas fir, spruce, redwood, cedar, cypress, fir, and hemlock. Wood pulp for use in paper making is also obtained from several conifers such as pines and spruces. Numerous other materials of economic value, such as turpentine, tannins, charcoal, tar, various oils, resin, and methanol, are derived from different conifers. Turpentine is the principal solvent for many paints and varnishes; tannins, obtained from hemlock, are used in the tanning industry, for ink making, and in the manufacture of certain drugs; and resins, extracted from pines, are used in the manufacture of linoleum, perfume, varnishes, and other products. Certain conifers such as spruce and fir are used extensively as Christmas trees.

CLASS ANGIOSPERMAE

The *angiosperms* are the most recently evolved and the most successful of all plant groups. Since their origin they have undergone a rapid and diverse evolutionary history which has culminated in the production of a specialized reproductive structure, the flower. The production of flowers, fruits, and seeds, and the ability to survive in practically every type of environment, are the two most important factors in the success of the angiosperms.

Although descended from a gymnospermous stock, angiosperms differ from gymnosperms in several ways. With respect to the sporophyte body, all gymnosperms are woody perennials, whereas many angiosperms are perennial or annual herbaceous plants. It is generally believed that herbaceous angiosperms evolved from woody ones. Angiosperms possess not only taproots typical of coniferous plants but also various root and stem modifications adapted to food accumulation and capable of vegetative propagation. Leaves of conifers are usually small and needle-like, whereas those of angiosperms exhibit enormous diversity in size and form depending upon variations in the environment. Internally, the xylem of angiosperms is structurally more advanced that that of gymnosperms. Conifer wood is composed principally of tracheids and ray cells; the wood of an angiosperm contains a greater variety of cells, including vessels, not found in conifers (Chapter 4).

Most certainly, the outstanding external characteristic which distinguishes angiosperms from gymnosperms, as well as all other plant groups, is the flower, which, through its activities, gives rise to a seed and a fruit (see Chapter 5). The seeds of gymnosperms are naked and not enclosed in a fruiting structure because the ovules are borne directly on the sporophyll as part of the sporangium. In angiosperms, the ovules and seeds are covered by the carpel(s) (megasporaphyll), which at maturity constitutes the fruit. The enclosure of the ovules prevents direct access of pollen to them, so pollen grains are received by a specialized portion of the pistil, the stigma. The pollen grains produced from microspores germinate on the stigma, and the pollen tubes grow through the tissues of the pistil until the sperm reach the egg. The process of fertilization is a double fertilization, unique to angiosperms. In both gymnosperms and angiosperms two nuclei are produced in each pollen tube, but in gymnosperms only one functions. In angiosperms both are functional; one fertilizes the egg and the other unites with other nuclei (polar) to form the endosperm. Archegonia, found in conifers, are not present in the ovules of angiosperms. Pollination in gymnosperms is by wind, whereas angiosperm pollination is by wind, insects, or other animals.

Origin and Classification

At present the origin of angiosperms is still unknown, largely because of inadequate fossil data. There is no evidence among the fossil remains that angiosperms lived at any time during the Paleozoic. It is assumed that they arose during the Lower Cretaceous of the Mesozoic. This view is supported not only by numerous angiosperm fossils from Cretaceous and Tertiary strata but also by the climate at that time, which was conducive to the rapid rise of a group such as angiosperms. Based upon comparative morphological studies, it appears that the remote ancestor of the angiosperms was an early seed fern or very primitive gymnosperm.

Angiosperms are divided into two subclasses, the *Dicotyledonae* (*dicots*) and the *Monocotyledonae* (*monocots*), and consist of 300 families and over 300,000 species (Table 20-2). The currently held view is that the dicots are the more primitive of the two groups and gave rise to the monocots at an early stage in evolution. Among the primitive features which relate dicots to gymnosperms are an active vascular cambium and a circular arrangement of fibrovascular bundles. Monocots probably evolved from primitive woody dicots through the loss of the vascular cambium, degeneration of one of the cotyledons, and rearrangement of the fibrovascular bundles from a cylindrical to a scattered arrangement. Some of the principal characteristics of monocots and dicots are shown in Figure 20-12 and need not be further enumerated here.

Each of the subclasses is divided into orders, which in turn consist of one or more families. The

TABLE 20-2. Classification of the Angiospermae

Kingdom Metaphyta
  Division Trachcophyta
    Subdivision Pteropsida
      Class Angiospermae (angiosperms)
        Subclass Dicotyledonae (dicots)
          Order Ranales (buttercup)
          Order Papaverales (poppy)
          Order Violales (violet)
          Order Caricales (papaya)
          Order Sarraceniales (pitcher plant)
          Order Euphorbiales (spurge)
          Order Geraniales (geranium)
          Order Rhamnales (grape)
          Order Urticales (nettle)
          Order Piperales (pepper)
          Order Primulales (primrose)
          Order Oleales (olive)
          Order Lamiales (mint)
          Order Scrophulariales (snapdragon)
          Order Rosales (rose)
          Order Cactales (cactus)
          Order Cornales (dogwood)
          Order Umbellales (parsley)
          Order Begoniales (begonia)
          Order Asterales (aster)
          Order Salicales (willow)
          Order Juglandales (walnut)
        Subclass Monocotyledonae (monocots)
          Order Lilliales (lily)
          Order Palmales (palm)
          Order Graminales (grass)
          Order Musales (banana)
          Order Orchidales (orchid)
          Order Hydrocharitales (elodea)
          Order Naiadales (pondweed)

families are further subdivided into genera. Among the principal characters used in classification into families are the position of the ovaries (inferior or superior), type of placentation, number and arrangement of floral structures, degree of fusion, if any, between floral parts, manner in which flowers are clustered on a branch, and structure of the seed and fruit (see Chapter 5).

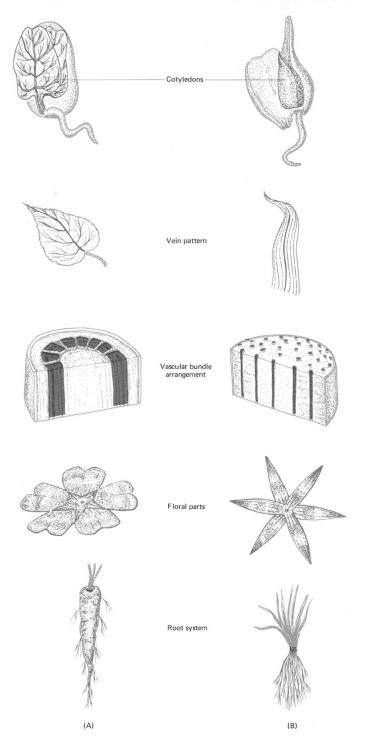

Cotyledons

Vein pattern

Vascular bundle
arrangement

Floral parts

Root system

(A)   (B)

Figure 20-12. Comparison of a few structural features of dicots and monocots. (**A**) In the dicots there are two cotyledons (seed leaves) in the embryo; the veins of the leaves are pinnately or palmately arranged; the vascular bundles in the stem are cylindrically arranged; floral parts are in fours or fives or multiples of these; the root system is typically a taproot; and there is an active cambium. (**B**) In the monocots there is one cotyledon in the embryo; the veins of the leaves are parallel; the vascular bundles in the stem are scattered; floral parts are in threes or multiples of three; the root system is typically fibrous; and there is no vascular cambium.

The general structure, function, and importance of the vegetative organs of an angiosperm have been discussed in Chapters 3 and 4 and the reproductive aspects have been treated in Chapter 5. At this point, therefore, attention will be turned to some of the major evolutionary trends exhibited by the angiosperms with particular emphasis on floral evolution.

Angiosperm Evolution

Perhaps the best indication of relationships among angiosperms and their subsequent course of evolution is found by analyzing the structure of the flower. Logically, the nature of these relationships and evolutionary changes is best revealed by comparison of advanced floral forms with a primitive flower. Unfortunately, the fossil record does not provide any flowers more primitive than any now living, so such a comparison must be restricted to flowers of the most primitive living angiosperms. Such flowers are found in the order Ranales, believed by most botanists to be the most primitive of extant flowering plants. Using the families of the Ranales as a starting point, the angiosperms have been analyzed and a list of modifications from a primitive type has been amassed. In general, angiosperm evolution has proceeded from simplicity to relative complexity. However, many aspects of angiosperm evolution, especially floral modifications, are characterized by regression in which simplicity and not complexity is the evolutionary product.

Assuming that the Ranalian theory is correct, the following principal evolutionary tendencies among angiosperms have been recognized:

1. Pollination by insects is considered to be a feature of the most primitive flowers with wind-pollinated flowers, representing a derived and reduced condition from earlier forms possessing more complex insect-pollinated flowers.

2. Flowers with spirally arranged sepals, petals, stamens, and carpels are primitive; those with cyclic or whorled parts are more advanced.

3. Flowers with many parts, the number of which may vary with the individual, are more primitive than those flowers with a definite number and few parts.

4. Flowers with separate parts (sepals, petals, stamens, and carpels) are more primitive than those with fused parts.

5. Fusion of one type of flower part with another (stamens fused to petals) is an advanced condition.

6. Flowers with superior ovaries are more primitive than those with inferior ovaries.

7. Plants with solitary flowers are more primitive than those with numerous flowers grouped into an inflorescence.

8. Monoecious plants are more primitive than dioecious ones.

9. Flowers that have one or more of their parts missing (incomplete) are derived from flowers with all their parts (complete).

10. Radial symmetry is a primitive characteristic; bilateral symmetry is advanced.

11. Woody angiosperms are more primitive than herbaceous plants.

12. Primitive plants tend to be perennial; biennials and annuals are more advanced.

CHAPTER IN RETROSPECT

Gymnosperms and angiosperms have developed two major evolutionary advances in their reproductive cycles: (1) water as a medium of fertilization is no longer required, and (2) the development of the seed habit has become firmly established. These characteristics are especially suited to the terrestrial environment in which most higher tracheophytes are found. Whereas gymnosperms typically produce naked seeds, the seeds of angiosperms are covered by the megasporophyll, which at maturity constitutes the fruit. Gymnosperms include two genetically unrelated subclasses—Cycadophytae and Coniferophytae. Perhaps the best known gymnosperm representatives are the conifers, which include pine, spruce, cedar, redwood, cypress, hemlock, and fir. These plants, sometimes referred to as cone-bearing plants, produce conelike strobili. The life cycle of pine (*Pinus*) characterizes the reproductive process of all conifers. In terms of an evolutionary sequence, pine, as a representative example of Coniferales, may be regarded as an intermediary between ferns and angiosperms. The economic value of gymnosperms ranks them high in their importance to man. They are important sources of lumber, particularly the conifers, as well as sources of turpentine, tannins, charcoal, tar, various oils, resin, and methanol.

By far the largest group of plants (over 300,000 species) and the most successful in modern times are the angiosperms. The appearance of a flower, more than any other external characteristic, distinguishes angiosperms from all other plant groups, including gymnosperms. Like gymnosperms, angiosperms are also divided into two subclasses. They are the Dicotyledonae (dicots), the more primitive group, and the Monocotyledonae (monocots). The relationship among angiosperms and their probable evolutionary history is revealed by analyzing the structure of the flower.

Questions and Problems

1. List the structural (external and internal) and reproductive differences of the following plant groups: Cycadofilicales, Bennettitales, Cycadales, and Ginkgoales. On the basis of this analysis, which are the most primitive and which are the most advanced plants?
2. Give a brief description of the function of each structure involved in the reproductive cycle of *Pinus*.
3. Using a series of diagrammatic sketches, illustrate the process of seed germination in pine.
4. Compare the process of alternation of generations, as seen in *Pinus*, with that of a true fern. What are the major differences?
5. What is the unique feature of *Ginkgo biloba* with respect to its evolutionary history?
6. Discuss the economic importance of gymnosperms.
7. Contrast the origin of angiosperms and gymnosperms. What evidence exists to show that one group of plants is declining and the other group is becoming more dominant?

Suggested Supplementary Readings

Andrews, N. H., *Studies in Paleobotany.* New York: John Wiley & Sons, Inc., 1961.

———, "How Old Are the Angiosperms?", *American Journal of Science,* vol. 259, pp. 447–459, 1961.

Axelrod, D. I., "A Theory of Angiosperm Evolution," *Evolution,* vol. 6, pp. 29–60, 1952.

Chamberlain, C. J., *Gymnosperms: Structure and Evolution.* Chicago: The University of Chicago Press, 1935.

Constance, L., "The Systematics of the Angiosperms," in *A Century of Progress in the Natural Sciences, 1853–1954.* San Francisco: California Academy of Sciences, 1955.

Eames, A. J., *Morphology of the Angiosperms.* New York: McGraw-Hill Book Company, Inc., 1961.

Foster, A. S., and E. M. Gifford, *Comparative Morphology of Vascular Plants.* San Francisco: W. H. Freeman and Company, 1959.

Meeuse, B. J. D., *The Story of Pollination.* New York: The Ronald Press Company, 1961.

Scott, R. A., E. S. Barghoorn, and E. B. Leopold, "How Old Are the Angiosperms?", *American Journal of Science,* vol. 258, pp. 284–299, 1960.

21
Relationships of Plants

INTRODUCTION

Plant ecology is a branch of botany which considers the effects of environmental factors on plant form and function, the migration and distribution of plants, and the formation of plant communities. Ecology examines the relationships of an organism or group of organisms acting and interacting with each other and with their environment. To examine the interrelations that exist between living organisms and their environment, all aspects of their activities must be considered. The sum of these relationships contributes to the development of an ecological system whereby both living and nonliving constituents of the environment function together and are interrelated in a level of organization called the **ecosystem.** For example, a deciduous forest consisting of maple, oak, and birch trees, various shrubs, and other ground-level plants in contact with the soil, water, minerals, atmospheric gases, and sunlight represents one type of ecosystem.

The highest level within an ecosystem is the **community** level of organization containing all the living members, both plant and animal. Within a community the organisms have a common relationship among themselves and with their environment. The members of a given community may consist of a mass of different algal types floating on a pond, a variety of slime molds on decaying logs, a field of wheat, or an oak–hickory forest. Communities may cover as little as 1 square inch of land, as in the case of a slime mold community, or as much as many thousands

of acres, as represented by an oak-hickory forest. A **community** may be viewed as several populations living together in the same area. **Populations** are groups of individuals of similar species existing within a given area delimited by certain restricting boundaries. Together the community, population, and individuals form the realm of life, the **biosphere.**

FUNCTIONING ECOSYSTEM

An ecosystem consists of four principal components: (1) nonliving substances, (2) producers, (3) consumers, and (4) decomposers. A freshwater pond, as illustrated in Figure 21-1, typically contains the basic constituents of an ecosystem. For example, the water and soil constitute the nonliving factors; the rooted, floating, and submerged vegetation, both microscopic and macroscopic, are examples of producers; and frogs, fish, insects and some microorganisms feed upon the green plants and upon each other and are therefore consumers. The organisms of decay—fungi, bacteria, and related microorganisms—are the decomposers.

Any given ecosystem will show evidence of change, both physical and chemical, as well as constant interrelations. These activities are the result of direct and indirect associations between **abiotic,** or nonliving, substances and **biotic,** or living, organisms. Each of these sets of factors represents a particular component of the ecosystem.

Abiotic Factors

The nonliving components, or **abiotic** factors, include a variety of naturally occurring chemical compounds and environmental influences. Some abiotic factors act upon plants through the atmosphere. These would include light, temperature, precipitation, and wind. Other abiotic factors influence plants through the soil. These include

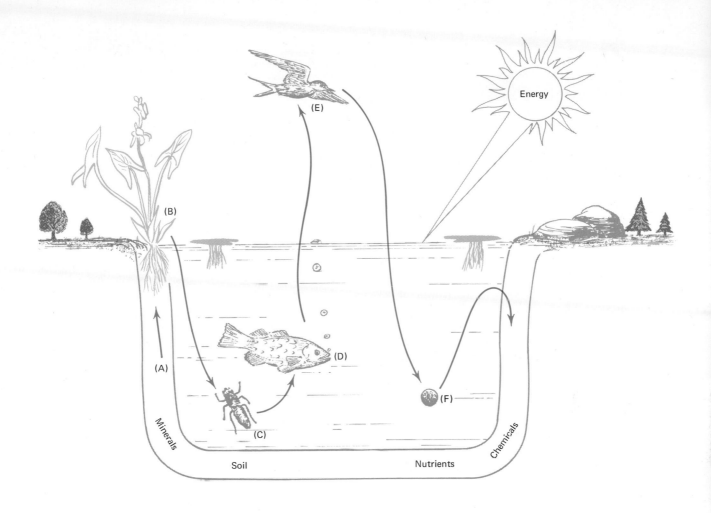

Figure 21-1. Cross-section of a pond ecosystem. The components are as follows: **(A)** Abiotic substances. **(B)** Producer organisms. **(C)** First order consumers. **(D)** Second order consumers. **(E)** Third order consumers. **(F)** Decomposers.

the physical properties of the soil, such as drainage, soil nutrients, and soil temperature.

Abiotic Factors of the Atmosphere. *The Influence of Light on Plant Growth.* The term *light* refers to the visible portion (wavelengths from about 390 to 760 mμ) of the electromagnetic radiation spectrum (Figure 21-2). As described earlier, light is an important factor that influences the growth and distribution of plants. Light is directly related to the processes of photosynthesis and transpiration and is a major factor in determining the time of flowering, the degree of water absorption, and the temperature of the soil. Different species of plants vary with regard to different light requirements. Some species, such as sugar maple, beech, hemlock, fir, and spruce, are able to survive in the

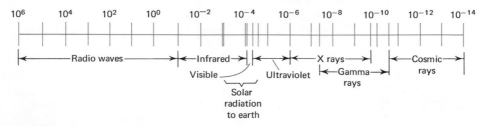

Figure 21-2. The electromagnetic spectrum from radio waves to cosmic rays.

shade of the forest floor, whereas other species, such as aspen, willows, birch, and some pines, will grow only when subjected to very high light intensity. A plant species which grows and develops in deep shade is said to be *tolerant;* one which cannot survive in the shade is said to be *intolerant.* In addition, light is an important variable in the enzyme mechanisms of cells. Inasmuch as light influences flower and seed production, it is a significant factor in determining, at least in part, the distribution of various plants.

Before it may be used in any biological process in the plant, light energy must be absorbed by receptor substances. These substances, known as *photoreceptors,* are pigments, and each has a characteristic *absorption spectrum* in which the maximum absorption of light is determined by the various wavelengths of light (the action spectrum). Some of the more important photoreceptor pigments found in plants are chlorophyll, carotenoids, and phytochrome. The visible wavelengths of radiation are not the only ones that have an effect on the growth of plants. One form of radiation with wavelengths longer than those of visible light is *infrared* or *heat radiation.* These frequencies of radiation cause an increase in the temperature of plants, but aside from that any other effects have gone unnoticed.

The radiation with wavelengths shorter than those of light exhibit considerable influence on the growth of plants. Ultraviolet light has been used to combat bacterial infections and may be injurious to the surface tissues of multicellular plants as well. Other shorter wavelengths, such as X rays and gamma rays, aid in promoting plant mutations and may be factors influencing plant development.

Temperature. One of the most critical factors affecting plants in their environment is temperature. A cursory observation of a variety of plant forms would show that some are able to endure high summer and low winter temperatures, whereas others are killed quickly due to fluctuating temperatures. In the vast majority of plants, the minimum and maximum temperatures at which physiological activities proceed are 32°F (0°C) and 110°F (43.3°C), respectively. This range is referred to as the *biokinetic range.* Within these limits, temperature has a marked effect upon the rates of growth, digestion, respiration, photosynthesis, and the absorption of water and minerals. Temperature is also one of the most important environmental factors in determining both the numbers and varieties of plants occupying a particular ecological setting. For example, in the United States cotton growing is restricted to the South because the cotton plant requires a warm growing period of at least 200 days. Other plants

(A) (B) (C)

Figure 21-3. (A) Prickly pear cactus in bloom, a typical xerophyte plant. (B) Water lilly, a common hydrophyte. (C) Hybrid corn, an example of a mesophyte plant. (Photos courtesy of U.S.D.A.)

that require a shorter growing period could be farmed as far north as Alaska, whereas some that require warmth yearround must be grown in a region with a 12-month growing period, such as is found in certain parts of California or Florida.

Cold-temperature preconditioning may be necessary to effect the breaking of both bud dormancy and certain types of seed dormancy. Such cold-temperature treatment is called **vernalization.** Many biennial plants, such as cabbage and beets, require vernalization treatment before flowering will occur. During the season of planting they grow vegetatively and, after a cold winter period in the ground, they produce seed the following year. Some plants, such as cotton, may require a high-temperature preconditioning of seeds.

Water. Precipitation represents the falling earthward of ice crystals or water droplets in the form of rain, hail, snow, sleet, or any combination of these. A percentage of this precipitation enters the soil and becomes part of the soil water. As with temperature, there is a direct relationship between the distribution of the various types of plants and the amount of soil moisture available. Some species of plants may exist in an area with very

little moisture. They have relatively small transpiring surfaces and extensive root systems. This type of plant is called a **xerophyte** and is usually found growing on deserts. An example is the cactus, illustrated in Figure 21-3A.

A plant that grows in or near water, partially or entirely submerged, is called a **hydrophyte** (Figure 21-3B). Other plants grow on land with their roots in soils which are neither excessively dry nor saturated with water. These are designated as **mesophytes.** They are the typical plants of the meadows and forests of the temperate zones (Figure 21-3C). In regions where there is a marked wet and dry season, many trees lose their leaves during the dry season and are physiologically active only during the wet season.

Wind. Of lesser importance, although significant to the growth of plants, is the factor of wind. The physical force of wind may place certain mechanical stresses upon plants by bending, twisting, and distorting organs (Figure 21-4). Wind is also an agent that affects the rate of transpiration, the loss of water by aerial portions of plants. Excessively high winds may alter soil types, thus may be a factor in determining plant distribution. Finally,

Figure 21-4. A windswept spruce deformed by wind. (Photo courtesy of U.S.D.A.)

These hairy outgrowths aid in the dissemination of the seed by wind.

Abiotic Factors of the Soil. Among the soil factors that affect plants are soil temperature, soil moisture, soil nutrients, and soil pH.

Temperature. As with atmospheric temperature, soil temperature greatly affects the ecological relationships of plants. For example, low temperatures tend to retard the mechanisms of absorption, and thus water and mineral absorption rates are altered and internal physiological processes may be modified.

Moisture. As described earlier, plants fall into three categories, depending upon the amount of soil moisture available: xerophytes, mesophytes, and hydrophytes. Soil moisture is important to plants in that the processes of absorption, photosynthesis, transpiration, and guttation are affected by the amount of water available from the soil at any given time.

Nutrients. The presence of soil minerals in the form of ions of salts is fundamental to the process of plant nutrition. Chemical elements such as nitrogen, phosphorus, potassium, sulfur, calcium, iron, magnesium, and copper are directly linked to the metabolic processes of plants as well as to reproduction, growth, and development. Some minerals are structural components of parts of plant cells, such as sulfur and nitrogen in proteins, magnesium in chlorophyll, and iron in cytochromes. Other minerals influence osmotic pressure, buffer activity, and cell permeability. Still others are basic components of enzyme systems.

pH. The degree of acidity or alkalinity of a soil influences the physical properties of the soil and consequently has an effect on plant growth. Some plants, such as azaleas, camellias, and cranberries, grow better in acidic soils, whereas other plants grow well in basic soils (Figure 21-5). Soils with a large amount of iron tend to be acidic, whereas

wind serves as a natural agent for pollination and seed dispersal in complex land plants. Wind pollination is common in both woody plants and herbs, but it is a relatively ineffective means of pollination because it relies on the chance that a pollen grain will be carried to another plant for fertilization. Plants that rely on wind pollination have certain structural adaptations that enable pollen grains to adhere and improve the efficiency of this method of pollination. Some seeds can be dispersed by the wind because they are small and light. They may develop outgrowths of their seed coats, such as the milkweed and cattail, or develop modified floral parts, such as in the dandelion.

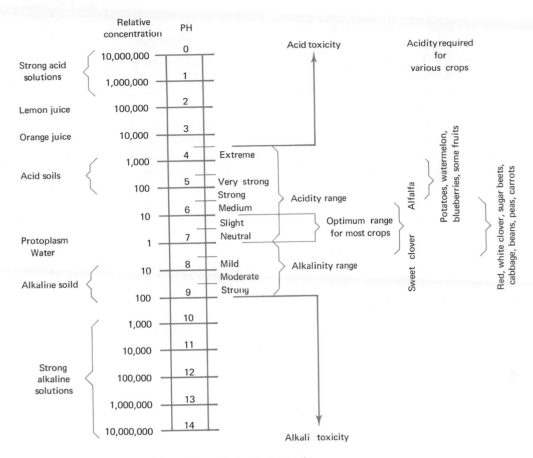

Relative
concentration   PH

	0	Acid toxicity
10,000,000		
Strong acid solutions {	1	
1,000,000		
Lemon juice	100,000	2
Orange juice	10,000	3
	1,000	4 — Extreme
Acid soils {	100	5 — Very strong
		Strong
	10	6 — Medium
		Slight
Protoplasm Water	1	7 — Neutral
	10	8 — Mild
Alkaline soild {		Moderate
	100	9 — Strong
	1,000	10
	10,000	11
Strong alkaline solutions {	100,000	12
	1,000,000	13
	10,000,000	14

Acidity range
Optimum range for most crops
Alkalinity range

Acidity required for various crops

Alfalfa
Sweet clover
Potatoes, watermelon, blueberries, some fruits
Red, white clover, sugar beets, cabbage, beans, peas, carrots

Alkali toxicity

Figure 21-5. The pH scale of soil and its relationship to plant growth.

soils with a high percentage of lime are generally basic. Most plants grow best in soil with a pH of 6 to 7, so it may be necessary to add certain elements to the soil to bring the pH to that level. For example, lime, usually in the form of finely ground limestone, is added to acid soils.

Biotic Factors

The ecology of plants cannot be considered solely in terms of abiotic factors. It must also be considered in terms of biotic factors and the interrelationships of these sets of factors in a dynamic functioning system.

Individuals. Plants must first be observed at the level of the *individual,* that is, as individual shrubs, individual ferns, individual trees, and so on, to provide a basic understanding of the patterns by which they associate. However, because plants do not function as isolated individuals, it is necessary to examine them in terms of their relationships with other plants.

Individual organisms of the same kind are grouped into series of higher organizational units. The next higher level of plant organization is referred to as a population.

Population. A **population** in a biological framework refers to a group of individuals of any one kind of organism in a definite ecological setting. For example, the twisted aspens of the "Crooked Forest" on the Grand Canyon's northern rim in Arizona represent a population. These trees have been bent and distorted by winter winds and snow drifts during formative stages. The dandelions in a lawn, the corn plants in a field, and the water lilies in a pond are other examples of local populations. Individual organisms in such a population reproduce and die, emigrate or immigrate, but as a whole the population continues to function and exist. The total of all populations of the same kind forms a **species.** For example, *all* corn plants, *all* water lilies, and *all* tulips represent separate species. A species is a group of interbreeding individuals that are self-perpetuating and do not interbreed with other groups.

Community. Several populations of many different species of plants living together within a limited area form the community level of organization. A **community** almost always consists of animals as well as plants; both are necessary for community survival. For example, a pond with its various animal and plant populations is a community; so is a forest, a meadow, or a town with its people, dogs, cats, and all other organisms. Most observers would note that certain types of plants grow together in a particular area. For example, oak and hickory trees are generally found together; dandelions and lawn grass are in close association, and maple and beech trees also occupy a similar habitat. A plant community, therefore, may be viewed as a particular group of plants in a specific habitat.

Community Relations

In every balanced community, green plants produce their own food and grow. They are the **producers** of the community. **Consumers,** by comparison, are organisms that lack the ability to synthesize their own organic nutrients. They must depend upon the producers to provide them with their basic nutritional requirements. Consumers utilize the chemical bond energy stored by the producers. For example, *herbivorous* animals eat plants; *carnivorous* animals, in turn, eat other animals. If there is a direct dependence of the consumer on a producer, the former is referred to as a *first-order consumer.* In the relationship, energy is exchanged directly between the producer and the consumer. If the consumer does not receive energy directly in the form of organic nutrients from the producer, but gets its energy from another consumer, the consumer is termed a *second-order consumer.* For example, a fish that feeds upon the insects of a pond receives the energy of the green plant through an intermediary; thus the fish is a second-order consumer. The insect represents a first-order consumer. In a similar fashion, the level of the consumer is determined by the number of intermediate organisms involved in the food chain (Figure 21-6).

The energy that is transferred from the producer to the consumer is eventually returned to the abiotic world by the activities of decomposers. For example, fungi and bacteria in their nutritional relationship decompose the remains of dead plants and animals, liberating the energy and allowing it to return to the abiotic world (Figure 21-7).

Plant Interactions

The various interactions that occur among plants may be classified as social and nutritive. In a social interaction neither plant receives food

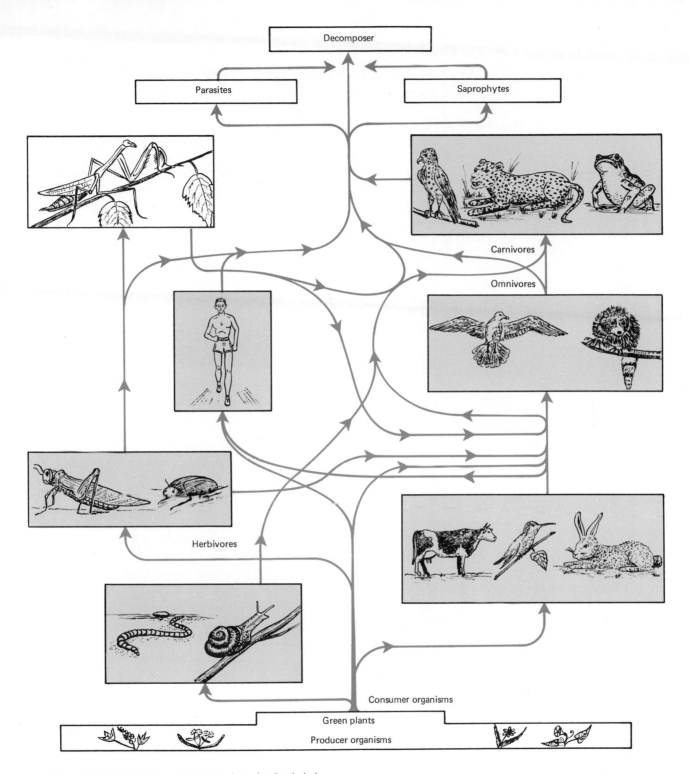

Figure 21-6. Diagrammatic representation of a food chain.

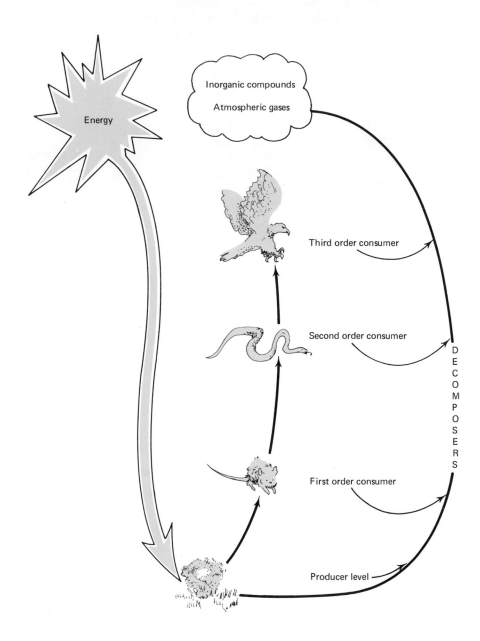

Figure 21-7. Producer-consumer relationships in an energy cycle.

from the other, whereas in a nutritive interaction one or both organisms may receive food.

Social Interaction. Plants such as Spanish moss (Figure 21-8) are classified as *epiphytes*. They grow upon other plants, usually trees, but do not obtain food from them. This type of plant is particularly common in the dense vegetation of the tropical rain forests. The epiphyte receives more light by this relationship than it would if it had to grow on the dense forest floor. Although the epiphyte does not receive nutrients from the supporting trees, it may cause harmful effects on the trees. For example, Spanish moss may become so thick that it shades the leaves of the trees, reducing photosynthesis to such an extent that the trees may die. Another harmful factor is the weight of the Spanish moss, which may be great enough to cause some limbs to break off the tree.

Vines may also grow on other plants, but, unlike epiphytes, vines have their roots in the soil. Vines benefit from their relationship by receiving more light than they would if they were confined to the ground. The vine may shade the leaves of the supporting plant or produce other harmful effects by coiling around the plant and restricting its growth.

Additional examples of social interactions may be viewed in the relationships between plants and animals as well as between two plants. Beavers may build dams, limiting the water supply to an area, resulting in a change in the physical environment. Man also has an effect on the natural environment of a region, as he has removed forests, built roads, cultivated the land, and constructed buildings. The future may bring additional changes as man continues to use radioisotopes and chemicals in his way of life.

Plants affect other plants in many ways. Trees in a forest may shade smaller plants, bring about a reduction in temperature, and change the atmospheric humidity of an area, as well as adding

Figure 21-8. Spanish moss growing on trees. (Photo courtesy of U.S.D.A.)

organic matter to the soil. Photosynthetic plants add oxygen to the atmosphere while removing toxic carbon dioxide. The environment is affected by plants almost as much as the environment influences plants.

Nutritive Interactions. Plants may be divided according to their nutritional relationships under two general categories: (1) **autotrophic** plants, which synthesize their own food molecules and in a sense are nutritionally independent; and (2) **heterotrophic** plants, which cannot synthesize their own organic nutrients and are therefore nutritionally dependent.

The autotrophs are chlorophyll-containing plants and a few species of chemosynthetic organisms. Because of their synthetic mechanisms, they occupy a pivotal position in the food chain, upon which all other feeders are dependent.

On the other hand, heterotrophs exhibit more diverse patterns of obtaining food. Some heterotrophs are closely associated in their nutritional patterns, whereas others are related in a more remote way. **Symbiosis** is typified by an association in which individuals of different species live together. One type of symbiosis is a nutritional relationship called **mutualism.** In this association both organisms benefit. Lichens, which consist of algae and fungi living in close association, represent a mutalistic relationship. The algae, by virtue of their photosynthetic capacity, provide food for the fungus, whereas the fungus plants afford protection for the algae by inhibiting excessive loss of water.

A second type of symbiosis is a relationship termed **commensalism.** In a commensal relationship one of the partners is benefited while the other is unaffected. Certain bacteria existing in the human intestine exemplify this type of relationship. These bacteria, which constitute part of the normal intestinal flora, live on waste materials within the intestine without doing any damage to the human being.

A final type of symbiosis is exemplified by **parasitism.** In parasitism one organism benefits while the other is harmed. An example of parasitism is seen in the activity of *Plasmodiophora brassicae,* a form of slime mold which causes a plant disease known as "club foot of cabbage." The slime mold, which may lie dormant in soil for many years, enters the root of cabbage plants and causes an overgrowth which results in an enlarged and malfunctioning root system. The slime mold obtains its nutrition while the cabbage plant is destroyed.

Heterotrophic plants may also live as **saprophytes.** These organisms live on dead or decaying organic matter. Common saprophytes are yeasts that ferment sugar; molds that live on fruit, leather, or bread; and fungi that live on dead trees.

SUCCESSION

The community level of plant organization represents a dynamic interacting system. These interactions cause certain changes to occur. These changes involve growth patterns which alter the physical makeup of the community. The end result in community change is *equilibrium,* a state of balance in which no further change takes place.

The sequence of changes within communities by which individuals replace each other in any one area in a given time is called **succession.** The individual rates of change vary from several years to many centuries, and each is referred to as a *seral stage.* The entire time required for succession to be completed is referred to as a *sere.* When succession no longer occurs, the community has reached a state of *climax.* Succession usually goes in one direction and therefore leads toward a climax community. For example, a lake community is slowly replaced by a soil community as a result of a deposition of sediments. This process is usually not reversible and the lake is doomed to extinction. All the diverse habitats of an area tend to move toward a single common climax. For example, if the history of most forests in the eastern United States was traced, their origin could be found in a barren field. During the year of development the annual herbs grew. The perennial herbs did not appear before three to five years. The next to follow were the conifers, which were succeeded by the hardwoods, which did not reach maturity for about 60 years. Because a climax forest takes approximately 60 years to

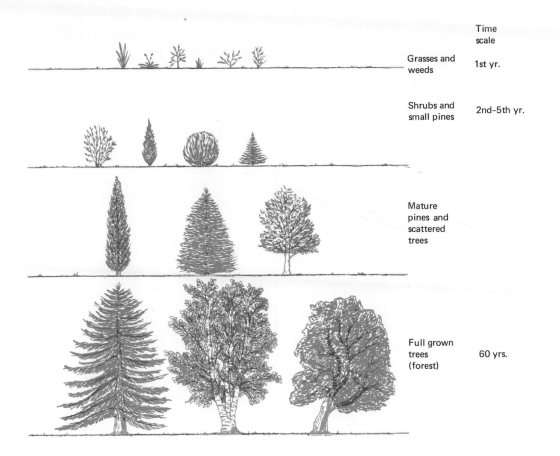

	Time scale
Grasses and weeds	1st yr.
Shrubs and small pines	2nd–5th yr.
Mature pines and scattered trees	
Full grown trees (forest)	60 yrs.

Figure 21-9. Old field succession.

develop, the damage done by fire to a forest is difficult to repair. Even with replanting, a productive forest would take almost a lifetime to develop (Figure 21-9). This type of development is known as *old field succession.*

In terms of available moisture, succession may be one of two types: (1) *xerarch succession,* and (2) *hydrarch succession.*

Xerarch succession begins in a region where water is not continuously available. The first evidence of seral changes occurs when lichens appear on the rock surface of the region. Lichens will grow only when moisture is available to them, so the amount of available moisture governs the rate of growth. With time the lichens grow large enough to trap dust particles carried by the wind. Eventually weathering causes the disintegration of the rock surface followed by the appearance of a very thin layer of soil. Seeds, also carried by winds, fall to the ground and begin to germinate.

Once again, the amount of water available determines the type of plant capable of establishing itself in this region. This type of succession is called *bare-rock succession.* In the climax community or the end of succession, the prior seral stages of succession are eliminated.

The first seral stage of hydrarch succession is represented by an open water system, that is, a pond, lake, stream, or river. Most open water systems are only temporary features of land topography and are soon replaced by other land forms. When the water is deep and the bottom sandy, the stage of succession is called the *bare-bottom stage.* With enough time the accumulation of sedimentary particles causes the lake to become shallow. The natural vegetation of the pond, including the plants growing on the bank as well as the floating and submerged plants, is now able to grow closer to the center of the pond. The filling-in process continues until, with time, the area becomes dry land, each step a seral stage closer to the climax community (Figure 21-10).

In addition to considering succession in terms of available water (xerarch and hydrarch), the process may also be viewed in terms of vegetation or the lack of it.

Primary succession begins in an area where there is no previous vegetation. Both xerarch and hydrarch are examples of this type of succession.

Secondary succession, by comparison, begins in an area where the natural vegetation has been destroyed. Agents of destruction include fire, floods, other natural catastrophies, and the influence of man.

A good example of the pattern of secondary succession was previously described in the discussion of an old field succession. At the outset, annual weeds invade the field, and in a few years are replaced by perennial herbs. In time, trees, shrubs, or other herbs invade the area and the climax is reached.

TYPICAL VEGETATION PATTERNS OF THE WORLD

The biosphere, or "world of life," is composed of a number of large geographical regions called *biomes.* Biomes are distinguished by the dominance of certain types of plants and animals. They are divided into two distinct types: (1) *land biomes,* and (2) *aquatic biomes.*

Land Biomes

Land biomes cover the land areas of the earth and differ vastly in character. However, no matter how diverse the environment, the fundamental requirements of all land life are air and soil. The annual temperature and rainfall of an area also plays a vital role in determining the type of life that may be sustained within the geographic realm. The land biomes, based on these physical factors, are divided into a number of distinct habitat zones: *grassland, desert, tundra, deciduous forest, coniferous forest,* and *tropical forest.*

Grasslands are found in the interior of continents. Their chief vegetation is grasses, because environmental conditions do not provide sufficient rainfall to support larger vegetation. The intermittent rainfall amounts to about 10 to 40 inches annually. The grasses are particularly adapted to irregular alternative periods of precipitation and dryness. Animals of this area include antelopes, prairie dogs, jack rabbits, coyotes, snakes, and other small animals (Figure 21-11).

A *desert* is an arid region usually having less than 10 inches of rainfall per year, and is characterized by a high rate of evaporation and low relative humidity. Xerophytic plants are well adapted to this environment. Plants such as cactus and yucca are capable of surviving the meager supply of water. Large animals are usually not found on deserts, but rodents and birds may be found in abundance (Figure 21-12).

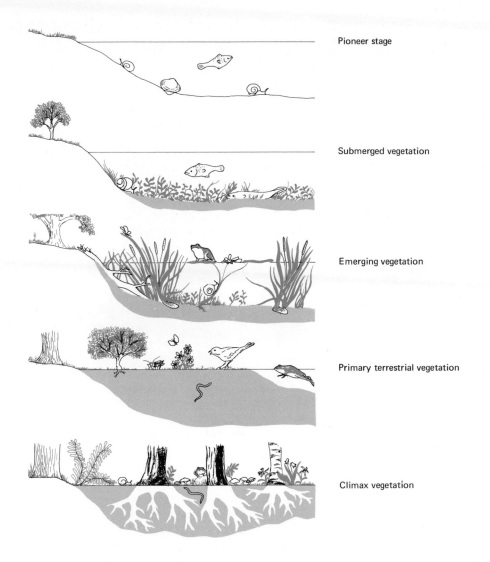

Figure 21-10. Pond succession.

A *tundra* is located in regions with frigid temperatures during the greater part of the year. The growing season is very short, lasting from a few weeks to a maximum of two months. In addition to low temperature, water supply is a limiting factor, because water may be frozen and thus unavailable to plants. The principal vegetation of the tundra consists of lichens, mosses, and some

Figure 21-11. A typical grassland as represented by Beaverhead National Forest, Montana. (Photo courtesy of U.S.D.A.)

grasses. Animals of tle tundra include polar bears, caribou, wolves, and foxes (Figure 21-13).

The *deciduous forest* is located in a temperate region, where seasonal temperatures vary from the extreme summer heat to the winter cold. Rainfall is well spaced, about 30 to 40 inches of precipitation per year. Deciduous trees lose their leaves in the fall and grow new leaves in the spring. They spend the winter in dormancy. Trees such as maples, oaks, elms, hickories, beech, and chestnut comprise the vegetation, and the animal life includes deer, fox, racoon, squirrel, and other related animals (Figure 21-14).

Coniferous forests are the evergreen forests of the world. They develop in areas typically known to have long severe winters with a very short summer growing season. Pine, spruce, fir, and hemlock predominate in the vegetation of coniferous forests. Animals found in those regions include elk, moose, deer, bear, beaver, and many small animals (Figure 21-15).

The *tropical forest* vegetation is supported by an abundance of rainfall and warmth. Rain falls practically every day and plant growth continues all year round. Several hundred different species of trees and plants grow in a dense tangle. They

Figure 21-12. (Left) Giant cacti growing in the desert. (Photo courtesy of U.S.D.A.)

Figure 21-13. (Below) Tundra. (Photo courtesy of U.S.D.A.)

Figure 21-14. (Below) Deciduous forest, Kenton Ranger District, Michigan. (Photo courtesy of U.S. Forest Service.)

Figure 21-15. Coniferous forest, Shoshone National Forest, Wyoming. (Photo courtesy of U.S. Forest Service.)

provide shelter for the myriad of animal life, which includes primates, snakes, and lizards (Figure 21-16).

Aquatic Biomes

Freshwater biomes include the rivers, streams, lakes, ponds, and other bodies of freshwater. Plants found in these regions would include the submerged as well as floating plants such as algae, pond weeds, water lilies, and other hydrophytes. Animal life is limited to those forms adapted to a watery habitat, including fish, water insects, amphibians, and some reptiles (Figure 21-17).

Marine biomes consist of the oceans and seas of the earth. The plants of the ocean are primarily algal types, many of which are microscopic. Animals found in the ocean are numerous and a partial list would include the many species of fish, shellfish, lobsters, marine mammals, and other kinds of animals which inhabit the seas.

CHAPTER IN RETROSPECT

Plants, like animals, interact with each other and also with the living and nonliving constituents of their surroundings. The sum of these relationships can be regarded as an ecological system, or ecosystem, which consists of and is governed by

Figure 21-16. (Above) Tropical forest.
(Photo courtesy of U.S.D.A.)

Figure 21-17. (Right) Fresh water biome
photographed at New Castle, Delaware.
(Photo courtesy of U.S.D.A.)

four principal components: (1) nonliving substances, (2) producers, (3) consumers, and (4) decomposers. Abiotic factors constitute the nonliving substances, and these consist of naturally occurring chemical compounds and environmental influences such as temperature, precipitation, and soil nutrients. Biotic factors, the living constituents of the environment, involve interreactions between the different organizational levels at which organisms occur, that is, individuals, populations, and communities. These reactions can be thought of as social, in which neither plant receives food from the other, or nutritive, whereby one or both plants may receive nutritive benefit from the other. Those organisms that can make their own food are regarded as producers and those which cannot (nongreen plants and most animals) are classified as consumers.

Within plant communities there is a sequence of gradual changes in which individuals replace each other in a given geographical area. This phenomenon is referred to as a plant succession, and it continues to occur until a state of equilibrium or climax is maintained.

Questions and Problems

1. What do the following terms signify?

ecosystem	biokinetic range
community	vernalization
population	succession
land biome	epiphyte
aquatic biome	mutualism
xerophyte	commensalism
hydrophyte	sere
mesophyte	

2. What are the principal elements of an ecosystem and how does each affect the stability of a plant community?

3. What are the characteristics of a typical hydrophyte and mesophyte? Can you identify one nearby?
4. Give examples of social and nutritive interactions among plants. How do these interrelationships contribute to the ecological stability of the biosphere?
5. What is meant by a first-order consumer and a second-order consumer? Consider energy transfer and conservation in your answer.
6. Briefly describe a xerarch succession and a hydrarch succession. Include the concept of primary and secondary successions in your description.
7. What are the elements of a typical land biome? Consider the following types of biomes: grassland, desert, and tropical forest.
8. How do aquatic and marine biomes differ?

Suggested Supplementary Readings

Carson, R., *The Sea Around Us.* New York: Oxford University Press, 1951.

Daubenmire, R. F., *Plants and Environment.* New York: John Wiley & Sons, Inc., 1947.

Deevey, E. S., "Life in the Depths of a Pond," *Scientific American,* October 1951.

Doyle, W. T., *Non-Vascular Plants—Form and Function.* Belmont, Calif.: Wadsworth Publishing Co., Inc., 1964.

Gleason, H. A., and A. Cronquist, *The Natural Geography of Plants.* New York: Columbia University Press, 1964.

Oosting, H. J., *The Study of Plant Communities.* San Francisco: W. H. Freeman and Company, 1956.

Platt, R., "Flowers in the Arctic," *Scientific American,* February 1956.

Salisbury, F. B., and R. V. Parks, *Vascular Plants—Form and Function.* Belmont, Calif.: Wadsworth Publishing Co., Inc., 1964.

Wardlaw, C. W., "The Study of Growth and Form in Plants," *Endeavour,* vol. 11, pp. 97–106, 1952.

Went, F. W., "The Ecology of Desert Plants," *Scientific American,* April 1955.

Appendix
Classification of the Plant Kingdom

The classification scheme that follows lists the principal groups of plants down to the level of orders, the approximate number of living species in each division, a brief description of each of the groups, and representative genera for each order. Although this scheme is fairly comprehensive, it does not contain a complete catalog of all known plants.

KINGDOM MONERA

Most Monera are unicellular forms, although there are some colonial and filamentous organisms. All contain procaryotic nuclei and therefore lack nuclear membranes and chromosomal organization. The cytoplasm of moneran cells lacks vacuoles and does not undergo cyclosis. Sexual reproduction is atypical, but when it occurs it proceeds according to a haplotonic life cycle. In photosynthetic forms, chromatophores, rather than organized chloroplasts, are the sites of synthetic reactions.

I. Division Schizophyta (fission plants). About 3100 species.

In general Schizophyta have the same characteristics listed for the Monera. In addition, chlorophylls and flagella, when present, are unique. The wall material is principally polysaccharides, proteins, and lipids. Glycogen represents the chief storage material.

Class Schizomycetes (bacteria). Unicellular; reproduction typically by fission; well-defined cell membranes; nutrition generally heterotrophic.

Order Pseudomonadales (pseudomonads). Rigid spherical, rod-shaped, or spiral cells; flagella when present are polar; includes sulfur and iron bacteria; most are gram negative; a few are human pathogens. *Nitrobacter, Nitrosomonas, Pseudomonas.*

Order Chlamydobacteriales (sheathed bacteria). Algalike branched or unbranched trichomes; often sheathed; motile forms have polar flagella; gram negative; freshwater or marine. *Sphaerotilus, Crenothrix.*

Order Hyphomicrobiales (budding bacteria). Single-celled, two celled, or colonial; pear-shaped or spherical cells with attached fine, threadlike stalk. *Hyphomicrobium, Rhodomicrobium.*

Order Eubacteriales (true bacteria). Spherical or rod-shaped; motile species have flagella arising all over the cell; many gram negative; free-living, parasitic, or saprophytic; includes many of the best known bacteria. *Azotobacter, Rhizobium, Clostridium, Salmonella, Escherichia, Streptococcus, Staphylococcus, Bacillus.*

Order Actinomycetales (branching bacteria). Rigid, elongated cells; many grow in branching colonies; similar in structure to Eubacteriales. *Mycobacterium, Streptomyces.*

Order Caryophanales (filamentous bacteria). DNA is arranged in a central body in the cells; found as parasites of animals and saprophytes. *Caryophanon.*

Order Beggiatoales (gliding bacteria). Multicelled filaments; nonflagellated; show gliding over surfaces; often contain sulfur granules. *Beggiatoa, Thiothrix.*

Order Myxobacteriales (slime bacteria). Flexible rods exhibiting gliding movement; mostly soil bacteria. *Myxococcus, Chondromyces.*

Order Spirochaetales (spiral bacteria). Flexible-walled, unicellular spirals motile; free-living, parasitic, saprophytic. *Treponema.*

Order Mycoplasmatales (pleuropneumonia-like organisms). Soft, fragile, nonmotile; reproduce by peculiar fragmentation into filterable cocci-like particles; saprophytic or pathogenic. *Mycoplasma.*

Order Rickettsiales (rickettsiae). Nonfilterable; rod-shaped or coccoid; many pathogenic. *Rickettsia, Coxiella.*

Order Virales (viruses). Unique ultramicroscopic intracellular parasites consisting of nucleic acid core surrounded by a protein coat; inert except when parasitizing host cells.

Class Schizophyceae (blue-green algae). Unicellular, colonial, or filamentous; nutrition autotrophic; reproduction by fission.

Order Chroococcales. Unicellular or nonfilamentous colonial forms; endospores usually absent. *Gloeocapsa, Chroococcus.*

Order Chamaesiphonales. Unicellular or colonial forms; capable of producing endospores. *Chamaesiphon.*

Order Oscillatoriales. Filamentous; hormogone-forming. *Nostoc, Oscillatoria.*

KINGDOM PROTISTA

Protista are unicellular, colonial, and filamentous forms (both plants and animals), all of which contain eucaryotic nuclei and a higher degree of cellular complexity than monerans. The genetic material exists as chromosomes delimited within a true nucleus bounded by a well-defined membrane. Nucleoli are also universally present. All protists are capable of mitotic and meiotic divisions of the nucleus. The cytoplasm, able to undergo cyclosis, contains definite chloroplasts and a variety of chlorophylls (*a, b, c, d,* and *e*). Prominent vacuoles are distributed throughout the cytoplasm. Reproduction is both asexual and sexual, and some forms exhibit alternation of generations in their life histories. In general, single cells function as reproductive cells (spores or gametes), and these cells are afforded little, if any, protection. Nutrition is both autotrophic and heterotrophic.

II. Division Euglenophyta (euglenoids). About 450 species.

Unicellular, green or colorless forms, with usually one or two flagella; most have plastids containing chlorophylls *a* and *b;* true cell wall absent; enclosed by a rigid or flexible membrane (periplast); principal storage products are lipids and paramylum; reproduction is typically by cell division in a longitudinal plane.

Class Euglenophyceae.

Order Euglenales. Unicelled; flagellate; flagellate motile cell is dominant phase of cycle; cell walls absent. *Euglena, Peranema, Astasia.*

Order Colaciales. Unicellular or colonial; sessile; cell walls present; encapsulated cell is dominant phase of life cycle. *Colacium.*

III. Division Chlorophyta (green algae). About 7000 species.

Unicellular, colonial, and multicellular forms; usually two or more whiplash flagella; chlorophylls *a* and *b;* cell walls of cellulose and pectins;

food stored as starch; asexual reproduction is by fission, fragmentation, or zoospores; sexual reproduction may be isogamous, heterogamous, or oogamous.

Class Chlorophyceae (green algae).

Order Volvocales. Unicellular or colonial; flagellated vegetative cells. *Chlamydomonas, Pandorina, Volvox.*

Order Tetrasporales. Nonmotile vegetative cells; mostly colonial; embedded in mucilage; biflagellated isogametes produced by dioecious colonies. *Tetraspora.*

Order Ulotrichales. Unicelled or filamentous; nonmotile; holdfasts present; biflagellated isogametes and zoospores with two or four flagella. *Ulothrix, Protococcus, Micrococcus, Coleochaeta, Stigeoclonium.*

Order Ulvales. Sheetlike, tubelike, or solid cylinders; many isogamous, some heterogamous. *Ulva, Schizomeris, Enteromorpha.*

Order Schizogoniales. Filamentous, platelike, or solid cylinders; nonmotile vegetative cells have stellate chloroplasts. *Schizogonium, Prasiola.*

Order Cladophorales. Filamentous; nonmotile vegetative cells; cells multinucleate, reproduction by zoospores or flagellated heterogametes. *Cladophora, Sphaeroplea, Pithophora.*

Order Oedogoniales. Nonmotile, filamentous vegetative cells; zoospores and microgametes with circlets of flagella; cell walls with terminal rings. *Oedogonium, Bulbochaete.*

Order Zygnematales. Unicelled or filamentous; reproduction by mitosis or by nonflagellated amoeboid isogametes. *Spirogyra, Zygnema,* desmids.

Order Chlorococcales. Nonmotile solitary or colonial vegetative cells; zoospores biflagellated. *Chlorococcum, Pediastrum, Hydrodictyon, Chlorella.*

Order Siphonales. Nonmotile vegetative cells are hollow, straight, or branched tubes; flagellated multinucleated zoospores; heterogamous; microgametes motile. *Acetabulum, Bryopsis, Codium.*

Order Siphonocladales. Multicellular, multinucleate thalli attached to substrate by rhizoids. *Valonia.*

Order Dasycladales. Multicellular, multinucleate thalli; branches in whorls. *Acetabularia.*

Class Charophyceae (stoneworts).

Order Charales. Growth by apical cells; upright stems, basal rhizoids, branches in whorls; heterogamous with biflagellate microgametes. *Chara, Nitella.*

IV. Division Xanthophyta (xanthophytes). About 400 species.

Mostly freshwater forms, a few are marine; characteristic color due to yellow-green plastids, many of which are found in each vegetative cell; most are unicellular, nonmotile forms, some are filamentous; motile cells (vegetative and reproductive) contain two anterior flagella of unequal length; principal food reserves are chrysolaminarin and sitosterol; asexual reproduction is by cell division, zoospore, and aplanospore production; sexual reproduction, although not widespread, is isogamous.

Class Xanthophyceae (yellow-green algae). Most are unicellular, nonmotile forms with uninucleate cells; zoospores have unequal flagellation. *Vaucheria, Botrydium, Tribonema.*

Class Chloromonadophyceae (chloromonads). Unicellular flagellates; many are small unusual forms. *Horniella, Rickertia, Gonyostomum.*

V. Division Chrystophyta (chrysophytes). About 5800 species.

Mostly unicellular of colonial forms, a few have simple, multicellular thalli, some motile; in motile cells, flagella are of unequal length; some have cell walls of silica, usually in two halves; principal food reserve is chrysolaminarin; chlorophylls *a* and *c,* xanthophylls, carotenes, lutein, and fucoxanthin comprise the principal pigments; asexual reproduction is by cell division, motile, and nonmotile spores; sexual reproduction, when it occurs, is generally oogamous.

Class Chrysophyceae (golden algae). Chlorophyll *a* plus undetermined types; lutein; fucoxanthin; usually one tinsel flagella; sexual reproduction rare.

Order Chrysomonadales. Unicelled and colonial; single tinsel flagellum. *Chromulina.*

Order Isochrysidales. Unicelled and colonial; one tinsel and one whiplash flagellum. *Isochrysis.*

Order Ochromonadales. Unicelled and colonial; one tinsel and one shorter whiplash flagellum. *Ochromonas.*

Order Rhizochrysidales. Unicellular and colonial. *Chrysamoeba.*

Order Chrysophaerales. Unicelled and colonial. *Epichrysis.*

Order Chrysocapsales. Colonial. *Hydrurus.*

Order Chrysotrichales. Filamentous. *Phaeothamnium.*

Class Bacillariophyceae (diatoms). Chlorophylls *a* and *c,* fucoxanthin; mostly unicellular; cell walls mainly of pectin and silica; walls conspicuously in two halves; sexual cycles common.

Order Centrales. Walls radial or concentric. *Coscinodiscus, Triceratium.*

Order Pennales. Walls have bilateral symmetry. *Navicula, Pinnularia.*

VI. Division Pyrrophyta (pyrrophytes). About 1000 species.

Chlorophylls *a* and *c;* food stored as lipids and carbohydrates, including starch; cell walls absent or, if present, composed of cellulose; usually biflagellate; reproduction is typically by cell division.

Class Dinophyceae (dinoflagellates). Mostly unicellular, some colonial; most have cellulose walls with interlocking plates; two anterior flagella; no transverse furrow.

Order Desmomonadales. Unicellular; no wall: *Desmomastix.*

Order Thecatales. Unicellular; wall in two vertical halves. *Exuviella.*

Order Gymnodiniales. Unicellular and colonial; wall absent or, if present, not divided into plates. *Gymnodinium.*

Order Peridiniales. Unicellular; wall in plates; no vertical separation into halves; transverse furrow equatorial. *Ceratium, Peridinium.*

Order Dinophysiales. Unicellular; wall in plates; vertical division into halves; transverse furrow above equatorial region. *Dinophysis.*

Order Rhizodiniales. Unicellular. *Dinamoebidium.*

Order Dinococcales. Unicellular. *Tetradinium.*

Order Dinocapsales. Unicellular or colonial. *Gloedinium.*

Order Dinotrichales. Filamentous. *Dinothrix.*

Class Cryptophyceae (cryptomonads). Unicellular forms with two anterior flagella; cell walls absent; cells slightly flattened; no transverse furrow.

Order Cryptomonadales. Unicellular, flagellate. *Cryptomonas.*

Order Thecatales. Unicellular; sessile. *Tetragonidium.*

VII. Division Phaeophyta (brown algae). About 1500 species.

All are sessile and multicellular; chlorophylls *a* and *c*, carotenes, and fucoxanthin; cell walls composed of cellulose and algin; food storage as laminarin and mannitol; exposed multicellular reproductive structures; biflagellate reproductive cells.

Class Isogeneratae. Gametophyte and sporophyte alike; diplohaplontic life cycle.

Order Ectocarpales. Filamentous, basal growth. *Ectocarpus.*

Order Sphacelariales. Cells form cylindrical columns; apical growth. *Sphacelaria.*

Order Tilopteridales. Bottom cylindrical, top filamentous; basal growth. *Haplospora.*

Order Cutleriales. Flattened blade; basal growth. *Zanardinia.*

Order Dictyotales. Flattened, parenchymatous; apical and marginal growth. *Dictyota.*

Class Heterogeneratae. Sporophyte dominant, gametophyte reduced to small, threadlike structure; diplohaplontic life cycle.

Order Chordariales. Filamentous; basal growth. *Leathesia.*

Order Sporochnales. Branch tips with tufts; basal growth. *Carpomitra.*

Order Desmarestiales. Parenchymatous thallus; apical growth. *Desmarestia.*

Order Punctariales. Parenchymatous thallus; intercalary growth. *Soranthera.*

Order Dictyosiphonales. Cylindrical, branched thallus; apical growth. *Dictyosiphon.*

Order Laminariales. Thallus differentiated into holdfast, stipe, and blade; basal growth. *Laminaria, Macrocystis.*

Class Cyclosporae. Sporophyte dominant, gametophyte much reduced; diplontic life cycle; oogamous.

Order Fucales. Parenchymatous thallus, apical growth. *Fucus, Sargassum.*

VIII. Division Rhodophyta (red algae). About 4500 species.

Multicellular thalli; cell walls contain cellulose and pectins; chlorophylls *a* and *b*, xanthophylls, phycocyanin, and phycoerythrin; food stored as floridean starch; flagella absent; asexual reproduction by nonmotile spores; sexual reproduction oogamous by nonmotile gametes.

Class Rhodophyceae (red algae). Same characteristics as in the Division Rhodophyta.

Subclass Bangiophycidae. Growth typically intercalary; plasmodesmata not apparent.

Order Porphyridiales. Unicellular or colonial. *Porphyridium.*

Order Bangiales. Filamentous, disclike, or leaflike; cells not enclosed in a gelatinous covering. *Porphyra.*

Order Goniotrichales. Filamentous; cells randomly scattered in a gelatinous matrix. *Goniotrichum.*

Order Compsopogonales. Filamentous; cortical layers surround a central row of large cells. *Compsopogon.*

Order Rhodochaetales. Filamentous; single row of cells; apical growth. *Rhodochaete.*

Subclass Florideae. Conspicuous plasmodesmata; terminal growth.

Order Nemalionales. Tetrasporophyte absent. *Nemalion.*

Order Gelidiales. Tetrasporophyte present; carpogonium produces carposporophyte directly. *Gelidium.*

Order Cryptonemiales. Tetrasporophyte present; carpogonium produces carposporophyte with auxiliary cells. *Cryptosiphonia.*

Order Gigartinales. Tetrasporophyte present; auxiliary cell is vegetative. *Iridaea.*

Order Rhodymeniales. Tetrasporophyte present; auxiliary cell differentiated before fertilization. *Rhodymenia.*

Order Ceramiales. Tetrasporophyte present; auxiliary cell formed after fertilization. *Polysiphonia.*

IX. Division Myxomycophyta (slime molds). About 500 species.

Photosynthetic pigments absent; similar to true fungi in nature of food reserves; diplohaplontic life cycles; haploid phases often as independent solitary cells; diploid phase as aggregated amoeboid mass.

Class Myxomycetes (true slime molds). Haploid phase flagellate cells; diploid phase a coenocytic, multinulceate plasmodium; no cell walls.

Order Ceratiomyxales. Spores borne externally; walls of spore and sporangium fused. *Ceratiomyxa.*

Order Liceales. Spores light in color; excretory threads in sporangium absent or poorly developed. *Licea.*

Order Trichiales. Excretory threads in sporangium well developed. *Trichia, Arcyria.*

Order Echinosteliales. Spores colorless or lightly pigmented; smallest slime molds. *Echinostelium.*

Order Stemonitales. Dark spores; no lime in excretory threads or sporangium; columella present in sporangium. *Stemonitis.*

Order Physarales. Excretory threads and/or sporangium contain lime. *Physarum, Didymium, Fuligo.*

Class Acrasiomycetes (cellular slime molds). Multicellular grouping of hundreds of uninucleate amoeboid cells; haploid phase amoeboid cells; diploid phase a multicellular plasmodium; fruiting bodies produce amoeboid swarmers.

Order Acrasiales. *Dictyostelium.*

Class Labyrinthulomycetes (net slime molds). Multicellular netlike plasmodium; fruiting bodies not formed; swarmers neither amoeboid nor flagellated.

Order Labyrinthulales. *Labyrinthula.*

X. Division Eumycophyta (true fungi). About 80,000 species.

Food stored as glycogen; cell walls of cellulose and chitin; typically filamentous organisms with thalli consisting of hyphae; nonphotosynthetic; reproduction is vegetative, by spore formation, and gametes.

Class Phycomycetes (algal fungi). Mycelia without septations; lower forms aquatic with flagellated reproductive cells.

Subclass Chytridiomycetidae. Reproductive cells generally with a single flagellum; cell walls of chitin.

Order Chytridiales. Thallus not a true mycelium; posterior flagellation; body unicellular or with rhizoid-like, sporangium-bearing outgrowth. *Chytridium.*

Order Blastocladiales. Thallus two-celled or a true mycelium with rhizoids. *Allomyces.*

Order Monoblepharidales. Well-developed mycelium; flagellate sperm; oogamous. *Monoblepharis.*

Subclass Oomycetidae. Sporangia of different types; asexual reproduction by biflagellate spores; cell walls of cellulose.

Order Lagenidiales. Mycelium simple; often multicellular; fertilization by tube or fusion of cells. *Lagenidium.*

Order Saprolegniales. Mycelial thallus extensive without cross walls; fertilization tube. *Saprolegnia, Achlya.*

Order Leptomitales. Mycelial thallus with holdfast; hyphae septate and constricted at intervals. *Sapromyces.*

Order Peronosporales. Mycelia well developed with branching hyphae. *Albugo, Pythium, Phytophthora.*

Subclass Zygomycetidae. Zygospores produced from the fusion of gametes; no motile cells; some produce conidia.

Order Mucorales. Fertilization by fusion of like gametangia; mostly saprophytic; usually coenocytic. *Rhizopus, Mucor.*

Order Entomophthorales. Fertilization by fusion of like gametangia; mycelia form septae that fragment into hyphal bodies; mostly parasitic. *Entomophthora.*

Order Zoopagales. Asexual reproduction by true conidia; parasitic. *Endocochlus, Cochlonema.*

Class Ascomycetes (sac fungi). Mycelium has septate hyphae; walls contain chitin; asci containing ascospores.

Subclass Hemiascomycetidae. Primitive; mycelium small or lacking; asci formed without ascogenous hyphae.

Order Endomycetales. Zygote results from fusion of two cells immediately forming an ascus. *Saccharomyces.*

Order Ascoideales. Coarse, septate mycelium; multinucleate cells; some produce conidia. *Ascoidea.*

Order Taphrinales. Cells binucleate; hyphal cell forms spore which germinates into an ascus; ascaspores reproduce by budding. *Taphrina.*

Subclass Euascomycetidae. Asci usually develop from ascogenous hyphae; distinct fruiting bodies; functional sex organs.

Order Eurotiales. Cleistothecial; asci irregularly arranged in poorly developed stroma. *Penicillium, Aspergillus, Talaromyces, Eurotium.*

Order Erysiphales. Closed ascocarp; large, globose, stalked asci. *Erysiphe, Phyllactinia.*

Order Xylariales. Globose, sessile, peritheca; perithecia dull-colored. *Neurospora, Xylaria.*

Order Hypocreales. Brightly colored perithecia. *Claviceps.*

Order Pleosporales. Stroma is small and one-chambered; asci contain multiseptate ascospores. *Pleospora, Venturia.*

Order Helotiales. Apothecia above ground; ascus opens by pore. *Pseudopezzia, Monilinia.*

Order Pezizales. Ascocarp an apothecium; operculum present. *Peziza, Morchella.*

Order Tuberales. Apothecia subterranean; asci not exposed. *Tuber.*

Class Basidiomycetes (club fungi). Basidiospores formed on basidia; septate hyphae.

Subclass Heterobasidiomycetidae. Basidia septate; basidiospores typically germinate by formation of conidia or by budding.

Order Tremellales. Fruiting bodies jelly-like; mycelium originates from uninucleate basidiospore or conidium. *Tremella, Dacrymyces.*

Order Uredinales. Basidiospores germinate by forming spore-bearing stalks; fruiting body represented by teliaspores. *Puccinia.*

Order Ustilaginales. Basidiospores germinate by budding; black spores. *Ustilago.*

Subclass Homobasidiomycetidae. Basidia usually nonseptate; basidiospores often germinate into mycelium.

Order Polyporales. Basidiocarps erect and resemble "cattails"; soft, fleshy texture; hymenium borne on smooth surfaces. *Clavulina.*

Order Agaricales. Basidiocarps with exposed basidia in layers. *Agaricus, Amanita, Coprinus.*

Order Hymenogastrales. Fruiting body closed; spore mass fleshy or waxy. *Gasterella.*

Order Lycoperdales. Spore mass powdery at maturity. *Lycoperdon, Arachnion.*

Order Phallales. Spore mass slimy and exposed. *Phallus, Mutinus.*

Order Nidulariales. Spore mass in chambers and waxy. *Nidularia, Crucibrilum.*

Class Deuteromycetes (imperfect fungi). No zygotes are formed; affinities uncertain.

Order Sphaeropsidales. Conidia borne in special fruiting structures. *Phoma.*

Order Melanconiales. Conidiophores subepidermal. *Pestalotia.*

Order Moniliales. Conidia borne directly on mycelia (undifferentiated). *Sporobolomyces.*

Order Mycelia sterilia. Spores unknown. *Sclerotium.*

Class Lichenes (lichens). Symbiotic combinations of algae and fungi.

Subclass Ascolichenes (fungal partners are Ascomycetes).

Subclass Basidiolichenes (fungal partners are Basidiomycetes).

KINGDOM METAPHYTA

The kingdom Metaphyta consists of two divisions—the Bryophyta and the Tracheophyta. The Bryophyta lack vascular tissues and true roots; the Tracheophyta have developed both of these features to a considerable extent. Members of the Metaphyta possess eggs that are retained and protected in the adult plant after fertilization and until a multicellular embryo develops; diplohaplontic life cycles; differentiated, multicellular gametophytes and sporophytes; chlorophylls *a* and *b;* starch as the reserve food material; cell walls of cellulose and pectin; subdivisions of the body into absorptive, conductive, and photosynthetic areas; adaptations for a terrestrial habitat.

XI. Division Bryophyta (mosses and allies). Sporophyte dependent upon and attached to dominant gametophyte; no vascular tissue; embryo not highly specialized; homosporous. About 23,600 species.

Class Hepaticae (liverworts). Prostrate gametophyte, foliose or thallose; compact dorsoventral thallus; leaves lack midrib and are arranged in two or three rows; elaters in spore capsule.

> Order Calobryales. Erect, foliose gametophyte; leaves in three vertical rows; no rhizoids; elongate spore capsule. *Calobryum, Takakia.*

> Order Jungermanniales. Foliose or thallose gametophyte; smooth-walled rhizoids; no scales. *Porella, Marsupella.*

> Order Metzgeriales. Thallose gametophyte; simple unicellular rhizoids; some contain fungal hyphae; some gametophytes contain marginal hairs. *Pellia, Riccardia.*

> Order Monocleales. Thallose gametophyte; some cells contain a single, brown oil body; rhizoids similar to Marchantiales. *Monoclea.*

> Order Sphaerocarpales. Thallose gametophyte; flash-shaped envelope around sex organs. *Sphaerocarpos, Geothallus.*

> Order Marchantiales. Gametophyte thallose and prostrate; considerable internal tissue differentiation; dichotomous branching; scales present; smooth rhizoids. *Marchantia, Riccia.*

Class Anthocerotae (horned liverworts). Cells contain one chloroplast each; gametophytes thallose with considerable tissue differentiation; scales absent; rhizoids smooth-walled; no air chambers or pores; elongated spore capsule; archegonia usually embedded in gametophyte; elaters present.

> Order Anthocerotales. Same characteristics as the class Anthocerotae. *Anthoceros.*

Class Musci (mosses). Gametophyte consists of protonema, erect leafy shoot-bearing sex organs, and multicellular rhizoids; sporophyte is comprised of foot, stalk, capsule (sporogenic tissue); no elaters.

Subclass Sphagnidae. Protonema thallose with a single erect shoot; no midribs; archegonia subterminal, antheridia lateral; spore capsule opens by removal of lid (explosive).

> Order Sphagnales. *Sphagnum.*

Subclass Andreaeidae. Filamentous protonema; midribs present; capsule discharges spores by splitting into four valves.

> Order Andreaeales. *Andreaea.*

Subclass Tetraphidae. Filamentous protonema; leaves contain midribs and are spirally arranged in three to five ranks; plants are bisexual. *Tetraphis, Tetrodontium.*

Subclass Polytrichidae. Filamentous protonema; erect gametophytes; most are unisexual; leaves typically spirally arranged in five ranks; very small spores. *Polytrichum, Dawsonia.*

Subclass Buxbaumiidae. Extremely simple gametophytes; filamentous protonema. *Buxbaumia.*

Subclass Bryidae. Filamentous protonema; midribs present; spore capsule opens by lid. The following orders (not a complete listing) are distinguished principally on the basis of the structure of the leaves and the mechanism and apparatus for discharge of the spores.

Order Fissidentales. *Fissidens.*

Order Dicranales. *Dicranella.*

Order Pottiales. *Hymenostylium.*

Order Grimmiales. *Grimmia.*

Order Funariales. *Funaria.*

Order Schistostegales. *Schistostega.*

Order Tetraphidales. *Tetraphis.*

Order Eubryales. *Mnium.*

Order Isobryales. *Fontinalis.*

Order Hookeriales. *Hookeria.*

Order Hypnobryales. *Thuidium.*

XII. Division Tracheophyta (vascular plants). About 300,000 species.

Vascular tissues present in sporophyte; sporophyte is independent and dominant, gametophyte is greatly reduced; most are differentiated into roots, stems, and leaves; mostly land plants; homosporous and heterosporous.

Subdivision Psilopsida (psilopsids).

Externally stem is dichotomously branched, internally it is protostelic or siphonostelic; leafy appendages without veins; no roots; rhizoids present; sporangia are homosporous, terminal, and bulbous; gametophyte is small and independent.

Class Psilopsidae. Same features as in subdivision.

Order Psilophytales. All genera extinct; rootless; protostelic stems. *Rhynia, Psilophyton, Hyenia, Horneophyton, Asteroxylon.*

Order Psilotales. Two living genera; sporophyte differentiated into rhizome and aerial shoot; protostelic to siphonostelic stem. *Psilotum, Tmesipteris.*

Subdivision Lycopsida (lycopsids).

Microphyllous leaves (often ligulate), stems, and roots present; sporangia on sporophyllus usually aggregated into strobili; protostelic to siphonostelic stem; some secondary growth; homosporous and heterosporous.

Class Lycopsidae. Features as in subdivision.

Order Lycopodiales. Homosporous; biflagellate sperms; no secondary growth; no ligules. *Lycopodium* (extant), *Phylloglossum* (extant), *Lycopodites* (extinct).

Order Selaginellales. Heterosporous; no secondary growth; ligulate microphylls; gametophytes retained in spore walls. *Selaginella* (extant), *Selaginellites* (extinct).

Order Lepidodendrales. All extinct; secondary growth; sporophytes treelike; ligules present; heterosporous; some with seedlike structure. *Lepidodendron, Sigillaria.*

Order Isoetales. Stem is cormlike; cambium present; ligulate microphylls; heterosporous; gametophytes retained in spore wall; sperm are multiflagellate. *Isoetes* (extant), *Stylites* (extant), *Isoetites* (extinct).

Order Pleuromeiales. All extinct; heterosporous; unbranched stem with apical cluster of ligulate leaves. *Pleuromeia.*

Subdivision Sphenopsida (sphenopsids).

Sporophytes differentiated into roots, stems, and leaves; stems ribbed externally, protostelic or siphonostelic internally; microphyllous leaves in nodal whorls; some contain a cambium; homosporous; sporangiophores borne in strobili.

Class Equisetinae. Features as in subdivision.

Order Equisetales. No cambium; stem solid at nodes, hollow at internodes; roots chiefly adventitious; gametophytes independent; green; sperm multiflagellate. *Equisetum* (extant), *Equisetites* (extinct).

Order Hyeniales. All extinct; shrublike, jointed stem; terminal sporangia. *Protolyenia, Calamophyton.*

Order Sphenophyllales. All extinct; protostelic stem; leaves wedge-shaped; homosporous. *Sphenophyllum.*

Order Calamitales. All extinct; large and much secondary growth; ribbed stems; lancet-shaped leaves; homosporous. *Calamites.*

Subdivision Pteropsida (ferns and seed plants).

Sporophyte differentiated into roots, stems, leaves; leaves megaphyllous and usually with petioles; leaf traces present; many have secondary growth; homosporous and heterosporous.

Class Filicineae (ferns). Stems protostelic or siphonostelic internally, rhizome type externally; no vessels; primary growth only; leaves usually large and pinnate; most homosporous, some heterosporous; sperms multiflagellate; spore germinates into independent gametophyte.

Order Coenopteridales. All extinct; protostelic stems; leaves imperfectly formed or pinnate; homosporous. *Stauropteris, Botryopteris, Ankryopteris.*

Order Ophioglossales. Siphonostelic stem; fronds simple or pinnate; subterranean nonchlorophyllous gametophytes; sporangia borne on a fertile spike. *Ophioglossum.*

Order Marattiales. Stem is dictyostelic; large pinnate fronds; circinate vernation; chlorophyllous gametophyte; sporangia arranged in rows. *Marattia, Danaea.*

Order Filicales. Protostelic to dictyostelic stems; homosporous; simple to compound pinnate fronds; circinate vernation; sporangia usually covered by indusium; spores discharged in catapult-like manner; chlorophyllous gametophytes. *Polypodium, Osmunda, Asplenium.*

Order Marsileales. Stems amphiphalic; heterosporous; fronds variable and usually borne in two rows; sporangia borne in sporocarps, covered by indusium, and arranged in two lateral rows; green gametophyte. *Pilularia, Marsilea, Regnellidium.*

Order Salviniales. Small, aquatic, heterosporous; sporangia borne in sporocarps; delicately branched stems. *Salvinia, Azolla.*

Class Gymnospermae (gymnosperms). Cambium present; vessels largely absent; heterosporous; exposed seeds; no endosperm; seeds borne in cones; all perennial, most evergreen.

Subclass Cycadophytae. Mostly extinct primitive forms; thick cortex, little wood and pith; flagellate sperms.

Order Cycadofilicales. All extinct; seeds exposed; veinlike or slender-trunked; leaves pinnate, fernlike; stem protostelic to siphonostelic. *Neuropteris, Lyginopteris, Medullosa.*

Order Bennettitales. All extinct; short column-like trunk; dictyostelic stem; fronds not compact; microsporangia borne on lateral branches. *Cycadella, Cycadeoidea, Weilandiella.*

Order Cycadales. Extant forms; compound pinnate leaves; dictyostelic stem; stems contain leaf scars; leaves arise in a tight spiral; sporophylls usually in strobili. *Cycas, Zamia, Dioon, Bowenia.*

Subclass Coniferophytae. Mostly extant forms; large amounts of wood and pith; relatively little cortex.

Order Cordaitales. All extinct; large, flattened leaves; stems tall and slender; larger than tree ferns; dictyostelic, sporangia in strobili on lateral branches. *Cordaites.*

Order Ginkgoales. Primitive; dichotomously veined; fan-shaped leaves; motile sperm and terminal seeds; no vessels; sporophylls not in strobili; a single living species. *Ginkgo.*

Order Coniferales. Apical dominance of sporophyte results in typical pyrimidal or conical form; no vessels; resin canals present; sperms lack flagella; dominant forest trees living today. *Pinus, Picea, Sequoia, Larix, Tsuga, Abies, Taxus.*

Order Gnetales. Root develops as a taproot; staminate strobili compound; no resin canals; sperms lack flagella; shoot apices superficially resemble angiosperms. *Ephedra, Gnetum, Welwitschia.*

Class Angiospermae (angiosperms). Vessels usually present; with or without secondary growth (cambium); heterosporous; sporophyte dominant; produce flowers; seeds usually contain endosperm and are enclosed in a fruit; most recently evolved plants.

Subclass Dicotyledonae (dicots). Usually two cotyledons in the embryo; woody and herbaceous; dictyostelic stems; net-veined leaves; floral parts in fours and fives; or multiples; cambium present.

Order Ranales (buttercups). Flowers have more than one pistil; leaves usually opposite; stipules generally absent. *Magnolia.*

Order Papaverales (poppy). Free stigmas; fused carpels. *Brassica.*

Order Violales (violet). Simple, alternate leaves; few stamens. *Viola.*

Order Caricales (papaya). Small trees; large alternate leaves on upper part of trunk; fruit is a berry. *Carica.*

Order Sarraceniales (pitcher plant). Leaves modified for insect catching; alternate leaves; numerous stamens. *Sarracenia.*

Order Euphorbiales (spurge). Simple, deeply indented leaves; no stipules; flowers unisexual. *Euphorbia.*

Order Geraniales (geranium). Pendulous ovules; stamens usually separate; flowers bisexual. *Erodium, Pelargonium.*

Order Rhamnales (grape). Simple leaves alternately or oppositely arranged; flowers usually bisexual; inverted ovules; fruit usually a berry. *Cissus.*

Order Urticales (nettle). Alternate leaves; stipules; flowers usually unisexual; fruit an achene. *Morus.*

Order Piperales (pepper). Opposite or alternate leaves; flowers in spikes; fruit a capsule or follicle. *Peperomia.*

Order Primulales (primrose). Simple leaves oppositely or alternately arranged; very small seeds. *Primula.*

Order Oleales (olive). Simple, pinnate leaves; flowers bisexual or unisexual; fruit usually a drupe or berry. *Ligustrum.*

Order Lamiales (mint). Opposite or whorled leaves; no stipules; square-shaped stem; fruits usually nuts. *Verbena.*

Order Scrophulariales (snapdragon). Opposite or whorled leaves; fruit typically a capsule. *Verbascum.*

Order Rosales (rose). Simple or compound leaves; fruits variable. *Rosa.*

Order Cactales (cactus). Stems fleshy and leafless (spines); epigynous flowers. *Opuntia.*

Order Cornales (dogwood). Simple leaves; flowers bisexual. *Cornus.*

Order Umbellales (parsley). Flowers usually bisexual; compound umbels; fruit a capsule or berry. *Aralia.*

Order Begoniales (begonia). Simple leaves, palmately veined; stipules; flowers unisexual; fruit typically a capsule. *Begonia.*

Order Asterales (aster). Flowers appear in heads; fruit an achene. *Helianthus.*

Order Salicales (willow). Simple, alternate leaves; flowers unisexual; flowers in catkins. *Salix.*

Order Juglandales (walnut). Pinnate leaves; monoecious; fruit is a nut. *Juglans.*

Subclass Monocotyledonae (monocots). Usually a single cotyledon; mostly herbaceous; atactostelic stems; leaves parallel-veined; floral parts in threes or multiples; no cambium; scattered vascular bundles.

Order Lilliales (lily). Simple leaves; flowers in threes; fruit typically a capsule or berry. *Allium.*

Order Palmales (palm). Alternate leaves; bisexual flowers; fruit typically a drupe. *Thrinax.*

Order Graminales (grass). Elongated leaves; bisexual flowers. *Carex.*

Order Musales (banana). Broad, pinnately veined leaves; bisexual flowers. *Maranta.*

Order Orchidales (orchid). Alternate leaves; fruit is a capsule. *Orchis.*

Order Hydrocharitales (elodea). Aquatic; opposite or whorled leaves; unisexual flowers. *Elodea.*

Order Naiadales (pondweed). Simple floating or immersed leaves; bisexual or unisexual flowers; fruit is nutlike. *Potamogeton.*

Glossary

Abiotic: Nonliving components of an environment.

Abscission: Separation of leaves, fruits, or other plant parts.

Abscission layer: Layer of thin-walled cells at the base of a petiole, the separation of which disjoins the leaf, fruit, or other plant parts from the stem.

Absorption: Materials taken into plant cells.

Achene: Dry, one-seeded fruit formed from a single carpel.

Activator: Forms a bridge between the enzyme and the substrate to facilitate transformation.

Adenosine diphosphate (ADP): Compound which forms high-energy phosphate bonds with the addition of inorganic phosphate; ATP is the resulting compound.

Adenosine triphosphate (ATP): Compound containing three high-energy phosphate groups; a phosphate group may be removed with the release of cellular energy; ADP is formed when such energy is liberated.

Aerial stem: Stem which grows above the ground.

Aggregate fruit: Fruit developed from a group of ovaries, produced in a single flower, all of which ripen together.

Akinete: Nonmotile thick-walled spore.

Allele: One of the two or more contrasting forms of a gene at a particular locus on a chromosome.

Allosteric interaction: Combination of a molecule with a protein that can influence the binding of another molecule.

Alternation of generations: Development of a sporophyte generation followed by the development of a gametophyte generation in a single life cycle of a plant.

Amino acid: Organic acid containing the NH_2 group (amino group); the units from which protein molecules are built.

Anabolism: Constructive or synthetic phases of metabolism, such as photosynthesis, assimilation, and the synthesis of proteins.

Anaphase: Stage in mitosis in which the chromatids of each chromosome separate and move to opposite poles of the cell.

Androecium: Male reproductive organs of a plant; stamens taken collectively.

Angiosperm: Group of flowering plants whose seeds are enclosed in fruits (ovaries).

Annual ring (*see* Growth ring).

Annulus: Ring of specialized cells in a fern sporangium; in mosses, thick-walled cells located along the edge of the sporangium.

Anther: Pollen-bearing portion of a stamen; the upper portion of a stamen.

Antheridium: Male sex organ in plants, contains sperm.

Anthocyanin: Natural pigment; a blue, purple, or red pigment; water-soluble; found in cell sap.

Aplanospore: Nonmotile true spores produced by a vegetative cell.

Apothecium: Cup-shaped structure containing asci in an open ascocarp.

Archegonium: Female sex organ in plants; produces egg.

Ascocarp: Structure in which asci are formed.

Ascogonium: Female gametangium of the ascomycetes.

Ascospore: Meiospore produced within an ascus.

Ascus: Characteristic spore-producing cell of ascomycetes; a saclike structure containing ascospores.

Assimilation: Transformation of nutrients into protoplasm and cell walls.

Atom: Smallest whole unit of a chemical element; made up of protons, neutrons, as well as other particles, which form an atomic nucleus, and a specific number of electrons which orbit around the nucleus.

Autotroph: Organism capable of manufacturing its own food, such as green plants and some bacteria.

Auxin: Plant-growth regulator that promotes cell elongation.

Axillary bud: Bud formed in the axil of a leaf.

Bark: Protective, corky tissue of a woody stem; all tissues outside the cambium.

Basidiocarp: Fruit body of Basidiomycetes.

Basidiospore: Spore of Basidiomycetes; produced on a basidium following a sexual process.

Basidium: Characteristic spore-producing structure of basidiomycetes; a club-shaped structure that bears basidiospores.

Berry: Simple fleshy fruit.

Binary fission: Division of a cell into two similar parts by the division of nucleus and cytoplasm.

Biological oxidation (*see* Respiration).

Biome: Major community of living organisms; the climax community of a major region, for example, grassland, tundra, and deciduous forest.

Biosphere: Part of the globe containing living organisms, including the land, water, and air.

Biotic: Living organisms of an environment.

Blade: Broad, expanded portion of a leaf.

Brownian movement: Passive movements of particles suspended in a fluid.

Bud: Undeveloped embryonic shoot.

Budding: Method of vegetative reproduction in yeasts; a form of grafting.

Bud mutation: Mutation which involves a single branch, different in some respect from the others on the same plant.

Bud scale: Modified protective leaf of a bud.

Bulb: Short, underground storage stem, composed of many fleshy scale leaves.

Calyx: Collective term for the sepals of a flower; outermost part of a flower.

Capsule: Simple, dry, dehiscent fruit that develops from a compound pistil and opens allowing the seeds to escape; the sporangium in liverworts and mosses.

Carbohydrate: Organic compound made up of carbon, hydrogen, and oxygen, the hydrogen/oxygen ratio being $2:1$.

Carotene: Reddish-orange pigment found in the plastids of plants.

Carpel: Ovule-bearing area of the ovary.

Casparian strip: Area that surrounds the radial and transverse walls of epidermal cells, making them impervious to water.

Catabolism: Destructive phases of metabolism in which complex organic molecules are broken down.

Cell: Structural unit of plants and animals; a unit of protoplasm surrounded by a cell wall.

Cell division: Division of the cytoplasm and nucleus of a cell into two.

Cell inclusion: Group of assorted materials that are produced as a result of cellular activities.

Cellulose: Complex carbohydrate, the major component of the cell wall in most plants.

Cell wall: Limiting layer of plant cells; formed by and completely surrounds the protoplasm.

Centromere: That part of the chromosome to which the spindle fiber is attached.

Chlorophyll: Green pigment found in plastids, important in the process of photosynthesis.

Chlorophyll *a:* One of the two pigments making up the mixture of green pigment called chlorophyll; the other is chlorophyll *b.*

Chlorophyll *b:* One of the two pigments making up the mixture of green pigment called chlorophyll; the other is chlorophyll *a.*

Chloroplast: Specialized body in the cytoplasm which contains chlorophyll.

Chlorosis: Condition characterized by the absence of chlorophyll due to its failure to develop because some physical condition, that is, light, proper nutrients, or a genetic factor, was missing.

Chromatid: One half of a doubled chromosome before the two portions separate.

Chromatin: Substance in the nucleus which readily accepts basic stains; forms the hereditary characters of the nucleus.

Chromoplast: Specialized body in the cytoplasm which contains pigments other than chlorophyll.

Chromosome: Group of structural bodies, located in the nucleus of the cell; the site of the hereditary determiners, the genes.

Chromosome mutation: May result from either changes in chromosome structure or chromosome numbers.

Cleistothecium: Closed spherical ascocarp; produces spores internally.

Codon: Three nitrogen bases together; triplets; same as gene.

Coenocytic: Multinucleate; the nuclei lack cross walls.

Collenchyma: Derived from parenchyma; supporting tissue in which cells have irregular wall thickenings.

Columella: Sterile central shaft of a sporangium or capsule.

Commensalism: Two organisms living in close association; one benefits from the relationship, the other is unaffected.

Community: Group of different organisms inhabiting a common environment.

Companion cell: Narrow, elongated cell with a prominent nucleus; derived from cells which also give rise to sieve-tube elements.

Compound: Two or more chemical elements so combined to form a new substance having definite properties of its own.

Compound umbel: Branches of the inflorescence grow from the tip of the main axis; each branch bears an umbel of flowers.

Concave cell: Dead cell which delimits the hormogonia of certain algae.

Conidiophore: Hypha which produces conidia.

Conidium: Asexual reproductive cell of fungi produced at the tip of a conidiophore.

Constitutive enzyme: Enzyme normally present in a cell.

Consumer: Organism that lacks the ability to synthesize its own organic nutrients.

Cork: External, secondary tissue; a protective tissue impermeable to water and gases.

Cork cambium: Secondary meristem tissue which gives rise to cork cells, also known as phellogen.

Corm: Short, enlarged, vertical, underground storage stem.

Corolla: Collective term given to the petals of a flower.

Cortex: Primary tissue of a stem or root, located between the epidermis and the stele; composed chiefly of parenchyma cells.

Corymb: Type of inflorescence; lower pedicels are elongated so that the top is nearly flat.

Cotyledon: Leaf that forms part of a seed embryo.

Cuticle: Waxy layer produced by epidermal cells on their outer walls; composed of cutin.

Cutin: Waxy substance impermeable to water and most gases.

Cutting: Artifically detached portion of a plant used for propagation.

Cyclosis: Circulation of protoplasm in cells.

Cytokinesis: Division of cytoplasm constituents resulting in the formation of a new cell wall.

Cytokinins: Group of plant hormones, structurally related to kinetin; influences cell enlargement and cell division.

Cytoplasm: All the protoplasm of a cell exclusive of the nucleus and plastids; a transparent viscous fluid with inclusions.

Cytoplasmic streaming (see Cyclosis).

Deciduous: Trees and shrubs which drop their

leaves at the end of the growing period.

Dehiscent: Spontaneous opening of a structure to liberate seeds.

Dehydrogenase: Enzyme which promotes the removal of hydrogen from certain organic compounds.

Deletion: Area or region missing from a chromosome.

Deoxyribonucleic acid (DNA): Hereditary material found in the chromosome.

Development: Progressive change in form and complexity.

Dictyosome (*see* Golgi body).

Diffusion: Movement of molecules or ions from a region of higher concentration to a region of lower concentration.

Diffusion pressure deficit (DPD): Difference between the osmotic pressure and the actual turgor pressure.

Digestion: Conversion of complex, insoluble substances into simpler soluble substances by means of enzymatic action.

Dihybrid cross: Cross between organisms differing in two pairs of genes.

Dioecious: Bearing male and female reproductive organs on separate individuals.

Diplobiontic life cycle: Both haploid and diploid plants are formed and these phases alternate with each other.

Diploid: Having a double set of chromosomes; characteristic of sporophyte generation.

Diplotonic life cycle: Haplobiontic life cycle characterized by a diploid adult.

Disaccharide: Compound sugar composed of two monosaccharides; usually 12 carbon sugars.

Dominant: Gene that completely dominates the presence of a contrasting gene.

Drupe: Simple fleshy fruit in which the endocarp develops into a hard, stony or woody, pit.

Dry fruit: Mature ovary wall is dry and paperlike, leathery, or woody.

Ecosystem: Interacting system existing between organisms and their environment.

Egg: Female gamete or sex cell.

Elater: Elongated, thickened, spindle-shaped hygroscopic cell in the capsule of spore-bearing plants; functions in spore dispersal.

Element: One of about 100 different natural or man-made types of matter which, singly or in combination, comprise all materials of the universe; a substance composed of only one kind of atom; an atom is the smallest unit of an element.

Embryo: Rudimentary plant formed from a zygote within an archegonium or an ovule.

Embryo sac: Female gametophyte of angiosperms; develops from a megaspore.

Endocarp: Innermost layer of the pericarp; usually hard.

Endodermis: One-celled layer of cells which surrounds the vascular tissue of certain plants; usually present in roots but may be lacking in some stems.

Endoplasmic reticulum: Cytoplasmic network of membranes adjacent to the nucleus which makes up a part of the submicroscopic structure of the protoplasm.

Endosperm: Triploid nutritive tissue formed within the enbryo sac of angiosperms; used by embryo and seedling at germination.

Enzyme: Complex protein molecule produced by living cells; speeds the rate of a particular transformation of materials in plants and animals.

Epidermis: Outermost protective layer of roots, stems, and leaves, the surface layer before cork is formed.

Ergastic substance (*see* Cell inclusion).

Etiolation: Condition involving stem elongation, poor leaf development, and lack of chlorophyll in plants grown in the dark or under low light intensities.

Eucaryotic: Applied to organisms having nuclei enclosed in membranes, Golgi apparatus, and mitochondria.

Exarch: Developmental pattern by which the outermost layers of xylem differentiate before any others.

Exocarp: Outermost layer of the pericarp.

F_1 (first filial) generation: Offspring resulting from the crossing of organisms of the parental generation.

F_2 (second filial) generation: Offspring resulting from crossing organisms of the F_1 generation among themselves.

False berry: Fruit in which both the floral tube and the ovary wall are fleshy at maturity.

Fat: Organic compound composed of carbon, hydrogen, and oxygen (the proportion of oxygen to carbon is considerably less than in carbohydrates), formed by union of glycerol with fatty acids.

Fatty acid: Organic aliphatic acid that may form a part of fat molecules.

Feedback inhibition: Type of self-regulatory mechanism; allows a cell to adjust the rate of synthesis of a metabolic intermediate.

Fermentation: Anaerobic respiration consisting of a series of complex oxidation–reduction reactions; hydrogen is released in glycolysis, combines with pyruvic acid to form alcohol, lactic acid, or other products.

Fiber: Elongated plant cell which serves to support and strengthen the plant.

Fibrovascular bundle: Bundle of vascular tissue surrounded by nonvascular tissue.

Filament: Stalk of a stamen bearing the anther at its apex; a threadlike row of cells, as in certain algae.

Fission: Asexual reproduction; a single cell divides into two equal daughter cells.

Flagellum: Long slender whiplike thread used as an organ of motility.

Fleshy fruit: Mature ovary wall is filled with sap and becomes fleshy.

Flower: Reproductive structure of angiosperms; consists of an axis on which are inserted sepals, petals, stamen, and pistil if all parts are present.

Follicle: Simple dry dehiscent fruit with one carpel; opens on one side only.

Fragmentation: Reproduction by isolating a part of an organism to form a new individual.

Frond: Leaf of a fern.

Fructose: Six-carbon sugar; fruit sugar; combined with glucose in sucrose.

Fruit: Mature ovary containing seeds.

Fucoxanthin: Brownish pigment found in brown algae.

Gametangium: Organ in which gametes are produced.

Gamete: Reproductive cell which fuses with another to form a zygote in the process of fertilization.

Gametophyte: Haploid, gamete-producing plant.

Gene: unit of hereditary characteristics located in chromosomes.

Gene mutation: Results from an alteration in the chemical structure of a gene.

Generative cell: One of two cells formed after the germination of the microspore in flowering plants; upon division it forms two male gametes.

Genotype: Genetic constitution of an organism.

Germination: Beginning of growth by an embryo or spore.

Gibberellin: Growth hormone, the effect of which produces elongation in stems.

Glucose: Six-carbon sugar.

Glycerol: Formed from the hydrolosis of fats and oils by the enzyme lipase; is readily changed to sugars and other compounds.

Glycine: Structural unit of various proteins; an amino acid; plays a part in the formation of creative and other compounds.

Glycolysis: Respiratory process during which glucose is anaerobically broken down to pyruvic acid.

Golgi body: Material present in cell cytoplasm; seems to play a role in the formation of certain cell secretions.

Grafting: To induce a union between two plants; to unite a scion to a stock.

Grain: One-seeded fruits in which seed is completely fused to the pericarp.

Ground meristem: Partially differentiated meristem which gives rise to cortex, pith rays, and pith.

Growth: Irreversible increase in size.

Growth regulator: Synthetic substance which influences growth and development in plants.

Growth ring: Amount of tissue formed during one year's growth, as seen in cross section.

Growth substance (*see* Hormone).

Guard cell: Two crescent-shaped cells surrounding stomata of aerial epidermis of plant tissue.

Guttation: Exudation of liquid water from plant leaves.

Gynoecium: Collective name for the carpels in the flower of a seed plant.

Haplobiontic life cycle: Single form of plant, either haploid or diploid; occurs in the life cycle.

Haploid: Having a single set of paired chromosomes; characteristic of gametophyte generation of plants.

Haplontic life cycle: Haplobiontic life cycle characterized by a haploid adult.

Haustorium: Specialized outgrowth of hyphae which acts as a penetrating and absorbing organ.

Head: Type of inflorescence whereby a dense cluster of flowers is grouped closely on a receptacle.

Herbaceous: Any nonwoody plant.

Heredity: Transmission of characteristics from parents to their offspring.

Heterocyst: Enlarged functionless cell which may occur in the filaments of cetain algae.

Heterogamete: Gametes which are unlike each other in size and behavior as egg and sperm.

Heterogamy: Reproduction involving two types of gametes.

Heterothallic: Condition in which an individual produces only one kind of gamete.

Heterotroph: Organisms which cannot manufacture their own food but must obtain nourishment from outside sources.

Heterozygous: Having two different genes of a Mendelian pair present in the same cell or organism.

Hilum: Scar on the seed coat which marks the place where the seed stalk attached the seed to the placenta.

Holozoic: Obtaining food in an animal-like manner.

Homologous: Organisms which are similar in structure and origin; chromosomes with the same sequence of genes.

Homothallic: Both sexes are produced by the same individual plant body.

Homozygous: Having members of a gene pair which are identical for a given characteristic.

Hormone: Specific organic product produced in one part of the plant and transported to another part, where it influences and controls physiological responses.

Hybridization: Process of crossing two individuals that are genetically unlike.

Hydrolysis: Conversion of compounds into simpler compounds by the addition of water.

Hydrophyte: Plant that grows submerged in water or in very wet soils.

Hypha: Threadlike filament of a fungus.

Hypocotyl: Part of the seedling stem below cotyledons.

Imbibition: Process by which certain solid substances, usually colloidal, absorb liquids and swell.

Incomplete dominance: Condition in which there is a lack of dominance between either member of a pair of contrasting characters.

Indehiscent: Fruit that does not split open when completely ripened.

Indusium: Membranous covering over fern sporangia.

Inflorescence: Flower cluster.

Integument: Covering of an ovule which becomes the future seed coat.

Internode: Portion of a stem found between two successive nodes.

Interphase: Stage between cell divisions in which the nucleus is not undergoing mitosis.

Inulin: Carbohydrate substance chemically similar to starch.

Inversion: Chromosome change in which the ends cross, resulting in a breakoff and exchange of positions.

Isogamete: Gamete that appears similar in form and size to the gamete with which it unites during fertilization.

Isogamy: Process of sexual reproduction among isogametes.

Karyolymph (*see* Nuclear sap).

Lamina (*see* Blade).

Leaf: Primary organ of most vascular plants; typically the principle photosynthetic organ.

Leaf scar: Scar left on a stem following leaf detachment.

Legume: Member of the pea family (Leguminosae) whose fruit is dry, dehiscent, and one-carpelled.

Lenticel: Small raised pore found mainly on the surfaces of roots and stems which permits gas exchange.

Leucoplast: Colorless plastid, in which starch is usually formed.

Lipid: Fatlike substance which occurs naturally in plants and animals and yields fatty acids upon hydrolysis.

Macronutrient: Plant nutrient required in relatively large amounts.

Maltose: Disaccharide sugar formed as a result of starch breakdown.

Megasporangium: Structure within which megaspores are developed; the ovule in flowering plants.

Megaspore: Larger of two kinds of spores which upon germination form the female gametophyte.

Megasporophyll: Leaf or modified leaf that bears one or more megasporangia.

Meiosis: Reduction division in which the chromosomes are changed from a diploid to a haploid condition.

Meiospore: Haploid spore produced during meiosis.

Meristematic tissue: Undifferentiated tissue, capable of further division and specialization.

Mesocarp: Middle layer of the fruit wall.

Mesophyll: Chlorophyll-bearing leaf cells found immediately below the upper epidermis.

Mesophyte: Plant growing under environmental conditions which are neither wet nor dry.

Metabolism: Sum total of catabolic and metabolic chemical processes occurring within a living organism.

Metaphase: Stage in mitosis or meiosis during which the chromosomes are at the central region of the spindle.

Metaxylem: Elements of primary xylem which mature last from the procambium.

Micronutrient: Plant nutrient (also called a trace element) required in relatively small quantities.

Micropyle: Small canal found in the integument(s) of an ovule which can later be recognized in a seed coat.

Microsporangium: Spore case within which microspores are produced.

Microspore: Smaller of two kinds of spores, which in flowering plants becomes pollen grain.

Microsporophyll: Leaf or modified leaf which bears a microsporangium; the stamen in flowering plants.

Middle lamella: Thin membrane separating two adjoining cells; serves to cement them together.

Mitochondrion: Small cytoplasmic bodies which occur in great numbers in all cells (except bacteria and blue-green algae); the site of the enzymes of the Krebs cycle and of oxidative phosphorylation.

Mitosis: Nuclear division during which process the chromosomes double and then split to form two identical daughter nuclei; cell division.

Molecule: Smallest particle of an element or compound that can maintain a stable, independent existence.

Monoecious: Condition in which both male and female reproductive organs are found on the same plant.

Monohybrid cross: Cross between parents which are the same but for one character.

Monosaccharide: Simple sugar which ceases to have the properties of a sugar when split into smaller molecules.

Multiple fruit: Fruit, such as pineapple and fig, which is formed from many flowers.

Mutualism: Two or more organisms living in an association which is mutually beneficial.

Mycelium: Mass of hyphae which forms the vegetative body of a fungus.

Node: Part of a plant stem where leaves are attached or may arise.

Nucellus: Central tissue of an ovule within which the megaspore develops.

Nuclear membrane: Outer membrane surrounding the nucleus.

Nuclear sap: Liquid portion of the nucleus.

Nucleic acid: Two kinds of nucleic acids occurring in cells are DNA and RNA, and these consist of many nucleotide molecules linked together.

Nucleolus: One or more small bodies, rich in protein and RNA, found in the nucleus.

Nucleus: Protoplasmic structure found suspended in the cytoplasm of most cells and containing the chromosomes.

Nut: Dry, indehiscent, one-seeded fruit with a hard woody pericarp, for example, walnut.

Nutrition: Process by which an organism secures and utilizes the substances required for metabolic activities.

Oogamy: Process of sexual reproduction in which a large nonmotile egg is fertilized by a smaller male gamete.

Oogonium: Unicellular specialized cell in which one or more eggs form.

Organ: Part of a plant body adapted for a special function or functions.

Organelle: Specialized part of a cell serving a specific function.

Organism: Living entity capable of reproducing itself.

Osmosis: Passing of a solvent, usually water, through a selectively permeable membrane from a region of low solute concentration to a region of high solute concentration.

Osmotic pressure: Force across a semipermeable membrane due to osmosis.

Ovary: Part of the basal portion of a pistil containing ovules or seeds.

Ovule: Structure found in gymnosperms and angiosperms which upon fertilization of an egg cell within it develops into a seed.

Oxidative phosphorylation: Formation of organic phosphate compounds using energy from respiration.

_P_ (parent generation): Initial generation in which all the males and females are uniform with regard to the allelomorphs under observation.

Panicle: Kind of inflorescence in which each of several branches bears a flower, for example, various grasses.

Parasitism: Method of obtaining nutrition from another organism (host) which is characteristic of a parasite.

Parenchyma: Cells that have thin walls and are loosely fitted together; retains meristematic capabilities.

Parthenocarpy: Development of fruit, usually seedless, in the absence of pollination and fertilization.

Pedicel: Stalk of an individual flower in an inflorescence.

Perianth: Calyx and the corolla considered collectively.

Pericarp: Ripened ovary wall after it has matured into a fruit.

Pericycle: Vascular cylinder located between the endodermis and the phloem, consisting mainly of parenchyma cells.

Perithecium: Flask-shaped or rounded fruit body in the ascomycete group having an opening and containing asci.

Petal: Single floral leaf or unit of the corolla, usually brightly colored.

Petiole: Leaf stalk.

Phenotype: Visible or external appearance of an organism.

Phloem: One of the vascular tissues that conducts foods through the plant; characterized by the presence of sieve tubes and companion cells.

Phloem fiber: Modified sclerenchyme cells, provide support.

Phospholipid: Fat containing phosphoric acid and a nitrogenous base.

Photoperiodism: Plant response to day length.

Photophosphorylation: Process of phosphorylation which is activated by the energy of light.

Photosynthesis: The synthesis of food, mainly sugar, from carbon dioxide and water in the presence of chlorophyll utilizing light energy and releasing oxygen.

Phycoerythrin: Photosynthetic red pigment usually occurring with chlorophylls in certain algae.

Pigment: Complex organic substance which produces the appearance of color.

Pinna: Basic or primary division of a pinnate leaf; a leaflet.

Pistil: Central reproductive structure of a flower consisting of a stigma, style, and ovary enclosing ovules.

Pit: Thin area in the cell wall; a cavity where the secondary wall fails to develop.

Placenta: Region of the ovary wall on which one or more ovules are borne.

Plasma (cell) membrane: Outermost layer of the cytoplasm of a cell.

Plasmodesma: Fine cytoplasmic threads which pass through the openings in the cell walls; connects the protoplast and adjacent cells.

Plasmolysis: Shrinkage of cell protoplasm away from the cell wall as a result of osmotic withdrawal from the central vacuole.

Plastid: Specialized cytoplasmic body; site of food manufacture and storage.

Plumule: Terminal bud of the embryo; also, the primary bud of a seedling.

Pollen grain: Microspore containing a developing male gametophyte.

Pollen tube: Outgrowth of a pollen grain following germination which carries the male gamete (sperm) to the egg.

Pollination: Transfer of pollen from anther to stigma (angiosperms) or from microsporangium to ovule (gymnosperms).

Polyploid: Organism (especially an angiosperm) carrying two or more sets of chromosomes in one cell.

Polysaccharide: Carbohydrate, such as starch or

cellulose, which consists of two or more chemically linked simple sugars.

Pome: Apple-like accessory fruit which is developed from a floral receptacle rather than from an ovary.

Population: Assemblage of interbreeding individuals which form a community by themselves.

Primary cell wall: First wall produced by a developing cell; usually associated with living protoplasts.

Primary pit field: Depression or cavity in a primary cell wall.

Primary plant tissue: Tissue which derives from the meristematic regions of roots and shoots.

Procambium: Tissue consisting of narrow, elongated cells which gives rise by further development to vascular tissue.

Procaryotic: Simple organisms, such as bacteria and blue-green algae, characterized by a lack of mitotic division and the absence of membranes surrounding various organelles, especially the nucleus.

Producer organism: Phototrophic organism.

Prophase: First stage in mitosis or meiosis; chromosomes double visibly.

Protein: Organic substance composed of numerous and complex amino acids.

Prothallus: Small, green thallus (ferns and other related vascular plants) bearing sex organs and usually attached to a substratum by means of rhizoids.

Protoderm: Outer cell layer of primary meristem; gives rise to epidermis.

Protonema: Branched filament of elongate cells produced by the germination of a moss spore.

Protoplasm: Living material of the cell.

Protoplast: Entire organized living unit of a cell.

Protostele: Simplest type of stele, present in certain ferns and club mosses; consists of a central xylem core surrounded by cylinder of phloem.

Protoxylem: Earliest xylem elements to be distinguished from procambium.

Pyrenoid: Center of starch formation found singly or in numbers embedded in the plastids of various algae.

Racene: Type of inflorescence having a main stalk with unbranched pedicels, each bearing flowers.

Radicle: Lowest portion of the plant embryo; develops into the primary root of the seedling.

Receptacle: Top of the flower stalk; bears flower parts.

Recessive: In genetics, a gene which has no effect on the phenotype; a masked gene.

Regulator gene: Gene produced by another gene that influences a metabolic reaction; a gene that controls the production of a number of structural genes.

Respiration: Process by which food is chemically oxidized, whereby organic material is broken down and energy is released.

Rhizoid: Hairlike structure, common in lower organisms which serves as a root.

Rhizome: Underground stem, usually horizontal, which serves as a means of vegetative propagation.

Ribonucleic acid (RNA): Nucleic acid which governs the synthesis of specific protein molecules.

Ribosome: Submicroscopic particles located in the protoplasm; contain RNA and are the site for protein synthesis.

Root: Descending portion of a plant; serves to fix it in soil; absorbs and conducts water and nutrients to other parts of the plant.

Root cap: Protective mass of cells at apex of root; develops from the apical meristem of the root.

Root hair: Thin delicate outgrowth of epidermal root cells which increase the absorptive surface of the root.

Root stock (*see* Rhizome).

Runner: Elongated prostrate aerial branch that roots at the tip, forming new plants.

Samara: Winged achene such as in a maple.

Saprophyte: Plant which obtains its nutriment from nonliving organic matter.

Sclereid (*see* Stone cell).

Sclerenchyma: Strengthening tissue; composed of thickened and hard cells of vessels.

Secondary cell wall: Innermost wall formed inside the primary cell wall.

Secondary tissue: Tissue formed by the cambium or cork cambium.

Seed: Characteristic reproductive structure of seed plants; the product of a fertilized ovule consisting of an embryo enclosed within a protective seed coat(s).

Semiconservative replication: Replication of a DNA molecule in which half of each parent molecule (one strand) is conserved in each daughter molecule; the daughter molecule consists of one strand from the parent and one strand synthesized from the surrounding medium.

Sepal: One of the modified leaves (usually green) which make up the calyx of dicotyledonous flowers.

Sessile: Lacking a stalk.

Sieve cell: Long and slender cell having perforated areas of cell wall.

Sieve plate: Perforated wall of a sieve cell through which pass strands of cytoplasm.

Sieve tube: Series of elongated cells placed end to end, forming a long tube for the conduction of food materials.

Silique: Special type of capsule or pod of a cruciferous plant developed from an ovary of two united carpels, which dehisces at maturity.

Simple fruit: Fruit derived from one ovary of one pistil without other structures attached to it, for example, peanut, corn, and rice.

Simple umbel: Umbrella-like floral cluster which in the parsley family (Umbelliferae) consists of five petals and five stamens.

Siphonostele: Stele in which phloem and xylem develop concentric cylinders which envelop a central core of pith.

Solitary flower: Flower born singly, as in tulip and lily.

Soredia: Organ of vegetative reproduction associated with lichens and consisting of algal cells and fungal hyphae.

Sorus: Group of fern sporangia located on the lower surface of a leaf.

Species: Basic unit of classification consisting of closely related individuals capable of interbreeding freely.

Sperm: Male gamete.

Spike: Kind of inflorescence in which the flowers are sessile and located upon an elongated axis.

Sporangiophore: Stalk which bears one or more spore cases (sporangia).

Sporangiospore: Spore produced in a sporangium.

Sporangium: Capsule or case in which spores develop.

Spore: Minute, asexual, reproductive structure, usually consisting of one cell and produced in a sporangium (except bacteria).

Spore-mother cell: Diploid cell which usually gives rise by meiosis to four haploid spores.

Sporophyll: Modified leaf which bears sporangia containing spores, for example, stamens and carpels of angiosperms and fertile fronds of ferns.

Sporophyte: Diploid phase of the life cycle of plants during which spores are produced.

Sporulation: Formation of spores.

Stamen: in angiosperms, the structure (microsporaphyll) that bears pollen, usually composed of an anther and filament.

Starch: Complex carbohydrate, not soluble in

water, consisting of numerous glucose molecules.

Stele: Cylinder of vascular tissue, found in roots and stems, in which the conducting tissues are located.

Stem: Normally the aerial part of vascular plants, which bears leaves and reproductive structures.

Stigma: Terminal part of a pistil, usually expanded, which receives pollen and permits it to germinate.

Stipe: Stalk which supports a fruiting body as found in certain higher fungi; also, the stalk of brown algae or the petiole of a fern frond.

Stipule: One or more appendages located at the base of a petiole in some species of higher plants.

Stolon (*see* Runner).

Stoma: Small opening between two guard cells in the epidermis of leaves and stems through which gases pass.

Stone cell: Short sclerenchyma cell having heavily lignified cell walls.

Strobilus: Cone or conelike structure bearing sporangia.

Structural gene: Gene that codes for messenger RNA, which in turn dictates the structure of an enzyme.

Style: That portion of the pistil located between the ovary and stigma.

Suberin: Complex waxy substance present in the walls of cork cells, rendering them waterproof.

Succession: Progressive and orderly sequence of plant communities from pioneer colonization to attainment of climax.

Sucrose: Disaccharide with 12 carbon atoms, which is commonly found in plants; cane sugar.

Symbiosis: Close association of two or more dissimilar organisms, resulting in mutualism or parasitism.

Telophase: Final phase of mitosis, in which daughter nuclei are reorganized and new cell wall begins to form.

Tendril: Modified stem, leaf, leaflet, or stipule used by many plants for attachment to a support.

Testa: Structure formed from integument(s), which provides a protective covering, usually hard and dry, for the embryo of seed plants; the seed coat.

Test cross: Cross of a homozygous recessive with a dominant.

Thallus: Simple vegetative plant showing no differentiation into true roots, stems, and leaves.

Thorn: Pointed, hard, woody outgrowth from a stem or other plant organ.

Tissue: Group of cells, usually of similar structure, which perform a common function.

Tracheid: Elongated, tapered cell, pitted with thick walls, which functions for conduction and support; contains no protoplasm.

Translocation: Movement of organic substances in the phloem; a process by which two non-homologous chromosomes exchange parts.

Transpiration: Loss of water vapor from the aerial parts of plants, principally through stomata and lenticels.

Trichome: Hairy outgrowth of plant epidermis.

Tube cell: In angiosperms, it is one of two cells, the other being the generative cell, which results from the germination of the microspore and directs the growth of the tube.

Tuber: Enlarged, fleshy, food-storing stem bearing rudimentary leaves and buds.

Turgor pressure: Force developed within a cell caused by the absorption of water.

Vacuole: Spaces within the cytoplasm containing air, sap, or a solution of various other substances.

Vascular bundle: Longitudinal strand of conducting tissue composed essentially of xylem and phloem.

Vascular cambium: Meristematic tissue which gives rise to secondary phloem and secondary xylem.

Vascular cylinder (*see* Stele).

Vegetative propagation: Asexual reproduction in plants by separation of vegetative structures.

Vein: Vascular bundle of a leaf; forms its basic framework.

Vernalization: Process in plants, usually related to day length, whereby flowering is affected by cold temperatures.

Vessel: Tubelike structure found in xylem which conducts water and mineral nutrients.

Vessel element: Portion of a vessel; an elongated, pitted, thick-walled cell, derived from the vascular cambium.

Wax: Complex hydrocarbon resembling beeswax which is often secreted by plants.

Wood: Principal water-conducting tissue in plants, technically known as xylem.

Woody (*see* Wood).

Xanthophyll: Yellow to orange carotenoid pigment associated with chlorophyll in chloroplasts.

Xerophyte: Plant which is both physiologically and structurally adapted to survival under conditions of prolonged drought.

Xylem: Plant tissue which conducts water from the roots upward; a woody tissue consisting of tracheids, vessels, parenchyma cells, and fibers.

Xylem fiber: Supporting cells of the wood, thick-walled and elongated.

Zoospore: Swimming spore common to algae and fungi.

Zygospore: Thick-walled resting spore produced by the union of similar gametes.

Zygote: Fertilized egg prior to any cleavage divisions.

Index

Cup fungi, 431
Cupressus, 489
Cuticle, 98
Cuticular transpiration, 193
Cutin, 45, 98
Cutleriales, 527
Cutting, reproduction by, 134
Cycadales, 485, 487, 533–34
Cycadofilicales, 485–86, 533
Cycadophytae, 485, 533–34
Cycas, 487
Cyclic photophosphorylation, 224–27
Cyclosis, 76–77
Cyclosporae, 527
Cytase, 245
Cytochrome, 209, 225, 256
Cytochrome system, 209, 255–57
Cytokinesis, 88–90
Cytokinin, 56
    effects of, 280
Cytology, study of, 11
Cytoplasm, 75–77

Darwinian evolution, 356–57
    See also Evolution
Dasycladales, 525
Day-neutral plant, 283
Deamination, 259
Decarboxylation, 253
Deciduous plants, 125, 518
Degradation, 165–66
Dehiscent fruits, 153
Dehydration synthesis, 39
Dehydrogenase, 209, 249
Dehydrogenation, 209, 249, 253, 255
Deletion, 308
Denaturation, 207
Denitrification, 176
Denitrifying bacteria, 176
Deoxyribonucleic acid (DNA)
    as agent of bacterial transformation,
        310
    in cell division, 90
    as genetic material, 310–12
    identification as genetic material,
        310–12
    as primary genetic material of bacterio-
        phages, 311
    replication of, 312–14

role in protein synthesis, 233–38
structure of, 48–52, 312–13
viruses and, 311, 341–42
Derbesia, 406
Desmarestiales, 527
Desmomonadales, 526
Deuteromycetes, 422, 530
Development, 167
    defined, 264
    See also Growth
Dextrin, 243
Diastase, 245
Diatomaceous earth, 410
Dicots, *see* Dicotyledonae
Dicotyledonae, 143, 499
    classification of, 534–35
    root of, 106–11
    seed germination of, 151
Dicranales, 532
Dictyosiphonales, 527
Dictyosome, 83–84
Dictyotales, 527
Differentially permeable membrane,
    184
Differentiation, 266–67
Diffuse root system, 106
Diffusion, 190
    definition of, 183
    direction and rate of, 184
    equilibrium in, 183
    relations to membranes, 184–85
Diffusion pressure, 184
Diffusion pressure deficit, 187
Digestion, 39, 165–66, 210, 242–47
Digitalis purpurea, 14
Dihybrid cross, 299–302
Dihydroxyacetone phosphate, 249
Dimorphism, 477
Dinocapsales, 527
Dinococcales, 526
Dinoflagellates, *see* Dinophyceae
Dinophyceae, 393, 410, 526–27
Dinophysiales, 526
Dinotrichales, 527
Dioecious, 142, 441, 484, 487, 489
Diplobiontic life cycle, 386, 404, 406
Diplococcus pneumoniae, 310
Diploid, 136
Diplotonic life cycle, 404

Disaccharide, 39
Divergence, 343
Diversity of plants, 2
Division, 327
DNA, *see* Deoxyribonucleic acid
DNA polymerase, 313
Dominance,
    apical, 277–78
    imperfect, 302–3
    incomplete, 302–3
    law of, 298
Dominant, 298
Double fertilization, 148
Drosophila melanogaster, 315
Drug products, 14–15
Drupe, 156
Dry fruit, 153
Dwarfism, genetic, 278–80

Echinosteliales, 528
Ecology, 11, 504–22
    abiotic factors in, 504–9
    biomes in, 516–20
    biotic factors in, 509–10
    community relations in, 510–14
    plant interactions in, 510
    succession in, 514–16
Economic botany, 12
Economic value
    of algae, 375–77
    of Bryophyta, 455
    of flower, 156
    of fruit, 157
    of Gymnospermae, 498
    of leaf, 128
    of root, 112–13
    of seeds, 157
    of stem, 121
Ecosystem, 504
Ectocarpales, 527
Egg, 402
Elater, 445, 474
Electron micrograph, 64
Electron microscope, 63–65
Electrophoresis, 69
Element, 25, 180
    atoms of, 25–26, 69
    chemical symbol of, 25
    commonly found in plants, 25*t.*

Floriculture, study of, 12
Florideae, 528
Florigen, 284
Flower, 140–48, 156
    accessory parts of, 141–42
    apopetalous, 143
    arrangement of, 143–45
    complete, 142
    definition of, 140
    economic importance of, 156
    epigynuus, 143
    essential parts of, 142
    hypogynous, 143
    imperfect, 142
    incomplete, 142
    perfect, 142
    regular, 143
    reproductive role, 145–48
    structure of, 140–45
    symmetry of, 143
    types of, 142–45
FMN, see Flavin mononucleotide
Follicle, 156
Fomes, 435
Foods, 12–14, 156–57
Foot, 441, 445, 448
Fossil, 345
Fragmentation, 132, 385
Fritschiella, 399
Frond, 475
Fructose, 39, 243
Fruit, 152–56
    aggregate, 156
    criteria for classification of, 153
    definition of, 152
    dehiscent, 153
    development of, 152, 276–77
    dropping off of, 275–76
    dry, 153
    economic importance of, 157
    fleshy, 155, 156
    indehiscent, 153
    kinds of, 153–56
    multiple, 156
    simple, 153
    structure of, 152
Fucales, 527
Fucoxanthin, 393
Fucus, 405, 459

Fungi, *see* Eumycophyta
Funariales, 532
Functional group, 206

β-Galactosidase, 318
Gametangium, 330, 406, 427
Gamete, 330
Gametogenic meiosis, 405
Gametophyte, 146, 331, 441, 446, 477
    of algae, 403–7
    of *Equisetum,* 473–74
    of Filicinae, 478–79
    of *Lycopodium,* 466
    of *Psilotum,* 462–64
    of *Selaginella,* 468–69
Gelidiales, 528
Gelidium, 409
Gemmae, 445
Gene, 295, 303–5, 307–9
    action on enzymes, 316
    crossing over of, 305
    linkage of, 305
    mechanism of function of, 314–22
    multiple, 303–5
    mutation of, 307
    operator, 321
    regulation of activity by, 317–22
    relationships to enzymes, 315
    role in inherited characteristics, 315
    structural, 320
Gene pool, 357
General Sherman tree, 2
Generative cell, 146, 490
Genetic dwarfism, 278–80
Genetic isolation, 358
Genetics, *see* Heredity
Genotype, 269, 296
Genus, 327
Geologic time scale, 347
Geosphere, 176–82
    definition of, 177
    minerals of, 180
    organic matter of, 182
    soil constituents of, 177–79
    soil organisms of, 176–77, 182
    soil water of, 179–80
Geotropism, 289, 291
Geraniales, 534
Germination, 150

Gibberella fujikuroi, 278
Gibberellic acid, 278
Gibberellin, 55–56, 278–80
Gigartinales, 528
Gill, 435
Ginkgo biloba, 488–89
Ginkgoales, 485, 488–89, 534
Girdling, 195
Gliding bacteria, *see* Beggiatoales
Gloeocapsa, 374
Glucose, 39, 243
Glycolysis, 249, 251
Gnetales, 485, 496–97, 534
Golden algae, *see* Chrysophyceae
Golgi body, 83–84
Goniotrichales, 528
Gonium, 396
Grafting, 134
Grain, 153
Graminales, 535
Green algae, *see* Chlorophyta
Grimmiales, 532
Ground meristem, 108
Group specificity, 206
Growth, 167, 264–87
    analysis of, 264–65
    control of, 271–87
        hormonal, 271–80
        light, 280–87
    definition of, 264
    distribution of, 267–68
    factors affecting, 268–71
    phases of, 265–66
    regulators of, 168
    of root, 111
    sigmoid curve of, 265–66
    stages of, 266–67
    of stem, 115
Growth movement, 288
Growth regulator, 168, 269
Growth ring, 117
Guard cell, 98, 124
Guttation, 191
Gymnodiniales, 526
Gymnospermae, 484–98
    classification of, 533–34
    economic importance of, 498
    general characteristics of, 484
    relationships of, 497

Nucleus (*Cont'd*)
  electron micrograph of, 85
  membrane of, 83
  nucleoli of, 84
  polar, 146
  procaryotic, 373, 382
Nut, 153
Nutation, 288
Nutrients, *see* Metabolites
Nutrition, types of, 163–64
  See also Metabolism; Metabolites

Obligate anaerobe, 369
Ochromonadales, 526
Oedogoniales, 525
Oedogonium, 401
Old field succession, 514–15
Oleales, 534
One-gene, one-enzyme hypothesis,
  316–17
Oogamy, 330, 402
Oogonium, 330
Oomycetidae, 529
Open tier, 494
Operator gene, 321
Operculum, 431, 452
Operon, 321
Operon model, 321
Ophioglossales, 533
Orchidales, 535
Order, 327
Organ, 87
Organelle, 77–84
Organic compound, 33, 37–56
Organism, defined, 162
Oscillatoria, 375
Oscillatoriales, 374–75, 524
Osmometer, 185–86
Osmosis, 185–86
Osmotic pressure, 186
Ovary, 141, 143
Ovule, 141, 484, 490–92
  development of, 146
Oxaloacetic acid, 253
Oxalosuccinic acid, 255
Oxidation, 225, 259
  beta (β), 259
  See also Respiration
Oxidative phosphorylation, 249

Oxidoreductase enzyme, 211
Oxygen cycle, 173–74

P generation, 295
Paleobotany, 245
Paleozoic era, 349–53
Palisade mesophyll, 125–26
Palmales, 535
Palmate net-venation, 123
Pandorina, 402
Panicle, 145
Panicum, 13
Papaver somniferum, 14
Papaverales, 534
Parasitic plant, 16
Parasitic root, 112
Parasitism as symbiotic relationship, 514
Paratonic movement, 288–91
Parenchyma tissue, 98, 104
Parthenocarpy
  defined, 152
  types of, 152, 276–77
Passive absorption, 188–89
Pathology, study of, 11
Pedicel, 140, 144
Peduncle, 144
Penicillium, 432
Pennales, 526
Pennisetum, 13
Peptide bond, 45
Peptone, 246
Perianth, 140
Pericarp, 152
Pericycles, 108
Periderm, 111
Peridiniales, 526
Perigynous flower, 143
Periplast, 388
Peristome, 452
Perithecium, 431, 433–34
Peronosporales, 425, 529
Petal, 140
Petiole, 121
Petrification, 345
Peziza, 431
Pezizales, 530
PGA, *see* Phosphoglyceric acid
PGAL, *see* Phosphoglyceraldehyde
pH, 35–37, 508–9

optimum, 212
Phaeophyta, 393, 527
Phallales, 530
Phellem, 111
Phelloderm, 111
Phellogen, 111
Phenotype, 296
Phloem fiber, 104
Phloem ray cell, 117
Phloem tissue, 117
  conduction by, 194–97
  in leaf, 125
  primary, 108, 115
  secondary
    root, 109–11
    stem, 116–17
  structure of, 102–4
Phosphoenolpyruvic acid, 251
Phosphofructokinase, 249
Phosphoglucoisomerase, 249
Phosphoglyceraldehyde (PGAL), 228–30,
  247, 249, 251
  synthesis of, 229–31
Phosphoglyceraldehyde dehydrogenase,
  249
Phosphoglyceric acid (PGA), 228
Phosphoglyceric kinase, 251
Phosphoglyceromutase, 251
Phospholipid, 45
Phosphotriose isomerase, 249
Photon, 221
Photoperiodism
  light and, 282–83
  mechanism of, 283–87
Photophosphorylation
  cyclic, 224–27
  noncyclic, 226–27
Photoreceptor, 506
Photosynthesis, 69, 164, 218–233
  CO_2 fixation, 228–29
  defined, 218
  discovery of, 219
  factors affecting, 221–27, 231–33
  fate of metabolites in, 219–20
  general equation for, 219
  importance in nature, 2
  limiting factors in, 231
  magnitude of, 218
  necessary conditions for, 220–21, 222

reactions of, 223–29
  dark, 227–29
  light, 223–27
  role of chlorophyll in, 220, 222, 223–27
  role of light in, 221–27
  significance of, 218
Photosynthetic autotroph, 164, 369
Photosynthetic Protista, 382–410
Phototaxis, 384
Phototropism, 289
Phycobilin, 220
Phycocyanin, 373
Phycoerythrin, 373, 393
Phycology, study of, 12
Phycomycetes, 421
  classification of, 529
  evolutionary trends in, 424–27
  form and function of, 424–27
  reproduction of, 424–27
Phyllid, 440–41
Physarales, 528
Physiology, plant, study of, 11, 18
Phytochrome, 286–87
Phytophthora infestans, 427
Picea, 489
Pigment, 53–55
Pileus, 435
Pinnae, 476
Pinnate net-venation, 123
Pinnus, 189–96
  evolutionary advances, 495–96
  life cycle of, 489–96
Piperales, 534
Pistil, 141
Pistillate flower, 142
Pisum sativum, 279
Pit in cell wall, 71
Pith, 116
Pitted tracheid, 101
Placenta, 141
Plant acid, 56
Plant ecology, *see* Ecology
Plant evolution, *see* Evolution
Plant form and environment, 2
Plant movements, 287–91
  autonomic, 288
  growth and, 288
  paratonic, 288–91

Plant organization levels, 24–25
Plant physiology, study of, 11, 18
Plant science, 16–20
  agricultural, 16–17
  current research in, 16–20
  molecular studies, 18
  physiological, 18–19
Plant taxonomy, *see* Taxonomy
Plasma membrane, 74–75
Plasmodesmata, 71
Plasmodium, 414
Plasmolysis, 188
Plastid, 77–78
Pleodorina, 396
Pleosporales, 530
Pleuromeiales, 532
Plumule, 149, 494–95
Poison hemlock, *see Conium masculatum*
Poison ivy, *see Rhus radicans*
Poison oak, *see Rhus toxicondendron*
Poison sumac, *see Rhus vernix*
Poisonous plants, 15–16
Polar nucleus, 146
Pollen cones, 145–46, 468, 488–89
Pollination
  cross-, 146–47
  self-, 147
  structures in, 146, 484, 490
Polymer, 42
Polypeptide, 246
Polyploidization, 308–9
Polyporales, 530
Polysaccharide, 41–42
Polytrichidae, 531
Polytrichum, reproduction by, 450–52
Pome, 156
Population
  defined, 504
  ecological setting of, 510
Pore space, 179
Porphyra, 17, 409
Porphyridiales, 527
Pottiales, 532
Primulales, 534
Probability and genetics, 298–99
Procambium, 108
Procaryotic nucleus, 373, 382
Producer, 510
Proembryo, 149, 494

Prophase, 89
Proplastid, 77
Prosthetic group, 208
Protease, 206, 246
Protein
  chemical composition of, 45
  digestion of, 245–47
  structure of, 45–49
  synthesis of, 233–38
Proteinase, 246
Proteose, 246
Proterozoic era, 248–49
Prothallial cell, 468, 490
Prothallus, 466
Protista
  classification of, 524–30
  evolution of, 364
  general features of, 382–83
Protoderm, 108
Protonema, 449–50
Protoplasm, 61
Protoplasmic streaming hypothesis, 197
Protoplast, 69, 73–74
Protopteridium, 352
Protostele, 460, 462, 476
Protoxylem, 465
Protozoa, *see* Protista
Pseudomonadales, 370, 523
Pseudoplasmodium, 417
Psilophytales, 461, 532
Psilophyton, 351, 461
Psilopsida, 460–64
  classification of, 461–63, 532
  general features of, 460–61
  importance and relationships of, 464
Psilopsidae, 532
Psilotales
  classification of, 532
  general features of, 461–64
Psilotum, reproduction by, 461–64
Pteropsida, 474–79, 533
  classification of, 533
  relationships and importance of, 479
  reproduction by, 477–79
  structure of, 474–77
Puccinia graminis, 435
Puffball, 436
Punctariales, 527
Pyrenoid, 384

Wall pressure, 186
Water, 189–97
  capacity of soil to hold, 179
  phloem transport of, 194–97
  photosynthesis and, 232
  plant growth and, 271
  specific heat of, 33
  xylem transport of, 189–94
Wave theory of light, 221
Wax, 45
Weeds, 16
Wilting, 186
Wood and wood derivatives, 14

Xanthophyceae, 389, 525
Xanthophyll, 54, 127
Xanthophyta, 389, 525–26
Xerarch succession, 515–16
Xerophyte, 507
Xylariales, 529
Xylem, 125
  function of, 101–2
  primary, 108, 115–16
  secondary, 109–11, 117
  water and nutrient transport by, 190–94
Xylem fibers, 102

Xylem ray cell, 117

Yeasts, *see Saccharomyces*

Zoopagales, 529
Zoosporangium, 425
Zoospore, 330, 385
Zygnematales, 525
Zygomycetidae, 529
Zygospore, 385, 425, 427
Zygote, 134, 148–49, 330, 441
Zygotic meiosis, 404